AF718595

The Mobilgas Economy Run

The Mobilgas Economy Run

A History of the Long Distance Fuel Efficiency Competition, 1936–1968

Dave Hermanson

McFarland & Company, Inc., Publishers
Jefferson, North Carolina

Unless otherwise noted, all photographs and advertisements appearing in this book are reproduced courtesy of ExxonMobil Corporation.

LIBRARY OF CONGRESS CATALOGUING-IN-PUBLICATION DATA

Hermanson, Dave, 1946–
The Mobilgas Economy Run : a history of the long distance fuel efficiency competition, 1936–1968 / Dave Hermanson.
p. cm.
Includes bibliographical references and index.

ISBN 978-0-7864-7562-9 (softcover : acid free paper) ∞
ISBN 978-1-4766-1322-2 (ebook)

1. Automobile rallies—California—History. 2. Automobiles—Fuel consumption—History. 3. Gasoline industry—United States—History. 4. Automobile industry and trade—United States—History. I. Title.

GV1029.2.H47 2014 796.7'309794—dc23 2014001063

BRITISH LIBRARY CATALOGUING DATA ARE AVAILABLE

On the cover: the Alsbury brothers cross the finish line simultaneously with Mel Jr. in the Imperial Crown (#6) and George in the Chrysler New Yorker (#1) during the 1958 Economy Run (courtesy of ExxonMobil Corporation)

Manufactured in the United States of America

McFarland & Company, Inc., Publishers
Box 611, Jefferson, North Carolina 28640
www.mcfarlandpub.com

Acknowledgments

This book would not be possible without the generous assistance of personnel affiliated with the Mobil Oil Company and United States Auto Club officials. These include Ken Downs, Roger Mahey, and John Parkinson, who supported my growing interest over the years by supplying press releases, photographs and other items for my constantly growing collection. These gentlemen introduced me to Arthur C. (A.C.) Pillsbury, chief steward of the Economy Run, who granted me access to the impound garage. There I observed some of the preparations for each entry while at the same time taking numerous slides and photographs of the competing cars. Other USAC officials, including 1967–1968's chief steward Art Rene, chief observer Russ Newton, and observer Smirle Morris, continued to support my continued interest providing me numerous items, including daily itinerary logs and final detailed results booklets. Today, these items are not commonly seen.

Several Los Angeles area dealers and drivers also supported my interest, especially the Alsbury family, Mel, Sr., Mel, Jr., and George. Also a special word of appreciation to American Motors engineer Les Viland, who always had a moment in his busy schedule for a short conversation and photo opportunity.

After the Run's cancellation in December 1968, my interest in the Economy Runs was put on the back burner, only to be rekindled in the early 1990s after I found some previously unknown to me Economy Run–related magazine advertisements in an antique shop in Maine. Thus began a new quest searching out missing information by researching the microfilm files of the *Los Angeles Times* and *San Diego Union* at the San Diego Public Library, along with Internet-based newspaper sites. Matthew Darby, archivist for the ExxonMobil Historical Collection at the University of Texas, provided additional assistance in locating items in their collections, especially with regards to the Mobil Mileage Rally. Several trips to the AACA Fall Meet at Hershey, Pennsylvania, always yielded additional items for my collection, as well as those found on eBay. Several collector friends of automobilia-related items have provided me with articles, advertisements and photographs over the years for my collection, and their assistance contributed greatly to documenting some earlier years. These include Jeremy Gilmore, John Heitmann, Alden Jewell, Frank Vogler, and Bob Zimmerman.

Table of Contents

Preface

This is the story of the Mobilgas Economy Run and its predecessor, the Gilmore Yosemite and Grand Canyon Economy Run, which were a part of American automotive and competition history from 1936 through 1968. While many articles have appeared in newspapers and magazines over the years, this is the first complete history to be written.

Based on newspaper and magazine articles of the day, along with some personal recollections from the Run's later years, each year is covered in detail from the announcement of the route, through preparations, to a daily record of the event culminating in the awards ceremony. Advertising and publicity efforts are also included. Over the years, I have been fortunate to acquire for my collection many of the official rules booklets, route descriptions with daily itineraries, and the final detailed result reports. These items provide additional details, some not readily known or reported, which provide a more thorough review of each year's Run.

A separate chapter covers the Mobil Mileage Rally, held from 1958 through 1961. Designed for imported models, these one-day affairs never gained the press's or public's attention and little was written about them at the time.

The results for each year are included, complete with each entry's mileage figures as well as driver and entrant information. Unfortunately, some of this information is incomplete for some of the early years. Preservation was not viewed as important in the thirties, forties and early fifties, with the result that some information, particularly regarding the Gilmore years and AAA Contest Board involvement, has been lost to history. It is this author's hope that some of these missing links will be located at some future date.

Notes for the Reader

As you read the history of the Economy Runs, you will encounter a few terms relating to procedures. To help you understand these more fully, the following descriptions are provided:

- **Ton Miles Per Gallon.** The ton miles per gallon formula was developed early in the 20th century in an effort to place all of the competing cars, regardless of their size or weight, on a more "level playing field," although it actually favored the heavier weight models. To calculate the ton miles per gallon, the weight of the car and passengers (in tons) was multiplied by the distance traveled, then divided by the amount of gasoline consumed. This formula was never completely understood by the motoring public and most advertising during this era focused more on the car's actual miles-per-gallon performance. The ton miles per gallon formula was discontinued following the 1958 Economy Run.
- **Topping Off.** This procedure was developed to accurately measure the amount of gasoline in each competing car. During the impound period, each car's gas tank was calibrated by leveling each car on a specially built platform. Special marks were then placed on each fender to indicate the proper level height and the tanks were filled to a predetermined mark set on the filler pipe, close to the gas cap. Prior to the start, each car was leveled, by use of hydraulic bumper jacks, and the gas tanks then filled to the set mark on the filler pipe. At each overnight stop (beginning in 1954) and at the finish, this same procedure was followed using the specially built platforms and portable gas pumps that were transported from stop to stop. These pumps had been previously calibrated to register in gallons and hundredths as opposed to the normal gallons and tenths. Beginning in 1959, specially designed 21-gallon gas tanks were installed in the trunk of each car, which resulted in a more accurate measurement of gasoline used. These tanks were hooked up to the regular gas line and included both leveling and temperature monitoring devices. These also eliminated the need for the specially constructed refueling platforms.
- **Highways.** During the later years of the Economy Run, today's Interstate Highway System was still being developed. All references in this book to route numbers and highway names are used as they were known at that time. As an example, in 1960 the drivers traveled over a highway named the "New Mexico Freeway," which today is known as Interstate 40.

Introduction

The purpose of the Mobilgas Economy Run was to provide the general public with an impartial and practical road test to determine the mileage potential of American passenger cars under driving conditions experienced in a year's time by the average motorist.

During the fifties and sixties, gasoline economy was not one of the most important aspects for car owners during what has been rightfully described as a "car crazy era." For those living in Southern California, and particularly in the Los Angeles area, it was difficult not to know about the Mobilgas Economy Run. Articles announcing the upcoming event, along with reports on the daily progress of the competitors, were common in area newspapers. Occasionally you would see some of the brightly lettered entries, with Flying Red Horse emblems on the hood and doors, on the streets as the drivers made various break-in and test runs. Immediately following each year's Run, articles and advertisements, some of which featured pictures of the competing cars, were featured in the Sunday automotive sections, and in the months following, articles appeared in various automotive-related magazines. After the completion of the Run, some of the class winning entries would be on display in dealers' showrooms along with their trophies. Adding to this exposure, Floyd Clymer, a well-known publisher of automobile-related books, began publishing a yearly book on the Economy Runs in 1950, and except for 1956 these books would appear yearly through 1959.

From humble beginnings in 1936, when Mobil's predecessor, the Gilmore Oil Company, began staging one-day trips to Yosemite National Park, the Economy Runs would evolve and grow in both size and scope, although basically remaining a western-states event through the fifties. As the Runs evolved into cross-country events beginning in 1964, traveling to the Mobil Pavilion at the New York's World Fair, national interest would continue, although at reduced levels. As the sixties progressed, individual new car dealers, who were the backbone of the earlier Runs, would begin to step aside as factory personnel took command of their programs, each determined to outshine the competition and profit from the resulting publicity. Competition for the public's attention in a car's gas economy would surface in 1960, when the Pure Oil Company began staging their Economy Trials, held at the Daytona International Speedway. Renamed and expanded in 1962, the Pure Oil Performance Trials were designed as a three-part test of new cars' acceleration, braking and economy. While never approaching the scope of the Mobil Economy Runs, the Pure trials did pit two well-known oil companies against each other in vying for the public's attention.

As imported models were becoming a more common sight on the roads by the latter half of the fifties, Mobil sponsored the Mobil Mileage Rally beginning in 1958, with some of the regular Economy Run drivers and USAC officials participating. Although some spectacular mileage figures were obtained, these rallies never captured the press's or public's attention, and were ultimately canceled in 1961 after a four-year run. Through the years, Mobil-sponsored Economy Runs would become part of the automobile landscape in Europe, Asia, and elsewhere, with competitions held in France, Great Britain, the Philippines, Malaya, Australia, New Zealand and South Africa, to name a few.

No history of the Economy Runs can be written without acknowledging several individuals who were instrumental in developing and guiding these events from the very beginning. Earl Gilmore, president of Gilmore Oil, revived in 1936 what was then known as the "Camp Curry Yosemite Economy Run." Gilmore provided the promotional backing for the Gilmore Yosemite Economy Runs, which evolved into the Gilmore Grand Canyon Run in 1941. Clarence S. Beesemyer, vice-president and general manager of Gilmore Oil, was also an early supporter of these Runs, and his leadership would continue into the mid–fifties after the merger of Gilmore Oil and General Petroleum in 1945. As the first postwar Economy Run took place, Frank Meunier, General Petroleum's advertising manager, began his active and enthusiastic involvement, which continued until the Run's cancellation in December 1968. Mobil Oil's special events and public relations personnel, including Ken Downs, Roger Mahey and John Parkinson, saw to it that members of the press were well treated, wined and dined in an effort to make sure factual, positive stories would be written, and for the most part they were.

Seeing a need for independent supervision of the Economy Runs, Gilmore Oil turned to the AAA Contest Board; directed by Chairman Col. Arthur W. Herrington, it was the sanctioning authority in the United States for automotive racing and endurance events. Under the direction of Arthur C. (A.C.) Pillsbury, West Coast regional director, rules and procedures were established which could guide the competition while attempting to ensure fairness and honesty in the conduct of the Run. Other AAA Contest Board officials who were prominent in supervising the Run in its early days included former race drivers Tommy Milton and Earl Cooper, Indianapolis Motor Speedway vice-president T.E. "Pop" Myers, and Los Angeles businessman J. Gordon Betz. In 1955, after several fatal racing accidents, the AAA Contest Board was dissolved. This group led the establishment of the United States Auto Club, which continued the sanctions and supervision of these events. A.C. Pillsbury was the chief steward responsible for the conduct of the Economy Runs from the beginnings in 1936 through 1966. After his death in September 1966, Art Rene, a former Economy Run driver, was appointed Chief Steward.

Created as an advertising promotion by Gilmore Oil, General Petroleum and their successor companies, the Economy Runs were generally considered to be a success. By the late sixties, performance was the name of the game, with gasoline economy becoming less of a factor. This ultimately led to Mobil Oil's decision to cancel the Economy Runs. The Runs were not without some controversy and skepticism from the public during their time, since the drivers possessed such skill in delivering excellent performance that the results were above those normally attained by the average motorist. However, the Economy Runs remain today an interesting and significant chapter in America's automotive history.

1

The Gilmore Years, 1936–1941

Since the automobile was first produced, various types of reliability and endurance runs were staged in an effort to publicize one manufacturer's success over all others; for the most part, these were individual efforts by the manufacturer. In 1917, there was a renewed interest in fuel economy. In an effort to publicize the mileage potential of their cars, the Los Angeles Motor Car Dealers Association began to participate in the annual Camp Curry Yosemite Economy Run and actively publicized their results. Held during the spring, these were two-day events with an overnight stop in Fresno. At first these were loosely organized; however, beginning with the 1924 Run, they were sanctioned and supervised by the Contest Board of the American Automobile Association with safeguards to minimize cheating. Price classes were established so competition would be between like models. In addition, a ton miles per gallon formula would be used to determine the overall sweepstakes (for attaining the highest ton miles per gallon figure) and price class winners. These Camp Curry Yosemite Economy Runs were sponsored by Standard Oil of California, so all entries used Standard's Red Crown gasoline. The rules were further tightened in 1925 with the requirement that "all cars were to be the same as a car that would be purchased off the showroom floor," and with "de-clutching" or coasting also being prohibited.[1] After 1926, these Economy Runs would fade into history. Then, in the early thirties, individual economy tests would be conducted by various manufacturers and oil companies, but nothing on the scale or organization of these former Camp Curry events. With the involvement of a major West Coast oil company and the AAA Contest Board, a revival of the Yosemite Economy Runs would become a reality.

Earl Gilmore, president of Gilmore Oil Company, led the company's involvement in various promotional and advertising activities, including sponsorship of automobile racing. Gilmore Oil, the largest independent oil company in the west, with its headquarters in Los Angeles, provided sponsorship for cars that set numerous land speed records in the 1930s. The company also participated in the Indianapolis 500 between 1932 and 1939 and again in 1941, winning the event in 1935 and 1937. Gilmore Oil's sponsored cars appeared at tracks all across the country and even in Vancouver, Canada. During the 1930s, the most popular and successful cars on the Pacific Coast tracks of California, Oregon and Washington were Gilmore sponsored. This success also led to the construction of Los Angeles' Gilmore Stadium in 1934. Located adjacent to the famed Farmers Market, it was the scene of popular midget races, which continued to be held through 1950. Earl, who passed away in 1964, was inducted into the National Midget Auto Racing Hall of Fame in 1989 and into the National Sprint Car Hall of Fame in 1992.

One of the operations of the American Automobile Association, since its organization in 1902, was the AAA Contest Board. Originally established to sanction attempts to establish world records for speed and endurance, the board was also involved in racing, hill climbs and reliability runs. Some of the best-known events sanctioned by the AAA Contest Board through the mid-fifties included the Glidden Tours, the Indianapolis 500, and attempts at land speed records at the Bonneville Salt Flats. Arthur C. (A.C.) Pillsbury was a construction engineer and developer who designed and constructed the majority of the board tracks across the country, including the Beverly Hills Speedway. By the mid-thirties he was a AAA Contest Board regional director, becoming the top official at the Bonneville Salt Flats Speed and Endurance Run. Well known for his strict, but fair, interpretation of the rules, he was appointed chief steward of the Gilmore Yosemite Economy Run and successor events. Considered the dean of the world's automotive contest officials, he was inducted into the National Sprint Car Hall of Fame in 1992.

1936

Los Angeles to Yosemite

January 9, 1936—352 Miles

In 1935, discussions were being held between various oil companies and Los Angeles area motor car dealers on a revival

Well-known racing driver Wilbur Shaw (left) is presented with his trophy by Earl Gilmore, president of the Gilmore Oil Company. Wilbur finished fourth in this price class with 46.40 ton miles per gallon in his Hudson Deluxe 8 on the 352-mile trip from Los Angeles (courtesy of Chrysler Group LLC).

of the Camp Curry Yosemite Economy Runs. At this time the Los Angeles–based Gilmore Oil Company, under the leadership of President Earl Gilmore and Vice-President Clarence S. Beesemyer, stepped in with a decision to revive this event with entries using their Red Lion brand gasoline and Lion Head motor oil. The AAA Contest Board was approached and agreed to sanction and supervise the Run under the direction of Arthur C. Pillsbury, its West Coast regional director. Based on criteria that were developed for the earlier Camp Curry Yosemite Economy Runs, rules and eligibility were established and the inaugural Gilmore Yosemite Economy Run was scheduled for January 9, 1936. The Los Angeles area's new car distributors and dealers became active participants. Each entry was prominently lettered on the sides and rear identifying it as a participant in the "GILMORE YOSEMITE ECONOMY RUN," together with the make and the entrant's name. This lettering style to promote the Run created "rolling billboards" and would continue to be used in future Gilmore Economy Runs, and with some modifications throughout the years of the Mobilgas events.

The inaugural Gilmore Yosemite Economy Run would be a 352-mile contest which required the drivers to complete the route within a 13-hour time limit. This time limit would require the drivers to average 27.1 miles per hour.

Entries were open to any stock 1936 metal-top sedan and most manufacturers were represented, although Buick, Cadillac, Dodge and Ford declined to participate. A total of 30 cars were entered, competing in seven price classes according to their delivered prices in Los Angeles. With the exception of Chevrolet, all of the entries were sponsored by distributors or dealers. Two of these, Ralph Hamlin Motors (Graham) and Maddux, Inc. (Lincoln), were veterans of the last Camp Curry Yosemite Economy Run held in 1926. Two well-known race drivers participated, with Wilbur Shaw driving a Hudson

Deluxe 8 and Babe Stapp driving a Terraplane Deluxe 6. Both were entered by Hudson's Los Angeles distributor, Earle C. Anthony, Inc.

Each entrant, after notifying AAA Contest Board officials of their intent to participate, was permitted to select from their inventory the actual car which would be competing. In the majority of cases, these represented the most popular-selling sedan models and were well broken in, each having been driven well over 1000 miles, and in some cases several thousand miles. On January 8, all of the competing cars were driven to Gilmore Stadium, where they were impounded and inspected for stock status by AAA Contest Board Officials. Each car was given a full tank of gasoline and was then weighed (along with all passengers and luggage) before being lined up in preparation for the next morning's start.[2] In addition to the driver and observer, each car also carried two or three passengers who were named by the entrant.

In the early morning hours of January 9, the drivers and observers were provided with detailed maps and instructions concerning the route, which involved two climbs to over 4000 feet in elevation. The rules covering the event were also presented, specifying that all cars must be kept in gear (declutching and coasting were forbidden) with the motor running at all times when the car was in motion.[3] Impartial observers, who were nominated by a competing entrant, rode beside the driver during the Run to ascertain that all rules were strictly adhered to. These procedures would remain in effect through 1941, although minor changes would be made reflecting current conditions.

Promptly at 5:00 a.m., the first car was flagged away by William Claus of the AAA, with the remaining cars leaving at one-minute intervals. Leaving Los Angeles, the drivers followed a familiar route crossing over the Ridge Route and into the San Joaquin Valley. In Bakersfield and Fresno, refueling stations under AAA Contest Board supervision "were provided for drivers, if they desired, to replenish their supply of gasoline, oil and water."[4] Arriving in Merced, each car was required to stop and check in with the AAA Contest Board before continuing to the finish line in Yosemite National Park. Here they would be provided with the latest road and weather conditions as they began the last leg of the trip. At all of these stops, no time allowances were provided, and the drivers were anxious go get back on the road. Throughout the day the drivers encountered headwinds along with a driving rainstorm and some snow in the mountainous terrain as they neared Yosemite. The first car to cross the finish line was a Willys "77" 4, driven by L.T. Sollender. After completing the Run, all cars were turned over to AAA Contest Board officials, "who under the watchful eyes of drivers, and observers, measured the gasoline, oil and water consumed with calibrated measurements."[5] Following this refueling procedure, each car was thoroughly inspected by AAA officials to verify its stock status; the officials then worked into the evening tabulating the results, which were scheduled to be announced on the morning of January 10. While this work was continuing, the competitors and others affiliated with the Run were guests of Gilmore Oil at the "Mileage Dinner and Dance" held at Yosemite's Ahwahnee Hotel.[6]

With all of the competing cars arranged in a semicircle in front of the Government Center Building, it was announced that Clay Moore, driving a Graham Supercharger 6, captured the coveted Sweepstakes Award with a figure of 55.47 ton miles per gallon. Coming in behind the Graham for the Sweepstakes Award was a Chrysler Airflow 8 with 53.35 ton miles per gallon. A Willys "77" 4 recorded the best miles per gallon figure with a record-setting 33.21, a record that would stand during the entire span of both the Gilmore and Mobilgas Economy Runs. The best miles per gallon performance for a six-cylinder car was the Sweepstakes-winning Graham, with a figure of 26.67, and the best performance for an eight-cylinder was a Hudson Deluxe 8 with 22.85. With 15.30 miles per gallon, the Reo 6 recorded the lowest average for all entries, and would stand until 1966. Gilmore Oil Company's President Earl Gilmore and Vice-President Clarence S. Beesemyer were among those from Gilmore Oil and the Yosemite Park Curry Company to present the winning drivers their trophies.

First-place trophies were presented to:

PRICE CLASS	MAKE and MODEL	TON MILES PER GALLON	MILES PER GALLON
1-A — Under $700	Willys "77" 4	49.48	33.21
2-A — $701 to $900	Graham Crusader 6	47.61	25.33
3-A — $901 to $1000	Pontiac Silver Streak 6	47.89	23.95
4-A — $1001 to $1100	Studebaker Dictator 6	50.98	24.28
5-A — $1101 to $1300	Graham Supercharger 6	55.47	26.67
6-A — $1301 to $1600	Studebaker President 8	48.63	20.35
7-A — $1601 to $1800	Chrysler Airflow 8	53.35	18.33

Final results for all competing cars can be found in the Appendix.

In announcing the results, the AAA Contest Board noted, "The thirty cars entered in the run achieved the composite average of 22.473 miles per gallon of Red Lion Gasoline," with all cars completing the Run in the specified time limit. In reporting the story and results, the *Los Angeles Times* noted, "This 'Stingy Run' was the biggest competitive stock car event ever held in the United States."[7] Over the next five years, these Runs would be referred to in some press reports as the "Stingy Driver's Run."

Longtime Los Angeles distributor Ralph Hamlin was especially pleased. Not only did his Graham Supercharger 6 capture the Sweepstakes Award, but a Graham Crusader placed first in its price class, despite being hampered by a minor accident en route to Yosemite.[8] Also highly pleased was Wallace Austin, vice-president of the Paul G. Hoffman Co., Inc., Los Angeles Distributor for Studebaker, whose entries both placed first in their respective price classes. A supporter of this type of events, Mr. Austin remarked, "While it is true that the new cars in the hands of owners have been building up excellent gasoline mileage records, we have felt the need for an official contest of some sort which would substantiate our claims and differentiate them in the minds of the motoring public."[9]

While there was some national advertising in newspapers, mostly by Graham and Studebaker, the bulk of the advertising was on the West Coast. In was not uncommon in the days following the Run for the winning dealers to advertise the results in Los Angeles area newspapers, with some of the cars being prominently displayed in their showrooms along with their trophies. Magazine advertising was quite limited. Graham noted it was "The Car That Beat Them All" in ads in both *Life* and the *Saturday Evening Post* that featured pictures of the winning cars.[10] Studebaker advertised their cars as being "Gas Economy Champions" throughout 1936, although most of these ads did not mention Gilmore, instead referring to "the National Economy Classic in California earlier this year."[11] Pontiac announced its winning performance in a two-page ad in the industry publication *Motor* illustrated with photographs of the Deluxe 6 sedan.[12] During the 1937 model year, Pontiac published a special eight-page brochure titled "1937 Automotive Economy News" that included a photograph of the class-winning 1936 Deluxe 6 sedan.[13]

Gilmore Oil ran newspaper advertisements noting "Thirty 1936 Model Cars Average 22.473 Miles Per Gallon" using Gilmore Red Lion.[14] As would be a feature of future advertising, these ads stressed the actual miles per gallon performance of the winning cars.

Dick Klein, driving a Hudson Deluxe 8, contends with icy road conditions on the All-Year Highway leading to Yosemite National Park. Weather conditions would continue to deteriorate, resulting in many of the drivers installing chains as they neared the finish at the Government Administration Building (courtesy of Chrysler Group LLC).

Newspaper articles were featured in Sunday automotive sections, and again these were primarily on the West Coast. Gilmore Oil's monthly publication, *The Gilmore Cub*, distributed by Gilmore dealers, featured stories and pictures in its February–March issue. For historians, in addition to newspaper articles, the feature articles found in *The Gilmore Cub* remain one of the prime sources for information on the Gilmore Runs through 1940. The few magazine articles that appeared were mainly found in industry publications, including *Motor* and *Motor West*.[15] Coverage of Studebaker's success was also found in their magazine, *The Studebaker Wheel*, distributed to Studebaker owners.

Motor West magazine, in its article covering the 1936 Gilmore Yosemite Economy Run, noted: "The affair was held under strict AAA Rules, and was ably managed. Much credit for its success goes to Earl Gilmore, President of the Gilmore Oil Co., which promoted the contest."[16] Based on the success of this inaugural event, plans were begun for the Gilmore Yosemite Economy Run to be held in 1937.

1937

Los Angeles to Yosemite

January 7, 1937—352 Miles

Early in December of 1936, plans were formally announced for the 1937 edition of the Gilmore Yosemite Economy Run during a presentation to manufacturers, dealers and newspapermen by officials of Gilmore Oil and the AAA Contest Board. Returning to Yosemite National Park, Gilmore Oil Company Vice-President Clarence S. Beesemyer remarked that early interest by the motor car dealers indicated that this year's Run would attract larger group of entries compared to last year's event. Arthur C. Pillsbury, AAA Contest Board West Coast regional director, would again supervise this year's Run, noting that greater emphasis would be made to ensure the stock status of each competing entry.[17]

On January 7, the drivers would face a 352-mile trip to Yosemite National Park, to be covered within a time limit of 11 hours. The highest elevation in this year's Run would be 4213 feet at Tejon Pass along the Ridge Route.

Entries were open to any stock 1937 metal-top sedan, with most manufacturers being represented. There were a few notable no-shows this year, including Buick, Cadillac, Ford and Studebaker. Studebaker's decision to sit out this year's Run was a surprise to most after its successful participation in 1936 and subsequent national advertising campaign. A total of 26 entries competed in eight price classes based on their delivered prices in Los Angeles. Although a larger group of entries had been predicted, economic conditions with various manufacturers, including Auburn, Hupmobile and Reo, kept the number of competitors under 1936's total. With the exception of the Chevrolet and Pontiac, all of the entries were sponsored by Los Angeles–area distributors or dealers, and most were veterans of last year's Run. Clay Moore, again driving a Graham Supercharger 6, returned to defend his Sweepstakes-winning performance.

In expanding the time for inspections, each entry was turned over to AAA Contest Board officials two days prior to the start and would remain under their control until after the completion. During this impound period, AAA technicians would conduct an expanded inspection of each entry to ensure that all components meet published stock specifications.[18] These inspections included a partial dismantling of the engine and carburetor. In its first year of participation, the Dodge 6 "failed to appear for checking and withdrew."[19]

On January 7, as the participants prepared for their trip to Yosemite, some of the coldest temperatures in years were being predicted. An overriding concern for the drivers was the constant threat of a frozen motor, since AAA Contest Board rules "specified that only water, without the addition of foreign compounds to prevent freezing, could be used."[20] The drivers knew very quickly the day was going to be a test of endurance, for them and their cars. At 5:00 a.m., A.C. Pillsbury sent the first car, a Willys 4 driven by Otis M. King, on its way, with the remaining entries leaving at one-minute intervals. Jack Stotts, in another Willys 4, made a wrong turn leaving Gilmore Stadium, driving off-course before discovering his error and getting back on track.[21] As the drivers made their way over the Ridge Route, they encountered their first weather challenge: snow developed into a blinding storm, slowing their progress. Two drivers encountered problems with their cars which resulted in brief delays. Roland L. Amberg had the radiator freeze on his Pontiac Deluxe 8, but after draining and refilling, he was able to continue. N.C. Williams had to stop to clean both carburetor jets in his Cord Beverly 8.[22] Descending into the San Joaquin Valley, the drivers had a brief respite from the snow as they headed to Bakersfield and Fresno for refueling stops at the option of the drivers. Leaving Bakersfield, they continued on their way to Fresno and Merced. After arriving in Merced, each driver was required to stop and check in with AAA Contest Board officials. They received the latest weather and road conditions before continuing on the final leg over the All-Year Highway into Yosemite.

Leaving Mariposa, as the drivers began their climb into the mountains, a heavy snowstorm and ice conditions made the road difficult to negotiate. For many of the drivers this made it necessary to install chains before continuing, which did nothing to improve their fuel economy. At several points along the way, non-competing cars skidded on the icy surface, blocking the road. One of these cars was carrying Gilmore

Oil Company's President Earl Gilmore, causing some impatience for the drivers attempting to stay within the 11-hour time limit.[23] Nearing the finish, a mechanical problem sidelined the Graham Custom Supercharger 6, driven by Erwin "Cannonball" Baker, when the engine stalled; "when attempting to restart [he] broke [the] distributor shaft."[24] Just before 3:00 p.m., the first car to cross the finish line at the Government Administration Building was a Willys 4 driven by Jack Stotts. By 5:00 p.m., with temperatures in the single digits, everyone had arrived safely, and the cars, with their fenders and radiators encrusted with ice, were again impounded by AAA Contest Board officials for refueling and inspections. At this time AAA officials announced: "Due to unusual road and weather conditions causing entrants to be delayed by non contestants and busses blocking the road, the time limit was extended to 12 hours."[25] At this time, Jack Stotts was notified that he was being disqualified under the rules for taking the wrong route leaving Gilmore Stadium.[26] During the final refueling process, it was noted that some drivers had elected not to add fuel along the way. This including Kenneth Dobbins, driving a DeSoto 6. After arriving at the finish he had approximately 12 ounces of gasoline remaining and later remarked, "I thought about getting gasoline at Merced, but I guess I didn't need it."[27] As AAA Contest Board officials worked well into the night tabulating the final mileage figures, Gilmore Oil hosted its annual "Mileage Dinner and Dance" at Yosemite's Ahwahnee Hotel.

In cold temperatures, with a good deal of snow visible on the ground, the awards ceremony was held at 11:00 a.m. on January 8 in front of the Government Center Building. With the competing cars arranged in a semicircle in front of the Government Center Building, Gilmore Oil Company's President Earl Gilmore and Vice-President Clarence S. Beesemyer were again on hand to present the trophies to the winning drivers. For the second year in a row, Clay Moore and his Graham (Series 116) Supercharger 6 captured the Sweepstakes Award with a figure of 53.4 ton miles per gallon. Placing second for the Sweepstakes Award was a Graham Cavalier 6 with 52.246 ton miles per gallon, followed closely by a Hudson Deluxe 8 with a mere .014 ton mile per gallon separating the two. Once again a Willys 4 recorded the highest miles per gallon figure for all entries with 28.16. This was 5.05 miles per gallon less than 1936's record-setting 33.21 miles per gallon, a result of this year's faster pace along with poor weather and road conditions. The highest miles per gallon performance for a six-cylinder car was 24.62 set by a Graham Cavalier, and the best performance for an eight-cylinder car was by a Hudson Deluxe 8 with 22.71.

First-place trophies were presented to:

PRICE CLASS	MAKE and MODEL	TON MILES PER GALLON	MILES PER GALLON
A	Willys 4	49.2	28.16
B	Chevrolet Master Deluxe 6	41.8	19.45
C	Terraplane Deluxe 6	48.4	22.00
D	Graham Cavalier 6	52.246	24.62
E	Graham 116 Supercharger 6	53.4	23.95
F	Nash Ambassador 8	49.6	20.35
G	Packard One-Twenty 8	43.6	18.05
H	Cord Beverly 8	45.6	18.24

As of publication date, the price class breakdown remains unknown. Final results for all competing cars can be found in the Appendix.

In announcing the results, the AAA Contest Board reported, "The twenty-three cars to finish the event made a combined average of 20.814 miles per gallon of Gilmore Red Lion gasoline. This average was a drop of 1.659 miles per gallon from 1936's average of 22.473. A further check-up revealed that every car had a full crankcase of Lion Head motor oil and not a drop of the lubricant was consumed by any of the cars during the course of the run."[28] With the extended time limit of 12 hours, the drivers averaged a speed of 32.8 miles per hour. Even with the challenging weather conditions, this was an increase of 5.7 miles per hour from last year's average. It was estimated that "at least one hundred miles of the 352 mile run from Los Angeles to Yosemite was made through falling snow, drifts, or over icy roads."[29] It was noted this year's Run "was by far the most grueling mileage test ever undertaken by such a representative group of cars in the annals of motordom."[30]

No one was more pleased with this year's results than longtime Los Angeles distributor Ralph Hamlin. Not only did his Graham (Series 116) Supercharger 6 capture its second consecutive Sweepstakes Award, a Graham Cavalier 6 recorded the best miles per gallon figure for a six-cylinder model. If there was any disappointment this year, it was the mechanical problem which sidelined a Graham (Series 120) Custom Supercharger 6 near Yosemite National Park.

In the midst of serious financial difficulties and declining sales, the one bright spot for the Auburn Automobile Company and its Los Angeles distributor, the Auburn Motor Sales Corporation, was the performance by its Cord Beverly 8. Competing against the Chrysler Airflow 8, last year's price class winner, the Cord had a strong first-place finish by a margin of two ton miles per gallon. By the end of the year, production of the Cord would come to an end.

While there was some national advertising in newspapers, the bulk of this was on the West Coast and was almost entirely by Graham. With the theme "Again! Graham Supercharger Wins Sweepstakes Championship in Gilmore-Yosemite Economy Run," accompanied by a photograph of the winning cars, Southern California newspaper advertising focused on the back-to-back wins by the Supercharger.[31] Outside of Southern California, Graham began using the "Official U.S. Economy Champion" theme in its advertising. Magazine advertising was also quite limited, with Graham the only manufacturer to advertise its achievements. In *Life* magazine, a photograph of the winning cars appeared under the headline "Graham again Captures National Economy Title!" along with a brief outline of this year's Run.[32] This achievement was also noted in Graham's showroom brochures for the 1938 models.

Graham also issued one of the most detailed publications covering the Economy Run with their January 1937 issue of *Graham Supercharger News*, a four-page newspaper-type folder. Featuring photographs and stories of Graham's achievements, the inside spread was a complete copy of the detailed final results issued by the AAA Contest Board. This document contained entry, timing and mileage calculations for all of the entries, providing a valuable source for historians researching the Gilmore Runs.

Gilmore Oil ran newspaper advertisements boasting "20.81 Miles Per Gallon of Red Lion — Average of Twenty-Three 1937 Motor Cars Under AAA Supervision," also stressing that every car competing also used Lion Head Motor Oil. Featuring a photograph of two competing cars battling snowy road conditions, these ads again stressed actual miles-per-gallon performance.[33] Additional advertisements appeared in their monthly publication *The Gilmore Cub*.

Newspaper articles again were featured in Sunday automotive sections and primarily found on the West Coast. *The Gilmore Cub* featured stories and pictures in its February issue, while the few magazine articles to appear were again found in industry publications such as *Automobile Topics*, *Motor* and *Motor West*.[34]

Despite the challenging weather and poor road conditions, everyone associated with this year's Run was pleased with the outcome. Within several months, plans were underway for a return to Yosemite in 1938.

1938

Los Angeles to Yosemite

January 13, 1938—314.5 Miles

It was becoming apparent that the Gilmore Yosemite Economy Runs were beginning to attract a larger nationwide audience, resulting in an increase in participation from various automobile manufacturers and their Southern California–based distributors and dealers. As plans were announced for a return visit to Yosemite National Park, A.C. Pillsbury, AAA Contest Board West Coast regional director, announced procedural changes relating to inspection of the competing cars.

To answer some criticism concerning the congestion which occurred during the two-day impound period, this time was extended to five days "in order that each entrant could be thoroughly checked by engineers for strict stock specifications." Additionally provisions were made for some inspections, which involved the removal of cylinder heads for checking the compression ratio and piston displacement, to be completed at the entrant's shop under supervision of an AAA official. After this was completed, seals would be placed on the cylinder heads to prevent any possible unauthorized modifications.[35] This effort to ensure that no unauthorized adjustments were made to the competing cars would continue to mark AAA's involvement with the Economy Runs.

This year's Run was scheduled for January 13, when the drivers would face a shorter 314.5-mile trip to Yosemite to be completed within an 11-hour time limit. Due to recent storm damage, caused by heavy rains, part of the All-Year Highway was closed for reconstruction work, resulting in the selection of a new route. The drivers would take the Wawona Road, a natural gateway into Yosemite for Southern California motorists. This route, "going directly from Fresno to Yosemite, is now partly a temporary road and in the Sierras it winds and twists through elevations up to and above 6,000 feet. A fair road for leisurely mountain touring, it discourages the driver who wants to wring record-setting mileage out of every [gallon] of gasoline."[36]

Entries were now restricted to any 1938 metal-top *four-door* sedan, with most manufacturers again being represented. For reasons not disclosed, Chrysler was a no-show this year, joining Buick and Cadillac who again declined to participate. After skipping the 1937 Run, both Hupmobile (which suspended manufacturing in 1937) and Studebaker were back. Ford would participate for the first time and would now create competition between the traditional "Low Price Three" models. A total of 26 entries would compete in ten price classes, based again on their delivered prices in Los Angeles. Two of the price classes, featuring luxury models from Lincoln and Packard, had a single entry each. Los Angeles area distributors and dealers again sponsored the bulk of the entries, with Chevrolet's and Pontiac's entries by their dealer associations. Several who were participating for the first time and included Hollywood Ford dealer Al Stuebing, Glendale Hudson dealer Boswell & Gratiot, and Los Angeles Oldsmobile dealer Kurtz Motor Company. These were joined by Hupmobile's longtime Los Angeles distributor Greer-Robbins Company. Returning in hopes of capturing his third Sweepstakes Award was Clay Moore in a Graham Supercharger 6.

Early in the morning of January 13, everyone connected with the Run gathered at Gilmore Stadium for the trip to Yosemite National Park.

Although the weather forecast looked favorable, the drivers knew that several challenges would be on tap as they traveled over a different route than in the previous two years. At 5:00 a.m., the first driver sent on his way was R.R. Rice, driving a Willys Deluxe 4, with the remaining entries leaving at the traditional one-minute intervals. The route out of Los Angeles was well known to the drivers as they headed over the Ridge Route and into the San Joaquin Valley, bound for Bakersfield and a refueling stop if the drivers so desired. Arriving in Bakersfield, Joseph Bozzani discovered a problem with his Dodge 6 as "he had lost a considerable amount of gasoline en route, caused by a leak in the gas gauge connection." Although test runs showed that he should get exceptionally good mileage, this problem eliminated any hopes of a top finish in his price class.[37] Continuing north, the drivers made their next stop for refueling in Fresno, where each entry was also required to check in with AAA officials. Leaving Fresno, the drivers headed in a northerly direction towards Yosemite, stopping in Coarse Gold to check in with AAA officials and receive weather and road condition reports for the last leg. Even with improved weather conditions compared to last year, the drivers had to contend with heavy fog and a slippery highway which led them through water up to the wheel hubs at one point. Crossing the finish line in front of the Government Administration Building, every one of the 26 entries arrived well within the 11-hour time limit. After last year's weather, everyone was pleased to see an absence of snow on the ground. At this time, AAA Contest Board officials began the task of refueling and performing detailed inspections for stock status. They then worked well into the night calculating the results to be announced at an awards ceremony the next morning. While this work was being done, Gilmore Oil again hosted its annual "Mileage Dinner and Dance" at Yosemite's Ahwahnee Hotel.

For an unprecedented third year in a row, Clay Moore and his Graham Supercharger 6 again captured the Sweepstakes Award with a figure of 55.927 ton miles per gallon. Placing second for the Sweepstakes Award was a Lincoln Zephyr 12 with 54.744 ton miles per gallon, followed by a Nash Ambassador 6 with 53.559. Once again a Willys Deluxe 4 recorded the highest miles per gallon figure for all entries with 30.534. With improved weather and driving conditions, this figure was a 2.374 miles-per-gallon improvement over the 1937 Willys' average of 28.16. The best performance by a six-cylinder car was 25.779 miles per gallon by a Hudson Terraplane Deluxe. The best performance by an eight-cylinder car was an impressive 28.850 miles per gallon set by a Ford "60."

First-place trophies were presented to:

PRICE CLASS	MAKE and MODEL	TON MILES PER GALLON	MILES PER GALLON
A	Willys Deluxe 4	48.900	30.534
B	Ford "60" 8	50.767	28.850
C	Hudson Terraplane Deluxe 6	50.461	25.779
D	Pontiac Deluxe 8	46.756	20.967
E	Studebaker Six	51.807	24.380
F	Nash Ambassador 6	53.559	23.647
G	Hudson Deluxe 8	49.409	22.790
H	Graham Supercharger 6	55.927	25.779
I	Lincoln Zephyr 12	54.744	23.470
J	Packard Eight	46.543	16.211

As of publication date, the price class breakdown remains unknown. Final results for all competing cars can be found in the Appendix.

AAA Contest Board officials reported that the field achieved a combined average of 22.88 miles per gallon, the highest average in all three years and just over a two-mile-per-gallon improvement over 1937's average. With this year's improved weather and road conditions, every competitor finished the Run "within the 11 hour time limit and the last arrival came in 24 minutes before the limit."[38] It was noted that these records "are all the more impressive and are considered to be an outstanding tribute to automotive and petroleum industries because the cars in the event had to follow a grueling temporary winter route for part of the run."[39]

Los Angeles distributor Ralph Hamlin celebrated his third consecutive Sweepstakes Award with a Graham Supercharger 6, thanks in part to the driving performance of Clay Moore. Even with its strong three-year record of performance, this would be Graham's last year of participation. Two Los Angeles distributors also scored their second consecutive price class wins, Nash's Homer C. Thompson and Packard's Earle C. Anthony, Inc. This year also marked Homer Thompson's second consecutive price class win as a driver. A Hudson Terraplane placed first in its price class for the second year in a row.

In its first year of competition, Ford surprised those associated with the Run with the strong performance of its "Thrifty 60" sedan. Not only did the Ford achieve the highest miles-per-gallon average for all eight-cylinder cars, it also was the third highest for all competing cars. The only two cars which scored higher in miles per gallon were the four-cylinder Willys Deluxe. This price class win would be the first of several for Hollywood Ford dealer Al Stuebing, who would remain active in the Economy Runs through the early fifties.

Newspaper advertising saw a very slight increase this year thanks to the efforts of both Ford and Lincoln. With the headline "Thrifty '60' Ford V-8 Averages 28.85 Miles Per Gallon!," photographs of the winning Ford were featured along with details of its winning performance.[40] Additional comments about this mileage performance continued during the remaining months of the 1938 model year. Lincoln followed the same theme, announcing the Lincoln Zephyr's performance with the headline "23.47 Miles Per Gallon With a V-12 Engine!," accompanied by a photograph of the winning model.[41] In *Ford News*, a magazine for owners of Ford Motor Company products, readers were told of Ford's and Lincoln's achievements.[42] Even with its third Sweepstakes Award, newspaper advertising for the Graham Supercharger was extremely limited, as Graham instead focused its efforts on magazine advertising. Continuing to refer to itself as the "Official A.A.A. Economy Champion," Graham advertised its Sweepstakes performance in general interest magazines such as *Life* and *The Saturday Evening Post* throughout 1938. Studebaker also noted the performance of its Six (later in the model year renamed Commander) and President models in various publications.

Gilmore Oil ran newspaper advertisements announcing "Proof of Record Breaking Mileage," noting: "Twenty-Six 1938 Stock Sedans in Gilmore Yosemite Run Average 22.88 Miles Per Gallon of Red Lion Gasoline," and continuing to stress that every car competing also used Lion Head Motor Oil.[43] Additional advertisements appeared in their publication *The Gilmore Cub*.

In its first advertising effort associated with the Economy Run, Champion Spark Plugs announced, "You Can Depend on Champions For Maximum Economy" in ads appearing in general interest magazines. Featuring a rendering of the Sweepstakes-winning Graham Supercharger during the awards ceremony, the text covered a brief overview of this year's event, noting that the winning entries were equipped with Champions. No direct mention was made of the Graham or other competing models.[44]

Newspaper articles continued to be featured in Sunday automotive sections, primarily on the West Coast. Gilmore Oil's monthly publication *The Gilmore Cub* featured stories, results and pictures in its March issue. The few magazine articles that appeared continued to be found mainly in industry publications such as *Automobile Topics* and *Motor West*.[45]

After three years, the Gilmore Yosemite Economy Runs had achieved a nationwide level of interest and acceptance both by various industry officials and by the motoring public. An example of this acceptance was a statement to Gilmore Oil Company's Earl Gilmore from Felix Doran, Jr., Chevrolet's general sales manager:

> In my opinion such an event as the annual Gilmore economy run is a highly constructive effort reflecting credit on all concerned. It focuses public attention on the constant progress which engineers are making not only in development of fuels and lubricants but in the improvement of the automobile. Please accept sincere congratulations on the originality and initiative it takes to evolve such a program and see it successfully through.[46]

The comments from Chevrolet's general sales manager shows the depth of support now evident from those participants whose models did not regularly place at the top of their respective classes. In the three years of the Gilmore Runs, Chevrolet's only first-place performance was in 1937. There was no doubt in anyone's mind that the Gilmore Yosemite Economy Runs would return once again in 1939. What was unknown were the challenges that would face the drivers and AAA Contest Board officials as well.

1939

Los Angeles to Yosemite

JANUARY 5, 1939—314.5 MILES

Pleased with the results and feedback from the participants in the 1938 Gilmore Yosemite Run, Gilmore Oil Company and AAA Contest Board officials announced another return to Yosemite National Park. Scheduled for January 5, the 1939 Run would see the drivers covering within an 11 hour time limit the same 314.5-mile route used in 1938 as repairs to the All-Year Highway had not been completed.

The only change AAA Contest Board officials felt necessary was a more enhanced inspection procedure. Now impounded a week prior to the Run, "AAA technicians literally took every car apart to ascertain that all parts conformed strictly to stock specifications. Heads of the motors were taken off, carburetors were taken apart, gas lines were checked, spark plugs were inspected. Grease and oil were removed from all parts of the lubrication system and every car was relubricated as specified by the Chek Chart system to be sure that each car was identical to that being driven by the average motorist."[47] Speaking with *Los Angeles Times* Auto Editor Lynn Rogers, Oldsmobile engineer J.W. Schiller remarked that he "had no idea the AAA people were so thorough in preparing for this event. Imagine their even looking at an air-cleaner to see if there could be any obstructions. No wonder this run is so successful, for they do everything possible to be sure that the cars entered are exactly the same as any motorist can buy."[48]

Entries were open to any 1939 metal-top four-door sedan with one of the widest ranges of makes participating. Only Buick and Cadillac continued their decision not to participate. The recently introduced Mercury made its Economy Run debut, along with Chrysler's return after a one-year absence. A total of 29 entries would compete in ten price classes based on the "Los Angeles prices according to rail freight delivery for standard five passenger, four door sedan, with standard equipment."[49] Continuing the practice from previous years,

Los Angeles area distributors and dealers sponsored the bulk of the entries. The exception to this were the five entries from General Motors. The Chevrolet entry continued to be sponsored by its Los Angeles Dealers Association, while the Oldsmobile and Pontiac entries were sponsored by their respective divisions. Reflecting on the increasing importance of this event, Oldsmobile's engineer J.W. Schiller traveled from its Lansing headquarters to drive the Oldsmobile "80" 8, becoming the first factory representative to have an active role.[50] Participating for the first time was Nash's new Southern California distributor, Nash California Company, joined by two area Chrysler dealers, Los Angeles' Lysle Greenman and Santa Monica's Ray E. Shafer. Three of the four joint sponsors for the new Mercury were first-time participants: Los Angeles dealers J.E. Coberly, Boyd Gibbons and Ham Nerney.

A surprise by its absence was the three-time Sweepstakes winner Graham Supercharger 6 and its driver Clay Moore. Although concrete documentation is lacking, it is presumed that this model was entered and then withdrawn during the impound period. This presumption is based on two factors. One is an article appearing in the *Los Angeles Times* noting, "Thirty stock sedans are undergoing minute examination by technical officials of the American Automobile Association."[51] The second factor is the starting position formula used by AAA Contest Board officials since 1937. Beginning with no. 1, for a Willys Overland competing in price class "A," these progressed through no. 31, a Lincoln Zephyr in Price Class "J." Starting position 13 was skipped in 1939. Based on available information, this starting position would have fallen in one of the price classes where the Graham had previously competed. Ralph Hamlin would continue his involvement, which would be his last, sponsoring three Hudson entries.

As the participants gathered at Gilmore Stadium in the early morning hours of January 5, weather reports indicated that this would be the most challenging in the short history of the Economy Runs. A.C. Pillsbury, AAA Contest Board regional director, "advised all drivers and observers to expect rain and snow in copious quantities before they reached the finish at the Government Administration Building in Yosemite. He also cautioned the group about the rules which demanded that the motor of the car be kept running at all times the car was in motion, and that no coasting or trick driving would be permitted."[52] This year there were no provisions for refueling at Bakersfield, with all entries being required to stop in Fresno and Coarse Gold. At 5:00 a.m., Bennett Hill in a Willys Overland Deluxe 4 was the first driver to depart, with the remaining drivers following at one-minute intervals. Light rain was falling as the drivers followed the usual route from Los Angeles to Bakersfield over the Ridge Route, where they would climb to an elevation of 4182 feet at Tejon Pass. Here the first of the weather issues surfaced as the drivers faced fog, then sleet mingled with rain, coupled with 45-mph wind conditions, making the driving treacherous on slippery roads until they reached Bakersfield. Already the drivers were facing delays in their planned schedule.[53] Continuing through the San Joaquin Valley, the drivers continued to experience slippery road conditions with high winds as they made their way to Fresno.[54] After refueling and checking in with AAA Contest Board officials, the drivers continued on their way following the Wawona Road to Coarse Gold, where they continued to encounter heavy rain, slippery roads and strong winds, although the weatherman in Fresno had reported earlier that the route was all clear.[55] The worst was yet to come.

At Coarse Gold, AAA officials notified the drivers of possible difficult road conditions on the route into Yosemite. Still battling rain and slippery road conditions, the drivers made their way along the Wawona Road and before reaching the 4000-foot elevation were in the midst of a sudden and unexpected snowstorm. With driving conditions now particularly difficult, AAA Contest Board officials held out as long as they could before requiring that all cars put on chains for the final 42 miles to the finish line. It was now becoming a race against time, since no time allowances were made for chain installation, as the drivers had to literally crawl around sharp curves as they climbed to an elevation of 6256 feet at Chinquapin Summit. At this point there were few straightaways available to make up for lost time. "Observers reported that most of the cars were forced to travel in low gear for miles after the order for chains was given and may of the drivers declared that had to stop repeatedly and dislodge snow and ice which wedged in between wheels and fenders."[56] Several drivers "reported that their cars slid off into snowdrifts and that they had to use up precious quantities of gasoline getting back on the highway."[57] Ray Logan, in a Hupmobile Senior 8, was one of the drivers who slid off the road at Chinquapin Summit, sustaining a broken wheel in the process. Ray was delayed for nearly 20 minutes waiting to be provided with a spare before he was able to continue.[58] Just prior to 3:00 p.m., Herb Ford in a Dodge Special 6 was the first to arrive at the finish line in front of the Government Administration Building, completing the Run in the fastest time for all entries, 9 hours 51 minutes. For the next hour and a half the remaining drivers, who had struggled through the snow and slush, arrived as the biggest crowd in the history of the Run was on hand to welcome them and take part in the various activities planned by Gilmore Oil Company. Norman Williams, driving a Hudson Country Club 8, crossed the finish line without a second to spare. Ray Logan, because of the minor accident with his Hupmobile Senior 8, arrived 20 minutes late and was disqualified. Legendary race driver Barney Oldfield, who was a passenger in the Nash Lafayette, commented, "In all my experience I have never seen more trying conditions for an economy run."[59] The drivers agreed that "this was the toughest battle against time and weather conditions they had ever encountered."[60] Following established procedures, each car was refueled and inspected for stock status before AAA Contest Board officials began the detailed task of computing the final results.

After plowing through the most difficult driving conditions in the four-year history of the Run, 12 of the competing cars crossing the finish line exceeded the 11-hour time limit. Those who ran over the time limit ranged from one minute for Fred Miller's Pontiac Deluxe 8 to nineteen minutes for Walter Jennings's Hudson 112 Deluxe 6. Six drivers arrived late by a margin of less than ten minutes. The matter of the delayed cars and the surrounding circumstances was now up for consideration, with serious discussions held between the participants and AAA Contest Board officials. While there were arguments on both sides, including one for having a time allowance for tire chain installation which would have enabled many to finish on time, strict interpretation of the rules was required. After deliberations, the AAA Contest Board's A.C. Pillsbury notified the late drivers that they had to abide by the rules and were disqualified. Although disappointed, everyone took the news in stride and made the best of a bad situation.[61] As the *Los Angeles Times* observed, "One mighty fine thing about the run is that the drivers always display a splendid brand of sportsmanship, and this year was no exception."[62]

With the awards ceremony scheduled for the following morning, AAA officials worked into the night calculating the results for all of the entries, including those who were disqualified. During the evening, Gilmore Oil hosted its annual "Mileage Dinner and Dance" with over 500 in attendance.

William C. Martin, driving a Studebaker Commander 6, captured the Sweepstakes Award with a figure of 55.875 ton miles per gallon. This was the second highest ton miles per gallon figure attained in the four-year history of the Run, only .052 ton mile per gallon behind that set by a Graham Supercharger 6 in 1938. Placing second for the Sweepstakes Award was a Packard One-Twenty 8, followed by a Studebaker President 8 by a slim margin of .051 ton mile per gallon. For the fourth consecutive year, a Willys Overland Deluxe 4 recorded the best mileage for all entries with 26.21 miles per gallon. This result was an astounding 4.324 miles per gallon less than the 30.534 average recorded by a Willys Deluxe 4 in 1938. The tough weather and driving conditions were main factors in this decline. As evidence of Studebaker's strong performance, its entries achieved the best miles-per-gallon figure for both six-cylinder and eight-cylinder cars. The Sweepstakes-winning Studebaker Commander 6 recorded a figure of 25.78 miles per gallon, and the Studebaker President 8 achieved 22.96 miles per gallon.

First-place trophies were presented to:

PRICE CLASS	MAKE and MODEL	TON MILES PER GALLON	MILES PER GALLON
A — Under $800	Willys Overland 4	42.720	26.21
B — $801 to $1000	All entries disqualified for not completing run in required time		
C — $1001 to $1060	Oldsmobile "60" 6	43.763	21.40
D — $1061 to $1130	Nash Lafayette Deluxe 6	47.388	21.25
E — $1131 to $1200	Chrysler Royal 6	42.808	20.03
F — $1201 to $1300	Studebaker Commander 6	55.875	25.78
G — $1301 to $1350	All entries disqualified for not completing run in required time		
H — $1351 to $1500	Studebaker President 8	52.914	22.96
I — $1501 to $1750	Packard One-Twenty 8	52.963	21.84
J — $1301 to $1350	Lincoln Zephyr 12	47.490	19.66

Final results for all competing cars can be found in the Appendix.

When the AAA Contest Board released the results, it reported that the field of 16 cars that crossed the finish line within the specified 11-hour time limit achieved an average of 21.9 miles per gallon. This figure was less than a mile per gallon below 1938's average of 22.88, a drop that was a reflection of the weather and road conditions.

Without a doubt, this year's results are the most confusing to understand at first glance. The 13 entries who were disqualified still had their ton miles per gallon and miles per gallon figures published by the AAA Contest Board. A.C. Pillsbury pointed out that "their results are not to be compared to those who finished on time."[63] In every price class, with the exception of two that had only single entries each, at least one entry was disqualified. Although most newspaper and magazine articles at the time focused on those entries who were awarded price class trophies, others did not. Beginning with the *Automotive News* 1939 Almanac, results were publicized without regard to whether or not a particular entry was disqualified. This type of listing continued throughout the years of the Economy Runs in background press information issued by Mobil Oil Corporation and its predecessor companies. As a result, most observers would conclude that a Ford "85" Deluxe 8, Pontiac Quality 6, Dodge Deluxe 6, Mercury 8 and Oldsmobile "80" 8 placed first in their respective price classes. This conclusion is incorrect, since these entries were among those who were disqualified. Through research, and confirmed in an article in the February 1939 issue of *Motor Age*, the actual standings are shown in the final results for 1939 in the Appendix.

Los Angeles Studebaker distributor Paul G. Hoffman Co. had plenty to celebrate. Not only did its Commander 6 capture the Sweepstakes Award for the first time, the President 8 was not far behind, placing third for the overall prize. With both models also placing first in their respective price classes, Paul G. Hoffman Co. Vice-President Wallace Austin exclaimed, "In the money four times with two entries."[64] Sweepstakes-winning driver William C. Martin, who also scored his third price class win, commented, "The contest ceased to be an economy run shortly after the start. Bad weather transformed it into a race against time."[65]

Packard and Lincoln scored back-to-back wins in their respective price classes, although it should be noted that they had no competition this year. For the first time competing in the Gilmore Runs, an Oldsmobile "60" 6 took home a first-place trophy.

Two entrants and drivers captured first-place trophies in their first year competing. Nash California Company saw its Lafayette Deluxe 6 take the top prize in its price class with Andy Henderson behind the wheel. This also marked the third year in a row a Nash took top honors in its price class. Santa Monica dealer Ray E. Shafer piloted his Chrysler Royal 6 into first place in its price class. However, both entries were the only ones in their respective classes to finish within the 11-hour time limit and remain eligible for awards.

Due to the unique circumstances surrounding this year's Run, newspaper advertising was basically dominated by Studebaker. Immediately after the results were announced, Los Angeles distributor Paul G. Hoffman Co. announced, "Studebaker Wins Grand Sweepstakes in Gilmore-Yosemite Economy Run" in an ad that also featured representations of the trophies awarded to the Commander and President models.[66] This was followed by a second advertisement, featuring a photograph of the winning cars and their drivers with the headline "Studebaker Wins Gilmore-Yosemite Economy Run," that told readers of the difficult driving conditions faced in this year's event.[67] Additional advertisements that appeared during the year, including those introducing the new Champion, continued to focus on Studebaker's economy. Nash's newspaper advertising was limited to noting that its Lafayette won Class D with 21.25 miles per gallon.

Studebaker also promoted its Sweepstakes win in advertisements appearing in both general interest and industry-related publications. A feature of most of these advertisements, which followed a craftsmanship theme, was a photograph of the winning Commander 6 along with its driver William C. Martin.[68] Featuring photographs of both winning cars and their drivers, a more detailed advertisement appeared in the February 1939 issue of *The Motor*, noting the difficult weather and driving conditions that were encountered.[69] Studebaker owners were told of this success in the May 1939 issue of *The Studebaker Wheel*, which also featured photographs of the winning cars and their drivers.[70]

One of the several magazine advertisements which appeared for the 1940 Nash, which usually featured the Ambassador Sedan, mentioned that it "was a class-winner in the Gilmore-Yosemite Run (21.25 miles to the gallon)" without noting this was achieved by the lower-priced Lafayette 6.[71] In showroom brochures for the 1940 model year, the achievement of the Lafayette 6 was discussed in more detail and included photographs of the competing car and its driver, Andy Henderson.[72] Oldsmobile advertised its success, proclaiming "Olds Wins 1st In Its Class In Gilmore Economy Run" in an ad that featured a photograph of the winning "60" 6 and its driver John Bodine.[73]

Gilmore Oil ran its usual advertisements in newspapers announcing it was the "Monarch of Mileage," and noting that "Sixteen Stock Cars Average 21.9 Miles Per Gallon of Red Lion Gasoline in the Gilmore Yosemite Run" despite the "Handicap of Snow, Rain, Mountain Grades and Blizzards."[74] During the year, additional advertisements appeared which were tailored to different price class winners. One of these focused on Studebaker's success using a comic book–style sketch showing a husband approaching his car carrying a recently filled gasoline can as his wife remarks, "Well! After This Use Red Lion! Studebaker Got 25.7 Miles Per Gallon In The Gilmore-Yosemite Run."[75]

Newspaper articles continued to be featured in Sunday automotive sections again found primarily on the West Coast. Gilmore Oil's monthly publication *The Gilmore Cub* also featured stories, results and pictures in its February issue. The magazine articles that did appear continued to be found in industry publications such as *Automobile Trade Journal*, *Motor Age* and *Motor West*.[76] The article in *Motor Age* featured one of the most detailed breakdowns of the results which included individual specifications of the competing cars and detailed those who were disqualified. In its publication *The Spirit of a Champion* for dealers and employees, Champion Spark Plug Company provided an overview of this year's Run, focusing on those winning cars equipped with their product.[77]

The most difficult Economy Run to date had come to a conclusion, and within a few months, plans were in the initial stages for a return to Yosemite in 1940. While there were still questions surrounding the strict interpretation of the rules, resulting in 12 entries being disqualified, all but two manufacturers would return in 1940. On the corporate side, changes were occurring with the acquisition of 75.56 percent of Gilmore Oil Company's stock by Socony-Vacuum Oil Company, known for its Mobilgas and Mobiloil brands. By 1945, Gilmore Oil would become a subsidiary of Socony-Vacuum.

1940

Los Angeles to Yosemite

January 4, 1940—306.5 Miles

Since the inaugural 1936 Gilmore Yosemite Economy Run, the increased interest from the manufacturers, press and driving public was becoming more and more evident as each year passed. While the majority of those participating were from the West Coast, primarily Southern California, more factory representatives from Michigan were on hand to witness the event. Much of the credit for this success goes to the AAA Contest Board's A.C. Pillsbury and his officials, who not only enforced the rules to the letter but also worked diligently, and fairly, insuring all entries were completely stock.

In December, officials from Gilmore Oil Company and AAA Contest Board met with participants and press representatives to outline the plans for the 1940 Gilmore Yosemite Run, scheduled for January 4. The drivers would face a 306.5-mile trip to be completed within a 10½-hour time limit. As in prior years, there would be no time allowance for the required stops in Fresno and Coarse Gold. After a discussion regarding weather conditions faced in last year's Run, the general consensus from the participants was that no additional time allowance would be made should the use of chains be required.

Entries were open to any 1940 four-door sedan with another wide range of makes participating. Joining Buick and Cadillac in not participating were Hudson and Hupmobile. Although it achieved some success with its Terraplane models, Hudson officials felt that these events were "not typical of cars with average drivers and common kinds of driving conditions" and had begun to sponsor their own tests.[78] Hudson's absence would only be for one year. Faced with a near collapse of sales, both Hupmobile and its Los Angeles distributor, Greer-Robbins Co., could not justify the expense of competing. A total of 28 entries would compete in nine price classes continuing to be based on the Los Angeles delivered prices for standard four-door sedans with standard equipment. Los Angeles area distributors and dealers continued to sponsor the bulk of the entries, with those from General Motors again being the exception. As was the case in 1939, the Chevrolet entry was sponsored by its Los Angeles Dealers Association, with Oldsmobile and Pontiac sponsored by their respective corporate divisions. Oldsmobile's engineer J.W. Schiller again this year traveled from its Lansing headquarters to drive the Oldsmobile "90" Custom Cruiser 8. Two Los Angeles dealers participating for the first time were Dodge's Bozzani Motors, Ltd. and Packard's Fleming Motor Co. The owners of both dealerships were Economy Run veterans who had previously driven cars entered by their Los Angeles distributors.

All entries were impounded on December 27 with "each car being subjected to a minimum of 234 separate examinations before being approved."[79] Instead of the assignment of starting positions based on price classes, a random drawing was held by AAA officials during the impound period. Starting position no. 1 was selected for Walter Hersley, driving a Willys Deluxe 4. Bringing up the rear, with no. 28, was a Chrysler New Yorker 8 with Eddie Seward behind the wheel. This year saw the widest variety of transmissions being used. Eleven entries were equipped with standard three-speed transmissions, and another fifteen with optional overdrive. The Lincoln Zephyr continued to be equipped with its Columbia two-speed axle plus overdrive. Two Oldsmobile entries were equipped with its all-new Hydra-Matic fully automatic transmission. This was the first appearance in the Runs for an automatic transmission–equipped car.

In the early morning hours of January 4, the participants arrived at Gilmore Stadium for the 5:30 a.m. start under a steady downpour of rain. Even with these wet conditions, weather forecasts were looking more favorable than last year's for the drive to Yosemite National Park. None of the drivers would take any chances of exceeding the 10½-hour time limit. Walter Hersley, driving a Willys Deluxe 4, led the participants north out of Los Angeles over the familiar Ridge Route, climbing to an elevation of 4182 feet at Tejon Pass Summit in rain and windy conditions. Descending into the San Joaquin Valley, passing through Bakersfield, the drivers encountered periods of hard rain. As they made their way north towards Fresno and Coarse Gold for their two required stops, the weather cleared, although they continued to battle head winds. As the drivers began their climb towards Yosemite National Park, temperatures dropped and they encountered light snow, up to 2 inches in depth on the road, crossing the 6256-foot Chinquapin Summit. Fortunately, the roads remained passable and chains were not required.[80] In Yosemite Valley, mist was falling as the drivers arrived at the finish line in front of the Government Administration Building, where traces of snow were on the ground ... a far cry from one year ago. At 3:11 p.m., Ray E. Shafer in a Chrysler Crown Imperial was the first to arrive, receiving a checkered-flag greeting from AAA's A.C. Pillsbury with several hundred spectators on hand to witness the event. All drivers arrived within the time limit, although Joseph Bozzani, driving a Dodge Deluxe 6, cut it pretty close, arriving with only a minute to spare. Gilmore Oil's traditional "Mileage Dinner and Dance" was held that evening while AAA officials worked well into the night calculating the results to be announced at the next morning's awards ceremony. During the evening's events "both drivers and observers agreed that the run was as close to actual conditions as possible."[81]

When the results were announced, one make stood out

from all the rest. That make was Studebaker. For the first time in the history of the Gilmore-Yosemite Runs, three cars from a single manufacturer placed first, second and third for the Sweepstakes while also placing first in their respective price classes. For the second consecutive year, William C. Martin and a Studebaker Commander 6 captured the Sweepstakes with 54.750 ton miles per gallon. Placing second for the Sweepstakes was the Studebaker President 8, with the Studebaker Champion 6 placing third. For the fifth consecutive year, a Willys Speedway 4 recorded the best mileage for all entries with 30.05 miles per gallon and a definite improvement on 1939's result of 26.21 miles per gallon. The Studebaker Champion recorded the best miles per gallon figure for a six-cylinder car with an impressive 29.19 miles per gallon, while the Ford "85" Deluxe recorded the best mileage for an eight-cylinder car with a figure of 24.92.

First-place trophies were presented to:

PRICE CLASS	MAKE and MODEL	TON MILES PER GALLON	MILES PER GALLON
A — Under $700	Willys Speedway 4	48.124	30.05
B — $701 to $850	Studebaker Champion 6	53.448	29.19
C — $851 to $950	Nash Lafayette Deluxe 6	52.699	23.76
D — $951 to $1000	Mercury 8	50.822	23.76
E — $1001 to $1050	Studebaker Commander 6	54.750	24.72
F — $1051 to $1200	Studebaker President 8	54.573	23.40
G — $1201 to $1400	Nash Ambassador 8	51.655	21.43
H — $1401 to $1600	Lincoln Zephyr 12	49.329	19.90
I — $1601 and Up	Packard One-Sixty 8	47.660	19.04

Final results for all competing cars can be found in the Appendix.

With the fastest running time in the five-year history of the Yosemite Runs, achieving an average speed of 30.32 miles per hour, the 28-car field also set a record for the best average miles-per-gallon performance. The overall figure of 22.96 miles per gallon eclipsed by a slim margin the 22.88 average set in 1938. Several drivers took top honors for the first time, including J.D. Bowerstock in a Nash Ambassador 8, Horace M. Jones in a Lincoln Zephyr 12, William R. Knopp in a Mercury 8, and Howard N. Kyser in a Studebaker President 8.

In addition to its three price class victories, the new Studebaker Champion gave the 4-cylinder Willys Speedway and Deluxe models a run for their money, "nearly becoming the stingiest car in the lineup" with an actual miles-per-gallon figure of 29.19. The total of 10.5 gallons of gasoline used for the entire trip was three-tenths of a gallon more than was consumed by the class-winning Willys Speedway 4.[82] In addition, the Champion was a runaway winner, finishing with a margin of 2.028 ton miles per gallon over the second place Ford "85" Deluxe 8.

Nash scored double wins this year with Andy Henderson in a Lafayette 6 and J.D. Bowerstock in an Ambassador 8 placing first in their respective price classes. This was the second year in a row that Andy and his Lafayette 6 took top honors in their class. In only its second year on the market, a Mercury 8 placed first in its price class by a comfortable margin over a DeSoto Custom 6.

In actual miles-per-gallon performance, both the Ford "85" Deluxe 8 and Mercury 8 finished with some of the best figures. The Ford recorded the best mileage for all eight-cylinder models with 24.92 miles per gallon with the Mercury not far behind at 23.76.

Packard and Lincoln both placed first in their respective price classes, and for the first time in two years, both faced some competition from Chrysler's Saratoga and Crown Imperial models. For Packard this was the fourth consecutive win, while it was the third consecutive win for Lincoln.

Newspaper advertising saw an increase this year and naturally Studebaker was in the forefront announcing its triple wins. Accompanying a photograph of its three winning models, a headline announced "Studebaker Wins! 1st 2nd 3rd Sweepstakes Prizes in Gilmore-Yosemite Economy Run!," which told readers the basic story of the event.[83] A second advertisement also appeared with the same headline, featuring a photograph of the Sweepstakes and price class trophies. In advertising specifically for the Champion, the model's actual miles-per-gallon result was featured, noting that this was superior performance over its competition from Chevrolet, Ford and Plymouth in the low-price field.

Although placing second to the Studebaker Champion in its price class, Ford found reasons to advertise its strong performance. "Again! Ford Leads in Economy" was the headline telling readers that in actual gasoline mileage, Ford showed the best results of all three volume sellers. Noting that the "85" V-8 achieved 24.92 miles per gallon, an accompanying chart also showed the results of the 2nd volume seller (Chevrolet) and 3rd volume seller (Plymouth). In addition, a separate panel reported the 1939 "95" V-8's figure of 24.57 miles per gallon. Naturally there was no mention of its disqualification for running over the 11-hour time limit.[84] Lincoln also announced the Zephyr's accomplishments with the headline "More Power — More Room — More Miles Per Gallon" above a photograph of the winning Zephyr at the finish line.[85]

Gilmore Oil continued this year to run advertisements in newspapers announcing "Red Lion Drives To Victory In Gilmore-Yosemite Run," pointing out that "28 Stock Sedans Average 22.9 Miles Per Gallon of Red Lion Gas" providing a brief thumbnail sketch of this year's event.[86]

Magazine advertising also saw an increase mainly in general interest publications such as *Life* and *The Saturday Evening*

Post. While the majority of these advertisements focused on the Champion models, they followed the same theme used in 1939. With a photograph of the winning Commander 6, along with its driver and trophies, went the claim that "*Studebaker Beats All Other Cars* in America's greatest gas economy classic." Details told readers of the mileage results for all three victorious models.[87] This same theme continued into the early months of the 1941 model year. Readers of the February 1940 issue *The Studebaker Wheel* were reminded of Studebaker's economy performance in an article which featured photographs of the winning cars and their drivers. A separate article, written by the Champion's driver, James E. Van Sant, provided tips for proper maintenance aimed at drivers wanting good mileage from their current vehicles.[88] In its 1941 showroom brochure, the "Studebaker Beat Them All" theme continued with a colorized photograph of the winning cars and drivers and noting the mileage figures achieved by each one.[89]

Magazine advertisements for the 1940 Nash, which again featured the Ambassador Sedan, mentioned that it "beat all others in its class in the Gilmore-Yosemite Run (23.76 miles to the gallon)," making no reference that this was by achieved by its lower-priced Lafayette 6.[90] While the majority of these ads only mentioned the mileage result, a couple did point out that this record was set by the Lafayette. In the features brochure titled "What Happened to Me ... When I Changed to Nash," one page was devoted to the achievements of the Lafayette 6 and Ambassador 8. This also included a photograph of the Ambassador Sedan at the Yosemite finish line.[91] With little fanfare, Ford told readers in one advertisement that "Ford topped all standard-equipped cars in its price class in the 1940 Gilmore-Yosemite Economy Run," in reference to those models competing without optional overdrive.[92] In *Ford News*, a magazine for owners of Ford Motor Company products, readers were told of the mileage achievements by the Ford, Lincoln and Mercury entries.[93]

Champion Spark Plugs made a brief mention of Studebaker's triple success along with the often-repeated photograph of its winning cars and drivers. This ad also noted that "Champion equipped cars won 7 of the 9 first class divisions."[94]

Los Angeles area dealers would usually display their winning cars and trophies in their showrooms. Two prominent displays were created by Ford and Nash dealers. Downtown Los Angeles' Nash California Co. displayed the winning Nash Lafayette with a large window banner proclaiming "Nash Wins Again!" that outlined the results attained by all three Nash entries. The trophies won by the three Nash models were also prominently displayed. Hollywood's Al Stuebing Ford displayed its Ford "85" Deluxe 8 with window artwork announcing "Ford Leads Other Two Volume Sellers," also noting its actual miles-per-gallon results.[95]

As has been the custom, newspaper articles were featured in Sunday automotive sections, primarily in West Coast publications. The magazine articles that did appear continued to be found in industry publications such as *Automobile Trade Journal*, *Motor Age* and *Motor West*.[96] Gilmore Oil's monthly publication *The Gilmore Cub* featured expanded coverage in its February issue. Not only did this include the standard stories and photographs, the center section contained a nearly complete copy of the detailed final results issued by the AAA Contest Board which contained timing and mileage calculations for all of the entries. Similar to the centerfold of the 1937 Graham *Supercharger News* issue, this is another valuable source of information for historians. Additional articles and photographs were featured in the January 1940 issue of *Gilmore Graphic*, a publication for Gilmore dealers and employees.[97]

Several manufacturers reported on their success in Gilmore Yosemite Run in publications designed for their dealers and employees. Studebaker highlighted its triple success in the January 19, 1940, issue of *The Studebaker News*, which included three pages of photographs of the winning cars and drivers along with background information on this year's Run.[98] Nash dealers also were told the story of its success, also with photographs of the winning cars and drivers, in the February 1940 issue of *Nash Merchandising News*.[99] In publications designed for dealers and their salesmen, Ford Motor Company reported on the success of its three entries, Ford "85" Deluxe 8, Mercury 8, and Lincoln Zephyr 12, also detailing the results of all competing models.[100]

As plans for the 1941 Run were progressing, a concern was being raised by General Motors Corporation and others over the growing use of overdrive, an extra-cost option available in some models that was designed to allow the engine to cruise at a sustained speed with reduced engine RPM. Their main argument was that competing cars should only carry standard equipment, since several makes, including Chevrolet, Dodge, Plymouth and Pontiac, did not have overdrive units available as options. The use of overdrive proved to be a factor in three price classes where overdrive-equipped units took the top prizes over those equipped with manual transmission. Studebaker was one of the first manufacturers to actively promote use of overdrive, and was the only make to have this option in those cars competing in 1936. An example of this advantage is evident when Studebaker's Economy Run results are analyzed. Between 1936 and 1940, ten Studebaker models competed, all but one of which were equipped with overdrive. Each of the nine overdrive models placed first in their price classes, along with capturing back-to-back Sweepstakes Awards in 1939 and 1940. The only non-overdrive–equipped model, a 1938 Studebaker Six, placed third (and last) in its price class. Not easily resolved, this issue would become an ongoing concern well into the 1950s.

1941
Los Angeles to Grand Canyon
January 9, 1941—599.3 Miles

After five years of having the Run's drivers travel to Yosemite National Park, both Gilmore Oil and AAA Contest Board officials felt a change was needed, as the drivers were getting overly familiar with the route. After looking at various possibilities, they selected the Grand Canyon as this year's destination in what would not only become a true test of both drivers and their cars but also provide a popular venue for all participants. Early indications were that a record number of individuals, representing automobile manufacturers, auto dealers and members of the press, would be on hand at the Grand Canyon as the cars crossed the finish line. To accommodate this group, arrangements were made for a special Santa Fe Railway train to travel to the Grand Canyon beginning a tradition that would continue well into the 1950s.

In December, the details for the 1941 Grand Canyon Run were announced, with a scheduled date of January 9. The route would cover a record 599.3 miles as the drivers, traveling through three states, to be completed within a 14½-hour time limit. This would result in the drivers averaging 43.01 miles per hour in covering the distance and "would afford an opportunity to attain higher driving speeds, resulting in fuel performance more closely corresponding to normal touring averages."[101] No time allowances would be made for required stops in San Bernardino, Las Vegas and Ash Fork, or for time necessary for chain installation should road conditions warrant. The rules would remain basically the same as in prior years, with the entries being impounded on January 2 for a thorough inspection by AAA Contest Board officials. In addition, those nominated as observers would be given a "rigid training schedule to insure [*sic*] adequate checks against any possible infraction of the rigid rules by drivers."[102]

Entries were open to any 1941 four-door sedan, although there fewer makes would participate. Continuing to voice their concern over allowing overdrive-equipped models to compete, General Motors made the decision not to enter any of its models. After sitting out last year's Run, Hudson was back with one of its employees, Glenn White, traveling from Detroit to handle driving duties in the Deluxe 6. A total of 24 entries, the smallest field in the six-year history of the Runs, were set to compete in ten price classes based on Los Angeles delivered prices for standard four-door sedans. Los Angeles area distributors and dealers continued to sponsor the bulk of the entries. Reflecting a growing interest by various manufacturers, the Hudson, Studebaker and Willys entries were factory sponsored.

After all entries were impounded on January 2, AAA Contest Board officials began checking each entry for stock status with a total of 234 separate examinations being made before all vital areas were sealed. During this impound period, a random drawing was held for starting positions, with no. 1 being selected for the Packard One-Twenty. Bringing up the rear at no. 24 was the Ford Super Deluxe 8. The growing use of optional overdrive was evident as sixteen entries were so equipped, compared to six with standard three-speed transmissions. The two Lincoln entries were equipped with its Columbia two-speed axle plus overdrive, and none were equipped with automatic transmissions.

In order for the cars to reach the South Rim of the Grand Canyon during daylight hours, the starting time was moved up to 1:00 a.m. on January 9. Les Butts, driving a Packard One-Twenty, was the first on their way from Gilmore Stadium, with the remaining cars leaving at the traditional one-minute intervals. As they made their way east from Los Angeles, the drivers faced a route that was later described as "particularly difficult for a mileage test because it started at an elevation of 300 feet at Gilmore Stadium and led over a series of twisting mountain climbs to an elevation of 7000 feet at Grand Canyon."[103] Fortunately the drivers would have clear skies during their trip. After a required stop in San Bernardino, the drivers made their way over the 4190-foot Cajon Pass and into the Mojave Desert as they continued towards Las Vegas on U.S. Route 91 while fighting stubborn headwinds along the way. Near Las Vegas, a problem surfaced with the Chrysler Saratoga when it coasted to a halt after running out of gasoline. At the start the Saratoga backed into a curb and damaged its gas tank, resulting in a leak that was not immediately noticed. After the Saratoga was supplied gasoline by a supporting tank car, it was determined that the puncture was too big to handle properly, forcing the car to withdraw.[104]

As the remaining participants arrived in Las Vegas in the early morning, the *Los Angeles Times* reported that "the drivers, observers and passengers were given a roaring welcome with all the warm hospitality for which the Nevada city is famed. The citizenry of the colorful frontier town, in 10-gallon hats and flashy western garb, swarmed about the competing cars as they pulled into the checking stations and served coffee and sandwiches."[105] Leaving Las Vegas, the drivers crossed over Hoover Dam and into Arizona as they made their way to Kingman, where they picked up U.S. Route 66 bound for Ash Fork and the last required stop before reaching the Grand Canyon. They continued to battle bitter head winds since leaving Las Vegas with no real relief in sight. "Just prior to the Run, a near-blizzard had covered the entire northern section of Arizona with snow so that as the drivers left Ash Fork and began their steep climb that leads into Williams they had to be constantly wary of ice on the sharp turns." They also encountered sub-freezing temperatures and slippery roads

At the Grand Canyon, Gilmore Oil Company's W.O. Headley (left) presents the first-place trophy to William Knoop for his winning performance driving the Ford Super Deluxe 8. This was the third time in the past four years that Ford came out on top in its price class, and was also William's second consecutive first-place finish.

with snow visible along the sides of the road.[106] On arriving at the Grand Canyon, the drivers reached an elevation of 6875 feet, which was the highest to date. Around 3:00 p.m. the first car arrived at the finish line, located at the El Tovar Hotel, receiving a checkered-flag greeting from AAA's A.C. Pillsbury along with nearly 500 spectators on hand to witness the event. Within 30 minutes the remaining 23 competitors had arrived, all within the 14½-hour time limit.[107]

As each entry arrived, it was immediately impounded by AAA Contest Board officials, who commenced their refueling and inspection procedures. As they continued to work well into the night calculating the results, to be announced the next morning, Gilmore Oil's annual "Mileage Dinner and Dance" was held at the El Tovar Hotel. That evening Charles Nash, veteran automobile manufacturer and founder of the Nash Motor Company, made the following remarks: "Both the automotive industry and the motoring public owe a deep debt of gratitude to Earl Gilmore for sponsoring such a renowned event as the Gilmore Grand Canyon Run. It is conducted on the highest possible plane, it encourages splendid American sportsmanship and good clean competition and it gives us all truly worth-while information about the improved performance of automotive products."[108]

A new record was set when it was announced that the Sweepstakes Award was captured by a Lincoln Custom 12 with 57.827 ton miles per gallon. Achieved by a first-time participant, Marshall Martin, this figure eclipsed the previous record of 55.927 set by a Graham Supercharger 6 in 1938. It was a close contest for the Sweepstakes Award: a Lincoln Zephyr 12 came in second with a mere .081 ton mile per gallon separating the two Lincoln entries. The 1939 and 1940 Sweepstakes champion Studebaker Commander 6 placed third, with each of the top three also placing first in its respective price class. For the sixth consecutive year, a Willys 4 recorded the best mileage for all entries, scoring 29.06 miles per gallon with its Americar Plainsman model. This figure is one mile per gallon lower than 1940s result, due in part to the higher speeds required to complete the trip. A Nash Ambassador 600 recorded the best mileage for a six-cylinder car with 25.81 miles per gallon, while a Ford Super Deluxe recorded the best mileage for an eight-cylinder car with a figure of 23.05. For the first time in the Run's history, two twelve-cylinder models competed, with the Lincoln Zephyr recording the best mileage at 22.96 miles per gallon.

First-place trophies were presented to:

PRICE CLASS	MAKE and MODEL	TON MILES PER GALLON	MILES PER GALLON
A — Four Cylinder	Willys Americar Plainsman 4	48.624	29.06
B — $750 to $809	Ford Super Deluxe 8	49.557	23.05
C — $810 to $914	Nash Lafayette Deluxe 6	48.651	25.81
D — $915 to $984	Hudson Deluxe 6	50.969	24.96
E — $985 to $1069	Studebaker Commander 6	52.975	24.36
F — $1070 to $1189	Studebaker President 8	51.988	22.53
G — $1190 to $1299	Chrysler Windsor 6	44.988	20.14
H — $1300 to $1389	Packard One-Twenty 8	46.846	19.78
I — $1390 to $1699	Lincoln Zephyr 12	57.749	22.96
J — $1700 to $4000	Lincoln Custom 12	57.827	21.03

Final results for all competing cars can be found in the Appendix.

Although one entry failed to finish, 1941 saw the fastest running time in the Run's six-year history, with the field averaging 43.01 miles per hour. The average for the 23-car field was 22.5 miles per gallon, which is the third highest average in the Run's history. Studebaker Pacific Corporation's President C.K. Whittaker made the following observation to the *Los Angeles Times* Auto Editor Lynn Rogers on the route and speed of this year's Run: "As compared to previous runs, I like the longer and more grueling drive to the south rim of the Grand Canyon because it gives a truer picture of the potential performance of modern motorcars. You really have to tread down on the throttle of a car to achieve an elapsed time average of 43 miles an hour on a long trip."[109]

In comparing mileage averages attained in 1940 and 1941, it appears that the goal of having these results "more closely corresponding to normal touring averages" was attained.[110] Some examples of class winners for both years:

	1940	1941
Chrysler Windsor 6	21.58	20.14
Ford Deluxe / Super Deluxe 8	24.92	23.05
Mercury Eight	23.76	23.35
Nash Ambassador 8	21.43	21.18
Packard One-Twenty 8	21.58	19.78
Studebaker Champion 6	29.19	24.61
Studebaker Commander 6	24.72	24.36
Studebaker President 8	23.40	22.53
Willys Deluxe / Plainsman 4	30.05	29.06

In addition to Sweepstakes Champion Marshall Martin, other drivers taking top honors for the first time included J.C. Marbre in a Willys Americar Plainsman 4, Glenn White in a Hudson Deluxe 6 and Lee Winchester in a Chrysler Windsor 6.

Ford Motor Company was the big winner this year. In addition to the Sweepstakes Award, the Ford Super Deluxe 8, Lincoln Zephyr 12 and Lincoln Custom 12 took top spots in their price classes. Placing second in its price class, the Mercury 8 came within .074 ton mile per gallon to overtaking the winning Hudson Deluxe 6. These results reflected the comments made by Edsel Ford prior to the start, who "characterized it as an outstanding event." He added, "I particularly like it because it is a survivor of the stock car tests which helped make motoring popular and helped build the automotive industry in its earlier days."[111]

After two years of Sweepstakes-winning performances, Studebaker was no match for Lincoln's strong ton-mile-per-gallon performance. Though Studebaker placed first in two price classes with its Commander and President models, the Champion's performance could only be termed as a disappointment. Placing a distant second to the Nash Ambassador 600, its actual miles per gallon figure was a drop of 4.58 miles per gallon compared to 1940s stellar performance. Part of this can be attributed to the Champion's new 169.6 cubic inch engine, although it was also noted that it experienced an unforeseen delay in Las Vegas. As reported in a Studebaker Confidential Sales Bulletin, the Champion "was delayed for 22 minutes because of congestion which the winner in this division did not experience," and under the rules this delay required it to make up part of this time in order to arrive at the finish within the time limit.[112]

For the third straight year, the smaller Nash Ambassador 600 (also known as the Nash 600) took top honors in its price class, continuing to match the performance of the previous Lafayette models. With Andy Henderson again behind the wheel for Los Angeles' Nash California Company, the Ambassador 600 not only outperformed the challenge from the Studebaker Champion but achieved the best mileage performance for all six-cylinder cars. The larger six-cylinder and eight-cylinder Ambassador models placed second in their respective price classes.

Packard continued to perform well in its price class by taking home a first-place trophy for the fifth consecutive year. However, in three of those years, it was the only entry in its price class. Les Butts, driving a One-Twenty Touring Sedan, held off a challenge from a Hudson Commodore in marking his third first-place finish. Although it was the only entry in its price class, the first-place finish by the Chrysler Windsor was the third time Chrysler came out on top after wins in 1936 with an Airflow 8 and in 1939 with a Royal 6.

Newspaper advertising saw another increase this year with a wider variety of manufacturers announcing their successes. Ford Motor Company took the opportunity to advertise the results of all four entries featuring photographs of the competing models with their drivers. Capitalizing on General Motors' nonparticipation, along with poor performance by Chrysler Corporation entries, the headline proclaimed, "Ford-Build Cars Score Smashing Economy Victory," telling readers of the Sweepstakes win by the Lincoln Custom along with first-place finishes by the Ford Super Deluxe and Lincoln Zephyr. Not to be left out, Mercury's second-place achievement was also mentioned. In referring to actual miles-per-gallon performances, the captions reminded readers that Ford "had the best gasoline record of all three volume sellers in the 1940 and 1939 Gilmore runs," while Mercury stressed it was a "close second in '41 to a 6-cylinder car with overdrive! First in '40—and first in '39 too."[113] Here again is evidence of the confusion surrounding the results from the 1939 Run, as both of these models were disqualified.

Studebaker told readers "Studebaker Wins Again! Two Firsts — One Second," emphasizing this was the third consecutive year for the Commander and President to take first place in their price classes.[114] A second advertisement followed, declaring "Studebaker Triumphs Again in Gilmore-Grand Canyon Run," and featuring a photograph of the three competing Studebakers with their drivers.[115] Both of these advertisements provided a brief overview of the Run while reporting on the actual miles per gallon achievements for all three models. An added feature this year was footnotes explaining "All Studebakers were equipped with extra-cost overdrive."[116]

Proclaiming "Nash Proves Its Amazing Economy," an ad with a photograph of the winning Ambassador 600 at the awards ceremony was featured with a second headline that noted it was "First in its Class" and "First in Miles Per Gallon" while providing an overview of the Run.[117] Although not as prominent as other manufacturers, Willys noted in an advertisement that its "Americar Plainsman with overdrive scored the highest mileage for all cars.[118] Although advertising on the Run was limited, the Plainsman's performance was discussed in the 1942 showroom brochure, which included a photograph of the class-winning car prior to leaving Gilmore Stadium, along with an elevation profile of the route.[119] Large posters were displayed in Willys showrooms announcing "Again! Willys Americar Beats All Other Cars" and featuring a photograph of the winning entry at the awards ceremony.[120]

Following tradition, Gilmore Oil's newspaper advertising announced the basic results, noting "23 cars average 22.5 miles per gallon with Red Lion," while adding, "Every car entered in this grueling run had its motor protected with Golden Lion Motor Oil."[121]

While newspaper advertising was prominent immediately following the Run, magazine advertising continued well into the summer months as the driving public was becoming more focused on their car's fuel economy and durability.

Hudson advertised the success of the Deluxe 6 throughout the year in its "All Cars Alike?" campaign, which focused on five areas including safety, economy, and driving ease. One advertisement featured a rendering of the awards ceremony, while others featured only the trophy.[122] The performance of both the Deluxe 6 and Commodore 8 were discussed in a special features brochure available from Hudson dealers. Continuing the theme "All Cars Alike?" the actual miles-per-gallon performance for the Hudson models was compared to others competing in their price classes, with photographs of both competing models and an elevation profile of the route.[123]

Nash advertised the performance of the Ambassador 600 in both general interest and industry-related publications. In a two-page spread in magazines such as *Life* and *The Saturday Evening Post*, a photograph of the winning car appeared with the headline "Nash Proves its Amazing Economy!" and noting that "the Nash Ambassador 600, Equipped with the Marvelous Fourth Speed Forward, *Officially Delivered More Miles to the Gallon* than any other 6, 8, or 12 cylinder car entered, regardless of Size, Price, Weight or Equipment."[124] A second advertisement, appearing in the spring, showed the winning car at the Grand Canyon finish line, while the text continued to tell about its overall mileage achievement.[125] Additional advertisements, appearing in industry related publications such as Motor and Motor Age, aimed at current or potential Nash dealers also reinforced this mileage record. Shortly after the Run, Nash reprinted its 1941 showroom

It's Talk Like This — THAT'S SWITCHING THOUSANDS TO NASH!

ACTUAL COLOR PHOTOGRAPH OF THE NASH DELUXE TRUNK-BACK, FOUR-DOOR SEDAN, ONLY $860*

A New Record! *From Los Angeles to the Grand Canyon . . . over 599 grueling miles of desert, mountain passes . . . Nash averaged 25.81 miles a gallon at 42.6 miles an hour!*

The "MOST MILES PER GALLON" Big Car

FACT! In the 1941 Official A.A.A. Gilmore Economy Run, Nash equipped with the marvelous Fourth Speed Forward, delivered more miles per gallon than any other competing 6 or 8 cylinder car!

SIX MONTHS ago if you heard someone say, "I just drove 500 miles without stopping for gas"—you wouldn't have believed it.

Now you can't go anywhere without hearing a Nash owner tell stories like these:

"*My new Nash is doing 25.6 miles a gallon over mountain roads. Power—perfect!*" P. A. S., CALIF.

Or—"*I've just completed a 9,897 mile trip in my new Nash—averaged 25.9 miles a gallon. Drove fast all the way.*" FROM J. H. P., RHODE ISLAND.

Or—"*Can't tell you how pleased I am with my new Nash. Rides like a dream. Doing over 26 miles a gallon every day.*" FROM G. R., NEW YORK.

It's not only this economy—but what these miles *are like* that is making Nash the fastest growing automobile in the lowest-price field.

You'll literally be amazed at every difference there is in this new kind of car.

You never felt *any* car—no matter how expensive—ride so smoothly, handle so easily.

Just five minutes behind the wheel will tell you why thousands of loyal owners of other low-price cars are now changing to Nash.

Here is the first low-price car with the comfort of coil springs *on all four wheels* . . . the "roadability" of Two-way Roller Steering.

The first low-price car with a Convertible Bed for vacation trips . . . the famous Weather Eye Conditioned Air System to end cold-weather discomfort, drafts and dust.

The very way this Nash is built—with body and frame solidly welded into a single twist-and-rattleproof unit—is different, too!

Before you buy any new car, ask yourself—*does it offer as much as this new Nash?*

Compare the differences at your Nash dealer's today—and you, too, will join America's most enthusiastic new car owners!

$745 Buys a big 6-passenger sedan. Other prices start at $710. *All prices, delivered at factory, include standard equipment, Federal tax. Weather Eye, Convertible Bed, White Side Wall Tires, Fourth Speed Forward, Two-tone paint, Deluxe Wing Bumpers or Bumper Guards at slight extra cost. *Prices subject to change without notice.*

Go NASH
AND SAVE MONEY EVERY MILE

1941 GILMORE GRAND CANYON RUN
Nash Ambassador 600 6
First Place Class C

brochure to include details on the Ambassador 600s winning performance.[126]

Although it didn't have a lot to shout about this year, Studebaker told readers of its "Fourth Straight Year of Triumphs in Gilmore Economy Runs" in the months following the Run. Featuring a photograph of the winning cars and drivers, the ad mentioned that the "Champion made the remarkable average of 24.61 miles per gallon and the President and Commander scored firsts in their divisions."[127] As the model year progressed, very little was mentioned about their Gilmore success in Studebaker's magazine advertising. Although its overall advertising was limited, Studebaker's performance continued to be stressed in its showroom brochures issued for both the 1941 and 1942 models. The emphasis in these brochures was on the class-winning Commander and President models, featuring photographs of the winning models and their drivers, reminding readers that its gas economy had been proved year after year.[128] This overall theme was contained in a brochure devoted to the 1942 Champion models, which also highlighted its 24.61 miles per gallon performance.[129]

One of the only advertisements covering Ford's performance appeared in industry-related publications like *Automobile Topics* and *Motor*. With the headline "Ford Economy Wins Again!" a photograph of Super Deluxe sedan at the awards ceremony was featured, along with an overview of the Run.[130] In its magazine for owners of Ford Motor Company products, *Ford News*, the achievements of the Ford and Mercury sedans were covered in a fairly detailed article. Interestingly, there was no mention of Lincoln's Sweepstakes Award in the article.[131]

Champion Spark Plugs exhorted drivers to "Choose your spark plugs for *greater gas mileage*," as a result of which, "you'll choose Champions!" in an advertisement appearing in general-interest publications. Featuring a rendering of the Sweepstakes-winning Lincoln Custom 12, the ad contained a brief overview of this year's event, also noting that the vast majority of the winning entries were equipped with Champions.[132] Although Lincoln captured the Sweepstakes Award, along with two price class victories, there was no mention of these achievements in Lincoln's general advertising campaign.

Following the trend of recent years, newspaper articles were in Sunday automotive sections, primarily on the West Coast. With one exception, magazine articles on the Run appeared in industry-related publications such as *Automobile Topics, Motor Age* and *Motor West*.[133] As a sign of the times, *Popular Science* reported on the Run for the first time, focusing on tips from winning drivers on "how to get more miles from a gallon of gasoline." With photographs ranging from preparation and inspection all the way to the awards ceremony, the article discussed ways to increase mileage along with steps in keeping your car in top mechanical condition in the uncertain times ahead. This article was a preview of *Popular Science*'s coverage of the Economy Runs that would continue well into the early 1960s.[134] Gilmore Oil's monthly publication *The Gilmore Cub* had apparently ceased publication before the Run was held, resulting in a loss of a reliable source of information and stories.

In publications for their dealers and employees, both Studebaker and Ford reported on their individual achievements. *Studebaker News* devoted two pages to highlighting its taking "honors in 4th consecutive Gilmore Run!," featuring photographs of the winning cars and drivers. In addition to covering the details of this year's Run, a chart showed Studebaker's winning performance from 1938 through 1941.[135] A separate "Confidential Sales Bulletin" was sent to its dealers which also included the results for all competitors.[136] Following the same format used in 1940, Ford dealers and salesmen were provided information on the Super Deluxe's performance along with detailed results for all competing models.[137] In its publication *The Spirit of a Champion* for dealers and employees, Champion Spark Plug Company provided an overview of this year's Run, focusing on those winning cars equipped with their product.[138]

Comments continued to be made regarding the use of overdrive by the majority of the competitors. Some of these comments were not at all favorable. In this year's Run, nine of the ten price classes saw overdrive-equipped models win over their non-overdrive–equipped competitors. A prime example showing the advantages of overdrive occurred in price class "A" for four-cylinder cars. Three Willys Americar models were entered, two equipped with overdrive and one without. The Americar equipped with the standard transmission placed third (and last), recording a figure of 25.81 miles per gallon, which was 3.88 miles per gallon behind the leader. One of the most critical comments was made in the January 20, 1941, issue of *Automobile Topics*:

> This run originally was designed for stock cars but the introduction of cars equipped with overdrive brought up a real question since in some cases the bulk of a company's production is sold with overdrive equipment. This year General Motors cars did not take any part in the event since none of the GM units employ overdrive as such. Unless some more definite rules are set up in the future the results of this run will have little value from the standpoint of establishing a yardstick of economy.[139]

Opposite: **Nash, along with Studebaker, regularly advertised their Gilmore Economy Run performance in both newspapers and magazines. In one of the few advertisements to feature a photograph of the winning car, Nash emphasized the Ambassador 600's 25.81 miles-per-gallon result (courtesy of Chrysler Group LLC).**

Although rules would be established permitting certain optional equipment for competing cars, the issue of overdrive-equipped models would continue until 1955. But there were more pressing concerns at this time with the conflicts occurring in both Europe and in the Pacific, and more and more production was now geared towards military and defense products.

Plans for a return to the Grand Canyon had been finalized by officials from Gilmore Oil and AAA Contest Board by early December 1941. Appearing in the *Seattle Times'* automotive section on December 7 was the announcement that the 1942 Gilmore Grand Canyon Run was scheduled for January 8 with various rule and route changes being implemented. Gilmore Oil's Vice-President Clarence S. Beesemyer noted, "At a time when national interest is focused on gasoline mileage, this test run should prove more valuable from a research standpoint than ever before."[140] Within hours of publication, the word was flashed around the world that there had been an attack on American naval forces in Pearl Harbor. Within days, the Gilmore Grand Canyon Run was postponed, joining other notable West Coast events whose cancellations were being termed as "war casualties," including the Pasadena (California) New Year's Rose Bowl game. Noting the importance of conserving rubber, gasoline and lubricants, an advertisement by Gilmore Oil Company pointed out that now "all energies must be directed to one goal — winning the war." The company expressed its hopes to again sponsor this test at a future date.[141]

Beginning in 1942, Socony-Vacuum began the transition of Gilmore Stations to the Mobil brand, which would be completed by 1945. As this transition neared completion, Gilmore Oil Company's President Earl Gilmore retired in September 1944 from active participation. Clarence S. Beesemyer, vice-president and general manager, also actively involved in the Gilmore Economy Runs since their inception, was elected president. The merger of Gilmore Oil Company into General Petroleum Corporation (the West Coast Affiliate of Socony-Vacuum Company) was announced on March 15, 1945, with Beesemyer named to the post of executive vice-president of General Petroleum.[142]

With the end of hostilities and a return to civilian automobile production, it was decided by Socony-Vacuum and General Petroleum to maintain the AAA Contest Board's sanction annually for a proposed Mobilgas Grand Canyon Economy Run as a comprehensive test to determine the efficiency of present-day engines and fuels. Spearheaded by Beesemyer, a survey of conditions in the automotive industry was conducted in early 1949. As a result of this survey and months of planning, a decision was made that resumption of an Economy Run was practical. Preparations commenced in the fall of 1949 for the 1950 Mobilgas Grand Canyon Run.

2
Grand Canyon

1950
Los Angeles to Grand Canyon Via Death Valley — Las Vegas — Hoover Dam
February 15 and 16, 1950—751.3 Miles

When a Willys Jeepster took the green flag in the early morning on February 15, it marked the end of months of planning and preparation by General Petroleum Corporation, led by Executive Vice-President Clarence S. Beesemyer and officials with the AAA Contest Board. It also marked a revival of earlier Economy Runs sponsored by Gilmore Oil Company a decade earlier. Arthur C. (A.C.) Pillsbury, regional director of the AAA Contest Board, again was in charge of all technical and mechanical details of the Run as chief steward, continuing his involvement dating back to the 1936 Gilmore Yosemite Run. Leaving Los Angeles, the competitors were faced with the most challenging test to date, which required them to cover the 751.3-mile route, ranging in altitude from 178 feet below sea level at Death Valley's Furnace Creek Ranch to 7005 feet at the Grand Canyon, and within an 18.5-hour time limit. This required them to achieve an average speed of 41.46 miles per hour to finish within the allowed time. As noted in the April 1950 issue of *Motor Trend*, "The route for this field test encompassed road and temperature conditions in a two-day period that the average motorist would only see in a year of driving. A more comprehensive test would be difficult to imagine."[1]

The rules generally followed those used in the 1941 Gilmore Grand Canyon Run, although slight changes were made to reflect current conditions. Entries were open to any stock 1950 four-door metal-top sedan of U.S.A. manufacture and design, not driven more than 5000 total vehicle miles, and complying with the specifications for that model as supplied by the manufacturer. Only one car of each make and model could be entered, although an exception to these rules was made for four-cylinder cars competing in price class A. The remaining price classes were established based on the December 1949 factory retail delivered price for the lowest-priced four-door sedan. Every car participating carried a payload of 750 pounds — which is the equivalent of four passengers plus their luggage.[2] The choice of transmission was left to each individual entrant, and of the 31 cars entered, 13 were equipped with overdrive or Fluid Drive, and 12 were equipped with automatic transmissions. For the first time, entrants were permitted to choose which grade of gasoline they preferred — regular or premium (Ethyl) — with 16 cars using regular grade and 15 using premium.

Every manufacturer, with the exception of Buick and Pontiac, was represented, with a total of 31 entries competing in 10 price classes. These were sponsored by either the manufacturer, local dealer associations, individual dealers, or in the case of the Chevrolet Fleetline Deluxe, private individuals. Several Los Angeles area dealers and distributors who participated in the Gilmore Runs were back in 1950, including Earle C. Anthony, Inc. (Packard), Bozzani Motors Ltd. (Dodge), Deaton Motor Company (Lincoln), Al Stuebing (Ford), and Willys-Overland Motors, Inc. (Willys).

Drivers and relief drivers were chosen by each entrant, and approved by the AAA Contest Board, based on their driving ability. The "granddaddy" of all stock car drivers, Erwin "Cannonball" Baker, piloted the Packard Super Eight. Four of the drivers who participated were veterans of earlier Gilmore Economy Runs, including Joe Bozzani (Dodge Coronet), Les Butts (Packard Eight), Andy Henderson (Nash Statesman), and Marshall Martin (Lincoln Cosmopolitan).

Preliminary inspection of all competing cars began on December 9, 1949, and subjected the entries to a total of 274 different inspections (which included a complete teardown of the engine), ensuring that all components were strictly stock. Three weeks prior to the start, all entries were delivered to officials of the AAA Contest Board for impounding, and from this time until completion of the Run, they remained under their constant supervision and control. During this period, all tires were completely deflated and refilled to specifications with bottled air. Cars could also be taken from the impound area and given no more than two test runs not to exceed 75

Bill Stroppe and his Sweepstakes-winning Mercury cross Hoover Dam as he heads to the Grand Canyon. The Art Deco–style Hoover Dam Monument is in the background.

miles each. At the same time the rules were published, a detailed description of the route was included, which enabled participants to make practice runs if they so desired. During the Run, an official AAA Contest Board observer rode beside each driver to ensure that all of the contest rules, including strict adherence to all traffic laws and regulations, were followed. No trick driving, coasting or other unusual aids to increase mileage were permitted at any time and none were reported.[3]

On the evening of February 14, all of the competing cars were driven from the impound area to the General Petroleum garage in downtown Los Angeles, where AAA Contest Board officials made their final inspections, then topped off each entry and sealed both the gas tanks and the hoods. Tires were also pressurized "according to a predetermined scale" which was "computed by averaging the pressures for that tire as recommended by the car manufacturer, the tire manufacturer and as listed in the Annual Issue of *Motor Magazine*."[4] Then they were moved by hand to their respective starting position on Flower Street. These same procedures would remain in place through 1958.

The start of the run was on Flower Street, adjacent to the General Petroleum garage in downtown Los Angeles, for the 452.51-mile, 11.25-hour trip to Las Vegas. The first car was flagged away on February 15 at 3:00 a.m. with the remaining cars leaving at one-minute intervals. With the aid of flagmen, the drivers navigated the nighttime streets through several Los Angeles area communities before they crossed the San Gabriel Mountains into the Antelope Valley. Continuing through the Mojave Desert, the cars arrived in Lone Pine for a refueling stop, under the continued supervision of the AAA Contest Board, and were greeted by members of the Lone Pine High School band along with residents of the community. This local enthusiasm would continue in future years. Here, as well as the other refueling stations along the route, drivers would order any amount of gaso-

Art Elliott guides his Cadillac Series 60 Special through rugged country in northern Arizona on his way to the Grand Canyon. This is one of the four Cadillac entries this year to take first-place honors in their respective price classes.

line they felt necessary to reach the next station. No time allowance was provided for these refueling stops, and the drivers were anxious to refuel as quickly as possible. In a surprise to both competitors and observers, Andy Henderson, driving a Nash Statesman, did not add any fuel.[5] Leaving Lone Pine, they continued their journey through Death Valley, dropping to an elevation of 178 feet below sea level at Furnace Creek Ranch, and by mid-afternoon had arrived Las Vegas for refueling. Following refueling, the cars were impounded overnight at the Motordome, adjacent to the Last Frontier Hotel. As reported in newspaper articles at the time, "the drivers, observers and passengers were given a roaring welcome featuring the warm hospitality for which the Nevada city is famous. The citizenry swarmed about the competing cars as they pulled into the refueling stations and served coffee and sandwiches to their drivers and passengers."[6]

On February 16, the first car to leave Las Vegas departed at 7:00 a.m. for the 298.79-mile, 5.25-hour trip to the Grand Canyon. Crossing the Hoover Dam into Arizona, the cars continued through northern Arizona, passing through Kingman before arriving in Williams for refueling, as needed, then continued on to the finish line at the Bright Angel Lodge. A large group of spectators were on hand, many of whom traveled on a special Santa Fe train from Los Angeles, as the cars arrived. The first to cross the finish line was a Cadillac Series 60 Special and all competitors completed the Run well within the 18.5-hour time limit. After all of the cars had arrived, they were turned over to AAA Contest Board officials, where they were topped off, using the same procedures used in Los Angeles before the start, and impounded overnight. While all competitors and others affiliated with the Run were guests of General Petroleum at the Mobilgas Mileage Dinner and Pegasus Party, AAA Contest

Board officials commenced computing final mileage figures and made final checks that stock status has been maintained. The results would be announced at an awards presentation on the morning of February 17.

During the Mobilgas Mileage Dinner, Clarence S. Beesemyer, Executive Vice-President of General Petroleum, remarked:

> In the 1950 Mobilgas Grand Canyon run we have had an event that gives us facts the customer can use now — today — in his own car. It has demonstrated what any car owner can get in the way of performance and economy in his own car if he maintains and drives it properly. This test has provided a demonstration of the capabilities of America's 1950 automobiles in that regard and, in addition, it has served as a guide to the selection of the fuel that will perform best and most economically in each customer's own motor car. We realize that there is no universal or absolute yardstick for determining which grade of gasoline a specific car needs, because our requirements change with age, mechanical condition, demand and need of the driver, and load variations. We also realize that the constant change in compression ratios is a most important factor. But we hoped to accomplish two things by the sponsorship of the 1950 Mobilgas Grand Canyon Run. First we hoped to be able to give car owners specific information that would show them how to get good performance and good economy when they buy gasoline, and, second, to dramatically bring to the attention of the public the great progress of the American automotive industry. Both aims have been accomplished.[7]

Bill Stroppe, participating in the Economy Run for the first time, wheeled his Mercury Sport Sedan into the winner's circle, capturing the coveted Sweepstakes Award with an impressive figure of 61.2711 ton miles per gallon. It was noted during the awards ceremony that this was the highest ton-mile-per-gallon figure attained in the seven-year history of the Economy Runs dating back to 1936. Following close behind the Mercury for the Sweepstakes Award was a Cadillac Series 60 Special, driven by Art Elliot, with 59.1179 ton miles per gallon. Although coming in second in its price class, a six-cylinder Studebaker Champion turned in the best miles per gallon figure for all entries with 26.551. The best miles per gallon figure for an eight-cylinder was the Sweepstakes Mercury with 26.524. In announcing the final results, the AAA Contest Board noted that ideal weather was encountered over the entire route, and no problems or accidents were reported. "Competition has been keen — and clean."[8]

In addition to the Sweepstakes Award, first-place trophies were presented to:

PRICE CLASS	MAKE and MODEL	TON MILES PER GALLON	MILES PER GALLON
A — Four Cylinder Cars	Willys Jeepster	44.3039	26.099
B — $1450 to $1750	Ford Deluxe "6"	48.5759	23.326
C — $1751 to $2100	Kaiser Special	53.0400	23.946
D — $2101 to $2275	Mercury	61.2711	26.524
E — $2276 to $2475	Studebaker Land Cruiser	55.6855	24.887
F — $2476 to $2950	Frazer Manhattan	54.3303	23.907
G — $2951 to $3150	Cadillac Series 61	58.5212	22.972
H — $3151 to $3750	Cadillac Series 62	58.5651	22.525
I — $3751 to $4500	Cadillac Series 60 Special	59.1179	22.080
J — $4501 to $6000	Cadillac Series 75	51.0874	17.245

Final results for all competing cars can be found in the Appendix.

All cars had a combined average of 22.0741 miles per gallon, which was reported by the AAA Contest Board as surprising, since it was thought that the 7,183-foot altitude span and fast pace (41.466 miles per hour average) would have precluded such a mileage figure. All price-class winners were equipped with overdrive (with the exception of Cadillac, equipped with Hydramatic), which continued to demonstrate its economy advantages over standard 3-speed units. This was especially evident in the results for price class B, in which the Ford Deluxe 6 was the only one of the three equipped with overdrive, an option that was not available on Chevrolet or Plymouth.

Hollywood Ford dealer Al Stuebing recorded his third consecutive class win with Ford, a streak dating back to the 1940 Yosemite Run. Henry J. Kaiser, Jr., was reported to be especially pleased with the overall performance of the Kaiser and Frazer entries. Not only did the Kaiser Special and Frazer Manhattan place first in their price classes, all three entries recorded miles-per-gallon figures that were very close to one another, ranging from 23.907 for the Frazer Manhattan, to 23.946 for the Kaiser Special, to 23.965 for the Kaiser Virginian.[9] Those associated with Cadillac were pleased with the performance of their models, although it should be noted that in two price classes the Cadillacs did not have any competition.

After the awards ceremony, advertising and promotion efforts went into high gear. Strict advertising guidelines were established by the AAA Contest Board to ensure clarity and accuracy and all advertising required AAA's approval prior to publication. These guidelines would continue throughout the history of the Runs, although they would be modified as events dictated. One specific requirement involved the type of transmission used. If a car was equipped with overdrive, semi-automatic or automatic transmission, this fact was to be prominently noted in all advertising copy.

Newspaper advertising was the most prominent, especially in Southern California where the Run originated. Sweepstakes-winning dealer Art Hall advertised his Mercury's

After receiving his Sweepstakes Award in recognition of the performance of his Mercury, driver Bill Stroppe (left) is joined by (left to right) Chief Steward Arthur C. Pillsbury, Long Beach Lincoln-Mercury dealer Art Hall, and General Petroleum Vice-President Clarence S. Beesemyer. In topping all other entries, Bill recorded a figure of 61.2711 ton miles per gallon.

success in both the *Long Beach Independent* and the *Long Beach Press Telegram* for several weeks following the Run. General Petroleum, under the direction of its advertising manager, Frank Meunier, ran both newspaper and magazine advertisements noting the overall results of the Grand Canyon Run. This included a full-color two-page spread proclaiming "Mobilgas Plus Good Car Care and Driving Can Increase Your Car's Gas Mileage," which appeared in both *Life* and *The Saturday Evening Post*.[10]

Mercury's success was noted in several magazine advertisements during 1950, although this was limited for the most part to footnotes in their regular copy. This also began Mercury's tagline as "America's No. 1 Economy Car," which would be used again in 1952. Although Nash did not place first in its price class, they did advertise the Ambassador's actual miles-per-gallon figures. Studebaker continued to stress the economy of their models, as they did during the Gilmore years, with emphasis on the Champion's actual miles-per-gallon performance.

In addition to newspaper and magazine advertising, Mercury produced a small postcard and a small 4-page folder for use by their dealers publicizing the Sweepstakes win. This folder also included tips for drivers on how to achieve the mileage figures in the Grand Canyon Run.[11] Mercury's success was further noted in a television commercial seen on Ed Sullivan's *Toast of the Town* during the summer, which also showed Mercury as the pace car for the 1950 Indianapolis 500. Potential buyers in the United States and Canada were told of the Champion's performance by the bold statement "Studebaker Champion beats 'em all in gas mileage!" in both the 1950 and 1951 Studebaker showroom brochures.

In the days following the Run, newspapers ran articles in their daily and Sunday automotive sections, most prominently in Southern California newspapers. One article featured

Sweepstakes-winning driver Bill Stroppe offering suggestions for saving gasoline, which included accelerating gradually, driving consistently while observing speed laws, and keeping your car mechanically sound, including brakes and tires.[12] In succeeding months, coverage of the Run appeared in *Time Magazine*, in several automobile-related magazines including *Motor Trend*, *Road & Track*, and *Speed Age*, and in industry publications including *Automotive News* and *Motor West*.[13] Through the fifties and into the early sixties articles would continue to appear in these publications. Even the British were told of the Run in an article that appeared in *The Motor* magazine.[14] Coverage of Ford Motor Company's successes were featured in their magazines *Ford Times* and *Lincoln-Mercury Times* distributed to owners.[15] Floyd Clymer, well-known publisher of automotive and racing publications, produced a yearbook covering the Run and would continue publishing these books through 1959.

In summarizing the Grand Canyon Run, Chief Steward A.C. Pillsbury remarked:

> The 1950 Mobilgas Grand Canyon Run is now history. The performance of the finest cars ever built is a proven fact. The responsibility of the American Automobile Association to the motoring public once again has been fulfilled. The AAA long has sanctioned the top automotive events. Such sanctions have not been given lightly. We have felt that any automotive event, regardless of scope, must have some purpose or aim to take it from the realm of mere competition. During the past six months, much planning and preparation has gone into the staging of the largest stock car event in the world. Those of you who competed know that the route was the toughest that could be devised.
>
> There was a purpose in that toughness. Had the course been laid out to enable the cars to set records of economy under perfect conditions, there would have been little point in this run. Instead, the course was designed to subject the cars to every sort of condition of altitude, weather, terrain and combined city-and-highway driving that would, in effect, run the whole range that the average motorist encounters in a full year. The outstanding records that you have set now give the public something to shoot at. In short, this run has provided the standard of performance that the average driver can expect to attain with his own automobile, of the make of his choice.[16]

These guidelines would continue to guide future Economy Runs.

1951

Los Angeles to Grand Canyon Via Death Valley — Las Vegas — Hoover Dam

March 6 and 7, 1951—840.05 Miles

After the successful completion of the 1950 Mobilgas Grand Canyon Run, the AAA Contest Board granted a sanction to General Petroleum Corporation for the 1951 Mobilgas Economy Run, which at that time was known as "the world's largest stock car event." In announcing this sanction, A.C. Pillsbury, AAA Contest regional board director and chief steward, reported: "Entries in the event will be under even more rigid controls than existed for the 1950 Run."[17] Gasoline conservation was again in the forefront of the public's mind due to the onset of the Korean War and fears were beginning to surface of gasoline rationing similar to what was experienced during World War II. Preparations began in September 1950 and involved more than 250 General Petroleum personnel, 40 AAA Contest Board officials, and during the Run, hundreds of personnel in related fields. Again with its destination the Grand Canyon, with an overnight stop in Las Vegas, the route was 90 miles longer than in 1950 and was designed to subject the cars, in two days' running, to every sort of altitude, weather and driving condition that the average motorist would encounter in a full year of driving. Leaving Los Angeles, the competitors were faced once again with a challenging test which required them to cover the 840.05-mile route, ranging in altitude from 279.6 feet below sea level at Bad Water in Death Valley (the lowest point in the Western Hemisphere) to 7005 feet at the Grand Canyon, within a 20-hour, 55-minute time limit. This required them to achieve an average speed of 40.6 miles per hour necessary to finish within the allowed time.[18]

Two significant changes were made to the rules based on lessons learned from the 1950 Run. The first was that all competing cars would now be selected by the AAA Contest Board, unannounced and without any advance notification, from dealers' showrooms, factory assembly lines, factory warehouses or storage yards. Entrants were required to specify their first and second preferences of the color of the car to be selected prior to car selection by the AAA Contest Board. Once a car was selected it remained under AAA Contest Board storage and supervision at all times.[19] The area for random selection was not confined to the Southern California area, as was pointed out in the March 1951 issue of *Nash News* (a publication for Nash dealers), where the Nash Ambassador was shown as being selected from the floor models at Sequoia Motor Co. in Oakland, California.[20]

Deciding that a new system of appointing impartial observers was in order, observers selected would from now on be "disinterested parties none of whom is in the employ of any automobile manufacturer," to serve "on any car in which he has any direct or indirect interest."[21] The California Institute of Technology (Caltech) was approached with a proposal to have senior engineering students to act as AAA Contest Board observers. Caltech approved this idea with the provision that students would participate on their own time. These students rode beside each driver to ensure that all contest rules, including all applicable speed, safety and traffic laws, were followed. No trick driving or coasting was permitted at any time and none was reported.[22]

Entries were open to any stock 1951 four-door metal-top sedan of U.S.A. manufacture and design, not driven more than 5000 total vehicle miles, and in compliance with the specifications for that model as supplied by the manufacturer. To make it possible that certain models not manufactured as four-door metal-top sedans might participate, two special lightweight classes were created. Only one car of each make and model could be entered. The competitive classifications (price classes) were established based on the December 1950 factory retail delivered price for the lowest-priced four-door sedan. Every car participating again carried a payload of 750 pounds.[23] The choice of transmission was left to each individual entrant, and of the 32 cars entered, 21 were equipped with overdrive or Fluid Drive, and 8 were equipped with automatic transmissions. Twenty-four of the competing cars elected to use Mobilgas Special.

Every manufacturer, with the exception of Buick, Dodge, Oldsmobile and Pontiac, was represented, with a total of 32 entries competing in 9 price classes, plus 2 lightweight categories. These lightweight categories featured the newly introduced Henry J and Nash Rambler. Bill Stroppe, 1950 Sweepstakes champion, returned to defend his winning performance, again driving a Mercury Sport Sedan for Long Beach Lincoln-Mercury dealer Art Hall.

At the same time the rules were published, a description of the route was included, which enabled participants to make practice runs if they so desired, and in most cases these were made. Several of the drivers made more than one practice run. Les Viland, who drove a Lincoln, said "his group made five complete trial runs, plus shorter ones"; most of the practice

Stan Van Horn, driving a Studebaker Commander, receives the green flag from well-known racing promoter and AAA official J.C. Agajanian (waving flag). Even with an early 3:00 a.m. start in light rain, a good-sized group of spectators were on hand to see the drivers on their way. Stan would place first in his price class with 58.173 ton miles per gallon.

After an overnight stop in Las Vegas, Arne Anderson's Henry J passes another car while crossing Hoover Dam. In its first appearance in the Economy Run, the Henry J would place first in the four-cylinder lightweight class with 49.153 ton miles per gallon. It also recorded an impressive actual miles per gallon figure of 30.109.

runs were made with models which were identical to the officially entered car.[24]

Promotion of this year's event began in late 1950, coordinated by Frank Meunier, General Petroleum's advertising and sales promotion manager, and West-Marquis, Inc., General Petroleum's advertising agency. This advance preparation included print advertising announcing the upcoming Run and meetings with the press in principal western cities to explain the Run. As a result of these efforts, 64 members of the press (including radio and television) turned out to cover the event first hand.[25]

Beginning in early January, after car selection and during preliminary storage, break-in runs could be completed under AAA Contest Board supervision. While the number of break-in miles was at the entrant's discretion, the total miles on the competing car could not exceed 5,000 miles. "With the official insignia already applied to the cars, many of them are daily sights on the highways around Los Angeles, making break-in runs with AAA observers accompanying them at every turn and returning them to the contest board's impound at the end of the run."[26] On February 24, all competing cars were impounded at the General Petroleum garage in downtown Los Angeles, where inspections were conducted to ensure all components were strictly stock. As outlined in the rules, "Evidence of tampering with the stock status of a competing car during impound periods shall be grounds for immediate disqualification of the car and entrant." During this impound period, the cars could be taken from the impound area and given no more than two check runs not to exceed 50 miles each. Seals were placed in several places of each competing car, including the hood and gasoline tank cap. One other change in procedures dealt with tire pressures. Continuing to use the predetermined scale formula from 1950, an additional two pounds of pressure was added "to compensate for the load and speed requirements of the contest."[27]

Prior to the start, Les Butts, a veteran of previous Economy Runs and slated to drive the Packard 300, became ill and was unable to participate. William "Parky" Parkinson, a salesman at Frost & French Packard in Los Angeles, was tapped to drive the Packard at the last minute without any prior knowledge of the route or road conditions.[28]

The start of the Run was on March 6, adjacent to the General Petroleum garage on Flower Street in downtown Los Angeles, a location that would continue to be used through 1958. Here the drivers began a 535-mile, 13.5-hour trip to Las Vegas, encountering rain for the first 50 miles. Art Rene, driving a Ford Deluxe 6, was the first car flagged away at 3:00 a.m. by racing promoter J.C. Agajanian with the remaining cars leaving at two-minute intervals. In light rain, the drivers navigated the nighttime streets through several Los Angeles area communities before they crossed the San Gabriel Mountains into the Antelope Valley. Continuing through the Mojave Desert, the cars arrived in Lone Pine for refueling. Repeating scenes from the 1950 Run, the contingent was greeted by members of the Lone Pine High School band, along with residents of the community. Leaving Lone Pine, they continued their journey through Death Valley, where the drivers encountered head winds of 40 miles per hour for approximately 50 miles, along with navigating a section of unpaved roadways.[29] Dropping to an elevation of 279.6 feet below sea level at Bad Water, the drivers continued on to Death Valley Junction for the second refueling stop of the day. Leaving Death Valley, they crossed the border into Nevada, and by late afternoon arrived Las Vegas, where the cars were impounded overnight at the Motor-Vu Drive-In Theater, adjacent to the Last Frontier Hotel.

On March 7, the first car to leave Las Vegas departed at 7:00 a.m. for the 299-mile, 7-hour and 25-minute trip to the Grand Canyon. Following virtually the same route as the 1950 Run, the contestants crossed the Hoover Dam into Ari-

Another price class winner crossing Hoover Dam is Melvin Lewis, driving a Chrysler Imperial. This year two Chrysler Imperials placed first in their price classes, with Bob Burns in a Chrysler Crown Imperial also placing first. This year would mark the beginning of a strong showing for Imperial models that would continue throughout the decade.

zona; they continued through Northern Arizona, passing through Kingman before arriving in Williams for refueling, as needed, then continued to the finish line at the Bright Angel Lodge. Although snow was on the ground, the roads remained dry. A large group of spectators were on hand when the first car, a Ford Deluxe 6, crossed the finish line followed, by all of the other competitors, completing the Run within the allotted time. After all of the cars had arrived, they were again topped off for a final time and impounded overnight. While all competitors and others affiliated with the Run were guests of General Petroleum at the Mobilgas Mileage Dinner and Pegasus Party, AAA Contest Board officials commenced computing final mileage figures and made final checks that stock status had been maintained. The results would be announced at an awards presentation held on the morning of March 8.

Les Viland, participating in his first Economy Run, piloted a Lincoln Sport Sedan into the winner's circle, capturing the coveted Sweepstakes Award with a record-setting 66.484 ton miles per gallon with an actual average of 25.448 miles per gallon. Second in the Sweepstakes Award was a Chrysler Crown Imperial, driven by Bob Burns, with 63.289 ton miles per gallon. In addition to placing first in the special lightweight six-cylinder class, a Nash Rambler, driven by Pierce Venable, turned in the best miles-per-gallon figure of 31.053. This was the highest miles per gallon achieved since the 1938 Gilmore Yosemite Run, when a 4-cylinder Willys recorded a figure of 30.534. This achievement would not be surpassed until 1961. The best miles-per-gallon figure for an eight-cylinder car was achieved by a Studebaker Commander, driven by Stanley Van Horn, with a figure of 28.001. At the awards ceremony, Studebaker's dealers and factory representatives on hand were sporting Scottish plaid caps and shirts with Studebaker banners declaring: "It's the Thrifty One for '51."[30]

In addition to the Sweepstakes Award, first-place trophies were presented to:

PRICE CLASS	MAKE and MODEL	TON MILES PER GALLON	MILES PER GALLON
A—$1400 to $1750	Ford Deluxe 8	54.587	25.994
B—$1751 to $1950	Studebaker Commander	58.173	28.001
C—$1951 to $2175	Mercury	59.868	25.945
D—$2176 to $2450	Packard 200	53.020	22.023
E—$2451 to $2700	Lincoln	66.484	25.448
F—$2701 to $3000	Cadillac Series 61	55.492	21.719
G—$3001 to $3300	Chrysler Imperial	59.457	21.178
H—$3301 to $4000	Cadillac Series 60 Special	58.795	21.979
I—$4001 to $6000	Chrysler Crown Imperial	63.289	19.208
Special Lightweight—4-Cylinder	Henry J	49.153	30.109
Special Lightweight—6-Cylinder	Nash Rambler	53.489	31.053

Detailed results for all competing cars can be found in the Appendix.

The entire field, from the four-cylinder Henry J to the huge Chrysler Crown Imperial, had a combined average of 23.923 miles per gallon, which would turn out to be the highest average recorded during the fifties. In releasing the performance statistics, the AAA Contest Board said the figure was surprisingly high since the Run "involved crossing of several mountain ranges, and a temperature range of 35 degrees plus the fact that for 50 miles of the trip headwinds of 40 miles per hour were encountered." The average speed for the entire field was 40.600 miles per hour, and driving at maximum legal speeds was required to finish within the stipulated time.[31]

Seven of the class winners, including the Sweepstakes Lincoln, were equipped with overdrive.

The Cadillacs and Chrysler Crown Imperial were equipped with automatic transmission, while the Chrysler Imperial was equipped with Fluid-Matic semi-automatic. As was the case in 1950, two of the most popular low-priced entries (Chevrolet and Plymouth) were among the three entries which had standard three-speed transmission without overdrive, and none of these were class winners.

Three of the class-winning drivers were veterans of the 1950 Grand Canyon Run, with A.E. "Bud" Ridings (Cadillac Series 60 Special) and Bill Stroppe (Mercury) repeating their previous performance and Melvin Lewis (Chrysler Imperial) being a first-time winner.

After the results were announced, a protest was filed by Andrew J. Crevolin, Alhambra (California) Chrysler-Plymouth dealer and entrant of the Chrysler Crown Imperial, charging that the Sweepstakes-winning Lincoln was not completely stock. His challenge involved both the rear axle gear ratio and speedometer drive pinion gear used on the Lincoln. Though the time for filing a protest had elapsed (the rules stated that any protests be made "within one hour after the last competing car has officially been refueled and all rights to protest are waived after this time"), Colonel Arthur W. Herrington, national chairman of the AAA Contest Board, ordered that the winning Lincoln undergo a tear-down inspection to verify that all parts were stock according to the parts catalogs of the Lincoln-Mercury Division. This inspection revealed that everything with the Lincoln was in order, and it was noted that it was picked at random from a group of production cars. The inspection also noted that this particular car was equipped with a 3.31-to-1 rear axle, which was also referred to as the "plains" rear end, designed to deliver fine economy and performance under high-speed driving conditions that were found in the Prairie States. As a result of the inspection, the protest was disal-

Admiring the Sweepstakes Award for the performance of the Lincoln, driver Les Viland (left) is joined by General Petroleum Vice-President Clarence S. Beesemyer (center) and Inglewood Lincoln-Mercury dealer Bob Estes (right). Les achieved a record-setting figure of 66.484 ton miles per gallon, which remains the highest achieved in Economy Run history.

lowed.[32] The *Los Angeles Herald-Examiner* noted, "It wasn't necessary for the AAA to hold the [protest] hearing as time of protest officially passed. It was done, however, to remove any doubt that Economy Run rules had been sidestepped and to assure the public that the event was carefully supervised and conducted."[33]

Advertising and publicity again were subject to strict guidelines established by the AAA Contest Board, and in an unusual twist, release dates for feature stories and advertising were staggered. In all states except California, publicity and advertising could begin on March 9. In California, press releases containing only essential facts could be released immediately after the awards ceremony; however, complete feature stories, pictures and all advertising had to be held until Sunday, March 18, or later. One exception to this rule was that each entrant was permitted to announce its achievements once in a news-type fashion on radio or television.[34] Newspaper advertising was again most prominent in the immediate weeks following the Run. Lincoln proclaimed a "Smashing Triumph!" in announcing its Sweepstakes Award, telling readers that this is "Proof again that Lincoln's dramatic engineering is a wise and economical investment for these uncertain times!"[35] Mercury declared that it is the "Economy Car of its Class," while Studebaker proudly announced that "Among 26 cars entered in Standard Classifications ... Three Studebakers Finished 1st, 2nd and 3rd in Actual Miles Per Gallon."[36]

General Petroleum's magazine advertisements featured the theme "All the Gas Mileage Your Car Can Deliver," featuring full-color one-page and two-page spreads appearing in general interest publications such as *Life* and *The Saturday Evening Post*. Noting that "32 Cars Averaged 23.92 Miles Per Gallon," this campaign ran through May 1951 and invited

This Nash Rambler achieved the highest miles per gallon for all entries with a figure of 31.053, topping the previous record set by a four-cylinder Willys Deluxe in 1938. General Petroleum's A.J. Donnelly (left) congratulates Nash Motors' Zone Manager R.D. Bolt on this achievement while driver Pierce Venable (holding trophy) looks on. The Rambler's record would stand until 1961.

readers to visit their Mobilgas Dealer for a free copy of the booklet "All the Gas Mileage Your Car Can Deliver."[37] This booklet featured hints on ways to maximize gas mileage, including regular servicing, good driving habits and using Mobilgas or Mobilgas Special, noting: "It takes all 3 to get Top Mileage under ALL Conditions." This also devoted space for the motorist to track his own mileage to see how it compared to the Economy Run entries.[38]

Mercury's magazine advertisements followed the same basic theme from 1950, as its success was limited for the most part to footnotes pointing out that it was the "Economy Winner Second Straight Time."[39] Lincoln followed the same style as Mercury, where the Sweepstakes Award was also found in footnotes concerning its "Proof of Economy!"[40] Ford also noted several times in their advertising text during 1951 that it was first in its class.

Lincoln-Mercury dealers, primarily in Southern California, prominently displayed duplicates of the winning models in their showrooms, produced in the same colors and lettering as the originals. A photograph in *Automotive News* showed 117 of these models (70 Tomah Ivory Mercurys and 47 Kent Gray-Blue Lincolns) lined up on the Santa Ana Freeway, forming a line nearly a half-mile long prior to delivery to the dealerships.

Nash proudly advertised the Rambler "First in Miles Per Gallon," claiming it had "Set a New All-Time High Record of 31.05 M.P.G."; this achievement continued to be noted throughout 1951.[41] This record-setting miles-per-gallon claim is accurate for the Mobilgas Economy Run. In the 1936 Gilmore Yosemite Economy Run, a 4-cylinder Willys "77" set the all-time record of 33.21 miles per gallon. A two-page advertisement in industry-related publications featured a pho-

Nash heavily advertised the 31.05 miles per gallon record set by the Rambler in 1951. Featured mainly in industry-related publications, this two-page spread featured photographs of the winning Rambler and driver, Pierce Venable (holding trophy, on left) (courtesy of Chrysler Group LLC).

tograph of the winning Rambler along with its driver, Pierce Venable. This was also noted in various Nash and Rambler showroom brochures throughout the 1952 model year. Henry J's achievement was part of an advertisement that proclaimed "All-round proof! Kaiser-Frazer's Henry J is America's all-round car!" and included a photograph of the winning Henry J.[42] Studebaker heavily promoted its models' having the best miles per gallon of all standard classification cars with taglines such as "Best by test in actual gas mileage!" and "The stand-out 8 in actual gas mileage!" Most of the emphasis was on the eight-cylinder Commander models.[43] Again, potential buyers in the United States and Canada were told of the Champion's performance by continuing the "Best by test in actual gas mileage!" theme in both the 1951 Studebaker showroom brochures. The Champion, Commander and Land Cruiser's performances were discussed in the 1952 Studebaker showroom brochure.

Borg-Warner, manufacturer of the overdrive transmission unit, also got into the act, proclaiming "First 15 Cars in Mobilgas Economy Run equipped with B-W Overdrive!" in magazine advertisements that were mainly found in automotive industry–related publications such as *Motor*.[44] This same advertising theme would continue through the 1954 Run.

In the days following the Run, articles appeared in newspapers, primarily in Sunday automotive sections, and once again these were most prominent in Southern California newspapers. Coverage of the Run also continued to appear in the months following the Run in both general interest and automobile-related magazines.[45] Economy driving tips from winning drivers, most notably Sweepstakes Winner Les Viland, appeared in *Motor Trend*, *Popular Science* and *The Lincoln-Mercury Times*. Since the vast majority of the entries were equipped with either 3-speed standard transmissions or overdrive, these driving tips were aimed specifically at the group concerned with getting into high gear as quickly as possible.[46] One of Les' tips was: "Be gentle with your car. Don't start it suddenly and stop suddenly, and don't have digout races with people. You never get good mileage while you're doing that."[47] In addition to newspaper and magazine advertising, both Lincoln and Mercury also produced small postcards used by their dealers publicizing their success.

In summarizing the 1951 Mobilgas Economy Run, Clarence S. Beesemyer, executive vice-president of General Petroleum, remarked:

> This one great comprehensive test of the performance capabilities of 1951 model cars is a proving ground for both motorists and the automobile industry and General Petroleum is proud to be leading the way to greater economies that inevitably result. Under conditions common to every motorist, the run demonstrates the economies that result from correct fuel selection, safe driving and proper maintenance of the automobile. On nature's own proving ground the event provides the automobile manufactures to test and compare their products. Many engineering improvements in 1951 cars are a direct result of technical lessons learned in last year's Economy Run. All are for the benefit and economy of the car owner, and we are pleased to be the one oil company that is helping make these economies and improvements possible.[48]

After two successful runs to the Grand Canyon, a new destination of Sun Valley, with longer days in both miles and time on the road, would be the choice in 1952.

3

Sun Valley

1952

Los Angeles to Sun Valley Via Grand Canyon — Salt Lake City

April 14–16, 1952—1415.4 Miles

Expanding the length of the route, along with some tightening of the rules, were key features in the 1952 Mobilgas Economy Run. In announcing the destination of Sun Valley, A.C. Pillsbury stated, "The Economy Run provides the one competitive opportunity for manufacturers to prove the exceptional performance and mileage capabilities of the 1952 cars, all under the same conditions and exacting supervision and on a common proving ground."[1] Facing the drivers would be another challenging test requiring them to cover a 1415.4-mile route in a 35-hour time limit over a three-day period. With overnight stops at the Grand Canyon and Salt Lake City, the route selected ranged in altitude from 70 feet below sea level at the All American Canal in Indio to 8010 feet at Jacob Lake Lodge in Arizona and included sections where the competitors had to climb above 6000' a dozen times.

Preparations for this year's Run began in July 1951 with nearly 450 personnel from General Petroleum and the AAA Contest Board involved in all facets, ranging from route selection and logistics to press relations and publicity. As noted in *Motorsport*, "The extensive preparations for the event by General Petroleum and AAA are at least as important as the actual running. Thorough planning in the early stages prevents last-minute errors and slip-ups that plague less-well planned events."[2] As evidence of this thorough planning, a set of detailed instructions was issued to all AAA Contest Board personnel involved in the actual Run. Covering ten pages (on legal-size paper), these instructions, which left nothing to chance, outlined each person's responsibilities beginning with the fueling activities prior to the start and continuing through the tabulation of the final results. Provisions for the use of an alternate route, should weather conditions dictate, were also included and these provisions would continued in future years.

Beginning in early 1952, General Petroleum began to aggressively publicize the upcoming Run and encouraged all participants to "exploit the event through every possible channel available to them."[3] With the expansion of the route, press parties, under the direction of General Petroleum's advertising manager Frank Meunier, were held in seven western states to create added interest. As a result of these efforts, reporters from newspapers in these states covered the Run along with those from Boston, Chicago, New York, and Dallas, to name a few.[4] Print advertising announcing the "World's Greatest Car Test ... 1952 Mobilgas Economy Run" ran in newspapers along the route of the Run and were seen in industry publications such as *Automotive News*.[5] These efforts to continually promote the Economy Runs would continue and expand in the future.

While remaining similar to those from previous years, two significant changes were made to the Official Rules designed to protect the validity of the Run. In announcing these changes, A.C. Pillsbury stated, "Exacting rules and regulations of the AAA allow for no loopholes. We will leave nothing undone to insure [*sic*] that this great, competitive mileage test is among regular stock automobiles and that it is planned and conducted with unquestioned fairness."[6] One change included specific production criteria for engine, transmission, and other options so that the competing models would accurately represent those being purchased by the driving public. This area was further expanded with regards to the axle ratio, which now must be "considered as 'standard' for the model selected and for the type of drive chosen or represents at least 50 percent of the national scheduled production for such model and drive."[7] Specially selected carburetors could no longer be used. Replacement carburetors could only be used when the original could be shown as being defective, and then the AAA would purchase one from a random source.[8] Competing cars would continue to be selected at random by the AAA Contest Board (a procedure that would continue throughout the duration of the Economy Runs), but entrants would no longer be able to specify a color preference of their entries. The other significant change was that the announcement of the actual route would be delayed until a date during final impounding, "thus eliminating practice over the course by the competing cars."[9]

Entries were open to any stock 1952 four-door metal-top sedan of U.S.A. manufacture and design and not driven more than 2500 total vehicle miles. As was the case in 1951, two special lightweight classes were created, and only one car of each make and model could be entered. Every car participating again carried a payload of 750 pounds, the equivalent of four passengers plus their luggage.[10]

A total of 29 cars competed in 9 price classes, based on the February 1952 factory retail delivered price for the lowest priced four-door sedan, plus 2 lightweight categories. The choice of transmission was again left to each individual entrant. Those equipped with overdrive or Fluid Drive represented the bulk of the entries; only four were equipped with an automatic transmission. Nineteen of the competing cars elected to use Mobilgas Special, with only the Chevrolet, Hudson and Studebaker entries using Regular. Cadillac, which enjoyed considerable success in the two previous Runs, decided not to participate, explaining that "they had tried the 1951 course for a month with their 1952 models and found that they had lost their economy."[11] In addition, Buick, Dodge, Oldsmobile, Pontiac and the newly-introduced Willys did not to participate. For unknown reasons, Les Viland did not participate to defend his 1951 Sweepstakes-winning performance.

Beginning in February, car selection and preliminary impounding commenced at a new location in Pasadena, California. This large building, originally designed for creating and decorating floats participating in the New Year's Tournament of Roses Parade, provided the necessary space for inspections and preparation of the competing cars. Random selection was not restricted to the Los Angeles area. As reported in the *Torrance* (California) *Herald*, "Without warning, AAA East Coast officials picked out two Henry J Corsairs and a Kaiser Deluxe at Willow Run (Michigan). Meanwhile, West Coast men went to Bakersfield (California) where a Hudson Commodore 8 was selected. Autos were driven to Los Angeles where they are impounded."[12] The impound was a hub of activity with representatives of each manufacturer servicing and conducting break-in runs on each competing car, all aspects under the constant supervision of the AAA Contest Board. Final impounding occurred on March 25, and after that date, two check runs not to exceed 50 miles each were allowed. Noting that the route would be fairly mountainous, the drivers of the three Hudson entries asked for and were granted permission to make a single 100-mile check run which enabled them to drive in higher altitude areas of Southern California with the competing cars to best duplicate the terrain of the route selected.[13] During this period, inspections were made to ensure that all components were strictly stock, and seals again were placed in some 60 different places on each competing car, including the hood and gasoline tank cap.[14] Based on their good performance in 1951, selected students from the California Institute of Technology (Caltech) were again selected as observers whose duties were to ensure that all Contest Board rules were followed, including all applicable speed, safety and traffic laws. As reported in the *Los Angeles Examiner*, A.C. Pillsbury noted that "the integrity and efficiency shown by Caltech students in the past have been major factors in the success of the run."[15] Caltech students would continue to act as observers through the 1959 Run.

After the actual route was announced, each competitor was provided with a route description booklet that in 6½ pages gave "turn-by-turn" instructions from Los Angeles to Sun Valley along with maximum elapsed time requirements for each day's leg. The importance of each competitor to make a practice run (using an identical model of the one participating in the Run) cannot be overstated. These runs provided insight to the terrain, road and traffic, which enabled entrants to develop various arrival times and required speeds during each day so as not to exceed the maximum time requirements. Failure to

Photographed at the Pasadena impound all lettered and ready to go is the Packard 200 entry sponsored by Los Angeles' Earle C. Anthony, Inc. Photographs of this nature were often taken to show a sample of the lettering format used on the sides, roof and trunk of each entry.

meet the time requirements set by the AAA Contest Board would result in disqualification.[16]

An hour prior to the start, the three Nash entries (Rambler, Statesman and Ambassador) were withdrawn without explanation. Later it was reported that even though the Ambassador, in particular, had performed well on practice runs, Nash officials "felt the ruggedness and added length of the course would prevent them from equaling their excellent 1951 results and the comparison would hurt sales."[17]

The start of the Run was on April 14. The drivers faced a 515.5-mile, 12-hour and 53-minute trip to the Grand Canyon. Les Butts, driving a Packard 300, was the first car flagged away at 3:00 a.m. by racing promoter J.C. Agajanian, with the remaining cars leaving at two-minute intervals. Heading east, they made their way through several Los Angeles area communities, then over the San Gorgonio Pass towards Palm Springs. Traveling through the Mojave Desert and Imperial Valley, the drivers had an advantage of a tail wind, which they used to their advantage until they arrived in Blythe for refueling.[18] As a result of the advance publicity, when the competitors arrived at the refueling station they were greeted by a large group of residents along with a local radio station broadcasting the event.[19] Crossing into Arizona, the drivers made their way through Prescott and into Williams, for the second refueling stop of the day. With a combination of mountain terrain and traffic, most of the field was reported to be running behind schedule as they approached Williams.[20] By late afternoon all competitors arrived at the Bright Angel Lodge at the South Rim of the Grand Canyon within the time requirements, although Willard McArthy, driving a Henry J Corsair 6, pulled in with a mere 10 seconds to spare. The cars were impounded overnight adjacent to the Bright Angel Lodge while the crews and officials got some well-needed rest.

On April 15, the first car to leave the Grand Canyon at 3:00 a.m. was the Packard 300, heading out on a long 575.3-mile, 14-hour trip to Salt Lake City. Traveling through the mountains and then into the Painted Desert, the drivers arrived at the Marble Canyon Trading Post for the first refueling stop of the day before continuing north to Zion National Park. Here drivers and observers enjoyed a 30-minute brunch stop, which gave everyone a chance to take a break and stretch his legs. In future years these stops would become a daily tradition. Leaving Zion National Park, they continued north to Beaver for the second refueling stop of the day. This leg between Zion and Provo was not kind to the competitors as they encountered strong head winds. Leaving Cedar City, the DeSoto Firedome 8-passenger sedan started to experience a mechanical problem with the carburetor, and 53 miles shy of the Beaver Refueling Station ran out of gas. After obtaining fuel from an unauthorized source, it was disqualified under the rules. A disappointed Randy Maahs discovered later that some foreign substance apparently had become lodged in the carburetor float and flooded the carburetor, resulting in a more than normal flow of gasoline.[21] Two other entries, a Chrysler Saratoga 8-passenger and another DeSoto Firedome, also ran out of gas. These drivers were able to catch a ride to the refueling station, where they obtained a sealed gasoline can from AAA officials, who returned to the two cars, broke the seals and added fuel to the stranded vehicles. This enabled the drivers to continue to the Beaver refueling station. Both entries were able to make up lost time and arrived at the Salt Lake City impound within the allowable time limits.[22] The drivers arrived in the late afternoon at the Capitol Grounds in Salt Lake City, where the cars were impounded overnight. They were met by a large group of spectators led by Utah Governor J. Bracken Lee along with Miss Utah, Bunny Reese. Although no official protests were received regarding the two entries that ran out of gas, there were discussions whether or not they would be able to continue to compete. Since no entry had run out of fuel in prior runs, and no rule specifically covered this situation, AAA Contest Board officials allowed both to continue.[23] A protest was received regarding the Mercury Monterey entry. As was noted by another competitor, the Mercury rolled approximately 40 feet at the Grand Canyon start before proceeding under its own power. After the protest was heard, it was decided by the AAA Contest Board officials that the Mercury should back up a total of 40 feet the next morning before leaving to compensate for the false start.[24]

On April 15, the first car to leave Salt Lake City at 8:00 a.m. was again the Packard 300, having been the first to arrive the previous day. With the finish line in sight, the drivers still faced a 324.6-mile, 8-hour and 7-minute trip to Sun Valley. Immediately after leaving the Capitol Grounds, the cars stopped for refueling before continuing on to Ogden. In Idaho, a second refueling stop was made in Twin Falls, where "each competing car was required to take on a minimum of five gallons of fuel" before continuing to the finish at the Sun Valley Lodge.[25] Leaving Twin Falls, a group of drivers encountered a roadblock caused by a herd of sheep, along with a warning sign "sheep on highway have right-of-way over motorists." Hudson Wasp driver Jim Abbott managed to get by using a smooth ditch on the side of the road, but some of the drivers were held up.[26] Les Butts, driving a Packard 300, was not only the first driver to leave Los Angeles, but was also the first to arrive in Sun Valley. Along with a large group of spectators, J.C. Agajanian was on hand greet the drivers, waving the checkered flag, and his flamboyant presence would be a fixture in future Economy Runs through 1966. All cars were again topped off for the final time, and while AAA Contest Board officials computed the final figures, General Petroleum officials hosted their annual Mobilgas Mileage Dinner for all competitors and others affiliated with the Run. The results

The Packard 300 driven by Les Butts is the first car sent on its way by AAA's J.C. Agajanian. Despite the early 3:00 a.m. starting time, a good-sized crowd of spectators were on hand to see the drivers leave on the first leg to the Grand Canyon.

would be announced at an awards ceremony on the morning of April 18.

Bill Stroppe captured the Sweepstakes Award for the second time in three years, piloting his Mercury Monterey into the winner's circle with a figure of 59.7118 ton miles per gallon and an actual miles-per-gallon figure of 25.4093. Unfortunately, he was unable to accept the Sweepstakes Award, having been hospitalized after arriving at Sun Valley with a case of the mumps complicated by a malaria relapse. Relief driver Clay Smith stood in for Stroppe at the awards ceremony, saying, "Bill complained of a 'sore throat' at the start of the run but refused to give up the wheel."[27] In addition to capturing the Sweepstakes Award, Bill Stroppe also scored his third consecutive first-place victory, while Lincoln-Mercury dealer Bob Estes took the top prize for the second consecutive year. Second in the Sweepstakes Award was a Lincoln Capri, driven by Art Rene, with 58.9085 ton miles per gallon. In addition to placing first in the special lightweight 4-cylinder class, a Henry J Corsair, driven by Woody Bell, achieved the best miles-per-gallon figure for all entries with 30.8558. The best miles-per-gallon figure for a six-cylinder car was 26.3676 set by another Henry J Corsair, driven by Willard McArthy, and the best figure for an eight-cylinder was 25.5968 set by a Studebaker Commander driven by John Van Trigt. Both of these models placed second in their respective price classes.

In addition to the Sweepstakes Award, first-place trophies were presented to:

PRICE CLASS	MAKE and MODEL	TON MILES PER GALLON	MILES PER GALLON
A—$1600 to $1950	Ford Mainline 6	53.8551	25.4634
B—$1951 to $2225	Kaiser Deluxe	53.6094	24.6480
C—$2226 to $2350	Mercury Monterey	59.7118	25.4093
D—$2351 to $2500	Hudson Wasp	47.7830	20.4638
E—$2501 to $2900	DeSoto Firedome	54.7368	21.2777
F—$2901 to $3450	Chrysler Saratoga	49.1762	19.0237
G—$3451 to $3750	Lincoln Capri	58.9085	22.3562
H—$3751 to $4500	Chrysler Saratoga 8-Passenger	50.5729	17.6520
I—$4501 to $7000	Chrysler Crown Imperial	53.1736	16.2362
Special Lightweight 4-Cylinder	Henry J Corsair	51.3749	30.8558
Special Lightweight 6-Cylinder	Plymouth Concord	47.0250	23.0797

Detailed results for all competing cars can be found in the Appendix.

The entire field recorded an average of 22.0057 miles per gallon, covering the 1415.4-mile route with an average speed of 40.846 miles per hour. The AAA Contest Board noted that these results were surprisingly good in view of the rigorous course that was used this year.[28] All of the price class winners were equipped with overdrive except for the Lincoln Capri and Chrysler Crown Imperial, which had automatic transmissions. A Chevrolet Styleline Deluxe, the only entry equipped with standard three-speed transmission, placed a disappointing fifth in its price class, demonstrating once again the gas-saving advantages of these optional overdrive units.

Officials from Kaiser-Frazer Corporation, faced with declining sales and struggling financially, had something to cheer about with the performance of two of their entries. Not only did a Henry J Corsair take top honors in the special lightweight 4-cylinder class, along with recording the highest actual miles per gallon for all entries, a Kaiser Deluxe also placed first in its Price Class.

Bob Burns repeated his 1951 class-winning performance, once again driving a Chrysler Crown Imperial. Also of note, six of the winning drivers (Jim Abbott, Bill Cameron, John Gerfen, Vern Holmquist, Bill Hooks and Lew Jabro) were first-time participants.

Vern Holmquist was especially pleased with his first-place finish in a DeSoto Firedome. This would be the only time DeSoto placed first in the years it competed in the Economy Run. As one of the competitors who ran out of fuel on the second day, the *Los Angeles Daily News* reported, Vern "wasn't filling his tank at the official stops in an effort to cut down on weight and fueling time. He figured eight gallons should take him to the next stop but ran into a strong headwind that ate into his gas supply faster than he expected. His win was all the more remarkable because of the added gasoline needed to start the car [and] he also had to drive at higher speeds to finish in the allotted time."[29]

After winning his price class in 1951 driving a Packard 200, Harold Michelson had a disappointing fourth-place finish. His Packard 200 had trouble early on when it "froze in overdrive and every time the car stopped, it had to start in high overdrive and this naturally took a lot of gas."[30]

While all advertising and publicity efforts continued under the strict guidelines established by the AAA Contest Board, participants were now free to begin these efforts immediately after announcement of the final results. There was a marked increase this year in both newspaper and magazine advertising, which showed the increased influence of the Economy Run for those participating. Once again, newspaper advertising would be the most prominent. Mercury proclaimed, "Again Mercury Wins Sweepstakes," making it the "Biggest three-time winner in history of Mobilgas Economy Run." Lincoln announced "Again Lincoln Wins Class G," adding "Lincoln takes top honors last year, wins in its class this year."[31] Studebaker continued to advertise, "Among 23 cars entered in standard classifications ... Studebaker finished 1st and 2nd in actual miles per gallon."[32]

"Ford Wins Again!" was the headline this year as Ford announced, "As in the past two years ... Ford was first in gas economy over all other cars in its price class."[33] This "first in gas economy" statement refers to Ford's top ton-miles-per-gallon performance due to the fact the Studebaker Champion (competing in the same price class) turned in a higher miles-per-gallon figure. Chrysler told readers that "180HP Chrysler V-8 *First in 3 Classes* in Economy Run!," which highlighted one of the winners, the price class F champ Saratoga.[34]

General Petroleum's magazine advertisements continued the theme "All the Gas Mileage Your Car Can Deliver," featuring full-color one- and two-page spreads. Continuing to appear through the summer months, these ads noted that "25 Different Makes and Models of U.S. Stock Cars Averaged Over 22 Miles Per Gallon with Mobilgas." Drivers were also reminded: "You too, can improve gas mileage if you do what the Economy Run drivers did — keep your car in tip-top shape ... drive carefully ... always fill up at a Flying Red Horse pump!"[35]

Mercury went all-out noting its second Sweepstakes Award, using the theme "America's No. 1 Economy Car!" in the majority of its advertising. Pictures of the Sweepstakes-winning car were featured in several of the magazine ads, appearing in general interest publications such as *Life* and *The Saturday Evening Post*. One proclaimed that the newly designed Mercury "can't be beaten for economy — won't be dated for years!" Another asked, "Why not get *all* that's coming to you!"[36] An advertisement that appeared in the July issue of *Motor Trend* announced, "Mercury Beats All Comers in Mobilgas Economy Run — *3 Wins in 3 Years*!"[37] Ford noted several times that its Mileage Maker Six won first place in its class. Henry J noted it was a Mobilgas Economy champ, highlighting its 30.85 miles per gallon performance by the 4-cylinder model.

While Lincoln made limited mention of its first-place performance in magazine advertising, Lincoln dealers were provided with a good deal of promotional material from Lincoln-Mercury Division's advertising and sales promotion department. Included in this packet was a detailed folder, "Lincoln Wins Over All Other Fine Cars," designed for a salesman's use to promote Lincoln's success, along with several suggested radio commercials for local use. Winning driver Art Rene also made an appearance on Ed Sullivan's *Toast of the Town* television show.

Studebaker went all-out in promoting the success in actual miles-per-gallon performance of its Champion and Com-

180 HP CHRYSLER V-8 FIRST IN 3 CLASSES IN ECONOMY RUN!

Here is truly dramatic proof that Chrysler's great new FirePower 180 HP V-8 engine design sets an entirely new standard of efficiency among American passenger car engines!

In this annual economy test, rigidly supervised, and limited strictly to stock car entrants, cars competing are divided, by price and size, into 11 standard classes. The route, from Los Angeles to Sun Valley, covered 1,415 miles of every possible kind of driving. Average speed for all cars was just under 41 miles per hour.

And in 3 of the 11 classes, the Chrysler V-8 engine was best for gasoline mileage.

In Class "F", this magnificent new engine won first place for a Saratoga 6-passenger sedan (125½-inch wheelbase, 4010 pounds). In Class "H", it won first honors for a Saratoga model 8-passenger sedan (139½-inch wheelbase, 4510 pounds). And in the top price and size, Class "I", it was again first, in a Chrysler Crown Imperial Sedan (145½-inch wheelbase, 5360 pounds).

To travel these substantial cars at this speed under these conditions gives additional proof, we believe, that here is the finest and most efficient engine ever put into an American passenger car!

WE INVITE YOU TO DRIVE THIS ENGINE, YOURSELF...

The same engine which has just scored these remarkable accomplishments is no farther from you than your own Chrysler Dealer. He will welcome the chance to let you take the wheel and feel for yourself what Chrysler has done . . . not only in FirePower engine performance, but in full-time power steering, power brakes, new Oriflow shock absorbers, and passenger comfort . . . to deserve the title: *"Finest Car America Has Yet Produced!"*

CHRYSLER *the finest car America has yet produced*

DRIVE A CHRYSLER Today! Learn the Difference!

This is one of the few advertisements created by Chrysler during the early fifties, when a Crown Imperial and two Saratoga sedans placed first in their classes. Advertisements of this nature were normally found in newspapers (courtesy of Chrysler Group LLC).

mander models in magazine advertisements appearing throughout the year. In a special two-page advertisement celebrating their centennial, Studebaker's success in both the Gilmore and Mobilgas Economy Runs was highlighted. Studebaker dealers on the West Coast heavily promoted these results, with dealers decorating their showrooms in a "Scotch Thrift" theme. Replicas of the winning Champion were built and lettered, noting Studebaker's success for display in dealer showrooms. This included a special "Drive Away" event for Studebaker dealers in the Los Angeles area. Again potential customers were reminded of the Champion's performance as "Best by test in actual gas mileage!" in the 1953 showroom brochures. Studebaker also produced a mailer for Studebaker owners featuring mileage-stretching tips from Studebaker Economy Run drivers and included an invitation for a "Thrifty Driving Demonstration."

Danny Eames takes his Dodge Coronet 8 on a break-in and check run during the impound period. After a two-year absence, the Dodge would go on to capture first place in the low-medium-price (standard and overdrive) class with 52.8565 ton miles per gallon (courtesy of Chrysler Group LLC).

During the Run, Studebaker provided one of the cars used by members of the press, a Starliner Hardtop, which was complete with a plaid-covered top and lettered with the slogan "Teilgeach Agat," which is Gaelic for "Your Economy." This promotion would be repeated in 1953.

In the days following the Run, articles appeared in newspapers, primarily in Sunday automotive sections, and once again these were most prominent in Southern California newspapers. Articles about the Run also continued to appear in automobile-related magazines. Some of the articles featured economy driving tips, as was the case in previous years. Two of the best articles written to date covered experiences of drivers who competed and gave somewhat detailed insight to the challenges they both faced. These were found in the July 1952 issues of *Motor Trend* ("Winning the Economy Run," by Bill Stroppe) and *Speed Age* ("Mobilgas Economy Run," by Jim Abbott and Bob Kiple).[38] Coverage of Mercury's Sweepstakes success was featured an article in "Lincoln-Mercury Times" which was distributed to Lincoln and Mercury owners.[39]

In summarizing the 1952 Mobilgas Economy Run, Robert L. Minckler, president of General Petroleum, remarked:

> I suppose that the feature of the Mobilgas Economy Run which creates the most business comment is the fact than an oil company will devote its advertising budget to telling its customers how to use less of its products. But that is consistent with the finest tradition and spirit of American business, because it is based upon the development of better service to our customers, and our customers are the most important people in our lives. It is that tradition and spirit of service, which has resulted in the steady and constant improvement in automobile design and in motor fuel. This is a contest, and in a contest there must be winners and there must be losers. But in the American tradition and spirit of which I have spoken, all participants who win, lose or draw in an event like this, contribute to the increase in knowledge which comes from such events as this. In this sense all participants in this Run are winners. The only losers are the automobiles who aren't in the Run.[40]

One of the benefits of the 1952 Run was the knowledge the automotive engineers gained during the event. Within a week after discovering the cause of the carburetor trouble that caused the two DeSoto entries to run out of gas, carburetors were being altered on Chrysler Corporation's assembly lines in an effort to prevent this problem from recurring.[41]

1953

Los Angeles to Sun Valley Via Reno — Boise

April 20–22, 1953—1206.1 Miles

Not long after the conclusion of the 1952 Run, officials from General Petroleum and the AAA Contest Board reviewed all of the details of the event and made the decision that the 1953 Mobilgas Economy Run would return to Sun Valley. The route, although 209.3 miles shorter this year, was described as the most rugged and toughest yet to face the drivers and cars. In announcing the destination, Clarence S. Beesemyer, who had retired as vice-president of General Petroleum and was now in a consultant position, remarked that "a course as tough and comprehensive as any ever mapped in the past will face this year's entries."[42] Added Chief Steward A.C. Pillsbury, "The idea of the Run is to put cars through every condition, weather, altitude and terrain that the average driver would hit in a full year's motoring. We want to show the public what it can expect of the new cars if they're driven wisely and kept in good condition. Nothing is done during the Run that can't be done by the average motorist when he drives the same model out of the showroom. Expect no trick driving. No special gadgets."[43] With overnight stops in Reno and Boise, the drivers would be required to cover the 1206.1-mile route in a 27-hour, 15-minute time limit on a route that ranged in altitude from 19 feet at Stockton, California, to 7383 feet at Echo Summit in the Sierra Nevada Mountain Range in California. This required the drivers to up their average speed to 44.7115 miles per hour, which would be the highest to date in Economy Run history.

Continuing to refine the rules, officials announced several important changes. Instead of the somewhat complicated price class groups, used since the 1936 Gilmore Yosemite Economy Run, new "classifications will be established for cars in price divisions known in the industry as Low Price, Low Medium, Upper Medium, and High, plus a Special Class for lightweight cars. Separate divisions for standard drive (including overdrive) and automatic transmissions were established in the Low, Low Medium and Upper Medium classes. Cars in the High-price class will compete with automatic transmissions only."[44] A.C. Pillsbury noted, "By limiting classifications, we hope to spur competition within the industry and give the public a better idea of the performance qualifications of each car."[45]

Additional rule changes were made as a result of events that occurred in 1952. One change involved the number of observers assigned. As explained in the Official Rules, "In order to insure [*sic*] better reporting and more alert observation, two observers will be assigned to each car." Sharing some of their responsibilities, one observer was now seated beside the driver and the second seated directly behind, and rest periods were allowed during each leg.[46] Another change was a result from three competing cars' running short of fuel in 1952, something that had not occurred previously in both the Gilmore and Mobilgas Runs. Beginning this year, "a car must reach the designated refueling station under its own power. A competing car cannot receive fuel or oil except when it is at an official refueling station."[47]

Entry eligibility mirrored those for 1952, although only a single special lightweight class (combining both 4-cylinder and 6-cylinder models) was established.[48] With the exception of Chevrolet, all entries were sponsored by individual dealers or dealer associations. As the public's knowledge of the Economy Run increased each year, it also continued to gain wide recognition along with a high degree of acceptance, resulting in additional support by some of the manufacturers. As one representative noted, "If our dealer wins, we go all-out on advertising; but if he loses, we can always claim the entry wasn't factory sponsored."[49]

A total of 26 cars competed in 7 classes,

Receiving a traditional green flag sendoff from AAA's J.C. Agajanian is the Studebaker Commander driven by Dick Griffith. Even with an earlier midnight starting time, a smaller crowd of spectators than in previous years was on hand to watch the drivers depart for the first overnight stop in Reno.

Pete Novotny guides his Chevrolet Two-Ten through the California Gold Rush town of Sutter Creek on the first leg of the 1953 Run. The next day, Pete would be disqualified after running out of gas before arriving at the Boise impound as a result of making a wrong turn outside of Winnemucca, Nevada.

plus a single lightweight class. Only the Chevrolet Two-Ten was not equipped with overdrive (an option that was not available on Chevrolet models), and 10 entries were equipped with automatic transmission. Nash, which in 1952 withdrew prior to the start, was back in full force with a Rambler, a Statesman and a pair of Ambassadors. Dodge also returned to competition after a 2-year absence. As was the case the previous year, Buick, Cadillac, Oldsmobile, Pontiac and Willys did not participate. Bill Stroppe returned, again driving a Mercury Monterey, to defend his 1952 Sweepstakes-winning performance.

Preparations, which began in late 1952, included plotting eight different routes between Los Angeles and Sun Valley. The Automobile Club of Southern California, the local AAA affiliate, was given the responsibility for route selection, and they covered each possible one in detail. Hotel accommodations were also secured for the several hundred who would travel the route, only to have the majority of these reservations canceled when the final route was announced.[50]

Car selection and preliminary impounding began in February, with random selection again being handled by the AAA Contest Board. While striving to maintain complete stock status of those cars competing, it appears that some manufacturers were beginning to try to "stretch the rules." Documents uncovered a number of years later, located in the Ford Archives at the Benson Ford Research Center in Dearborn, revealed that "some 50 specially-readied cars, with engine tolerances modified and engines balanced, were sent to Los Angeles showrooms and the Long Beach factory. At Ford factories, even production and tire sizes were changed so that more economical versions could be homologated." There is no evidence available that would indicate conclusively that one of these specially readied cars was selected by the AAA Contest Board to participate.[51] As in prior years, the impound remained a hub of activity with the servicing and break-in runs, under the watchful eyes of the AAA Contest Board, being handled on each competing car.

Final impounding occurred on March 31, with no mechanical work on the cars being permitted during this period, and the announcement of the final route occurring at this time.[52] Maximum net running time (not to include stops for refueling or the 30-minute brunch stop) for each day was also announced, and failure to meet this time would result in disqualification.[53] This rule would continue to be enforced through 1956.

The start of the Run was on April 20, with the drivers facing a long and fairly mountainous 548.3-mile, 13-hour trip to Reno. With stops for refueling and brunch along the way, the drivers would be on the road for well over 14 hours. Andy Henderson, piloting a Nash Ambassador Custom, was the first car flagged on its way at 12:01 a.m., with the remaining cars following at two-minute intervals.

As the drivers left downtown Los Angeles in a light rain, AAA Contest Board flagmen were once again stationed at various intersections to guide the drivers through Hollywood and the San Fernando Valley, and then over the Ridge Route into the San Joaquin Valley. In addition to rainy conditions, some dense fog was encountered on this leg, though everyone came through without any problems or appreciable loss of time.[54] The first refueling stop of the day was at Merced, and the scene here would be repeated at every stop over the next three days. As noted in *Speed Age*,

As each car pulled into the station, AAA officials broke the seals on the fuel tank and hood; observers recorded the stopping time in their logs. Representatives of every entrant recorded the amount of fuel each car took and each team maintained a complete set of records on all of their competitors. The drivers seemed oblivious to the greetings of the local welcoming committee, the strains of the village high school band and to the coffee, juices and doughnuts provided by the ladies auxiliary. It was noted that, in the case of Bill Stroppe's Mercury and Clay Smith's Lincoln, every inch of the grille, hood, fenders and windshield was wiped clean on the theory that dirt and every bug could affect the streaming air flow. Even the tires and wheels were wiped clean.[55]

Heading north through California's Mother Lode country, the drivers traveled through a number of historical mining towns before arriving in Amador for a 30-minute brunch stop. For the first time since leaving Los Angeles, the drivers could relax. Here entertainment was provided by a local school band, and this would become routine during the trip to Sun Valley. Encountering rain showers off and on during the day, the drivers crossed the Sierra Nevada mountain range at Echo Summit, the highest point on the Run, and into Carson City for the second refueling stop of the day. With the end of the first day's drive in sight, traffic conges-

Traveling along U.S. Route 50 at Lake Tahoe, California, is the Chrysler New Yorker, driven by Jim Ickes. Although the drivers encountered rain showers on and off during the day, the roads remained clear as they crossed the Sierra Nevada Mountains en route to Reno.

tion coupled with road construction greeted the drivers as they arrived in Reno. "Most of the cars arrived well in advance of the deadline, but Verne Holmquist's DeSoto Firedome had less than two minutes to spare upon arrival at the impound area."[56] The weary crews were ready for a well-earned overnight break after a long trek from Los Angeles.

On April 21, the first car to leave, a Nash Ambassador that had been the first to arrive the previous afternoon, left at 6:00 a.m. for the 431.5-mile, 9-hour and 15-minute trip to Boise. With weather conditions clearing, the drivers faced a route through largely open and undeveloped areas. Vern Holmquist, after encountering delays the previous afternoon arriving in Reno, was once again delayed at a railroad crossing as he headed out of town. Leaving Reno, the drivers headed for Winnemucca through an area where large cattle ranches dotted the landscape. During this leg, drivers had to contend with cattle roaming freely on the road, causing some competitors brief delays. The drivers arrived in Winnemucca for a brunch stop, followed by the only refueling stop of the day.

Here is where the Chevrolet Two-Ten crew had some tough luck. Outside Winnemucca Pete Novotny missed the sign for Route 95, which was reported to have been blocked by a large trailer truck, and continued on the wrong road. Both his relief driver, Wally Tietzel, and the AAA observers were working on paperwork related to the refueling and did not have their eyes on the road. After traveling about 10 miles the crew discovered their error, and Pete turned around and retraced his steps. He drove at speeds up to 75 miles per hour to make up his lost time, and along with the added miles driven, he ran out of fuel 15 miles from Boise and was disqualified.[57] Later it was noted that the Chevrolet was doing fine up to this point as unofficial calculations showed that it was "doing better than 21 miles per gallon."[58]

In Oregon, a good deal of hilly and winding roads were present, and the drivers were challenged with strong head winds during this leg.[59] Between Winnemucca and Jordan Valley, the Studebaker Land Cruiser's crew had to contend with a flat tire, which took 7 minutes to change. Since they were running well ahead of their planned schedule, they continued on as normal and were able to arrive in Boise on time. In Idaho, some delays were reported as the drivers had to be alert crossing the Snake River's one-lane bridge, where traffic control by the Idaho Highway Patrol was in place. Arriving in Boise in late afternoon, the drivers made their way to Boise Junior College, where they were greeted by a good-sized crowd of spectators and the cars were impounded overnight.

On April 22, everyone enjoyed a relaxing 9:00 a.m. start as they faced a 226.3-mile, 5-hour trip to the finish line at Sun Valley. The first car to leave was once again a Nash Ambassador and, although disqualified, the Chevrolet crew continued on to Sun Valley, following behind the competing cars. Immediately after leaving the Boise Junior College impound, they made a refueling stop before heading to Twin Falls, traveling through Idaho's rugged terrain and following the Snake River for part of the journey. The Twin Falls police were on hand, along with onlookers lining the highway, to help direct traffic as the drivers headed north to Sun Valley and the finish line near the Opera House.[60] A flock of sheep blocked the road after leaving Twin Falls, which delayed a couple of the competitors, including Clay Smith (Lincoln Capri) and Danny Eames (Dodge Coronet). Bill Stroppe, driving a Mercury Monterey, "pulled around the side of the road and edged through the flock," saving both time and fuel.[61] Led by Idaho Governor Len Jordan, a group of nearly 600 spectators, many traveling to Sun Valley from Los Angeles on the "Mobilgas Special Train" operated by Union Pacific, were on hand at the finish line to greet the competitors along with members of Studebaker's Highland Lassies and Bagpipe Band. Long Beach (California) Hudson dealer Joe Rickets, driving a Hudson Jet, was the first car to cross the finish line, where well-known race driver Ralph DePalma greeted each of the drivers with the checkered flag. Staying close to the maximum allowable time, three of the competitors cut their cushion so close that they arrived at the finish line with only seconds to spare. The Kaiser Manhattan crossed the line with a margin of two seconds to spare, the Lincoln Capri came in with three seconds left, followed by the Dodge Coronet with a "comfortable" five seconds.[62] All competing cars were immediately impounded and topped off for the final time under the watchful eyes of the drivers and AAA Contest Board officials. AAA officials worked well into the night computing the final figures, with the results being announced at an awards ceremony the following morning.[63]

Les Viland returned to the winner's circle piloting his Ford Mainline 6 to capture his second Sweepstakes Award in three years. The winning figure was 56.7028 ton miles per gallon, with an actual miles-per-gallon figure of 27.0335. This was the first "light car" (weighing in at 4195 pounds, which included the 750-pound payload) to take the top prize that had generally been won in previous years by heavier cars. Hollywood Ford dealer Al Stuebing celebrated his first Sweepstakes Award as an entrant and also marked the sixth time since 1938 his entries have placed first in their class. This win also marked the fourth victory in a row for a Ford Motor Company product to take top honors, Mercury having won in 1950 and 1952 and Lincoln in 1951.[64] Coming in second for the Sweepstakes was a Dodge Coronet, driven by first-time participant Danny Eames, with 52.8565 ton miles per gallon. For the second year in a row a Henry J Corsair 4, driven by Woody Bell, achieved the best miles-per-gallon figure for all entries with 28.2587. The best miles-per-gallon figure for a six-cylinder car was 27.0335, set by the Sweepstakes-winning Ford Mainline, and the best figure for an eight-cylinder was 24.5080, set once again by a Studebaker Commander, this year driven by Dick Griffith. Although scoring well in actual miles per gallon, the Studebaker Commander placed third in its class. All of these models were equipped with overdrive. The best miles-per-gallon result for those entries equipped with automatic transmission was set by a Studebaker Land Cruiser with a figure of 22.8885.

In addition to the Sweepstakes Award, first-place trophies were presented to:

CLASS	MAKE and MODEL	TON MILES PER GALLON	MILES PER GALLON
A — Low Price (Std & Overdrive)	Ford Mainline 6	56.7028	27.0335
B — Low Price (Automatic)	Hudson Jet	42.4606	22.0575
C — Low Medium Price (Std & Overdrive)	Dodge Coronet	52.8565	23.4189
D — Low Medium Price (Automatic)	Studebaker Land Cruiser	49.3476	22.8885
E — Upper Medium Price (Std & Overdrive)	Nash Ambassador Custom	51.3140	22.5457
F — Upper Medium Price (Automatic)	Nash Ambassador Custom	48.9037	21.1156
G — High Price (Automatic)	Lincoln Capri	52.3467	19.9416
Special Lightweight	Henry J Corsair 4	48.5768	28.2587

Detailed results for all competing cars can be found in the Appendix.

In announcing the results, the AAA Contest Board noted that the entries averaged a startling 22.2869 miles per gallon over the 1206.1-mile course. This resulted in the "second best postwar mileage increase the Run has seen," topped only by 1951's average of 23.92 miles per gallon, achieved when the cars traveled a shorter, less trying two-day route to the Grand Canyon.[65] Of particular interest this year was the miles-per-gallon performance of those cars equipped with overdrive compared to those with automatic transmissions. The 15 entries equipped with overdrive averaged 23.8130 miles per gallon, while the 10 with automatic transmission averaged 19.9979. No accidents or mechanical problems were reported.[66]

After an absence of two years, Dodge came roaring back, marking the first time it captured a first-place trophy, ending a drought dating back to the inaugural 1936 Gilmore Yosemite Economy Run. After sitting out the 1952 Run, Nash was the only make this year to score victories in two Economy Run classes, with Ambassador Customs taking top honors for models competing in both upper medium price classes. Andy Henderson, a veteran of both the Gilmore and Mobilgas Economy Runs, placed first in class "E" in his Ambassador equipped with overdrive. Andy's previous class wins, driving Nash entries, dated back to the Gilmore Runs in 1939, 1940 and 1941. Archie Bowen, making his first appearance as a driver, placed first in class "F" in his Ambassador equipped with Hydramatic. These wins masked the disappointing second-place performance by the Rambler in the special lightweight class, competing against the 4-cylinder Henry J

Henry J. Kaiser, Jr., vice-president of Kaiser-Frazer Corporation, was on hand to see Woody Bell repeat his 1952 special lightweight class–winning performance in a Henry J Corsair. This marked the third year in a row a Henry J "4" took top honors in this class. Because of declining sales of its products and continuing financial problems, this would be the last year Kaiser would compete.

Clay Smith, driving a Lincoln Capri, scored his first class win as a driver, having previously participated as a relief driver while Beverly Hills Lincoln dealer John Deaton celebrated his seventh win since the 1938 Gilmore Yosemite Run. Three of the winning drivers (Archie Bowen, Danny Eames and Clarke Sanders) were first time participants.

After enjoying considerable success in the three previous Runs, Bill Stroppe placed a close second to the Dodge Coronet in his Mercury Monterey. The difference between the two was .2936 ton miles per gallon and .2635 actual miles per gallon.

The increased interest by the general public, and the Run's overall importance to the various manufacturers participating, were noted by those attending the activities at Sun Valley. In addition to executives from Socony-Vacuum and General Petroleum, attendees included automobile executives traveling from home offices of the Ford Motor Company, Hudson Motor Company, Kaiser-Frazer Corporation, Nash Motors and Studebaker Corporation. As a result of the efforts of General Petroleum's advertising manager, Frank Meunier, over 70 members of the press, representing newspapers stretching from Los Angeles to New York, were on hand to report the results. The majority of the press attending were based in Southern California, with many of these travelling along the route.

Advertising and publicity commenced immediately after the announcement of the final results. Newspaper advertising once again was the most prominent, as a decline was seen in magazine advertising. Featuring a picture of the winning Mainline 6 Sedan, "Ford Economy Wins Again!" was the headline in newspaper ads noting that it was an "Economy Winner 4 Years in a Row!"[67] Henry J proclaimed that it "beats all comers with 28.2 miles per gallon!," adding: "For the

Parked at the Reno impound after a long 548.3-mile trip from Los Angeles, the competing cars await their turn to be topped off on special refueling platforms. After being topped off, each car was pushed into starting position for the next day's leg to Boise.

second year in a row, Kaiser-Frazer's Henry J scored the highest mileage in the famous Mobilgas Economy Run"; the ad also featured a picture of the winning entry.[68] "Dodge Sweeps the Field" was the headline announcing that it "Tops all 8's in famous 1206-mile economy test!" in an ad also featuring a picture of the winning car.[69] Here is where the use of the ton-miles-per-gallon formula began to cause confusion with the public. While it is true that Dodge did top the field of 8-cylinder models in ton miles per gallon, a Studebaker Commander scored higher in actual miles-per-gallon performance. All three of these ads also appeared in the magazine *Automobile Topics*.

Lincoln noted its three-year winning streak with the headline "Winner! Lincoln Again Sweeps its Class," detailing the previous 1951 and 1952 victories.[70] Studebaker didn't have a lot to cheer about, only being able to headline that a "Commander V-8 with Overdrive finished 1st in actual miles per gallon among all competing 8-cylinder cars," as well as mentioning the Automatic Drive–equipped Land Cruiser's class win. Studebaker also noted that the Champion "scores amazing 26.86 miles per gallon," although this was not the highest figure achieved this year.[71]

In addition to announcing the winners and results, General Petroleum's advertising theme for 1953 stressed "High Quality Economy Gasoline" as "Proved in the Annual Mobilgas Economy Runs," featured in full-color single-page spreads. These began a new focus on ingredients and refining programs designed to deliver top economy. Drivers were reminded that "25 new U.S. stock cars, using Mobilgas or Mobilgas Special, averaged an amazing 22.3 miles per gallon over a punishing 1206-mile route."[72]

Champion Spark Plugs claimed, "Champions That Deliver Mileage Like This ... Will Make Your Car a Gas Miser, Too!" in ads featuring a picture of the Sweepstakes-winning Ford Mainline 6. Champion also noted, "For the fourth straight year Champion Spark Plugs were used by the Sweepstakes Winner of the famed Mobilgas Economy Run."[73] A different style magazine ad appeared in Spanish-language magazines with the theme "Champion Campeon de Economica." Champion's advertising efforts would continue through 1955.[74]

Dodge advertised heavily its Economy Run success with the new Red Ram V8 engine, not only in the 1953 model year but well into 1954. Magazine advertising noted Dodge's class-winning performance "tops all other 8's in the famous 1206-mile Mobilgas Economy Run" in ads that were part of its overall "Dependable Dodge" campaign.[75] Dodge owners were told of this achievement in an article in *Dodge News*, along with two-page color advertisements highlighting this "Proof of Performance!"[76] During the year, Dodge produced several special promotional materials which included folders and postcards highlighting this winning performance that was mailed to potential customers.

Ford made only brief mention of its Sweepstakes-winning performance throughout the remainder of the model year, stressing the performance of its "Mileage Maker Six" engine. Lincoln continued to make limited mention of its first-place performance. One exception was in an advertisement featuring a replica of the winning Capri Sedan next to a replica of the winning Pan American Road Race models. Nash made only a brief mention of its Hydra-Matic equipped Ambassador's success.

Studebaker's Economy Run performance in advertising was limited this year, with only brief reference to its being "the car that officially proved itself a gasoline mileage star" along with its "stand-out thrift." The company's ads instead focused more on the new styling of its new Starlight coupe models. Potential customers were reminded of Studebaker's mileage performance in its 1954 showroom brochure, which noted that a Commander with overdrive was "Thriftiest V-8 in all the world" that "for the third straight year, beat all other eights that competed in the contest in actual gas mileage." Another ad trumpeted, "The Champion proves its economy again in the Mobilgas Run," accompanied with pictures of the cars and drivers.[77]

Although advertising surrounding the Economy Run was seen less than in previous years, articles were more numerous, no doubt due to the fact one of the popular "Low-Price Three" models took top honors.[78] Newspapers in the western states provided daily reports on the progress of the Run, even though results would not be known until the finish in Sun Valley. In ads appearing primarily in automobile-related magazines, emphasis was placed on Les Viland's Sweepstakes-winning performance with the Ford Mainline 6. Interviews with Les, which covered both the preparation of the Ford along with his driving techniques that were used to achieve his winning result, were the theme of this ad campaign. These interviews were found in the June 1953 issue *Motor Trend* ("How Did You Do It?") and in the May 22, 1953, issue of *Motor World* ("How Ford Got More Miles Per Gallon"). The British also learned of this year's results in an article appearing in the May 15, 1953, issue of *The Autocar*.[79]

One of the more comprehensive articles to appear was in the July 1953 issue of *Speed Age*. Featuring a photograph of Les Viland and the winning Ford Mainline 6 on the cover, the magazine asked, "Is the Economy Run Honest?"—

At Sun Valley, Sweepstakes winner Les Viland crosses the finish line in his Ford Mainline 6. In capturing his second Sweepstakes, Les also achieved the best mileage performance for a six-cylinder car with an impressive 27.0335 miles per gallon. If this picture looks familiar, that's because it was used by Champion Spark Plugs in an advertisement that appeared in several magazines.

After being topped off at the General Petroleum garage in downtown Los Angeles, a Studebaker Land Cruiser is pushed by AAA Contest Board personnel into its starting position on Flower Street.

the title of an article written by *Speed Age*'s editor Don O'Reilly.

In more than 3½ pages of 3-column text, he covered every aspect (and in much more detail than is covered here) of the Run from selection, preparation and actual events through the final results. Don wrote:

> It is obvious that every mechanical and engineering trick in the books were tried.... AAA claims none got by ... probably some did, but not many. In our opinion, AAA officials make every effort to keep the Mobilgas Economy Run honest and, of course, they are backed up by General Petroleum Corp. which maintains a "hands off" policy in that department. In our opinion, the Economy Run is not won as a result of the pre-race mechanical work. True, the cars are tuned to absolute perfection. The big score is made during the driving. The expert driver knows the course well enough to drive it blindfolded. He knows precisely where to ease off the accelerator going into a turn; he gauges his traffic signals and he knows the timing of those lights.

While making some recommendations regarding the timing of the course announcement, which would be adopted several years later, he concluded: "In summation, we believe that the Mobilgas Economy Run is honest ... as honest as anything of this scope can be; that it does prove the maximum economy you can get under normal highway conditions; that it is being improved every year; and that it can be improved still more."[80]

At the conclusion of the Run, Dodge announced that it would attempt to set a new AAA National Stock Car Record (Class C Closed Car Division), currently held by a 1939 Hudson Eight Sedan with 94.73 miles per hour, with the same car that had just competed. Immediately after the conclusion of the Run, A.C. Pillsbury had the Dodge Coronet entry impounded and kept under the watchful eyes of the AAA Contest Board. In early May, Danny Eames drove through the speed traps at El Mirage Dry Lake, California, recording 102.622 miles per hour for the flying-start mile and averaging 72.325 for the measured mile from a standing start, breaking the old record by a wide margin.[81]

1954

Los Angeles to Sun Valley Via San Francisco — Elko

April 5–7, 1954—1335 Miles

After reviewing the previous two Runs to Sun Valley, officials made a decision to return for a third time to this well-known Idaho resort. One of the main reasons for this selection was due to the "many varied routes available from Los Angeles to Sun Valley, and the wide range of weather and driving conditions, not to mention the basic attractions of the resort itself."[82] A rugged 1335-mile route, ranging in altitude from sea level at San Francisco to 7135 feet at Donner Summit in the Sierra Nevada Mountain Range in California, faced the drivers as they traversed three states, making overnight stops in San Francisco and Elko. In order to cover the route within the 31-hour, 45-minute time limit the drivers had to average a speed of 41.1001 miles per hours.

While the rules were basically similar to those for 1953, a few changes were made to further protect the integrity of the cars competing. The changes mainly dealt with the

Just before midnight, AAA's J.C. Agajanian prepares to send Archie Bowen and his Nash Ambassador Super on his way for the 554.4-mile trip to San Francisco. A larger crowd of spectators was on hand to witness the event compared to previous years.

handling of competing cars both before and during the run. The most significant change involved the servicing and adjustments which could be made to the competing entries. The only changes permitted were "adjustment of carburetor and ignition system, and other normal servicing adjustments as specified by the car manufacturer. Cylinder heads and oil pans may not be removed unless tests prove the existence of mechanical defects which can be corrected only by removal of these parts. A cylinder head or oil pan may be removed only after all other entrants in that class have been notified and while the representative of at least one competitive entrant and an official of the Contest Board, A.A.A., are on hand to observe the work being done."[83] Entrants were no longer given the option of using regular or premium grade gasoline, as all competitors would be using only Mobilgas Special. The option to use regular grade gasoline would not return until 1959.

Competitive classifications followed those used in 1953 with a total of 20 cars competing, the smallest number of cars participating in the history of the Economy Runs. With its popularity starting to show an increase, eleven entries were equipped with automatic transmission. Except for Chevrolet and Oldsmobile, all of the entries were sponsored by either individual dealers or dealer associations. Acknowledging a wider interest in the Runs, Studebaker dealers from outside Southern California participated for the first time, with Paramount Motors in San Francisco and Don Rasmussen Co. in Portland, Oregon, sponsoring Land Cruiser entries. Although provisions were made for a special lightweight class, no entries were received. Faced with continuing production, sales and financial struggles, the recently merged Kaiser-Willys Sales Corporation and its Southern California Dealer Association members declined to participate. Buick, Cadillac and Pontiac also did not participate.

Random car selection and preliminary impounding began in early March. "To make sure there are no 'spotted' cars, AAA officials make their selections throughout Southern California and in one case, the event went as far as Salt Lake City to pick up one make."[84] Les Viland, hoping to repeat his 1953 Sweepstakes victory, encountered a problem with his Ford Mainline 6 early on. The problem was that the Ford had a blowby (caused by a small amount of unburned fuel and exhaust gases escaping around the piston rings and entering the crankcase in one cylinder). He asked for a new engine, but the AAA told him to work with the car. "Normally," said Viland, "a blowby is nothing serious. It will work itself out in 5,000 miles or so." But the competing cars may be broken in for only 2,000 miles. After 800 miles, Viland still had his blowby, and got a new engine only after demanding that AAA officials put their refusal in writing. However, with 800 miles already used up, AAA officials would give him no more than the 1,200 miles still left for break-in.[85]

Among the challenges that faced the drivers on the first day was road construction and a detour on Highway 140 between Merced and Westley. All of the drivers, including Bill Hooks in a Ford Mainline 8, had to contend with gravel road surfaces and heavy construction equipment for several miles.

Working within the rules, competitors experimented with techniques aimed to increase the mileage performance of their models. One of these, which came to light in 2007, involved the Studebaker entries. Jack Smith, a Studebaker engineer who was responsible for the general conduct and strategy, noted that one way to gain an advantage involved tire pressure. "Studebaker printed the owner's manual with an asterisk next to the recommended tire pressure. The related footnote said that the tire pressure should be increased by 2 psi when carrying loads of more than 700 lbs. While the Studebaker tire pressure footnote has been added especially for the Economy Run, it made good sense for normal driving. So Studebaker gained a 2 psi tire pressure advantage over the competition."[86] While helpful, it is doubtful this additional tire pressure solely contributed to Studebaker's success, which resulted more from full factory support and detailed preparations going back to the previous November.

Final impounding occurred on March 19, along with the announcement of the final route and maximum net running time. In addition to open country highways, the route included 188.5 miles within city limits and on restricted-speed roads in Yosemite National Park. Chilton's *Motor Age* noted, "A recent government survey indicates that 38 per cent of all automobile trips in this country are less than three miles in length, [so] the Economy Run has logically placed more emphasis on this phase of driving conditions."[87] Teams began making practice runs over the course, using test cars to gain knowledge of the terrain and road conditions that would be encountered, in addition to recording their timing requirements for each segment. One of the more detailed articles on the preparation for this year's event was in the August 1954 issue of *Hot Rod*. "Mobilgas Economy Run: How it is Won," by Don Francisco, covered each team's preparations so "that the driver and his observer tried to develop a driving technique that would enable them to travel the course with a minimum of fuel." One of the more critical tests involved the driver's technique in climbing hills, finding the most economical method possible.[88]

As evidence of the increasing interest in the Economy Runs, the AAA Contest Board announced that starting this year each entry would be topped off at the end of each day's leg, with the individual results and standings being released daily. Refueling platforms with portable gas pumps, designed to accommodate two cars at a time, were installed at each impound location. This daily topping-off procedure would continue pretty much without any modifications through 1958.

On April 5, with the 20 entries lined up "double file" on Flower Street, the drivers were ready for a long 554.4-mile, 13-hour and 35-minute (not including refueling or brunch stops) trip to San Francisco, which would consist of a fair amount of mountain driving. With stops for brunch and refueling, they would be have nearly a 15-hour day ahead of them. With a large group of spectators on hand, Archie Bowen, driving a Nash Ambassador Super, was the first car flagged away at 12:01 a.m. As in prior years, the remaining cars followed at two-minute intervals, headed for California's San Joaquin Valley and the first refueling stop of the day at Fresno. Leaving Fresno, the drivers headed for Yosemite National Park, driving through some rain and heavy mist with snow alongside the road, arriving around 8:00 a.m. at the historic Alhwahnee Hotel for a well-earned one-hour brunch stop. Known for being the site of the terminus of the Gilmore Economy Runs from 1936 to 1940, this stop would be the Economy Run's last visit to Yosemite National Park. The Nash Ambassador Super was the first to arrive, and its driver, looking tired, remarked: "Lord, I'm hungry!"[89] Leaving Merced, although the rain cleared, the drivers now had to contend not only with continuing strong head winds but with road construction and detours as they headed to Westley for the second refueling stop of the day. Here everyone was greeted by the local high school band and a group of well-wishers. By midafternoon, the drivers had made their way through Bay Area traffic to San Francisco's St. Francis Yacht Club, where the entries were impounded overnight.

Later that evening, the first day's results were announced: Dick Griffith in a Studebaker Land Cruiser was leading the field with 59.1295 ton miles per gallon. Behind the Studebaker was Danny Eames in a Dodge Royal with a figure of 55.9766 ton miles per gallon. Last year's Sweepstakes champion, Les Viland in a Ford Mainline 6, was fourth overall in ton miles per gallon with 53.9187. This was a surprise to most, as Les and his Ford were thought to have the best chance for a repeat win. The best performance in actual miles per gallon was 28.8423 recorded by Dick Johnson driving a Studebaker Champion Custom. All of the leading entries were equipped with overdrive.

The most challenging day since the 1939 Gilmore Yosemite Run faced the drivers as they began the second leg on April 6 to Elko, Nevada. Leaving San Francisco at 3:00 a.m., the drivers encountered foggy conditions for the first 20 miles of their 532-mile trip, which was to be covered within a 12-hour and 10-minute time limit. Crossing the Golden Gate Bridge, one of the AAA Contest Board officials was on hand "to pay their tolls so they could roll on without stopping." The drivers headed for Sacramento following U.S. 40 and the first refueling stop of the day.[90] With weather conditions already deteriorating over Donner Summit (the highest elevation on the Run), AAA Contest Board officials were faced with the decision whether or not to proceed beyond Sacramento. A.C. Pillsbury, riding an hour ahead of the competitors surveying road conditions, made the decision that it was safe to proceed over the summit, although tire chains would probably be required.

Snow conditions and icy road surfaces greeted the drivers in the High Sierras, resulting in the need to use chains for 24 miles over Donner Summit. No time allowances were made by AAA officials for chain installation. In 28-degree temperatures, Chrysler Imperial driver John Gerfen works to finish the job before continuing on his way.

Outside of Sacramento, rain turned to sleet and then to snow, and as the drivers arrived at Emigrant Gap, the California Highway Patrol mandated that chains would be required for the next 24 miles. With no time allowance made for chain installation, drivers were under pressure to make the installation in 28-degree temperatures and continue on their way. The drivers were given a choice of installing chains or using spare tires (each car competing was required to carry two space tires) with chains already in place.[91] In an unofficial contest, the fastest time for chain installation was set by Dick Griffith, who installed the chains on his Studebaker Land Cruiser in 2 minutes and 2 seconds. John Gerfen was the slowest, taking 18 minutes and 10 seconds to install the chains on his Chrysler Imperial.

Drivers then cautiously made their way over Donner Summit, where the speed limit was reduced to 30 miles per hour due to slick road conditions. Everyone made it safely to Donner Lake, where chains were removed. Again there was an unofficial contest to see who could remove the chains the quickest. Verne Holmquist (DeSoto Firedome) took 40 seconds, while Clay Smith (Lincoln Capri) was the slowest at 6 minutes. With a sigh of relief, the drivers drove into clearing skies and continued to Lawton Hot Springs Resort, just across the Nevada state line, for a 30-minute brunch stop. Then, heading east through northern Nevada, the drivers continued to Lovelock for the second refueling stop of the day before arriving in Elko and the impound at the city recreational center. Tired from nearly 14 hours on the road, the competitors received a warm welcome from the residents of the community, and the entries were impounded overnight.

At each overnight stop, special refueling platforms were constructed so each entry could be topped off. After being leveled by jacks, the Ford Mainline 8 is ready for AAA Contest Board officials to begin the refueling process.

When the second day's results were announced, Dick Griffith continued to lead the field in his Studebaker Land Cruiser with an impressive 60.9801 ton miles per gallon. There were no changes in the overall standings with Danny Eames, in a Dodge Royal, second overall with 58.5411 ton miles per gallon. The Studebaker Champion Custom continued with the best performance in actual miles per gallon with Dick Johnson turning in a figure of 29.5689. It was also announced by AAA Contest Board officials that one hour would be added to the time limit to compensate for the adverse weather conditions that were encountered.

On April 7, the drivers had a relaxing start, with the first car flagged away at 8:00 a.m. for the 248.6-mile, 6-hour trek to the finish line in Sun Valley. Leaving Elko in crisp but clear weather conditions, the drivers headed for Twin Falls, traveling along good straight roads through open grazing land. Just north of Elko, some of the competitors were delayed when a herd of cattle blocked the road until some cowboys were able to clear the road. As the drivers were leaving Wells, AAA officials reported, "A favorable tailwind into Idaho aided some of the lighter cars."[92] They arrived in Twin Falls, which was celebrating its Golden Jubilee, for the only refueling stop of the day, where the best turnout of residents on the entire Run greeted the drivers, led by Archie Bowen in his Nash Ambassador Super. The Chamber of Commerce arranged for Miss Twin Falls (Shirley Fullmer) along with six hostesses to greet the competitors, presenting them with commemorative Golden Potatoes and other locally produced items while everyone was entertained by the local high school band.[93] In future years, cities and towns making specialized presentations to Economy Run participants would become a common practice. Leaving Twin Falls, the drivers followed the same route used in 1953 to Sun Valley and the finish line near the Opera House. A large group of spectators, many again traveling to Sun Valley from Los Angeles on the "Mobilgas Special Train" operated by Union Pacific, were on hand at the finish line to greet the competitors as John Gerfen, driving a Chrysler Imperial, was the first car to cross the Finish Line with the rest arriving in a steady procession.[94] All competing cars completed the Run within the time limits and were immediately impounded and topped off for the final time. After crossing the finish line, "Bill Stroppe, the two-time (sweepstakes) winner but an also-ran this year, shook his head as he climbed from his Mercury. 'Who said this economy run is a daisy chain,' he muttered, 'I'm three times more bushed than I was after the Mexican Road Race!'"[95] Although everyone had a good idea

Top: Dick Griffith guides his Sweepstakes-winning Studebaker Land Cruiser across the finish line at Sun Valley. Dick's strong performance, recording 60.8465 ton miles per gallon, marked the third Sweepstakes Award for Studebaker dating back to the 1939 and 1940 Gilmore Yosemite Runs. ***Bottom:*** With a wave of the checkered flag from AAA's J.C. Agajanian, Willard McArthy's Oldsmobile Super 88 crosses the Sun Valley finish line, placing first in the low-medium (automatic drive) price class. This was Oldsmobile's first victory since the 1939 Gilmore Yosemite Run.

which entry would take the Sweepstakes and individual class trophies, AAA officials worked late computing the final mileage figures in preparation for the awards ceremony on April 8.

Dick Griffith led what was termed "Studebaker's Clean Sweep," guiding his Land Cruiser to capture the Sweepstakes Award with an impressive figure of 60.8465 ton miles per gallon. His actual miles-per-gallon figure was an equally impressive 28.1046, which was the best mark set by any Sweepstakes winner. This win not only marked the third Sweepstakes victory for Studebaker, dating back to the 1939 and 1940 Gilmore Yosemite Runs, but was the third highest ton mile score achieved by a Sweepstakes winner since 1936. Danny Eames placed second for the Sweepstakes in his Dodge Royal with 58.0456 ton miles per gallon and was considered a serious contender for the top prize until the final day's results were computed. The best miles-per-gallon figure for all entries was set by Dick Johnson in a Studebaker Champion Custom with 28.2587. The Sweepstakes-winning Studebaker Land Cruiser also recorded the best miles-per-gallon figure for an eight-cylinder with 28.1046. The best miles per gallon result for those entries equipped with automatic transmission was set by Robert Donkin in a Studebaker Land Cruiser with a figure of 24.5763. Those associated with Ford were disappointed at the performance this year after a string of successes in capturing a first-place trophy dating back to the 1940 Gilmore Yosemite Run.

For the second year in a row, Dodge placed first in the low-medium (standard and overdrive) price class with Danny Eames behind the wheel. Danny also placed second for the Sweepstakes Award behind Dick Griffith's Studebaker Land Cruiser.

In addition to the Sweepstakes Award, first-place trophies were presented to:

CLASS	MAKE and MODEL	TON MILES PER GALLON	MILES PER GALLON
A — Low Price (Std & Overdrive)	Studebaker Champion 6	56.5138	29.5806
B — Low Price (Automatic)	Hudson Jet-Liner	43.1249	21.6382
C — Low Medium Price (Std & Overdrive)	Dodge Royal	58.0456	25.3973
D — Low Medium Price (Automatic)	Oldsmobile Super 88	49.7122	19.7506
E — Upper Medium Price (Std & Overdrive)	Studebaker Land Cruiser	60.8465	28.1046
F — Upper Medium Price (Automatic)	Studebaker Land Cruiser	53.9942	24.5763
G — High Price (Automatic)	Lincoln Capri	52.2894	19.7542

Detailed results for all competing cars can be found in the Appendix.

The AAA Contest Board reported that despite the adverse weather conditions encountered, the entries averaged 21.8466 miles per gallon over the 1335-mile route. The rugged driving conditions contributed to a lower average than was recorded in 1953, when a figure of 22.2869 miles per gallon was recorded. Comparing the miles-per-gallon performance of those equipped with overdrive to those with automatic transmissions continued to be a topic of interest. The nine entries equipped with overdrive averaged 24.3561 miles per gallon, while the eleven with automatic averaged 19.7934. Chilton's *Motor Age* noted, "The overall results were surprising in view of the fact, first, the Run was made through heavier traffic and poorer weather conditions than last year; and second, it was thought the limit has been reached in economy driving in 1953. Considering that the only changes were those that the manufacturer incorporated in the 1954 models, these facts indicate that the economy of gasoline, automobile and motor oil have been improved."[96] Although tire chains were used over Donner Pass, no accidents or other mechanical problems were reported.

Studebaker scored "three for three" with all three entries placing first in a feat described by members of the motoring press as impressive. As noted by Floyd Clymer in his book on the 1954 Run:

> The Studebaker victories came as little surprise to me, or to most Economy Run followers. Studebakers have participated in these events since the original Gilmore Run in 1936, and have always placed high in the final standings. The team maintained class leads throughout the three-day trip. Since the inauguration of Economy Runs, no Sweepstakes winner has ever turned in an actual-miles-per-gallon mark equaling the Studebaker Land Cruiser's performance of 28.1046. This is not to imply that Studebaker had no competition. In fact, the Dodge V8 (overdrive), which has been in the news lately for top-speed records, was a close second-place in the overall Sweepstakes standings and a Dodge won first place in the Automatic Drive class.[97]

Other observers noted that Studebaker's success was due in part to its "aerodynamic design and excellent organization for the Run. The Studebaker team, according to reports here, has been practicing on various possible routes to Sun Valley since last November."[98] "Upon learning of his success as driver of the Sweepstakes winner, Griffith said that careful tuning of his car, plus 'my constant thought that there was an unbroken egg beneath the accelerator,' enabled him to complete the difficult route with the highest ton-mile rating of any of the 20 competing cars."[99] Dick's "unbroken egg" comment would remain a topic of conversation for years to come.

San Fernando Hudson dealer Lee Hamer scored his first first-place trophy in four years of competing, driving a Hudson Jet-Liner to victory in the low-price cars (with automatic transmission) class. Danny Eames and Bill Stroppe continued their friendly rivalry from 1953 in the low-medium price class, with Danny driving his Dodge Royal (with overdrive) once again to a first-place victory by a wide margin over Bill's Mercury Monterey. Mercury's new overhead valve V8 returned a disappointing 21.4781 miles per gallon compared to 1953's result of 23.1554, which used the older L-Head V8.

In what was described by *Automotive News* as "an unusual victory," Oldsmobile took first place in the low-medium-price (with automatic transmission) class. The Super 88, driven by Willard McCarthy, was a private entry, along with the Chevrolet Two-Ten of Los Angeles businessman Leonard DeBell, who had long sought a victory. "Oldsmobile officials feigned no interest in the private entry, but unofficially called Mobilgas headquarters each evening to see how it was doing."[100] This was Oldsmobile's first win since the 1939 Gilmore Yosemite Run.

Clay Smith repeated his 1953 first-place win in the high-price class, once again driving a Lincoln Capri. After four years of participation, either as driver or relief driver in a Sweepstakes or class-winning team, 1954 would mark his last appearance. On September 6, Clay lost his life in a tragic accident during the running of the Red Horn Memorial Race at DuQuoin, Illinois.

Celebrating Studebaker's Sweepstakes win are (left to right) driver Dick Griffith, Paramount Motors' L.E. Von Schiltheis, General Petroleum Corporation Consultant Clarence S. Beesemyer, AAA Contest Board Referee Wilbur Shaw, and relief driver Oscar Allfie. The Studebaker Land Cruiser achieved the third highest ton miles per gallon in the history of the Economy Runs.

Executives from Socony-Vacuum, General Petroleum and various manufacturers were once again in attendance at Sun Valley along with a large press contingent representing all forms of media. In addition to those guests traveling on the special Union Pacific train was a select group of Mobilgas dealers from Southern California who followed the route on a specially chartered bus. One special guest in attendance was Wilbur Shaw, president of Indianapolis Motor Speedway, who was serving as an AAA Contest Board referee for the Run. His experiences with the Economy Runs dated back to the inaugural 1936 Gilmore event, in which he drove a Hudson Deluxe 8.

For the third year in a row, a Lincoln Capri took top honors in the high-price class. Admiring the trophy are (from left) Beverly Hills Lincoln-Mercury dealer John Deaton, driver Clay Smith, Lincoln-Mercury Dealers Association representative H.B. Daniels, and General Petroleum Vice-President Arthur Connelly. This was Clay's second consecutive year taking first-place honors and would be his last year to participate.

Advertising and publicity commenced immediately after the announcement of the final results. As in previous years, newspaper advertising continued to be most prominent, although a noticeable decline was evident. With its corporate advertising in 1954 focusing more on the new Mobilgas Special blend, General Petroleum's Economy Run advertising effort was restricted mainly to the announcement of the final results. Found in newspapers and automotive-related publications, it noted that the Run "was the final and conclusive test in an exhaustive series of research and road tests of New Mobilgas Special."[101] Even though its own advertising was limited, the Mobilgas name was in the forefront, both in articles covering the run and in individual efforts by the winning manufacturers and this would continue for the duration of the Economy Runs.

As would be expected, Studebaker went all-out advertising its success. In newspapers, Studebaker proudly announced, "Studebaker Wins 1954 Mobilgas Economy Run Sweepstakes," but also highlighted records set, including the Champion's being first in actual miles per gallon of all cars competing.[102] Advertisements in the months following proclaimed Studebaker as "America's No. 1 Economy Car," noting that its advanced styling contributed to this success.[103] In one of their first magazine ads, "Another triumph! Advance style Studebakers win the '54 Mobilgas Economy Run!" was the headline. The text, accompanied by a rendering of the Sweepstakes-winning Land Cruiser, and told readers of the performance of both the Champion and Land Cruiser entries.[104] Another exhorted, "Cut your driving costs with an out-ahead '54 Studebaker," and also featured designer Raymond Loewy.[105] Canadian readers were told of the Economy Run wins in an ad focusing on actual miles-per-gallon performance. It featured a photo of the class-winning Champion sedan and noted that Studebaker is "Canada's No. 1 Economy Car."[106] Potential customers were reminded of Studebaker's success in its 1955 showroom brochure that claimed, "Studebaker Tops All Cars in America in Mobilgas Economy Run Victories" and added "Studebaker's 1954 triumph the greatest of all!"[107]

Featuring pictures of their class-winning entries, both Dodge and Lincoln used newspapers to announce their victories. "Dodge Wins Again!" was the headline of an ad noting "Dodge V-8 Scores 2nd Straight Victory in Its Class!" The ad further pointed out that the other Dodge entry, "equipped with Powerflite, automatic-drive, beats nearest competitor by 1.6 miles per gallon!"[108] Lincoln announced, "Lincoln Wins Again! Class G Mobilgas Economy Run," noting that Lincoln "is America's pace-setter in the fine car field." Ads also made reference to Lincoln's success in the Pan-American Road Races.[109] Both Dodge and Lincoln's magazine advertising that focused on the Economy Run achievements was limited, and only in brief text comments did Lincoln remind readers that it "took first in its class for the 4th straight year."[110] Dodge produced a special "Winner and Still Champ!" folder for potential customers that highlighted various aspects of the Run, noting that this was "proof again that Dodge gives you top operating economy!"[111] Both Hudson and Oldsmobile ignored their first-place victories in advertising.

Newspaper coverage was noticeably expanded, including the release of results at the completion of each day's drive, and was found primarily in locales that were on the route between Los Angeles and Sun Valley. Articles continued to be found in enthusiast magazines, although the depth of reporting was less than in 1953.[112] Specialized articles in company publications *Dodge News* and *Lincoln-Mercury Times* focused on their models' winning performance.[113] Some of the articles dealt with the preparation of each team's entry and specialized driving techniques used by the drivers in an attempt to answer the recurring question from motorists: "Why can't I get the same kind of mileage as the Economy Run drivers?" One answer was in Sweepstakes-winning driver Dick Griffith's discussion of economical driving techniques used, which included avoiding jackrabbit starts, driving at an even and moderate speed, and anticipating the road ahead, along with traffic light changes. Other hints included using low and second gears to gain momentum and shifting into overdrive as soon as the overdrive cut-in speed is reached.[114] Walt Woron, editor of *Motor Trend*, concluded, "This isn't the first time it's been said that the average car owner can't duplicate the kind of mileage that's obtained in the Mobilgas Economy Run. And yet, we got comparable mileage in a Ford Six, merely by observing restraint and consistency in driving."[115]

After five years of growth, a new destination of Colorado Springs was on the horizon, along with keener competition.

4

Colorado Springs

1955

Los Angeles to Colorado Springs Via Tucson — Albuquerque

March 14–16, 1955—1323 Miles

Desiring a destination at a point east of the Pacific Coast region, officials of both General Petroleum and the AAA Contest Board had little difficulty when it came to selecting Colorado Springs for the 1955 Run. Before reaching the finish line at the famed Broadmoor Hotel, the drivers would travel 1323 miles over a three-day period along a route that was referred to as the "Indian Trail" in the American Southwest. With overnight stops in Tucson and Albuquerque, the route ranged in altitude from 218 feet below sea level at the Salton Sea in California to 9382 feet at LaVeta Pass Summit in Colorado. The drivers would average a speed of 40.6543 miles per hour in order to stay within a 32-hour and 50-minute time limit set by the AAA Contest Board.

One of the biggest changes for 1955 was the requirement that the field be limited to models equipped with automatic transmission. As explained by the AAA Contest Board, "Now that the car with automatic drive represents the typical, rather than the exception in new car purchase, the economy performance of cars so equipped calls for new study and attention."[1] *Automobile Topics* reported that in 1954, "58.3 per cent of cars manufactured in the United States were equipped with automatic and this figure was projected to exceed 60 per cent in 1955."[2] Remaining basically similar to prior years, the rules were refined as situations warranted which would continue to enhance the stock status of competing cars. Randall Gould, columnist for the *Denver Post,* further noted that the rules for 1955 were "developed through experience since the run began in 1936."[3] One other significant rule change was that a maximum of 24 man-hours would be allowed for servicing and checking each entry during the period from car selection through pre-start procedures. In the event of a major mechanical defect, provisions were made for the AAA Contest Board to allow additional time for servicing and correcting.[4] All competing cars would use Mobilgas Special exclusively, as the option to use regular grade gasoline would not be available for several years.

Entry eligibility continued to follow those established in prior years. With all entries only equipped with automatic transmissions, competitive classifications were reduced to four, following the well-known groups referred to as low, low-medium, upper-medium and high-price. A total of 22 cars competed, sponsored by individual dealers or dealer associations, with the exception of the Chevrolet Bel-Air 6 as the lone private entry. Continuing a trend that would increase in future years, dealers from outside California participated. Sponsoring Dodge and Studebaker entries, these dealers were located in Clinton, Indiana; Las Vegas, Nevada; and Portland, Oregon. With production winding down, it was not surprising that Kaiser and Willys were not represented. Cadillac continued its previous practice not to participate. General Motors was well represented, with Pontiac returning to competition for the first time since 1940, and Buick making its inaugural Economy Run appearance. Dealers now saw an opportunity to compete on a level playing field once the overdrive-equipped models were eliminated.[5]

There was quite a bit of interest in class "A" for low-price cars, with the Chevrolet Bel-Air and Plymouth Belvedere models, equipped with their new V-8 engines, competing against the Ford Fairlane V-8 along with the 6-cylinder Chevrolet, Ford and the smaller Rambler Super. Two-time Sweepstakes champion Les Viland, who had recently become an engineer with American Motors Corporation after leaving Ford, was named to drive the Rambler Super. Marshall Martin, Sweepstakes-winning driver in the 1941 Gilmore Grand Canyon Run, returned to competition for the first time since 1950, this year piloting a Ford Fairlane 6. In addition to dealer or factory personnel handling driving and navigation duties, familiar names from racing and enthusiast publications made their first appearance. These included 1953 and 1954 Mexican Road Race participant Ackton (Ak) Miller, driving the Plymouth Belvedere, with Ray Brock from *Hot Rod Magazine* as his relief driver; Don Francisco, also from *Hot Rod*

Magazine, as relief driver for the Mercury Monterey; and Bob Russo from *Speed Age* as relief driver for the Dodge Coronet. For reasons unknown, 1954's Sweepstakes champion Dick Griffith did not participate. Two-time Sweepstakes champion Bill Stroppe chose not participate in order to focus his efforts on preparing Mercury entries for the stock car racing circuit.

Random car selection and preliminary impounding began in February as AAA officials continued to make their selections throughout Southern California. Under the continuing leadership of Chief Steward A.C. Pillsbury, the competing cars would remain under the watchful eyes of AAA Contest Board officials until the conclusion of the Run. As in prior years, the competing cars could only be broken in for 2,000 miles; most of these trips were reportedly made to Death Valley or Las Vegas, as the rules stipulated that the competing cars could not be away from the impound for more than 24 hours.

Final impounding occurred on March 4, at which time the final route and maximum net running time were announced. The route once again was selected by the Automobile Club of Southern California and included 110.3 miles of restricted city driving. With only 10 days before the start of the Run, the teams were quick to begin practice runs over the course, using test cars similar to the ones on the actual Run to gauge terrain and road conditions and establish their timing requirements for each segment. As in prior years, these test cars were equipped with fuel metering equipment and other testing devices to determine the most economical driving style for automatic transmissions over the course. Interviewed by *Automotive News*, drivers said that "elimination of standard transmission and overdrive cut down on special techniques which have been a feature of past runs."[6] Bob Russo, *Speed Age*'s racing editor, gave readers an in-depth look at this year's Run in its July 1955 issue. "I Drove in Mobilgas Economy Run" covered the detailed preparations of the two Dodge and one Plymouth teams as they met and planned their approach to tackling the 3-day event. Their practice runs were as identical as possible, down to timing the first leg to coincide with their scheduled starting time in Los Angeles. So intense were the preparations, the teams arrived back in Los Angeles less than 24 hours before the start. For a first-time participant in the Economy Run, Bob soon discovered "this contest is anything but a pleasure drive," and he classed it as "one of the finest and fairest tests of the American Automobile."[7]

On March 14, the 22 brightly painted and lettered entries were lined up "double file" on Flower Street ready for the 519.9-mile, 12-hour and 50-minute (not including refueling or brunch stops) trek to Tucson. This stretch would feature mostly open-road desert driving, and at times temperatures would reach close to 90 degrees. With a large group of spectators on hand and searchlights lighting up the night sky, Pete Novotny in a Chevrolet Bel-Air 6 was the first car flagged away at 12:01 a.m., with the remaining entries leaving at two-minute intervals. For the next three years these post-midnight starts would be common to accommodate reporters for newspapers in the Midwest and on the East Coast in meeting their deadlines for reporting the daily results. Under cloudy skies the drivers headed east from Los Angeles, where they encountered their first challenge of the day. Crossing the San Gorgonio Mountains, east of Redlands, they entered a section of thick fog that slowed them almost to a crawl. This was followed by strong head winds past Indio which a Chrysler engineer noted "could affect mileage by as much as 5 miles per gallon."[8] Crossing California's Imperial Valley road construction for nearly a dozen miles hampered their progress as they headed for Yuma for a refueling stop, immediately followed by a brunch stop at Loo's Restaurant.[9] Evidence of advance

All lettered and ready for the trip to Colorado Springs is the eight-cylinder Chevrolet Bel-Air entry. This is the first year Gardena Chevrolet dealer James Rush would compete as both an entrant and a driver.

planning by General Petroleum's Frank Meunier and his staff was noted as Yuma television station KIVA provided on-the-spot coverage of the refueling operations, along with a good turnout of local residents to greet the participants.[10] Leaving Yuma in hot, dry weather the drivers passed by several Indian reservations, and by early afternoon arrived in Tucson under sunny skies to a colorful and enthusiastic welcome at University of Arizona's Athletic Field, where the cars were impounded overnight. Everyone was ready for a well-earned break after enduring several hours of uncomfortable desert driving where, for economy's sake, windows remained tightly closed to lessen wind resistance.

With his trademark leap into the air, AAA's J.C. Agajanian sends Archie Bowen in his Pontiac Chieftain on his way from the start in downtown Los Angeles. Another good-sized crowd of spectators was on hand again this year for the midnight start. This was Pontiac's first appearance in the Economy Runs since the 1940 Gilmore Yosemite Run.

On the outskirts of Tucson, the unthinkable happened to Bill Losher and Bob Russo in the Dodge Coronet 8: the engine coughed and then quit. They had run out of gas three miles short of the impound. Sitting on the side of the road, the Dodge was refueled by a AAA Contest Board truck, then continued on to the impound. Upon arrival, A.C. Pillsbury reviewed the observer's report and announced, "You're disqualified for taking on fuel at an unauthorized stop. I'm sorry."[11] At the refueling stop, Bill had decided to order only 8 gallons (as opposed to the others, who were ordering between 10 and 12 gallons), reasoning that that would be sufficient for the 240-mile trek to Tucson, and reckoning that each additional gallon meant more weight.[12] As Bob Russo later explained, "Bill reasoned that something was wrong somewhere, possibly in the fuel tank. Our [gas] gauge indicated that the tank was not empty and it just did not seem reasonable. Later, upon checking, Bill found that the pickup line, which draws fuel from the tank and feeds it to the fuel pump and carburetor, was slightly higher than it should have been. Although we still had fuel left, we were not getting the use of it since the level was below the pickup line."[13] Disappointed, they continued although no longer in competition.

The results for the first day showed Dick Johnson in a Studebaker Commander Custom 8 leading the field with 58.1604 ton miles per gallon. Lurking close behind was Pierce Venable in a Nash Ambassador Super 6 with 57.8873 ton miles per gallon. The margin between the two was a slim .2731 and would get closer as the Run progressed. The best performance in actual miles per gallon was turned in by Les Viland in a Rambler Super 6 with 27.6965. A surprise to everyone was the performance of the Buick Roadmaster, leading the field in the high-price class with 53.2700 ton miles per gallon. The Buick's actual miles-per-gallon figure was a respectable 19.6026.

Leaving Tucson, beginning at 2:00 a.m. on March 15, the drivers were on the road for a 455.1-mile, 11-hour and 20-minute trip to Albuquerque over relatively smooth desert highways in clear weather with little wind. Heading eastward across the Arizona desert, they crossed the Continental Divide before crossing into New Mexico for a refueling stop in Deming. As was becoming commonplace for the duration of the Run, everyone was greeted by the local high school band along with an enthusiastic crowd of local residents. The drivers were here only for a short period of time, and to the surprise of everyone, three drivers—Les Viland (Rambler Super), Hart Fullerton (Hudson Hornet) and Lee Hamer (Hudson

Wasp)—elected not to take on any fuel.[14] Heading north across barren desert landscape, they would next stop in Truth or Consequences for brunch at Our Lady of Perpetual Help Catholic Church's Community Hall. A staff of volunteers served brunch to the participants. The participation of local community and church groups would become commonplace in future years, especially in smaller population centers, as the Run expanded in size and scope.

It would now be a direct route to Albuquerque, where the impound was located at Tingley Field, home of the Albuquerque Dukes baseball team. Shortly after 1:30 p.m., Pete Novotny, driving a Chevrolet Bel-Air 6, was the first to arrive, and another good-sized crowd of spectators were on hand to welcome the competitors. This included around 200 special guests of General Petroleum traveling to Colorado Springs on the "Santa Fe Mobilgas Special," which made a stop in Albuquerque for several hours. Interviewed after his arrival, John Hartman, relief driver for Jim Rush in the Chevrolet Bel-Air 8, voiced a slight complaint about the condition of some of New Mexico's roads: "The paint is worn off the school zone signs and caused us some discomfort and trouble about the time school was letting out." John also noted one of his jobs was to keep Jim aware of approaching school zones and that under the rules any violation of traffic regulations, including speeding in marked speed zones, would result in penalties being assessed.[15] The drivers and observers were ready for a break from their nearly 13-hour day on the road before tackling the final leg the next morning. Snow and freezing temperatures had been reported in the mountain passes of Colorado.

Anyone who thought Studebaker had a lock on the Sweepstakes Award was in for a surprise when it was announced, "We have a new leader." Pierce Venable in his Nash Ambassador Super 6 recorded the best ton miles per gallon of all entries with a figure of 59.9730.[16]

Following close behind was Dick Johnson with his Studebaker Commander Custom 8 with a figure of 59.2643. Les Viland in his Rambler Super 6 continued to record the best performance in actual miles per gallon with 28.4706.

On March 16, the drivers faced the most difficult day of the Run as they headed north to Colorado Springs. The first to leave at 4:00 a.m. was Pete Novotny in his Chevrolet Bel-Air 6 for the 348-mile, 8-hour and 40-minute trip that began with temperatures hovering near zero. Leaving Albuquerque, the drivers were buffeted by strong side winds that resulted in a "blinding dust storm [that] cut visibility to less than 100 feet at one point."[17] As they made their way north, strong headwinds and snow flurries were encountered on U.S. Route 85 through the Rio Grande Valley, an area known as the location of several Indian reservations. In Santa Fe and Taos, local residents lined the road as the cars passed through their downtown sections. Near the Colorado border, Verne Holmquist, driving a DeSoto Firedome, had to contend with a herd of sheep blocking the road, making it difficult to stay on schedule.[18] Arriving in San Luis, Colorado, the participants had a break at Centennial Union High School's cafeteria for brunch. AAA Contest Board officials had temporary gas pumps installed at this location for refueling, which was accomplished before the drivers headed toward La Veta Pass Summit. At this point, drivers encountered winter driving conditions with snow and slush covering the ground, but the wet highway surface was safe for travel and did not require the use of chains. Everyone was pleased to see level terrain return as they headed north through Pueblo and into Colorado Springs. Bundled up against the near-freezing temperatures with snow falling intermittently, a good-sized group of spectators were on hand as the drivers arrived at the Broadmoor Hotel. Well out of contention in the low-price class, Ak Miller in his Plymouth Belvedere decided to "speed it up" and took the honor of being the first to cross the finish line.

Dick Johnson's Studebaker Commander Custom travels through open country near San Luis, Colorado, en route to the finish at the Broadmoor Hotel in Colorado Springs. In a close contest, Dick would capture Studebaker's second consecutive Sweepstakes Award, duplicating its back-to-back successes in the 1939 and 1940 Gilmore Yosemite Runs.

Despite the weather conditions and livestock obstacles, all of the drivers completed the Run within the specified time limits, and the cars were topped off for the final time and impounded overnight. For the next few hours, AAA Contest Board officials would be computing the final results that were announced in the late afternoon. The next morning, on the grounds of the Broadmoor in 16-degree temperatures, a formal awards ceremony was held.

In the closest competition for the Sweepstakes Award in the history of the Run, Dick Johnson, guiding his Studebaker Commander Custom 8, edged out Pierce Venable in his Nash Ambassador Super 6 by the slimmest of margins: .0760 ton miles per gallon. The Studebaker's ton-mile-per-gallon average was an impressive 58.6320 compared to the Nash's average of 58.5560. "Actually, by standard methods of measurement, it is understood that the cars wound up in a dead heat, which was broken only by the condensation report that was brought into play for the first time during the Runs. The condensation report is an intricate measurement based on the temperature of the gas tank, which is taken by an electronic device. AAA officials had to go back and compute the condensation factor for all three days."[19] This handed the victory to Studebaker. Studebaker engineer Jack Smith noted "a couple of traffic-light stops spelled the difference between victory and second place."[20] For Studebaker, this marked its second consecutive Sweepstakes win dating back to the Gilmore Yosemite Runs. The best miles-per-gallon figure for all entries was set by Les Viland in a Rambler Super 6 with 27.4733, also the best figure achieved by an automatic transmission–equipped car in the history of the Run. This record would stand until 1961. The Sweepstakes-winning Studebaker Commander Custom also recorded the best miles-per-gallon figure for an eight-cylinder with an impressive 27.4366 — and was within a mere .0367 miles per gallon of the Rambler's final mark. All four of the class-winning drivers recorded class victories in previous Runs.

In addition to the Sweepstakes Award, first-place trophies were presented to:

CLASS	MAKE and MODEL	TON MILES PER GALLON	MILES PER GALLON
A — Low Price	Rambler Super 6	51.2789	27.4733
B — Low Medium Price	Studebaker Commander Custom 8	58.6320	27.4366
C — Upper Medium Price	Nash Ambassador Super 6	58.5560	25.1854
D — High Price	Buick Roadmaster	53.7596	19.7827

Detailed results for all competing cars can be found in the Appendix.

In announcing the results, the AAA Contest Board noted the entries averaged 21.5615 miles per gallon over the route. Those associated with the Run were amazed, since "it was not anticipated that the figure would be so high since the run was limited to cars equipped with automatic transmissions for the first time." Robert L. Minckler, president of General Petroleum, added: "The mileage marks established in the Economy Run are a tribute to the drivers, the American automobile industry and the quality petroleum products used by all entrants."[21] The AAA Contest Board added that "this had been the cleanest run in experience. The Official stated, 'You can read every one of our observer's reports for each of the three days and I defy you to find one single insertion that even suggests unfair driving or rule infraction. This has been a good run.'"[22] With the exception of one entry running out of fuel on the first day, no accidents or mechanical problems were encountered.

> Drivers were in agreement in praise of the new course, along with the completeness and efficiency of arrangements. In contrast with the course of 1954 and several previous years to Sun Valley, Idaho, the 1955 route was somewhat more level, grades were less steep and weather conditions were better. The biggest adverse factor, shared by all alike and therefore not a ground for complaint, was the requirement that every car be equipped with automatic transmission which in general was regarded as less efficient than the manual transmissions previously allowed, but, on the other hand, in line with the trend of American buying (habits).[23]

It was referred to as "a year for the Independents," who scored almost a clean sweep, with only Buick representing "The Big Three" in the winner's circle. Three of last year's class-winning cars, including Dodge, Lincoln and Oldsmobile, did not do as well, finishing in either second or third position. For struggling Studebaker, the Sweepstakes Award was a bright spot, a result of a total team effort headed by Jack Smith.[24] Despite the fact the Commander's engine had substantially greater horsepower, observers noted it turned in a more economical performance than last year's entry. With a figure of 27.4366 miles per gallon, the Commander (with a 162 hp engine) outpaced the 1954 Land Cruiser (with a 127 hp engine) by 2.86 miles per gallon.[25]

While the smaller Rambler Super 6 took the top honors in the low-price class, 1941's Sweepstakes-winning driver Marshall Martin took the second in a Ford Fairlane 6. Pete Novotny, driving a Chevrolet Bel-Air 6, was pleased this year with his third-place performance in the low-price class now that competition was limited to automatic transmission–equipped models. A participant since 1950, always in a Chevrolet, along with Leonard DeBell's Southern California Plating Company as his sponsor, Pete was usually found toward the bottom standings in the low-price class, hampered by the fact all of his competition was equipped with overdrive, an option that was unavailable in a Chevrolet.

Buick's first participation was not only successful, taking first place in the high-price class with 19.7827 miles per gallon, but resulted in an interesting story as well. As a AAA Contest Board official related, "When the Buick was entered the dealer told us this car would get exactly 19.5 mpg. On that basis it

In a year where the independent manufactures dominated the Economy Run, Nash was quick to advertise its two class wins. You have to read the text to note that the Rambler's all-time record is for automatic transmission-equipped cars. Nash continued to note the Rambler's record set in 1951 (courtesy of Chrysler Group LLC).

One of the more unusual advertisements by General Petroleum announced the results of the 1955 Economy Run. With photographs representing each make competing, the best actual miles per gallon result was shown.

would win its class. After impounding that particular car was hardly touched. It was unchanged in every department and we even had to call the entrant to come in and take advantage of his break-in allowance. It is safe to say that particular car was closer to showroom delivery than any car in the run. Its performance was normal!"[26] Added Al Cottle, driver of the Buick, the Roadmaster's V8 was powerful enough to "eliminate lugging on hills," which is an asset in maintaining good mileage performance.[27]

Packard engineers were pleased with the performance of their Patrician, which placed second behind the Buick Roadmaster. As noted in *Automotive News* prior to the start, "This is the first year in which a car with torsion-bar suspension has been entered. Packard Engineers are hopeful that the balanced ride may be an important factor in the contest."[28] The engineers' theory appears to have been proven with the 1955 Patrician achieving a figure of 18.7515 miles per gallon compared to 17.8957 set in 1954 by a Clipper Deluxe. Both models were equipped with Ultramatic automatic transmission and were piloted by Harold Michelson.

With corporate advertising continuing to focus on the new Mobilgas Special blend of gasoline and motor oil, General Petroleum's print efforts focused on publicizing the results. In one of the more unusual advertising layouts, found in newspapers and automotive-related publications, this ad featured a photograph of each make's best mileage entry, noting: "When American car manufacturers added horsepower, they didn't sacrifice mileage." The ad reminded readers that a "regular stop at the Flying Red Horse sign of Mobilgas will help you obtain 'par' mileage on any course you run."[29]

After several years of advertising the success of its overdrive units, Borg-Warner highlighted the performance of its automatic transmission, used by the Sweepstakes-winning Studebaker, noting that it "wins over all other automatics for economy."[30]

With the domination of Studebaker and American Motors in Sweepstakes and first-place victories, advertising by Detroit's "Big Three" was virtually nonexistent. Limited to newspapers immediately following the Run, Buick told readers, "Buick Roadmaster with Variable Pitch Dynaflow Wins First In Its Class in Famed Mobilgas Economy Run," and mentioned the benefits of its transmission over other cars competing in Buick's class.[31] Ford, with a rendering of its entry in the Run, proclaimed, "Ford Tops 'Big 3' in Mobilgas Economy Run," and went on to highlight its performance over Chevrolet and Plymouth.[32]

Studebaker, as expected, went all-out advertising its second consecutive Sweepstakes victory. Newspaper ads proclaimed, "Again Studebaker wins Mobilgas Run Sweepstakes!," also noting that it scored the best actual mileage of all eight-cylinder models competing. Brief mention was made of the President's being "a close second among all eights with an actual 25.29 miles per gallon," although it placed a disappointing third in the upper-medium price class.[33] A simplified version of this appeared in both *Automobile Topics* and *Automotive News*. Featuring a photograph of the winning Commander, Studebaker proclaimed its Sweepstakes win and advanced style, noting this "victory is another terrific publicity bonanza for Studebaker dealers," and extending an invitation for "successful automobile men anxious to be a dealer" to contact Studebaker's general sales manager.[34] Readers of *Life* and *The Saturday Evening Post* were reminded throughout the summer of Studebaker's victory with pictures of the winning Commander highlighting "Top honors again in Mobilgas Economy Run."[35] In an effort to capitalize on this success, dealers could order a "factory-lettered replica of the '55 Commander Custom 4-door" with the individual dealer's name and address appearing prominently on the front fenders. Dealers were also encouraged to stage their own local economy run using this replica.

American Motors ran several newspaper ads focusing on the success of both the Rambler and Nash Ambassador. Featuring photographs of the winning cars, "Nash Wins!" announced that the "Nash Ambassador Wins 'Big Car' Class C," along with "Rambler Sweeps 'Low-Price Field' Class A." Readers were invited to "Drive America's Top Economy Cars at Your Nash Dealer Today!"[36] Aimed primarily at dealers, or prospective dealers, an ad in *Automotive News* proclaimed, "Only Nash Dealers Have 2 Winners To Sell!" With photographs of both entries leaving Los Angeles at the start of the Run, the ad outlined these results as new proof of the value of a Nash franchise.[37] In a second ad in *Automotive News*, a two-page spread was devoted to the Rambler's success with the headline "Rambler Smashes Miles Per Gallon Record," highlighting its performance over the low-price Three. This also marked the first use of slogan "Economy King" when referring to Rambler.[38] American Motors also revived their "X-Ray" Series Comparison Catalog (last issued in 1940), which compared Nash to other models competing in the medium-price range. A full page was devoted to the Ambassador's achievements and noted, "While Studebaker won by a hair, the difference in gasoline amounted to a running distance of only 110 *yards* over the 1,323 *mile* course from Los Angeles to Colorado Springs." The same theme was continued in the 1956 "X-Ray" Comparison Catalog.[39] A similar publication was issued comparing the Rambler to Chevrolet, Ford, Plymouth and Studebaker models, and highlighting the Rambler's performance over these competing models.

Newspaper coverage continued to be the primary means of reporting on the Run. This was especially true in Southern California, but also in the cities and towns along the route. Covering the events each day, readers were also kept abreast of the close contest for the Sweepstakes Award. Expanded articles and pictures in the Sunday automotive sections appeared in the weeks following. With the horsepower race in full swing, economy driving was not in the forefront of people's minds, and this was evident in enthusiast magazines. The feature articles that did appear were far smaller in scope due primarily to the victories of Studebaker and American Motors over the more popular models from the Big Three, along with the requirement that automatic transmissions be used.[40] *Road & Track*, in an article titled "1955 Economy Run in which the Independents show the Big 3 a thing or three," focused more on the technical aspects of both the transmission and axle ratios used than on the event itself. *Motor Trend* followed suit in an article titled "Slushpumps aren't Spendthrifts. That's the surprising lesson of the automatics in the ... Mobilgas Economy Run," which also noted some of the driving techniques used by the winning drivers. Sweepstakes-winning driver Dick Johnson provided guidance on how to get the best mileage with time-tested techniques. These tips included: "On cars with automatic transmission, accelerate evenly. The transmission will shift into regular drive sooner and thus reduce consumption. With any kind of transmission avoid jackrabbit starts."[41]

Pierce Venable guides his Nash Ambassador Super through La Veta Pass on his way to the Colorado Springs finish. Pierce placed first in the low-medium-price class, although he would be edged out of the Sweepstakes Award by Dick Johnson's Studebaker by a slim 0.760 ton mile per gallon.

Months later the question, "What happens to these Run cars [and] where is the Studebaker that won the Sweepstakes?" was raised to Russ Catlin, press coordinator for the Economy Run. Russ reported later:

> A diligent search turned up the immortal piece of machinery. The car had been entered by a San Diego Studebaker dealer (J.R. Townsend Co., Inc.). Following its 1955 triumph the vehicle was put on display in the dealer's showroom. In time the car was painted and placed on a used car lot as "slightly used." A sailor bought the car for his wife. It is now being driven to and from the market and wherever else a sailor's wife motors while hubby is at sea. Her reaction was, "I get tremendous mileage with this car yet I didn't know it was the economy winner!"[42]

Changes were on the horizon that would have an impact on the sanction and supervision of future Economy Runs. Within a month of the tragic loss of Bill Vukovich during the AAA-sanctioned 1955 Indianapolis 500 Mile Race, a horrific crash during the running of the 24 Hours of LeMans Race in France claimed the lives of 83 spectators, which has been called "the most catastrophic in motorsports history."[43] As a result the American Automobile Association decided to dissolve its Contest Board so it could better focus on the services it provided to the motoring public. Along with members of the AAA Contest Board, A.C. Pillsbury became a prime mover and financier in the creation of the United States Auto Club, which would now assume this sanctioning authority.

In 1955, the name of General Petroleum's parent company, Socony-Vacuum Inc., was changed to Socony-Mobil Oil Company.

1956

Los Angeles to Colorado Springs Via Las Vegas — Salt Lake City — Grand Junction

MARCH 19–22, 1956—1468.8 MILES

Ushering in a new era was the announcement that the Sports Commission of the United States Auto Club (USAC) would now be the sanctioning organization assuming the duties of the former Contest Board of the American Automobile Association, dissolved in mid–1955. In addition it was announced that USAC was now affiliated with the Federation Internationale de l'Automobile, an international supervising organization of automotive sporting competition headquartered in Paris, giving the Economy Runs international recognition. A.C. Pillsbury, named a regional director of the Sports Commission, continued his role as chief steward in charge of all technical and mechanical details of the Run.[44] Other than the name change, few would have realized that a change had occurred; this transition was virtually seamless, with long-time AAA officials who had worked on the event for years transferring over to USAC. As it grew in scope, "nearly 300 different employees or officials worked in the event and it is estimated that more than 1200 persons had something to do with the contest in one way or another."[45] While General Petroleum continued to sponsor and underwrite the Economy Run, publicity emphasis by Socony-Mobil Oil Company was becoming more evident.

Returning for a second time to the Broadmoor Hotel in Colorado Springs, this year the drivers would tackle a 1468.8-mile route over a four-day period, with overnight stops in Las Vegas, Salt Lake City and Grand Junction. With an altitude range of 203 feet below sea level in Death Valley to 11,312 feet at Monarch Pass Summit in Colorado, this route would reach the highest elevation to date, and this mark would not be surpassed until 1959. In order to keep within the 36-hour, 20-minute time limit set by USAC, drivers would average a speed of 40.9905 miles per hour.

Pierce Venable, this time driving a Pontiac Chieftain 870, travels through Fishback Canyon in Utah en route to Salt Lake City. Placing first in the low-medium-price class, Pierce scored Pontiac's first class win since the 1938 Gilmore Yosemite Economy Run.

Taking note of the popularity of pillarless hardtop models (in 1955 the top-selling model for Buick, Mercury and Pontiac were two-door hardtops), rule changes were made that either a conventional four-door sedan or a two-door or four-door hardtop model could be entered. Just over half of the entries were four-door hardtop models, while only two two-door models competed. Competitive classifications were increased in number, with those for medium-priced models expanded to three, covering the low-medium, medium and upper-medium price ranges. In response to the smaller Rambler's substantial victory in 1955 over Chevrolet, Ford and Plymouth in the low-price class, a new special limited displacement class was added. Aimed primarily at American Motors' Rambler, this new class was restricted to those cars having a piston displacement of less than 200 cubic inches. This also relegated the previously successful six-cylinder Studebaker models into this special class. The only other significant rule change of note was a reduction in the maximum number of man-hours allowed for servicing of competing cars to 16, down from 24 in 1955.[46] As Randall Gould of the *Denver Post* noted, "Rules developed through the years have steadily worn down any special gimmick giving certain drivers advantages beyond the average. The whole idea of the run is to put the main emphasis on performance of the individual car."[47]

Not only were the rules developed and enhanced over the years, procedures followed by USAC personnel overseeing the various aspects of the Run were constantly developed as well. To ensure that the conduct of the Run was done as correctly and smoothly as possible, "General Instructions to U.S.A.C. Officials" were issued in February (covering 25 pages on legal-sized paper), so everyone involved with start, refueling, brunch, impound and calculating crews knew without questions their tasks and responsibilities. These instructions even detailed when and by which route individual crews would travel from point to point. Should adverse weather conditions occur, alternate routes were discussed so that all contingencies could be addressed.

Continuing to be primarily sponsored by individual dealers or dealer associations, a total of 21 cars competed, representing every make except for Cadillac. For reasons unknown, Chrysler did not enter either their Windsor or New Yorker models, placing their hopes instead on the Imperial Southampton hardtop. Joining the Chevrolet Bel-Air 6 as a private entry was the Studebaker Champion 6, which reflected Studebaker's corporate decision not to provide factory support. Three Southern California area dealers who made their inaugural appearance this year — Bill Murphy Buick in Culver City, Paola Oldsmobile in LaCrescenta, and W.R. Shadoff Plymouth in Pomona — would continue their involvement in the Economy Runs well into the 1960s. Ray Brock (from *Hot Rod Magazine*) and Bob Russo moved up to the driver's seat, with Ray piloting the Oldsmobile "98" Holiday and Bob piloting the Studebaker Champion 6. Assisting Ray Brock as relief driver was Wally Parks, also from *Hot Rod Magazine* and well-known founder of the National Hot Rod Association. Dick Johnson returned to defend his 1955 Sweepstakes-winning performance this year, piloting a Packard Four-Hundred hardtop.

Random car selection by USAC officials and preliminary impounding commenced in early February. At the time of selection, cars were inspected from bumper to bumper to ensure stock status, as had been the rule in prior years. This year it was announced that USAC officials may check "cylinder compressions to ensure that the selected car will meet factory specifications."[48] Rules regarding servicing, break-in runs or the total mileage allowed on the entries did not change from previous years.

Final impounding occurred on March 9, coupled with the announcement of the final route and maximum net running time. The route selected by USAC, covering 22½ pages of turn-by-turn instructions, was the longest in the event's history, and the rugged terrain was designed to challenge both the car and more importantly, the driver. "From the day the course was announced, veteran drivers said they could make outstanding economy records only under the most favorable weather conditions."[49] The crews only had enough time to make a single practice run and soon discovered the challenges facing them. In order to maintain the time schedule, driving at near-maximum speeds would be required, as the possibility of less than ideal weather conditions could easily cause them to run over their allotted time, resulting in their being disqualified. Other challenges facing the drivers would be narrow winding roads coming out of Death Valley, traffic in the Salt Lake City area, and a wintertime climb over Monarch Pass at the Continental Divide. As Wally Parks noted in *Hot Rod Magazine*, "To say that this year's course was a tough one is the understatement of the season — it was downright rugged. Credit for selecting the route over which the '56 run was conducted goes to Art Pillsbury and his capable crew of hard-working USAC officials. Always a stickler for proof of performance, Art made certain that no one was going to win in a breeze — certainly not with a tail wind."[50]

During this time, the Lincoln Premiere entry was withdrawn for reasons not made public. Observers made an interesting observation after its withdrawal: that in the random drawing for starting positions, the Lincoln had drawn no. 13.[51]

On March 19, the remaining 21 brightly lettered and decorated entries were lined up "double file" on Flower Street, ready for the first leg across the Southern California desert region to Las Vegas, covering 422.5 miles. A 10-hour, 30-minute time limit was set, and this time did not include stops for brunch or refueling. At this point the drivers knew they were facing a stiff test as some of the mountainous sections had been hit by blizzard conditions which resulted in "hordes of deer being driven down to the highways by the storms."[52] With advance publicity by General Petroleum's advertising staff, along with pre-run articles in Los Angeles area newspapers, a large group of spectators were on hand when Tommy Thomas, driving a Rambler Deluxe 6, took the green flag at 12:01 a.m. As in prior years, the remaining cars left at two-minute intervals heading eastbound for San Bernardino, then over Cajon Pass and across the Mojave Desert on straight, narrow roads with many dips along the way. Later in their travels these roads would become steep and winding. It was here the drivers faced their first weather obstacle as head winds, referred to as "wind flurries of the desert," plagued them for most of the trip. North of the town of Trona in Panamint Valley, the first accident in the 13-year history of the Economy Runs occurred when Les Viland's Nash Ambassador 6 overturned at the bottom of a grade, marring the opening day. As reported by USAC, "On the second turn going into Panamint Valley, the left wheels went off onto the gravel shoulder and when the brakes were applied the car went into a slide and partially turned over. It then hit some rocks which threw it back onto its wheels."[53] Although Les escaped injury, his relief driver (Loren Adkins) and both USAC observers (Richard Fiddler and James Robins from Caltech) suffered minor injuries and lacerations which required treatment at Trona Hospital. This was the first time anyone had been injured in the history of the Runs. As Les later explained, he lost control of his car when it was hit by a gust of desert wind on the twisting narrow road and in fast subsequent maneuvers the Nash overturned.[54] This was a disappointing ending to what had been predicted by other competing drivers as a possible third Sweepstakes victory for Les based on "his meticulous approach to his job. He studies the rival cars scientifically (and) does not overlook a single factor than can help him win."[55]

Arriving at the Stovepipe Wells Hotel around 6:15 a.m., everyone was ready for a 30-minute break for brunch after battling winds and dust for several hours. USAC officials had temporarily installed gas pumps in the hotel's parking lot to facilitate refueling, after which the drivers continued their trek through Death Valley and on to Las Vegas. Arriving around 11:00 a.m. at New Frontier Sports Drome, just off the Las Vegas Strip, the drivers were ready for a well-earned break as entries were impounded overnight. USAC officials began calculating the first day's results which were due to be released by 2:00 p.m. to accommodate press representatives from the East Coast in meeting their deadlines.

When the results were posted, Mel Alsbury, Jr., in his Imperial Southampton had a commanding lead with 63.0665 ton miles per gallon. During the calculations it was reported that the USAC calculating crew refused at first to believe their own figures for the Imperial's performance. As told by Russ Catlin, USAC press coordinator:

> When a car has completed its run seven of these operators compute the result. Each works independent of the other and all seven must agree. In the case of the Imperial, George Inglis, chief of the crew, ordered a second run through the production run. When this trial also turned up the same phenomenal 63.0665 ton-mile figure he ordered a third check. Only then would he admit that the Imperial had a ten ton-mile lead over its nearest competitor.[56]

In actual miles per gallon performance the Imperial also made an excellent showing with a figure of 21.6203. Well back, with 53.4972 ton miles per gallon, was last year's Sweepstakes contender Pierce Venable, this year driving a Pontiac Chieftain 870 Catalina. The best performance in actual miles per gallon was 25.5264 set by Tommy Thomas in his Rambler Deluxe 6. In addition to these three cars, both Oldsmobile entries, along with the Ford Customline 8, were leading their respective classes. Competition was extremely close in the popular low-price class, with Marshall Martin, piloting a Ford Customline 8, leading Art Rene, in a Plymouth Belvedere 8, by only three-tenths of a ton mile per gallon. The field achieved an overall average of 20.70 miles per gallon, which was lower than everyone expected. This lower result was due to the strong desert winds encountered throughout the day.

Leaving Las Vegas beginning at 12:01 a.m., the drivers faced a tough 466-mile, 11-hour and 10-minute trip to Salt Lake City, returning to this area for the first

Various Socony-Mobil distributors would erect welcome signs for the Economy Run drivers traveling through their territories. Danny Eames and his DeSoto Fireflite pass one of these on U.S. Route 50 in Colorado as he heads for Grand Junction.

time since 1952. USAC officials inaugurated an "inverted start" procedure, whereby the last competing car to arrive would be the first out the following morning. This would become standard practice in all future Runs. Based on reports of heavy traffic conditions between Provo and Salt Lake, the Utah State Police recommended a time extension, and the USAC added an additional 20 minutes to the net running time.[57] Facing a long, mostly uphill trip to Salt Lake City, the drivers had to contend with some head winds, along with some snow flurries, as they headed north. Passing through areas of typical Nevada grazing land, the drivers were cautioned to watch for cattle on the roadway. Crossing into Utah they passed through St. George and continued over several mountain ranges, arriving in Beaver just past 7:00 a.m. for refueling. This year's run continued not being kind to the Nash Ambassador entries. Carl Chakmakian, driving an Ambassador Special 8, hit a post near Beaver, causing minor damage to the left rear fender and bumper. Leaving Beaver, the drivers continued traveling on winding highways in deer and elk country to Richmond and a brunch stop at the Rainbow Café. Continuing north, they headed along the Wasatch Mountain Range towards Provo. Arriving around 12:30 p.m., they made their way to the impound at the Utah State Fair Grounds. Here is where heavy traffic conditions played havoc for the drivers, with four arriving with less than two minutes to spare. Bob Russo, driving a Studebaker Champion 6, not only came in with a mere 34 seconds to spare, but his gas tank was close to empty, the "engine dragging instead of pulling" him into the impound area.[58] Bill Losher, driving a Dodge Coronet 500, was planning on an uneventful trip after his experience running out of fuel last year. Driving through Provo, he was stopped twice by police to pose for pictures with the local high school band. Already running late, he had to "speed it up," arriving at the Salt Lake City impound with 13 seconds to spare.[59] Bill Hooks, driving a Ford Fairlane 6, was not as fortunate, reaching the impound 2 minutes and 20 seconds late; he was disqualified, thus reducing the field to 19. On the opposite end of the spectrum, "for some unexplained reason the Mercury crew subtracted twenty minutes rather than added and arrived in Salt Lake after a day of blinding speed, 22 minutes early."[60] All entries were impounded overnight while the crews, tense from a long day's drive, were ready for another well-earned break.

After the posting of the results for the second day, it appeared that the Imperial Southampton was pretty much a runaway for the Sweepstakes Award with a figure of 62.6704 ton miles per gallon. Trailing in the overall Sweepstakes Award continued to be Pierce Venable in his Pontiac Chieftain 870 Catalina with 52.89 ton miles per gallon. Following closely behind with 52.56 ton miles per gallon was last year's Sweepstakes winner Dick Johnson in his Packard Four Hundred. The best performance in actual miles per gallon continued to be Tommy Thomas in his Rambler Deluxe 6 with 25.0583. There were no changes in the individual class-leading entries.

On March 21, everyone had the chance to sleep in, as the first car headed to Grand Junction at 4:30 a.m. Covering 288.8 miles within a 7-hour time limit, the route would be a fairly straightforward climb through mountainous terrain. For a change the drivers had relatively little wind to contend with. Passing once again through Provo, the drivers headed southeastward on U.S. Routes 6 and 50 toward their first stop of the day in Green River. As they crossed Soldier Summit, with an elevation of 7463 feet, herds of deer were spotted and snow 4 feet deep lined the highway. In Price and Green River, schools were dismissed so students could watch and cheer on the competitors as they passed through town.[61] With the first car arriving at the Green River Union School around 8:25 a.m., brunch was served by community volunteers in the school cafeteria. Immediately following brunch was a stop for refueling where each competitor was required to take a minimum of 5 gallons of Mobilgas Special. Continuing on U.S. Routes 6 and 50, they crossed into Colorado, arriving around noon at Lincoln Park Stadium in Grand Junction, where the entries were impounded overnight. Watching the weather forecasts, "a number of drivers expressed concern over Monarch Pass, [but] they were advised the highway had been scraped and clear weather could be expected."[62]

When the results were posted, it was apparent that barring any unforeseen difficulty on the final leg, the Sweepstakes trophy for Mel Alsbury and his Imperial Southampton was "almost a sure thing."[63] With a figure of 62.5654, he had a commanding lead of over 10 ton miles per gallon over Pierce Venable in his Pontiac Chieftain 870 Catalina and Dick Johnson in his Packard Four Hundred. Only a mere .07 ton miles per gallon separated the Pontiac and Packard entries. Tommy Thomas in his Rambler Deluxe 6 continued to record the best performance in actual miles per gallon with a figure of 24.6535. There were no changes in individual class leaders, although competition continued to be close in the popular low-price class. Marshall Martin, in a Ford Customline 8, continued an extremely slim lead over Art Rene in a Plymouth Belvedere 8.

In reviewing the results, it was noted that all but two of the competitors showed a drop in their mileage figures for the third day of competition. Searching for an answer to this drop, Randall Gould of the *Denver Post* interviewed several drivers including the Chevrolet Bel-Air 6 driver Pete Novotny. Pete explained:

> For the first three hours out of Salt Lake City, the drivers had been plagued by a multitude of "restrict speed" zones. The day's distance seemed refreshingly short, a mere 288.38 miles, but at the same time only seven hours were allotted and the required average was relatively high, close to 42 miles an

> hour. With four hours to go, each driver in turn discovered that he had a seemingly disproportionate amount of total distance still to go. So they "whipped up the horses" and began cruising as high as 58 miles per hour to make up time. This actually brought most of them in with more time to spare than they'd intended — close to 13 minutes in one instance, 12 minutes in another.

Pete concluded that the drivers "hit higher speeds than customary and the immediate penalty was a greater fuel consumption for the seemingly easiest leg."[64]

On March 22, the drivers were ready for the final and most rugged leg of the Run. The first car left Grand Junction at 3:30 a.m. in comfortable 46-degree weather for the 291.5-mile, 7-hour and 20-minute trip to Colorado Springs. Heading east, the drivers faced winding highways on U.S. Route 50 through Colorado's beautiful mountains towards Gunnison, where deer grazing alongside the road became a common sight. Climbing to an altitude 11,312 feet at Monarch Pass Summit and Continental Divide, the drivers were ready for a brunch stop at the Monarch Crest Café, with the first car arriving around 7:00 a.m. The drivers were reported to be relieved that the roads were dry over Monarch Pass although a good deal of snow lined the sides of the road.

Heading downhill, encountering some steep grades along the way, their next stop was in Salida for refueling at the Warner Mobilgas service station. Here each competitor was required to take a minimum of 5 gallons of Mobilgas Special before continuing to Cañon City and the finish line at the Broadmoor Hotel in Colorado Springs. In much more milder temperatures than encountered last year, a good-sized group of spectators were on hand when Art Elliott, driving a Mercury Montclair Pheaton, was the first car to take the checkered flag, arriving around 11:00 a.m. One by one, the drivers crossed the finish line within the specified time limits, and the cars were topped off for the final time. The cars were kept under USAC supervision while the final results were computed. Later in the afternoon, with the competing cars arranged in a semicircle in the front of the hotel, the Awards Ceremony was held. The drivers, "like their storm-scarred cars, showed the wear and tear of the grueling test they had just undergone."[65]

After experiencing the closest competition for the Sweepstakes Award in 1955, this year turned into a runaway success for 26-year-old Mel Alsbury, Jr., and his Imperial Southampton. This win also marked the first time a Chrysler Corporation automobile took the top prize. In only his second year competing, Mel led the field from the beginning, achieving a ton-miles-per-gallon average of 61.3796. This average was second only to the 66.484 ton miles per gallon mark set in 1951 by Les Viland driving a Lincoln. Equally impressive was the Imperial's actual miles per gallon figure of 21.0420. Pierce Venable placed second for the Sweepstakes in his Pontiac Chieftain 870 Catalina with 51.7297 ton miles per gallon, as well as recording the best miles-per-gallon figure for eight-cylinder cars with 21.1098. The best miles-per-gallon figure for all entries was set by Tommy Thomas in a Rambler Deluxe 6 with 24.3545. A surprise to most observers was the drop in miles-per-gallon performance (compared to 1955's results) of Rambler's new "high efficiency" engine.[66]

After his disappointing sixth-place finish in 1955 driving a Chrysler Windsor, Mel was never considered to be a contender for the Sweepstakes Award. Chrysler officials were reported to be disappointed when Mel was named driver of the Imperial, as a more experienced driver was preferred. The Imperial's entrant, Hollywood Chrysler-Plymouth Dealer Mel Alsbury, Sr., stood his ground and prevailed in his selection of Mel Jr. as driver. After learning of the Imperial's spectacular performance when the first day's results were announced, some of these officials decided to travel to Colorado Springs for the finish.[67] When Mel Jr. accepted the Sweepstakes trophy, his father was heard to shout, "That's my boy!"[68]

Marshall Martin returned to the winner's circle for the first time since his Sweepstakes-winning performance in 1941, this year driving a Ford Customline 8. Two of the winning drivers (Ray Brock and Tommy Thomas) were first-time participants as drivers, having held relief driver duties previously.

In addition to the Sweepstakes Award, first-place trophies were presented to:

CLASS	MAKE and MODEL	TON MILES PER GALLON	MILES PER GALLON
Special Limited Displacement	Rambler Deluxe 6	48.2218	24.3545
A — Low Price	Ford Customline 8	47.7643	20.5217
B — Low Medium Price	Pontiac Chieftain 870	51.7297	21.1098
C — Medium Price	Oldsmobile 88 Holiday	50.8077	19.7044
D — Upper Medium Price	Oldsmobile 98 Holiday	50.9488	18.6694
E — High Price	Imperial Southampton	61.3796	21.0420

Detailed results for all competing cars can be found in the Appendix.

In announcing the results, the USAC reported that all entries averaged 19.9536 miles per gallon, which was a drop of 1.6079 from 1955's average. This was the lowest average miles-per-gallon figure since the Economy Run's inception in 1936. The rugged route from Los Angeles, coupled with high altitude over the Continental Divide, contributed to this drop. Chief Steward A.C. Pillsbury noted that the results "demonstrate the efficiency potential of 98 percent of America's stock cars. Today's finish demonstrated this with convincing proof of the mileage possibilities when cars are properly maintained and driven."[69] Other than the opening day's accident involving the Nash Ambassador 6, which eliminated it from competition, no entries encountered any mechanical or tire problems during the Run. Weather conditions were described as generally good.

USAC's J.C. Agajanian waves the checkered flag for the Sweepstakes-winning Imperial Southampton driven by Mel Alsbury, Jr. In only his second year competing, Mel achieved a figure of 61.3796 ton miles per gallon, the second highest in Economy Run history.

In the popular low-price class, a battle for first place developed from the beginning between the Ford Customline 8 driven by Marshall Martin and Plymouth Belvedere 8 driven by Art Rene. They were close for all four days, but in the end Marshall edged out Art by a slim .0262 ton miles per gallon in one of the closest competitions in the event's history, referred to by one reporter as "a dog fight from beginning to end."[70] In actual miles-per-gallon performance, Plymouth outperformed Ford; however, both were behind the two Chevrolet entries. Pete Novotny in his Chevrolet Bel-Air 6 achieved 21.1715 miles per gallon, which was the top figure for standard size six-cylinder models. Prior to the Run, factory interest was focused on four entries equipped with new engines designed more for performance than economy. One of these was Ford's 292-cubic-inch "Thunderbird V-8" powering the Fairlane Hardtop, which placed a disappointing fifth with 18.7496 miles per gallon. The class-winning Customline was powered by the smaller 272-cubic-inch engine.

General Motors, who in previous years had a rather indifferent attitude toward the Economy Runs, could not have been more pleased with the success of both Oldsmobile and Pontiac dominating the medium-price classes. Oldsmobile scored a double win. The "88" Holiday, driven by Woody Bell, took the first-place trophy in the medium-price class, and the "98" Holiday, driven by Ray Brock, took first in the upper-medium-price class. Prior to the start, observers pegged Woody as one of the drivers most likely to take the top prize.[71] Pontiac also scored a class victory with the Chieftain 870 Catalina, driven by Pierce Venable, in the low-medium-price class, as well as coming in second to the Imperial for the best ton-miles-per-gallon figure. This marked Pierce's second year in a row placing second in the Sweepstakes, narrowly losing out to a Studebaker Commander in 1955. Both Buick and Chevrolet entries also placed well in the final standings.

Disappointing this year was the performance of the Studebaker Champion 6, which turned in the lowest ton-miles-per-gallon-figure of all competing cars with 39.7426. Its actual miles-per-gallon figure of 20.0467 was significantly lower than any automatic transmission–equipped Studebaker previously recorded, whether a six- or eight-cylinder model. This result was explained in part by Studebaker's decision not to provide factory participation and support, along with Bob Russo's first experience as a driver.

In what would be their last year of competition, Hudson, Nash and Packard saw mixed results. American Motors' newly developed 350-cubic-inch V8 engine, powering both the Nash Ambassador Special (driven by AMC engineer Carl Chakmakian) and the Hudson Hornet Special (driven by San Fernando Hudson dealer Lee Hamer) placed second and third in the medium-price class, although both models achieved higher actual miles-per-gallon results than the class-winning Oldsmobile "88" Holiday. This second-place result was especially pleasing to Carl Chakmakian, who tangled with a post near Beaver, Utah, causing minor damage to his car. Both Hudson and Nash closed out their competition; having participated in twelve of the thirteen Economy Runs held since

1936 (Hudson missed the 1940 Run, while Nash missed the 1952 Run), they accumulated a total of fifteen first-place awards — ten for Nash and five for Hudson. Last year's Sweepstakes champion Dick Johnson, driving a Packard Four Hundred, placed distant second to the Sweepstakes champion Imperial Southampton in the high-price class. Competing in all of the Economy Runs, Packard recorded six first-place awards. Also of note was the steady participation of Packard's Southern California distributor, Earle C. Anthony, Inc., having been involved as an entrant each year since 1936.

General Petroleum Officials continued to be pleased with this year's events, and longtime officials, including Robert L. Minckler and Clarence S. Beesemyer, were again heavily involved. Advertising Manager Frank Meunier and Assistant Advertising Manager Roger Mahey worked tirelessly throughout, getting the word out to members of the press and ensuring good follow-up publicity nationwide. Noticeably absent this year was popular racing driver Ralph DePalma. A regular from General Petroleum's public relations department since 1950, also serving as an official with both the AAA Contest Board and USAC, he was seriously ill, battling cancer and unable to travel. Ralph died at age 73 shortly after the Run was completed.

Any history of the Economy Runs could not be complete without mentioning one of the more colorful automotive editors of the time, Slim Barnard of the *Los Angeles Examiner*. A regular covering the Run since 1950, Slim always found a way to entertain the spectators at the finish line, whether driving a Roman chariot or riding a burro. For the second year in a row, everyone from the press to the competitors who entered the grounds of the Broadmoor were greeted by two large signs. One read "Los Angeles City Limits" while the other said "Read Your *Los Angeles Examiner* Daily." As part of Slim's antics, copies of the *Examiner* were flown to Colorado Springs for distribution to all of the participants at the Awards Banquet. He would continue to entertain the participants in one way or another through the end of the decade.[72]

Although the class-winning models came from various manufacturers, advertising was quite limited this year, and those ads that appeared were found mostly in newspapers and in industry publications such as *Automotive News*. Socony-Mobil's corporate print advertising continued to focus primarily on publicizing the results, using a more simplified format this year. On a border featuring Mobil's famous Flying Red Horse, "Official Winners Mobilgas Economy Run" were announced, along with a brief description of the event, noting that "the Mobilgas Economy Run is a vigorous mileage test of cars that compete against others in the same price class." Further noted was that the Run "is annually sponsored by Socony Mobil's western affiliate General Petroleum Corporation as a public service to motorists."[73] Another ad for Mobilgas Special, appearing mainly in enthusiast publications, noted that the Run was a "testing ground of performance and mileage."[74] A unique ad, appearing in Spanish-language editions of *Life*, "Economia Mobiloil 1956," featured a photo of the Sweepstakes-winning Imperial and route details, noting the benefits of using Mobiloil Special.[75]

Chrysler advertised the Sweepstakes win for the Imperial, proclaiming "Chrysler-Built Imperial Beats All Cars In Mobilgas Economy Run!," further noting that it won "with the greatest margin in sweepstakes history." Featuring a picture of the winning Imperial Southampton, this ad was found in newspapers and *Automotive News*.[76]

Ford's newspaper advertising headlined "Ford Wins in Economy Run too," featuring a picture of the low-price class-winning Customline 8, and tied in this success to Ford's stock car racing performance. Ford had won the Manufacturer's Trophy at NASCAR's National Speed Week events in Daytona Beach in February.[77] Aimed primarily at Ford dealers, a two-page spread in *Automotive News* proclaimed, "Ford wins all 3 of The Industry's Most coveted awards!," showing the NASCAR Trophy (for performance), the Mobilgas Economy Run Trophy (for economy) and *Motor Trend*'s Car of the Year Award (for safety). Using the tagline "The car that goes *first* in Performance ... Safety ... Economy!,"[78] Ford revised their regular sales brochure to further publicize their recent success.

While Chevrolet did not publicize its performance in national advertising campaigns, a small tri-fold folder was issued to dealers and salesmen emphasizing both the six-cylinder and eight-cylinder Bel-Air's performance: "The Hot One's Thriftier, Too!" Focusing on actual miles-per-gallon performance, this folder noted that Chevrolet scored the best mileage in the popular low-price class, beating out both Ford and Chevrolet, emphasizing that "Chevrolet's figures are the ones that actually mean something to your prospects."[79] This was the first time since 1950 that Chevrolet acknowledged its participation in the Run.

Oldsmobile and Pontiac featured simple print advertising in both newspapers and in *Automotive News*. Noting Oldsmobile's double class wins with the "88" and "98" models, their ad proclaimed, "Oldsmobile Rockets To Two First Places In Mobilgas Economy Run," telling the readers that "without fanfare, two strictly stock Oldsmobiles were entered. In a blaze of glory, both won decisive victories."[80] Pontiac not only announced its class win, but also announced that it scored the most miles per gallon of all eight-cylinder models competing with the headline "Here's The Big *Fact* You Need To Know."[81] Individual manufacturers continued to rely on these miles-per-gallon figures over the more confusing ton miles formula.

Naturally, American Motors heavily advertised the fact that again this year a Rambler Deluxe 6 delivered the best actual miles-per-gallon performance of all entries. Featuring a

Imperial's first Sweepstakes Award resulted in advertising appearing in both newspapers and industry-related publications. These advertisements followed an established trend of showing either photographs or renderings of the winning models (courtesy of Chrysler Group LLC).

photo of the Rambler at the start, this ad noted, "Rambler still holds the all-time record for economy!," in reference to the overdrive-equipped and slightly smaller Rambler Convertible's 1951 Economy Run performance.[82] A two-page ad, aimed primarily at American Motors dealers, also featured a photo of the winning Rambler, declaring that "Rambler Tops The Field In '56 Mobilgas Economy Run!"[83] This success was also highlighted in the 1956 "X-Ray" Series Comparison Catalog, available at American Motors dealers, which compared the Rambler to other low-price cars from Chevrolet, Ford, Plymouth and Studebaker. The results were repeated in the 1957 "X-Ray," although more as a footnote to their NASCAR-supervised Coast-to-Coast Run with a 1956 overdrive-equipped Custom Station Wagon.

The only mentions of the Economy Run in ads in general interest magazines were for the newly introduced Nash Ambassador Special models. Found in publications such as *Life* and *The Saturday Evening Post*, these ads proclaimed the economy of the new Torque-flo V8 Engine, noting these engines had power with traditional Nash economy. Hudson, who rarely advertised its participation in the Economy Runs, advertised the Hornet Special's performance in the newsletter *Hudson Owners Report*, which was mailed to Hudson owners. Featuring a photo of the Hornet Special crossing the finish line, the article focused on actual miles-per-gallon performance. It noted that "V8 Specials Finish 1st and 2nd in Class 'C'" without actually stating that the Hornet finished behind the Ambassador Special by just over two-tenths.[84]

Newspaper coverage continued again this year to be the prominent means of reporting on the Run. Articles about each day's events allowed readers to follow the progress of the competitors along with the runaway performance by Mel Alsbury, Jr., and his Imperial Southampton for the Sweepstakes award. Expanded articles and pictures in the Sunday automotive sections also appeared in the weeks following. As in prior years, Floyd Clymer's *Automobile Topics* provided the most in-depth coverage of any publication, devoting 13 pages in its April issue to the Run. Other enthusiast magazines mainly reported on the results, and most of these articles were focused on the drop in the average miles-per-gallon performance compared to previous Runs. Noting the continuing "horsepower race" by American manufacturers, *Motor Life* observed, "Comparisons between economy runs by year is not entirely fair. For instance, the route always varies, wind and road conditions naturally differ, and in the past couple of years automatic transmissions have been required equipment. Nevertheless, it is pretty evident that engines of higher power output hooked to gear boxes of reduced efficiency are costing the driver who insists upon them plenty of dough."[85] Wally Parks, writing in *Hot Rod Magazine*, made the same basic observation as he provided some additional insight of the Run, due in large part to his participation as Ray Brock's relief driver. *Popular Science* reported on the results, continuing its practice of providing its readers with tips on getting the best gas mileage from their cars. These tips followed those from previous years and were starting to become redundant. The story of the Ford Customline's class-winning performance, along with a color photograph, was featured in *Ford Times*, a publication for owners of Ford Motor Company products.[86]

Pleased with the United States Auto Club's handling of this year's Run, both Socony-Mobil Oil and General Petroleum announced continuation of the event. Looking for ways to broaden the scope and interest in the Economy Runs, the next three years would usher in major changes to accomplish this goal.

5

Sun Valley and Galveston

1957

Los Angeles to Sun Valley Via San Francisco — Winnemucca — Ogden

April 14–18, 1957—1568.3 Miles

In announcing a return to the Challenger Inn in Sun Valley, officials with General Petroleum and the USAC also had a surprise up their sleeve. The rules, published in late February, included one paragraph which stood out by making a significant change in eligibility. This read:

> An entrant may exercise an option to enter a second car of the same make and model providing that second car is to be operated by a qualified woman driver and woman relief driver. Official score for that make and model will be determined by whichever of the two cars achieves the most ton-miles per gallon over the entire course. Except as provided in this option, all drivers and relief drivers will be of the male sex.[1]

What had previously been a male-dominated event now took on a new look. Not only would women be competing for Sweepstakes and class victories, they would be in direct competition with male drivers handling the same make and model, although this rule would cause confusion when the results were announced. This subject had been under serious discussion, and at General Petroleum's suggestion the new rule was drafted by the USAC. Chief Steward A.C. Pillsbury remarked, "There's hardly a woman in the country who doesn't think she can drive better than most men. Now we can really find out."[2]

This year the drivers would face the longest route in history, covering 1568.3 miles over a four-day period, with overnight stops in San Francisco, Winnemucca and Ogden. A fairly moderate altitude range from sea level just north of Ventura, California, to 7382 feet at Echo Summit in the Sierra Nevada Mountains of California would be encountered. This year drivers would average 41.4524 miles per hour to keep within a 38-hour and 30-minute time limit set by the USAC.

Following the same basic criteria used in the 1956 Run, eligibility was tightened with regards to engine size with a requirement that entries be powered by an engine of 200-cubic-inch displacement or larger. This had a direct effect on participation of six-cylinder models manufactured by American Motors and Studebaker, since the special limited displacement class was also discontinued. Competitive classifications were reduced to four this year, covering low, low-medium, upper-medium and high-price classes. This reduction was due in part to the decline in sales and availability of medium-priced models. The only other rule change of significance was a further reduction in the maximum number of man-hours allowed for servicing of competing cars to 14, down from 16 in 1956.[3] After reviewing the events surrounding last year's disqualification for exceeding the time limits for each day's drive, the USAC issued a bulletin stating that no car would be disqualified for being late, but a penalty would be imposed.[4]

With one exception, all entries were sponsored by individual dealers or dealer associations. The only private entry this year was Leonard DeBell, owner of the Southern California Plating Company in Los Angeles. A regular participant since 1950, always with a six-cylinder Chevrolet, this year DeBell sponsored a Studebaker President. With a total of 24 cars entered, several makes were not represented, including Cadillac, Hudson, Nash and Packard, and their absence had been expected. A surprise was the absence of Lincoln and Mercury for reasons that were not disclosed. The new faces this year included Vince Piggins, an engineer with Chevrolet Motor Division, in his first Economy Run as a driver; he had been responsible for training and preparations of the entire Chevrolet team. Mel Alsbury, Jr., returned, again behind the wheel of an Imperial Crown, to defend his Sweepstakes-winning performance. Three previous Sweepstakes-winning drivers, Dick Griffith, Marshall Martin and Les Viland, also participated.

Nine entrants exercised the option to enter a second car, with women as drivers and relief drivers piloting entries from Chevrolet, DeSoto, Dodge, Ford, Imperial, Oldsmobile and Plymouth. Selected based on their driving abilities, most had racing or rallying experiences in their background. Well-known names included sports car rally participant Myra Buchanan, sports car racer Mary Davis thrill show driver Patricia Jones, and sports car and aviation record holder Betty Skelton. The Oldsmobile team featured the wives of veteran Economy Run drivers, Lorraine Bell (Mrs. Woody Bell) in a Golden Rocket "88" and Marilyn Miller (Mrs. Ak Miller) in a Starfire "98." Mildred Alsbury, wife of Hollywood Chrysler-Plymouth dealer Mel Alsbury, Sr., drove an Imperial Crown, with Rae-Nell Alsbury (Mrs. Mel, Jr.) making their participation an all-family affair. The women's participation was not just for show, as they covered hundreds of highway miles in practice runs. Also on their agenda were practicing changing a flat tire or installing tire chains in the fastest time possible. A few would compete only this year, while others would be active participants well into the 1960s.

Advance publicity and promotion began in March under the continued direction of Frank Meunier. Press releases were issued several times each week covering differing aspects of the Run. One release concerning the route was a "teaser" describing four of the possible routes the drivers would cover during the event, although USAC officials emphasized that the final selection would not follow any of these four.[5] After the actual route was announced, additional releases included information and travelogue-style stories on the route selected, noting that "one thing is certain about the Mobilgas Economy Run — the entries will never run the same course in two succeeding years."[6] Press releases frequently covered the preparations involved, with some featuring an emphasis on the women's participation. To show this importance, Nancy Reynolds of the Los Angeles CBS television station's morning show *Panorama Pacific* was one of two women reporting on the Run, following the competitors over the route in one of the press cars.[7]

New this year was a Mobilgas Economy Run Fact Book, published and distributed to the press and other interested parties. This 23-page spiral-bound book covered the important aspects of the Run including route and timing information, entry list and biographical sketches of each driver and USAC official. These Fact Books would continue to provide an important source for press and historical information through 1963.

Random car selection and preliminary impounding commenced in March following established procedures from previous years, which included placing seals on 64 parts to prevent tampering. In an attempt to answer skeptics who "believe there 'must be a catch to it' because such amazing mileage figures are obtained," two press releases were devoted to addressing this issue in some detail. Reviewing the rules and procedures established by USAC, these releases emphasized: "In other words, there can be no gadgets, gimmicks, devices or parts which in any way differ from the model on the showroom floor."[8]

With an enthusiastic send-off by USAC's J.C. Agajanian, Marshall Martin leaves the Los Angeles starting line in his Ford Fairlane 500 6 bound for San Francisco. Another large crowd of spectators was on hand for the 11:00 p.m. start. Marshall, the Sweepstakes winner in the 1941 Gilmore Grand Canyon Run, would place second in the low-price class.

While most of the competing cars were selected in the Southern California area, the Rambler Rebel "8" entry was taken from the factory in Kenosha, Wisconsin.[9] A press release covered the various steps taken in the selection of one of the Oldsmobile Starfire "98" entries:

> A short while after the entry blank is received, A.C. Pillsbury, regional director of the USAC Sports Commission, telephones dealers in Southern California until he finds a car of the model specified by the entrant. Within an hour after locating the car, two of Pillsbury's aides walk into the dealer's showroom. It might be in Whittier, San Diego, or

> Cucamonga. But it is never at the dealer who made the entry. The USAC officials show him their papers, pick out the car they want and drive off to the official impound in Pasadena, California. There the car is inspected to be sure it is strictly stock. After going over it with a fine tooth comb, USAC staff technicians seal 64 major parts. Next Pillsbury calls the entrant and for the first time informs him that his official entry is on the line.[10]

Bill Losher guides his Dodge Coronet 500 through picturesque Skyline Drive in the Big Basin San Mateo County Park heading for San Francisco. Bill placed first in the low-medium-price class, the only time he would take top honors.

During this time the impound continued to be a hub of activity, as had become common. Under the supervision of USAC officials, servicing and break-in runs, limited as usual to 2000 miles, were the order of the day.[11]

Taking a different approach to reporting on this year's Run, *Motor Trend* took readers inside the impound and covered some of the details on preparation and break-in of each entry, noting: "When it comes down to automotive interest, the most important thing in any Economy Run seems to be the car."[12] One of the observations made in this article included selection of the Chrysler Saratoga, driven by George Alsbury, and the Oldsmobile Starfire "98," driven by Marilyn Miller. When inspected by USAC, it was discovered that both cars that had been selected had the wrong axle ratios which, under the rules, had to be changed to conform to regular production standards. Duplicate models were then picked at random by USAC and the rear ends were swapped.[13] The participation of the women drivers was highlighted during this period with various publicity sessions which introduced the concept of special uniforms used by the Chevrolet, Chrysler Corporation and Oldsmobile teams.[14]

Final impounding occurred on April 5, with announcement of the final route and maximum net running time. After several months of planning, personnel from USAC charted several possible routes, not only to avoid duplication from previous years, but taking into consideration that adequate hotel accommodations would be available.[15] This 1568.3-mile route selected, while not as mountainous as last year's, would challenge the drivers with 225 miles of restricted speed zones, 274 traffic signals, along with 29 stop signs. With the Route Description (a booklet covering 36 pages of turn-by-turn instructions) in hand, the crews only had time for a single practice run to familiarize themselves with the course, making notes concerning terrain and road conditions. Most made their practice runs during the same hours that would be encountered during the actual run.

On April 14, the scene from prior years was repeated as 24 brightly lettered and decorated entries were lined up on Flower Street ready to head north towards San Francisco for a 453.7-mile trip. Not including stops for refueling or brunch, this leg was to be covered within an 11-hour, 15-minute time limit. In front of a large group of spectators, with searchlights lighting up the night sky, Mel Alsbury, Jr., in his Imperial Crown took the green flag at 11:00 p.m. from starter J.C. Agajanian, with the others following at two-minute intervals. Lori Nelson, of movie and television fame, served as honorary starter and stepped in to send the first woman driver, Mildred

Alsbury, in another Imperial Crown, on her way.[16] Traveling through the suburban communities of Los Angeles' San Fernando Valley, they followed California's Pacific Coast Highway (U.S. Route 101) towards Paso Robles with pleasant weather forecast for the entire day. Bob Russo, *Speed Age*'s racing editor, gave a detailed example of the conditions drivers faced as they left Los Angeles:

> Getting out of Los Angeles onto the Hollywood Freeway was the first problem. There were six [traffic] signals, and we made all but the last, at the bottom of a steep upgrade leading to the Freeway. Missing it unnerved us a bit, but once on the Freeway, we concentrated on the task ahead. Heading out on Route 101 toward Ventura, we passed through another congested zone of thirty traffic signals, with restricted speed zones ranging from 25 to 35 mph. While we reduced our speed in accordance with the posted signs, cars all around us zipped by at a fast clip. It was practically impossible to find another car that stayed within the speed limits with us. They accelerated quickly from the signals and raced from one to another, only to jam on their brakes and wait until the lights changed. Meanwhile, maintaining a steady, even pace, we eased through them all with little effort. Once out of the congested area, we had an open highway, except for a few small towns until our fuel and brunch stop in Paso Robles.[17]

Arriving in Paso Robles just before sunrise, the drivers made their first refueling stop of the run at Chuck Frazier Mobilgas Service, immediately followed by a 30-minute brunch stop at the Paso Robles Inn.

Continuing north through the Salinas Valley, they then followed California's beautiful Skyline Drive into San Francisco. When the cars began to arrive at the Presidio of San Francisco's Crissy Field, everyone was greeted by music from a Chinese children's orchestra. In an effort to speed up the topping-off process each day, additional refueling pads were constructed and installed to enable handling four entries at once. The cars were then impounded overnight.

When the first day's results were posted, Mel Alsbury,

It was a busy scene at Walt Orchard Mobil Service in Reno as several drivers arrived at the same time. While USAC officials finished refueling four competing cars, Betty Skelton's Chevrolet Bel-Air 8, Don Bridges's Buick Roadmaster and Pierce Venable's Pontiac Chieftain waited their turn.

Jr., and his Imperial Crown picked up where he had left off last year, leading all entries with 61.4078 ton miles per gallon. Mel's actual miles-per-gallon performance was 19.9376. Second for the Sweepstakes Award was Santa Monica DeSoto dealer Hart Fullerton, recording 53.9067 ton miles per gallon in his Firedome Sportsman Hardtop. In what was to become a sweep of all classes by Chrysler Corporation, individual class leaders were:

Low Price	Plymouth Belvedere 8 50.6500 Ton Miles Per Gallon
Low Medium Price	Dodge Coronet 500 52.2776 Ton Miles Per Gallon
Upper Medium Price	DeSoto Firedome 53.9067 Ton Miles Per Gallon
High Price	Imperial Crown 61.4078 Ton Miles Per Gallon

Mary Davis became the first woman to place first, topping all entries in the low-price class in her Plymouth Belvedere 8, which also recorded an actual miles per gallon figure of 20.5909. Later Mary recounted, "Skyline Drive is a narrow winding road and we were late so I was cornering like crazy. Our timing of thirty signals thru San Francisco brought us in ten minutes earlier than scheduled and both Ginny [relief driver] and I were practically in tears because we figured we wasted too much gas being ten minutes early."[18]

As Ray Brock, driving an Oldsmobile Starfire "98," later observed, "It was obvious after the first day's finish in San Francisco that although the '57 Olds was better than the '56 model, chances of competing with the Chrysler products were very poor. The Torque-Flight transmission used on all '57 Chrysler products completely released on downgrades and permitted them to coast freely while the Hydra-Matic equipped Oldsmobiles had to maintain partial throttle on downgrades to keep compression from slowing the car down."[19]

Leaving San Francisco on April 16, beginning at 12:01 a.m., the drivers faced a 419.3-mile, 10-hour and 25-minute trip to Winnemucca, Nevada. Although there were concerns of snow in the Sierra Nevada Mountains, balmy spring weather and dry road conditions prevailed in this section, to the relief of everyone. Climbing to the highest elevation on the Run at Echo Summit, they made their first stop in Carson City for brunch, arriving just before 6:00 a.m. at the Nugget Café. An hour later in Reno was the day's refueling stop at Walt Orchard Mobilgas Service, where crowded conditions gave the drivers a few extra minutes to unwind.[20] With the weather turning cold, the drivers encountered strong head winds along with some road construction through the Truckee River Valley as they headed to the impound at Winnemucca's Municipal Ball Park. Compared to the events of the first two days in the 1956 Run, this year's Run appeared to be relatively a breeze. Interest by members of the press centered on the women drivers; it was reported, "The girls, for the most part, seemed to be standing the strain of long hours and tension with remarkable ease."[21]

When the results were released for the second day's drive, it appeared that not only the Imperial Crown was on target for a second consecutive Sweepstakes victory, but Chrysler Corporation was set to make history by capturing all classes. With a ton-miles-per-gallon figure of 61.8964, Mel Alsbury, Jr., continued his comfortable lead over all entries. Hart Fullerton, driving a DeSoto Firedome, continued second in the Sweepstakes with 54.9803 ton miles per gallon. At this point it looked like Plymouth and DeSoto were on their way to their first wins since 1952. Class leaders continued to be:

Low Price	Plymouth Belvedere 8 51.2501 Ton Miles Per Gallon
Low Medium Price	Dodge Coronet 500 53.4804 Ton Miles Per Gallon
Upper Medium Price	DeSoto Firedome 54.9803 Ton Miles Per Gallon
High Price	Imperial Crown 61.8964 Ton Miles Per Gallon

There were no other changes with regards to the class leaders, although two contests were becoming close. In the low-price class, Mary Davis in her Plymouth Belvedere 8 held a .4070 lead in ton miles per gallon over Marshall Martin driving a Ford Fairlane 500 6, while in the upper-medium-price class, Hart Fullerton in a DeSoto Firedome had a margin of .6186 over George Alsbury in a Chrysler Saratoga. These slim leads would tighten in the coming two days.

On April 17, another midnight start faced the drivers as they headed for Ogden on a 396.2-mile, 9-hour and 35-minute trek. Following U.S. Route 40 across Nevada, they headed for their first stop in Wendover for brunch at the State Line Café. The stop at the State Line Café could not be missed, as it featured a 65-foot-high cowboy sign proclaiming, "This is the Place." Arriving here, the drivers were ready for a break after encountering slow-moving cattle trucks over the 6967-foot Pequop Summit. Immediately after brunch was a stop for refueling at Elden Williams Mobilgas Service before crossing into Utah, passing the Bonneville Salt Flats, where many recognized international land speed records were set, before continuing on to Salt Lake City. From there they headed north to Ogden and the impound at the Municipal Stadium. The Miss Ogden Contest was in progress during this time and the 15 candidates competing were official greeters. The drivers began to arrive around 11:15 a.m. after a fairly uneventful trip, although a few obstacles delayed the drivers. Lorraine Bell, driving an Oldsmobile Golden Rocket "88," was forced to wait while a freight train blocked the road.

While nearing the impound, several other competitors were delayed by a traffic accident blocking part of the roadway.[22] Pierce Venable, driving a Pontiac Chieftain Catalina, arrived at the impound 1 minute and 57 seconds late and was penalized, although the exact nature of the penalty is unknown.[23] It was here that an entry was withdrawn due to a mechanical problem for first time since the 1937 Gilmore Yosemite Run. Link Paola, driving an Oldsmobile Golden Rocket "88," reported that a vibration developed early on this leg, and during the day this vibration increased in severity. The cause was undetermined, but the Oldsmobile was withdrawn as a safety precaution.[24]

When the results were released, a Chrysler Corporation sweep was almost assured with three drivers continuing to lead their respective classes. Mel Alsbury, Jr., and his Imperial Crown maintained a comfortable lead over all entries with 63.9099 ton miles per gallon. Improving on their ton-miles-per-gallon performance, the class leaders were:

Low Price	Plymouth Belvedere 8 52.2749 Ton Miles Per Gallon
Low Medium Price	Dodge Coronet 500 55.3464 Ton Miles Per Gallon
Upper Medium Price	Chrysler Saratoga 56.0679 Ton Miles Per Gallon
High Price	Imperial Crown 63.9099 Ton Miles Per Gallon

Celebrating his 20th birthday, Mel's brother George took over first place in the upper-medium-price class in his Chrysler Saratoga, edging Hart Fullerton in a DeSoto Firedome by a very slim margin of .0621 ton miles per gallon. Mary Davis continued to hold a slim lead in the low-price class over Marshall Martin.

Ready for the final leg, the 23 remaining drivers faced a relatively easy 299.1-mile, 7-hour and 15-minute trip to the finish line. With the first car departing at 2:30 a.m. on April 18, the drivers headed north to Pocatello. Just before 5:00 a.m., their first stop was for refueling at John M. Jackson Mobilgas Station, where there was a minimum requirement of 3 gallons. Despite the early hour, one of Pocatello's radio stations provided live coverage, interviewing several drivers while refueling was completed. After refueling, they made a stop at the Green Triangle Club for brunch, then the drivers headed towards the Craters of the Moon National Park and on to Sun Valley, where traces of snow remained on the mountainsides. A good-sized group of spectators, some of whom traveled on the Union Pacific's "Mobilgas Economy Run Special," were on hand at the Challenger Inn as the drivers began crossing the finish line at 10:30 a.m., led by Patricia Jones driving a Dodge Coronet 500.[25] After being topped off for the final time, the cars were kept under USAC supervision while the final results were computed. In the mid–afternoon, with the competing cars

Driving a Plymouth Belvedere, Mary Davis becomes the first woman to capture a first-place trophy in the Economy Runs. Mary managed to hold off last year's low-price class winner, Marshall Martin in a Ford Fairlane 500 6, by a slim .1030 ton mile per gallon.

Part of Chrysler Corporation's domination of the 1957 Economy Run was the performance of George Alsbury and his Chrysler Saratoga. George, who turned twenty during the Run, placed first in the upper-medium-price class, becoming the youngest driver to accomplish this feat.

parked on the lawn adjacent to the Inn, the Awards Ceremony was held, followed by a banquet that evening.

Two names dominated this year's Economy Run — Chrysler and Alsbury — with Chrysler Corporation automobiles dominating the field. "With a victory blast of the horn on his big, white Imperial, Mel Alsbury, Jr. swept across the finish line" to capture his second consecutive Sweepstakes Award with 64.5153 ton miles per gallon.[26] In addition, Mel's actual miles-per-gallon figure of 20.9465 was almost identical to last year's mark. Second in the Sweepstakes was Mel's brother George in a Chrysler Saratoga with 56.7267 ton miles per gallon. George, competing in his first Economy Run, also took the first-place trophy in the upper-medium-price class, becoming the youngest driver to achieve this feat. Although he was penalized for being 1 minute and 22 seconds late, this penalty did not affect the outcome. Mary Davis had the distinction of being the first woman to capture a first-place trophy when she topped all entries in the low-price class in her Plymouth Belvedere 8. *Automobile Topics* noted, "Never before has one manufacturer swept the field [and] never before has one family garnered so many gleaming trophies, symbolic of automobile supremacy, this famed event."[27] The best performance in actual miles per gallon for all entries was set by Marshall Martin in a Ford Fairlane 500 6 with 22.2534, while the best performance by an eight-cylinder model was 22.0047 by Bill Losher in a Dodge Coronet 500.

Remarking on Chrysler Corporation's success, Bob Russo wrote, "Undoubtedly, much of the success of the Chrysler products may be attributed to torsion-bar suspension and exceptionally good handling. True, the transmissions and engines had a lot to do with the economy figures turned in, but handling helped a great deal, especially in the many hard corners. We were able to take the downhill corners easier and faster. Where other drivers had to brake, we had a decided advantage that definitely figures in the final mpg payoff."[28] Added Ray Brock, "The final results bore out the Chrysler superiority in the transmission department and also proved that women are here to stay in economy driving."[29]

In addition to the Sweepstakes Award, first-place trophies were presented to:

CLASS	MAKE and MODEL	TON MILES PER GALLON	MILES PER GALLON
A — Low Price	Plymouth Belvedere 8	52.6211	21.3907
B — Low Medium Price	Dodge Coronet 500	55.8920	22.0047
C — Upper Medium Price	Chrysler Saratoga	56.7267	20.7032
D — High Price	Imperial Crown	64.5153	20.9465

Detailed results for all competing cars can be found in the Appendix.

The USAC reported that all entries averaged 20.4865 miles per gallon, which was an improvement over 1956's average. In the "battle of the sexes," the men averaged 20.6461 miles per gallon, only fractionally higher than the women, who averaged 20.2382. The overall performance of women drivers was commendable. In addition to Mary Davis's class win, Lorraine Bell, driving an Oldsmobile Golden Rocket "88," placed second in the low-medium-price class, while Marilyn Miller, in an Oldsmobile Starfire "98," placed third in the upper-medium-price class. Mildred Alsbury, running an unofficial second in the Sweepstakes Competition, remarked, "I hope this helps to prove that women are just as good drivers as men."[30] Other than the withdrawal of Link Paola's Oldsmobile Golden Rocket "88," no other problems were reported. Weather and road conditions were described as being very good.

A battle that began in 1956 for top honors in the popular low-price class between Ford and Plymouth continued this year with the results reversed. Close for all four days, Mary Davis in her Plymouth Belvedere 8 managed to hold off Marshall Martin in a Ford Fairlane 500 6 for the first-place trophy by a margin of .1030 ton mile per gallon. *Automotive News* reported, "If Martin had not run into headwinds up around the Crater of the Winds, he might have caught her."[31] Mary's class-winning performance was described as "sweet indeed," since she was competing not only against three previous Sweepstakes champions (Dick Griffith, Marshall Martin and Les Viland), but against Chevrolet's Betty Skelton as well.[32]

Those associated with American Motors were no doubt disappointed at the fifth-place finish of its new Rambler Rebel 8 piloted by Les Viland. Although the Rebel's engine was the largest (327 cubic inches) competing in the low-price class, it managed to outperform all 8-cylinder models' actual miles per gallon with a figure of 21.6214.

The third time was a charm for Bill Losher and his Dodge Coronet 500. Bill, who had to endure the disappointment of running out of gas in 1955, along with last year's close call arriving in Salt Lake just under the wire, had clear sailing for all four days, winning the low-medium-price class by a comfortable margin. He also had the distinction of achieving the best actual miles-per-gallon figure for all 8-cylinder cars competing. This year would mark Bill's last appearance competing as a driver.

Another close contest occurred in the upper-medium-price class between George Alsbury, piloting a Chrysler Saratoga, and Santa Monica DeSoto dealer Hart Fullerton in a DeSoto Firedome. Hart, a regular competitor since 1950, had a fairly comfortable lead for the first two days and was looking for his first class win. As Jack Weed, writing in *Automotive News*, reported: "On the almost-level run [from Winnemucca] to Ogden he lost out to Alsbury who came into Ogden .0621 ton-miles ahead. Many predicted that Fullerton would gain this back in the hills of the last lap, but he finished .2803 ton-miles behind the Saratoga. One of his observers said Fullerton experienced trouble in the right bank of his engine during this lap, and the records indicates this might have happened as the Firedome outeconomized the Saratoga in the hills prior to this lap."[33]

Any review and analysis of the results of the 1957 Economy Run cannot be complete without a discussion of the impact on the rules for the women competing. As outlined earlier, under a somewhat complicated rule, whichever of the two identical cars attained the better ton-mileage, this figure would be used in the final results.[34] This had a direct impact on three women competing. As various newspapers reported on the results daily, Mildred Alsbury was shown in second place for the Sweepstakes Award and tops among the women participating. This achievement was further noted in a Chrysler Division press release issued after the final results were announced.[35] Since she was competing against her son in an identical Imperial, this was an unofficial score under the rules, and this fact was not explained in the various articles or press releases. When the final results were released, Mildred was shown unofficially in third place behind Don Bridges in a Buick Roadmaster, who scored a poor 52.3466 ton miles per gallon. This rule relegated Ina Mae Overman, driving a Ford Fairlane 500 6, to an unofficial seventh-place finish in the low-price class, although she achieved the third best ton-miles-per-gallon figure. Patricia Jones in a Dodge Coronet 500 was also relegated to an unofficial fifth-place finish in the low-medium-price class, although she achieved the second best figure. As Jack Weed pointed out in his article "Pro-Male Rule 'Robs' 3 Women in Economy Run" in *Automotive News*, "Six of the nine teams of women entered in this year's run actually placed first, second, or third in their division." He further speculated, "While Chief Steward Art Pillsbury, of the U.S. Auto Club, and the Mobilgas people wanted to boost in-

With its second Sweepstakes Award, Chrysler continued to advertise its success in newspapers and industry-related publications. While also noting the class win for the Chrysler Saratoga, credit for this year's success for both models was given to the Alsbury family (courtesy of Chrysler Group LLC).

terest in the run by the addition of women drivers this year, it is unlikely that they ever dreamed the women would score as well as they did."[36] A review of the events and an analysis of this year's Run would result in rule changes for women participants in 1958.

With Chrysler Corporation's complete domination of this year's Run, advertising in turn was limited and was once again found primarily in newspapers and in industry publications such as *Automotive News*. Socony-Mobil's corporate print advertising followed the same theme and format used in 1956 with only an update of the results and routing. Without reference to the addition of women drivers and the scoring system that was used, the ad stressed that "the Mobilgas Economy Run is an impartial test of cars competing against others in the same price class."[37]

As expected, Chrysler Corporation advertised its clean sweep of all classes, declaring that "Forward Look Engineering and Design Write Performance History," while also giving credit to DeSoto's second-place finish in the upper-medium-price class. Other advertisements, appearing in general-interest publications such as *Life* and *The Saturday Evening Post*, made only a brief mention of Chrysler Corporation's success, noting that "for the first time ever, *one* manufacturer took top honors — in every class — in this year's Mobilgas Economy Run."[38] Featuring a picture of the Sweepstakes-winning Imperial Crown, a separate ad appearing in *Automotive News* proclaimed "Imperial Wins," and also made note that Mildred Alsbury "topped all women drivers with 61.75 ton-miles per gallon also in a 1957 Imperial Crown."[39]

Plymouth's performance was limited to newspaper advertising. Designed specifically for the Southern California market, one featured a Belvedere convertible participating in

Two names dominated the 1957 Mobilgas Economy Run — Chrysler Corporation and the Alsbury family. In addition to their second Sweepstakes Award, the Alsburys placed first in two of the four classes. Standing behind the Sweepstakes trophy are George Alsbury, Mel Alsbury, Sr., Rae-Nell (Mrs. Mel, Jr.) Alsbury, Mildred (Mrs. Mel, Sr.) Alsbury, and Mel Alsbury, Jr. The winning Imperial Crown is in the background.

a ticker-tape parade. Under the headline "Southern California Loves The Mobilgas Economy Run Winner," this ad further noted, "Mary Davis, the first woman to win this event, swept to victory in Class A over all cars in the low price field."[40] Dodge advertised that not only did its Coronet 500 place first in its class, but it also achieved the best actual miles per gallon of all V-8s competing. Dodge owners were told of this achievement in an article, "Here's how we won the Economy Run!" which appeared in *Dodge News*. Included in this article were gas-saving tips from the Dodge team members.[41]

In one of this year's largest print advertising programs, appearing in both general interest and enthusiast publications, Auto-Lite pointed out that their "Power Tip" spark plug "fires up winners in 1957 Mobilgas Economy Run." Featuring photographs of the class-winning models, the ad told readers that this "Power Tip design gives better initiation of combustion for top performance and economy at all speeds."[42]

Although placing a close second in the low-price class, Ford advertised its performance in achieving the best miles per gallon of all entries. Appearing in newspapers, featuring a photograph of one of the Fairlane 500 models competing, it announced, "Ford Beats All Mobilgas Entries in Actual Miles Per Gallon," pointing out that "Ford Costs less to own ... less to drive."[43]

In the 1958 showroom brochures issued by Chrysler, Ford, Imperial and Plymouth, only brief mention was made of their Economy Run performance.

Despite its disappointing performance, American Motors still pointed out the performance of its new eight-cylinder engine powering the Rambler Rebel. In newspapers, "Rambler Wins!" was the headline telling readers that the Rebel V-8 topped all low-price V-8s in actual miles per gallon with a "phenomenal record of 21.62 miles per gallon."[44] This achievement was also noted in a footnote to an advertisement appearing in publications such as *Life* and *The Saturday Evening Post*. Appearing in *Automotive News* and aimed primarily at potential dealer operators, "Rambler Does It Again!" was the headline that announced not only its recent achievement but told of Rambler's record in previous Economy Runs dating back to 1951.[45] In a separate newspaper advertisement publicizing their private, NASCAR-supervised Border-to-Border Run (using an overdrive-equipped Rambler Custom "6" hardtop), American Motors made pointed criticism of USAC's eligibility rules regarding engine size: "The Rambler 6, holder of the all-time mileage record in this event, was barred by a change of rules."[46] Press releases announcing the Rebel's performance also pointed out this change of rules.

Newspaper coverage was prominent, especially in the western states along the route of the Run. Readers were able to follow the progress of Chrysler Corporation and the Alsbury family's success each day with emphasis in the early stages on the mother-and-son competition for the Sweepstakes Award. *Time* magazine emphasized Chrysler's sweeping the field along with the participation of women drivers for the first time, making special note of Mary Davis's taking home a first-place trophy.[47] This same theme was found in enthusiast and industry-related magazines in their articles covering the Run. In *Flying Red Horse*, a publication for Socony-Mobil employees and shareholders, the women's accomplishments was the theme of their article on the Run, with special emphasis on the participation of Mildred and Rae-Nell Alsbury.[48] Among enthusiast magazines, *Speed Age*'s Bob Russo again provided some of the most detailed insight into the preparation and competition faced as relief driver for the class-winning Dodge Coronet 500.[49]

In its article covering the 1957 Economy Run, *Road & Track* observed:

> Those who have watched the 21-year-old Mobilgas Economy Run from its start have seen plenty of changes. From a purely local Western affair that took a whole day in 1936 and went all the way from Los Angeles to Yosemite National Park, the Run has grown in every way. A nationwide survey has showed that the public now looks on it as the top test of stock models. A justified complaint of years past — that almost every make somehow managed to come away with a price — is now outdated. Only cars with automatic transmissions may not enter, a rule which we certainly object to on the grounds of its ruling out a large portion of the skill involved in top economy driving; at the same time it gives a more nearly accurate view of what U.S. cars are likely to do under expert hands.[50]

For those who had witnessed these changes over the years, they would see that more changes were in store.

1958

Los Angeles to Galveston Via Phoenix — El Paso — Lubbock — Dallas

April 13–17, 1958—1883.2 Miles

When Robert L. Minckler, General Petroleum's president, announced that Galveston, Texas, would be the terminus of the 1958 Economy Run, it would once again become the longest to date. With this announcement, a surprise was in store for the competitors, as USAC officials would be withholding the details of the route, including overnight stops, until April 11, less than 36 hours before the first car would be sent on its way from Los Angeles.[51] This rule eliminated practice runs over the route which "had been rather extensively indulged in during the past few years."[52] This also eliminated the final impound period, although specific dates were specified for servicing and test runs.

Following the success of the women's competition in 1957, although not without some controversy, USAC officials worked during the ensuing months to change the rules in an attempt to level the playing field. Retaining the basic eligibility criteria that "two cars of the same make and model may be entered [provided that] one car is operated by a male driver and relief driver team, and the other is operated by a team of women," the women would now compete for awards on their own merits and would not be required to defeat their male counterparts. Otherwise, only one model could be entered with no restriction on having a male or female driving team.[53] This change had a positive reaction from all of the competitors and press.

Eligibility was further tightened with regards to engine size with the requirement that entries be powered by an engine of 220-cubic-inch displacement or larger. An increase from the minimum 200-cubic-inch displacement in 1957, this rule was designed to represent the majority of cars currently being purchased by the driving public. This prevented the participation of six-cylinder models manufactured by American Motors and Studebaker. This had no effect on Studebaker's participation, which continued to reflect its corporate decision not to provide factory support. American Motors, on the other hand, voiced its displeasure in both press releases and a two-page advertisement titled "Why the Rambler 6, America's No. 1 Economy Car, Is Not in the 1958 Mobilgas Economy Run!," appearing in *Automotive News*.[54]

In a statement issued by American Motors:

> The Rambler Six will not participate in the Mobilgas Economy Run because of a rule change made by the sponsoring organization, according to Roy D. Chapin, Jr., Automotive division executive vice-president, American Motors Corp. Official rules now bar cars with engines of less than 220 cubic-inch piston displacement. The Rambler Six has a dis-

From 1952 to 1966, the Pasadena impound was in a large Quonset hut, which provided ample storage of the competing cars along with space for lubrication and tire alignment equipment. This building was also used to construct floats for the annual New Year's Tournament of Roses Parade.

placement of 195.6 cubic inches, which, Chapin said, "results in maximum efficiency and economy of operation. We at American Motors feel that the change in rules defeats the purpose of such an event, which is to show economy of operation."[55]

This would not be the last battle American Motors would have with USAC officials over the rules governing the Run.

Additional requirements were made regarding optional equipment required on competing cars, with all eight-cylinder models required to be equipped with power steering and power brakes. While four-door sedan, two-door or four-door hardtop models continued to be eligible, only four-door sedan models could be entered when equipped with six-cylinder engines. Competitive classifications remained as they were in 1957, covering the low, low-medium, upper-medium and high-price classes.

With a total of 29 cars competing, the largest field since 1952, over half of the entries were sponsored by individual dealers, with the remainder by dealer associations. The Studebaker President returned as the lone private entry by Leonard DeBell's Southern California Plating Company. Edsel made its Economy Run debut, joining both Lincoln and Mercury models, which were back with Bill Stroppe, returning after a three-year absence to lead the team's training and preparation responsibilities. In addition to American Motors' Rambler, Cadillac and Packard chose not to participate. Two dealers participating for the first time were from outside of California — Al Parker Buick in Houston, Texas, and Holiday Chevrolet in Chandler, Arizona. Gus Retzlaff, owner of Holiday Chevrolet, would drive a Chevrolet Delray 6, while his wife, Dorothy, was behind the wheel of a Chevrolet Impala 8 in what was reported to be a result of a "flip of a coin" to decide who would drive each entry.[56] Gus's son Jim, a Pontiac dealer in Barstow, California, also participated as an entrant and relief driver in a Pontiac Super Chief. Over the next eleven years the Retzlaff name would become well known in Economy Run competition. *Hot Rod Magazine*'s Ray Brock and Wally Parks returned for a second year in an Oldsmobile "98" Hardtop. Mel Alsbury, Jr., driving an Imperial Crown, was back to enhance his two consecutive Sweepstakes-winning performances. Previous Sweepstakes-winning drivers Marshall Martin and Bill Stroppe also returned.

Seven entrants entered cars with women as drivers and relief drivers piloting entries from Chevrolet, DeSoto, Dodge, Edsel, Lincoln and Plymouth. The women teams in the Chevrolet Impala and DeSoto Firedome entries had the distinction of not competing directly against their male counterparts. Mary Davis and Myra Buchanan were back for a second year, joined by two other well-known names from Southern California's sports car circles, Ruth Doushkess and Joan Fischal.

Random car selection began in March, continuing to follow established procedures from prior years. All entries were selected from various locations in California and were immediately inspected for stock status, with USAC staff continuing to place seals on 64 major parts including the hood and trunk lid. There was an unusual twist this year regarding the selection of the Chrysler, Imperial and Lincoln entries, as later explained in an article in *Motor Trend*:

> Several Chrysler Corp. and Lincoln cars had been secretly tagged as satisfactory at dealer establishments in Berkeley, Calif., by a United States Auto Club representative. Typical of the intense rivalry between contestants, a private deal was made between the Alsburys and Bill Stroppe, representing the Lincoln-Mercury Dealers Association, to approve each other's final selection of the cars.
>
> Stroppe spent three days carefully examining each of the cars which the Alsburys were authorized to select. He wanted to make sure that the cars were strictly stock and that there was no break-in mileage concealed by a disconnected speedometer. Mel was much more casual about the affair. He had complete confidence in the Chrysler and Imperial winning potential. In 15 minutes Stroppe had Mel Jr.'s stamp of approval on two Lincoln Continentals.[57]

The rules this year contained a further reduction in the maximum number of man-hours allowed for servicing of competing cars to 12 (down from 14 in 1957), while the total mileage on the competing cars remained a maximum 2000 miles.[58] Break-in and servicing of the competing cars, under the continuous supervision of USAC officials, were the order of the day. When asked how his Imperial Crown was prepared during the impound period, Mel Alsbury, Jr., explained: "Actually little more than new car service. We installed one-size-leaner carburetor jets and advanced the spark 12 degrees as allowed by the rules. Beyond that we made sure the brakes did not drag and the cars were tuned and adjusted as we would do for our customers. We had no big staff of mechanics; the work was done by John Justice, who works in our service department."[59]

With the earlier announcement of Galveston as the destination of the Run, advance publicity and promotional efforts commenced, with press releases being the main source of information. Arrangements were also made for radio and television coverage along the route and included daily reports back to Los Angeles television viewers by local CBS affiliate KNXT. Under the guidance of General Petroleum's Frank Meunier and his staff, additional promotional efforts were undertaken by Magnolia Petroleum Company, which was Socony-Mobil's affiliated company in New Mexico and Texas. With the new timing on the announcement of the actual route and elimination of practice runs, the drivers and relief drivers were more readily available for various promotional and publicity activities. One of these activities was the drawing of

Art Rene pulls out of the Safford, Arizona, refueling station heading towards El Paso on the second day of the Economy Run. In its inaugural appearance, Art's Edsel Citation was plagued by mechanical problems which contributed to his placing well down in the standings.

starting positions, which had previously been handled without fanfare by USAC officials. In front of two rows of competing cars, each driver walked up to a table and selected a rolled pennant marked with his or her starting position. George Alsbury, driving a Chrysler New Yorker, selected No. 1. As was the case in 1957, special uniforms were designed for the teams from Chrysler Corporation and Ford Motor Company. Originally from Texas, the DeSoto Firedome team of Myra Buchanan and Betty Shutes were outfitted in typical western costumes complete with cowboy hats.[60]

When the route was announced, the competitors discovered they would cover a distance of 1883.2 miles through the American Southwest over a five-day period. Overnight stops would be in Phoenix, El Paso, Lubbock and Dallas before arriving at Galveston's Galvez Hotel. With an altitude range from 22 feet below sea level at Indio, California, to 7540 feet at Apache Summit, New Mexico, they would encounter 350.2 miles of restricted speed zones through 141 towns and cities along the route. Another obstacle the drivers would have to contend with were 316 traffic signals. In order to stay within the 42-hour time limit set by the USAC, drivers would average 45.2224 miles per hour. This average would be the highest to date, surpassing the previous mark of 44.7115 miles per hour set in 1953.

Instead of basic turn-by-turn instructions used in previous years, new route books (spiral bound, measuring 14" × 8½") were provided. Surprisingly detailed, these books eliminated the need for drivers and relief drivers to develop their own time parameters for completing each leg. As described by Bill Doushkess, a newcomer to the Economy Run and husband of DeSoto Fireflite driver Ruth Doushkess:

> General Petroleum gave each crew three of the most completely detailed, expensively finished route books I have ever seen or of which I have heard. Seventy pages in length, the book gave every conceivable bit of information to the navigator — highway numbers, odometer readings, speed limits, suggested average speeds, elevations, percentages of grades — even the traffic signal timing. The back of practically every page contained a map outlining the route of a particular point, either a difficult intersection, or perhaps the correct way to enter a brunch stop. In addition each leg of the route had a

Chandler, Arizona, Chevrolet dealer Gus Retzlaff receives a warm welcome as he crosses the finish line in Galveston, Texas, in his Chevrolet Delray 6. Placing third in the low-price class, Gus also achieved the second highest miles-per-gallon figure for all entries with 20.6441.

Rand McNally map showing the correct roads and the location of each brunch and refueling stop.[61]

After the route was announced, Magnolia Petroleum kicked off a newspaper advertising campaign in New Mexico and Texas. Using the same format personalized for each city along the route, the ads announced the approximate arrival time, noting that "for the first time in its history, you can watch one of the nation's major performance tests of 1958 cars as the Mobilgas Economy Run arrived in your area."[62] As the competitors would see over a four-day period, this campaign was successful. "As the cars rolled through cities and towns they were greeted with waiting throngs everywhere. No matter whether a stop was scheduled or not, people lined the curbs."[63]

Houston Buick Dealer Al Parker was pleased to see the Run passing through that city on the final leg. After entering both a Buick Special (driven by Assistant General Manager Joe Hill) and a Buick Century (driven by General Sales Manager R.M. Flanagan), he was quoted as saying he "hopes his two entrants will pass through his home town on the last leg to Galveston."[64]

For the final time, the General Petroleum garage in downtown Los Angeles was the scene for the start of the Run. In the early morning hours on April 13, a large crowd of spectators including television and camera crews were on hand as the 29 lettered and decorated entries were angle-parked against the curb, ready for the 425-mile trek to Phoenix. Driving time was set at 9 hours, 25 minutes, which as usual did not include stops for refueling or brunch. With searchlights lighting up the night sky, George Alsbury in his Chrysler New Yorker was the first to take the green flag at 2:00 a.m., with the rest of the field following at two-minute intervals. Following a new route out of Los Angeles, the drivers headed southeast to Riverside and Hemet. In observing their driving styles, Bill Doushkess observed:

> After only twenty minutes on the road it was already evident that the drivers of the Run were a different breed. No weaving in and out of freeway traffic indiscriminately; instead a steady pace keeping out of traffic. This was to be the pattern for four and a half days. Windows up, electric accessories off, no smoking, two Cal Tech observers per car to insure [*sic*] that each and every traffic law and rule of the Run was obeyed to the letter; truly a great test of driver skill. Not so much a test of physical endurance (although certainly a goodly amount is helpful), as one of mental alertness.
>
> Coming through the first series of hills past Riverside on our way to Hemet we began to see cars bunching together. Each driver sets his own pace and if the car ahead on a hill is slower than those following, particularly in a no passing zone, then a procession develops. It was interesting to watch Bill Stroppe negotiate turns in his Mercury Turnpike Cruiser. Each bend was straightened out as much as possible; high on the outside, then dipping toward the apex, then out wide again. Anything, in order not to vary engine speed.[65]

After crossing the San Jacinto Mountains, they continued traveling through Indio and across the desert to Blythe for refueling at Saunders & Peterson Mobilgas Service. Quickly arriving in Ehrenberg for brunch at the A-1 Restaurant, the drivers then continued on their way towards Phoenix, and with the cooperation of state officials, they were not required to stop at the Arizona Inspection Station. Shortly after leaving Ehrenburg, Art Rene's Edsel Citation developed mechanical trouble, which at first was thought to be either carburetor- or ignition-related. It was discovered that the ignition switch was shorting out, and Art made five stops trying to correct this problem before arriving in Phoenix. Unable to fully correct the problem, he elected to continue. Crossing the desert, several drivers complained about encountering strong head winds, which hampered their mileage performance.[66] A few minutes before noon, Loretta Colange in an Edsel Pacer was the first to arrive at the Arizona State Fair Grounds, where a Mexican band was on hand to entertain the competitors. With ideal weather conditions and temperatures in the 80s, a small crowd of onlookers were on hand as the cars were topped off and then impounded overnight.

When the first day's results were announced, two names were prominent once again — Mel Alsbury, Jr., and Chrysler Corporation. For the third year running, Mel led all entries, posting a figure of 63.1156 ton miles per gallon, with an actual miles-per-gallon figure of 20.7123. Second for the Sweepstakes

The Alsbury brothers cross the finish line simultaneously with Mel, Jr. in the Imperial Crown (#6) and George in the Chrysler New Yorker (#1). In addition to Mel Jr.'s third consecutive Sweepstakes Award, George also achieved the highest miles per gallon for all entries with a figure of 21.0217.

Award was Mel's brother George with 60.0173 ton miles per gallon in his Chrysler New Yorker. The best performance in actual miles per gallon was also set by the Chrysler New Yorker with an impressive figure of 21.5696. Mary Davis also picked up where she left off last year, topping the field in the low-price class. Spoiling another sweep by Chrysler Corporation, an Oldsmobile Dynamic 88 led in the low-medium-price class. Faced with disappointing sales and questionable build quality, those associated with Edsel were pleased to see their Pacer in second place behind the Oldsmobile Dynamic 88. Individual class leaders were:

Low Price	Plymouth Belvedere 8 49.8710 Ton Miles Per Gallon
Low Medium Price	Oldsmobile Dynamic 88 50.9465 Ton Miles Per Gallon
Upper Medium Price	Chrysler New Yorker 60.0173 Ton Miles Per Gallon
High Price	Imperial Crown 63.1176 Ton Miles Per Gallon

Leaving Phoenix on April 14, Don Rice was the first driver on his way at 3:00 a.m. Pointing his Edsel Pacer towards El Paso, he was ready to cover a distance of 408.6 miles within a 9-hour time limit. In the darkness the drivers headed east through the Arizona desert to Safford for a refueling stop at D and D Mobilgas Service, followed by brunch at the Manor House Café. When she was ready to leave the brunch stop, Joan Fiscal's Dodge Custom Royal would not start. Mercury driver Bill Stroppe stepped in and found the trouble, and Joan was on her way.[67]

As the competitors entered Lordsburg, New Mexico, they were greeted by a large banner, "Welcome to Magnolialand," in reference to Magnolia Petroleum Company's operating area. The drivers then faced several areas of road construction which slowed down their progress. Upon arriving in Deming, the cars were halted momentarily for traffic signals and were greeted by young girls handing out cups of coffee and doughnuts.[68] From there it was on to El Paso and the impound at Sun Bowl Stadium on the grounds of Texas Western College. As the competitors arrived, they were greeted by the Texas College band and cheerleaders, along with other students in typical Indian dress. Here, as in other stops along the way, General Petroleum personnel announced the arrival of each entry over a public address system. When the announcement was made of the arrival by the two Buick entries from Al Parker Buick in Houston, the drivers received a big ovation from the spectators. Three of the drivers — Robert Flanagan in a Buick Century, Nicky Griffen in a Lincoln Continental, and Dorothy Retzlaff in a Chevrolet Impala 8 — arrived late and were penalized. After spending close to ten hours on the road, their day was not finished while they observed the topping-off process, and for a number of drivers, including Mel Alsbury, Jr., and Mary Davis, there were interviews with members of the press.[69]

With release of the results for the second day, it appeared that Mel Alsbury, Jr., and his Imperial Crown were well on their way to a third consecutive Sweepstakes Award. With a figure of 63.0354 ton miles per gallon, he maintained a good-sized lead over all entries. George Alsbury and his Chrysler New Yorker continued second in the Sweepstakes with 59.7998 ton miles per gallon while continuing to lead all entries in actual miles per gallon with 21.1320. Mary Davis continued to be the only woman placing first in her class, with two sitting in third in their respective classes — Myra Buchanan (DeSoto Firedome) in the low-medium-price, and Nicky Griffen (Lincoln Continental) in the high-price class. Class Leaders continued to be:

Low Price	Plymouth Belvedere 8 49.1923 Ton Miles Per Gallon
Low Medium Price	Oldsmobile Dynamic 88 50.9151 Ton Miles Per Gallon
Upper Medium Price	Chrysler New Yorker 59.7998 Ton Miles Per Gallon
High Price	Imperial Crown 63.0354 Ton Miles Per Gallon

Comments made by some of the drivers indicated that the next leg would be one of the toughest, with over 60 miles of restricted speed zones and construction areas, coupled with the threat of rain near Lubbock.[70]

On April 16, another 3:00 a.m. start faced the drivers as they headed back through New Mexico on their way to Lubbock. Loretta Colange, driving an Edsel Pacer, was the first driver on her way for the 383.7-mile, 8-hour and 30-minute trip. Heading north out of El Paso, the drivers followed a route to Alamogordo, then headed east crossing Apache Summit (the highest elevation on the Run) at daybreak, continuing to Roswell for a brunch stop at the Holiday Inn. Prior to their arrival at Roswell, some confusion developed between George Alsbury and one of his USAC observers:

> One of the things [the USAC observers] check is the entering of restricted speed zones. Unless you want to incur the penalty of running your engine for five minutes while at a standstill or having five minutes deducted from your driving time, you had better be down to the stated speed before and not after you pass the zone sign. The observers cooperate with the drivers in calling their attention to signs. Between Hondo and Roswell, N.M., the interpretation of a sign by an observer almost got George in trouble. George, busy with other matters, failed to spot a sign posted on a construction zone. The observer saw it, however, and told George he was entering a 15 miles-per-hour zone. What the observer failed to see was that the sign was a suggested limit sign and there was a second sign placing the legal limit at 40 miles per hour. As a result, George plodded along through nine long miles at 15

On the grounds of Galveston's Hotel Galvez, Mel Alsbury, Jr., proudly displays the Sweepstakes Award sitting on the hood of his Imperial Crown. With this win, Mel became only the second driver in Economy Run history to score three consecutive Sweepstakes Awards (courtesy of Chrysler Group LLC).

miles per hour and then had to sprint to make the stop at Lubbock, Texas, within the time limit.[71]

Leaving Roswell, their next stop was in Tatum for refueling at Phil Murph Mobilgas Service before crossing back into Texas, traveling through sparsely populated country, where in spots ranchers and schoolchildren lined the side of the road greeting the competitors. Road construction again hampered the drivers as they continued to Lubbock and the impound at the Municipal Auditorium and Coliseum. Arriving just after 1:00 in the afternoon, everyone was greeted by the Lubbock Championship Band, along with members of the Lubbock County Sheriff's Posse. Continuing to reflect the advance publicity efforts, radio and television crews were on hand to report on all of the activities and this would be common during the remaining days of the Run. Five competing cars were late in arriving, and after a review of the observers' reports, an extension of time was granted "due to confusion on signs at certain road construction" sites. In addition to George Alsbury (Chrysler New Yorker), these drivers included Don Bridges (Buick Roadmaster), Loretta Colange (Edsel Pacer), Nicky Griffen (Lincoln Continental) and Bill Stroppe (Mercury Turnpike Cruiser). Loretta Colange's challenges continued after reaching Lubbock when her relief driver, Marguerite Monnett, became too ill to continue. Due to these circumstances, USAC "bent the rules," and Les Gough stepped in as Loretta's relief driver for the remainder of the Run.[72]

When the results were released, it appeared that Chrysler and the Alsburys continued on their way to repeat victories. Mel Alsbury, Jr., and his Imperial Crown maintained a comfortable lead over all entries with 63.8194 ton miles per gallon. George Alsbury and his Chrysler New Yorker continued second in the Sweepstakes with 59.1119 ton miles per gallon, continuing to lead all entries in actual miles per gallon with 21.2422. A close contest was shaping up in the low-price class between Plymouth Belvedere 8 drivers Mary Davis and Pierce Venable. After the former had held a fairly comfortable lead, Pierce had now closed to within .2551 ton miles per gallon.[73] With Link Paola and his Oldsmobile Dynamic 88 preventing an all–Chrysler Corporation sweep, class leaders continued to be:

Low Price	Plymouth Belvedere 8 49.5497 Ton Miles Per Gallon
Low Medium Price	Oldsmobile Dynamic 88 51.3290 Ton Miles Per Gallon
Upper Medium Price	Chrysler New Yorker 59.1119 Ton Miles Per Gallon
High Price	Imperial Crown 63.8194 Ton Miles Per Gallon

With the strain beginning to show after covering just over 1217 miles, along with the early morning starting times, the drivers headed for the "Big D," covering a distance of 383.7 miles within an 8-hour, 25-minute time limit. With reports of last-minute changes to posted speed limits due to road construction, USAC officials added 10 minutes to the original published time limit. Don Rice, driving an Edsel Pacer, led the competitors, leaving at 4:00 a.m., bound for Abilene. Arriving at the Sands Hotel Café for today's brunch stop, everyone was greeted by various groups providing entertainment, including members of the Citizens Square Dancing Group along with the Hardin-Simmons Band & Drill Team. Before leaving Abilene, the drivers made a refueling stop at Roy Owens Service. From here it was a fairly straight shot to Fort Worth and Dallas, where the freeway between these cities was a welcome relief after the road construction encountered in the previous two days.[74] The first car to arrive at Dallas's Cotton Bowl Stadium was Ray Brock in his Oldsmobile Ninety-Eight, pulling in just prior to 1:00 p.m. Among the group of spectators on hand to greet the competitors with cold drinks were students from Texas College. Seven competing cars were late in arriving, their tardiness ranging from as little as 29 seconds to 17 minutes and 59 seconds. These drivers, who included Mary Davis (Plymouth Belvedere), Ruth Doushkess (DeSoto Fireflite), Fran Hernandez (Mercury Montclair), Hart Fullerton (DeSoto Fireflite), Ivan Lewis (Ford Custom 300 6), Lynn Paulette (Pontiac Super Chief), and Gus Retzlaff (Chevrolet Delray 6), would be required to make up their delinquent time on the final leg to Galveston.[75] This requirement would have a direct impact on Mary Davis.

At this point, it was almost certain that Mel Alsbury, Jr., and his Imperial Crown would capture a third consecutive Sweepstakes Award with a ton-miles-per-gallon figure of 63.2455. George Alsbury's Chrysler New Yorker continued to place second in the Sweepstakes with 58.8545 ton miles per gallon, and he was virtually assured of his second consecutive class win. In the low-price class, Mary Davis barely held onto her lead over Pierce Venable, with the margin between the two shrinking to a mere 13/100th of a ton mile per gallon. The only change of note was in the low-medium-price class, where Myra Buchanan, driving a DeSoto Firedome, took over second place from Don Rice, driving an Edsel Pacer.[76] Continued interest was on the highest figure for actual miles per gallon among the entries. George Alsbury continuing to set the pace with an average of 21.1516 in his Chrysler New Yorker.[77] Class leaders continued to be:

Low Price	Plymouth Belvedere 8 48.9353 Ton Miles Per Gallon
Low Medium Price	Oldsmobile Dynamic 88 50.8714 Ton Miles Per Gallon
Upper Medium Price	Chrysler New Yorker 58.8545 Ton Miles Per Gallon
High Price	Imperial Crown 63.2455 Ton Miles Per Gallon

While the competitors were getting some well-needed rest and relaxation before the final leg, the city of Galveston was busy getting everything ready for their arrival, including installation of the finish line banner and flags at the Galvez Hotel. "Hundreds of checkered flags bearing the insignia of the flying red horse have been distributed to beachfront business houses for display," according to Don Ward of the National Hotel Co. and one of the local area representatives.[78]

Ready for the final leg to Galveston, the drivers would face their shortest day of the Run, covering 297.37 miles within a 6-hour, 40-minute time limit. Bill Stroppe and his Mercury Turnpike Cruiser was the first on its way at 5:00 a.m. on a stretch that would turn out "to be the most difficult of all [with] highway repair and reduced speed zones through much more densely-populated sections."[79] Heading south out of Dallas bound for Madisonville, the drivers made a refueling stop at Burr Brothers Mobilgas Service, then it was on to Houston, where heavy traffic was encountered. The drivers were ready for a breather on the freeway between Houston and Galveston, but a final obstacle arose when the drawbridge over the Galveston Causeway was raised, holding the competitors up for approximately 15 minutes.[80] USAC officials deducted this time and no late penalties were assessed.

During this delay, Joan Fischal in her Dodge Custom

Royal and Bill Stroppe in his Mercury Turnpike Cruiser were working to see who would have the honor of being the first to arrive at the Galvez Hotel. They were ahead of the other competitors by a good 15 minutes, so no one was ready for their arrival, including USAC Official J.C. Agajanian with the checkered flag, as Bill crossed the finish line first by a hood length in front of a crowd estimated at 2500. This estimate included nearly 200 members of the press, including those from radio and television.[81] As the competitors continued to arrive, the Alsbury brothers crossed the finish line side-by-side in a well-choreographed move that had been planned earlier, with both drivers adjusting their speed during the trip. This side-by-side arrival was a sight never seen before in Economy Run competition ... and one which would not be repeated in future years. Under overcast skies with temperatures in the 70s, a group of swimsuit-clad girls, contestants in the "Miss Explorer" beauty contest, were on hand at the finish line to greet all of the competitors. The large turnout of spectators at times hampered USAC Officials in their final topping-off and inspection of the cars at the adjacent Moody Center.

The announcement of the final results was held up for two hours as protests were filed against the Chrysler New Yorker and Imperial Crown, causing some members of the press to miss their deadlines. "Filed by Ford, Lincoln, Mercury, Chevrolet and Oldsmobile entries, the protests alleged the Chryslers were not actual 'stock' cars, that their carburetors and air cleaners had been 'fixed' and that Mel's Crown Imperial had trick wiring in the power seat rigged up to the transmission."[82] A thorough analysis of the challenged parts showed that all of them were fully within the rules and the protests were disallowed. In announcing the decision, Chief Steward A.C. Pillsbury remarked that "such protests in the economy run are 'rare' and added that this was the first such instance in about eight years."[83] Following this announcement, the awards ceremony was scheduled to begin at 4:30 p.m., followed by a banquet that evening at the Moody Center. It was interesting to note that sales and marketing executives from Socony-Mobil and Magnolia Petroleum, and not those from General Petroleum, were prominent in presenting the various awards.

After several years driving both Nash and Pontiac entries, Pierce Venable moved over to Plymouth and placed first in the low-price class with a Belvedere hardtop. This was Plymouth's second consecutive year for Plymouth to win its class.

Spoiling a second sweep of all classes by Chrysler Corporation was the performance of the Oldsmobile Dynamic 88, placing first in the low-medium-price class. Standing with their trophy are driver Link Paola (left) and co-driver Bill Likes.

It was another big day for Chrysler Corporation and the Alsbury Family. For the third consecutive year, Mel Alsbury, Jr., captured the Sweepstakes Award in his light yellow Imperial Crown with 62.7188 ton miles per gallon. His actual miles-per-gallon performance was 20.5821, which was close to last year's figure of 20.9465. What makes this figure impressive for such a heavy car was the fact that the route was covered at a much faster pace, averaging 45.2224 miles per hour. This win put Mel, Jr., as the only three-time Sweepstakes winner in the history of the Mobilgas Economy Run, and he joined Clay Moore as a three-time champion. Clay captured his awards driving a Graham Supercharger 6 the Gilmore Yosemite Economy Runs. Second in the Sweepstakes was Mel's brother George in his Chrysler New Yorker, who celebrated his 21st birthday on the final day of the Run. Both Mel, Jr., and George won their respective classes by wide margins. Most notable about George's performance was his achieving the highest actual miles-per-gallon figure of all entries of 21.0217 miles per gallon with the eighth heaviest car in the run.[84] Overshadowed by the Chrysler New Yorker's actual miles-per-gallon performance was the battle between the two six-cylinder entries — a Chevrolet Delray and a Ford Custom 300. In a close contest from the start, Ivan Lewis in a Ford Custom 300 managed to edge out Gus Retzlaff and his Chevrolet Delray by a mere .2429 miles per gallon.

In addition to the Sweepstakes Award, first-place trophies were presented to:

CLASS	MAKE and MODEL	TON MILES PER GALLON	MILES PER GALLON
A — Low Price	Plymouth Belvedere 8	48.3264	20.0088
B — Low Medium Price	Oldsmobile Dynamic 88	50.5951	19.2377
C — Upper Medium Price	Chrysler New Yorker	58.4928	21.0217
D — High Price	Imperial Crown	62.7188	20.5821

Detailed results for all competing cars can be found in the Appendix.

The average miles per gallon for the entire field was 18.7529, down 1¾ miles per gallon from last year. This drop was largely due to the higher average speed required this year, 45.2224 miles per hour, compared to 1957's average of 41.4524. Also contributing to this mileage drop were some 60 miles of unexpected construction, which slowed their run across Texas, along with a good deal of traffic on the final day. The men averaged 18.8598 miles per gallon, edging out the women drivers, who averaged 18.4171. All entries completed the Run, and other than Art Rene's mechanical problems with his Edsel Citation, no problems were reported. Weather conditions were very good with warm temperatures over most of the route, and except for areas of road construction conditions, were also described as very good.

The biggest upset occurred in the low-price class, as Pierce Venable, driving a Plymouth Belvedere 8, overtook Mary Davis on the last leg to take first place by a mere .1463 ton miles per gallon. Competition between the two was close from the start, with Mary hoping for a repeat win, but Pierce narrowed the distance between the two each day. What may have enabled Pierce's move into first place was Mary's late arrival in Dallas by 5 minutes and 50 seconds. This time had to be made up on the last leg, which therefore resulted in gas-consuming higher speeds and driving techniques. "Mary, although heart-broken, accepted her loss like the champion she is."[85]

Spoiling a sweep by Chrysler Corporation was the Oldsmobile Dynamic 88's victory in the low-medium-price class, driven by La Crescenta (California) Oldsmobile dealer Link Paola. In only his second time competing, he maintained a comfortable lead from the start over Myra Buchanan in a DeSoto Firedome and Don Rice in an Edsel Pacer. This was especially pleasing for Link after his having had to withdraw from competition last year due to a vibration problem with his Oldsmobile. A surprise was the disappointing performance of both Dodge Custom Royal entries. After a strong first place performance in 1957, this year these models struggled throughout the run, placing eighth and ninth in a field of ten entries. With different driving teams competing this year, these results were viewed as a temporary setback in Dodge's overall Economy Run performance.

With the women drivers competing on their own for awards, their overall results were quite good. In addition to Mary Davis and Myra Buchanan each placing second in their respective classes, Nicky Griffen placed third in the high-price class, piloting a Lincoln Continental. Although they placed last in the low-price class, the Chevrolet Impala 8 team of Dorothy Retzlaff and Lillian Vann were said to have had more fun than anyone else competing.[86]

With Chrysler Corporation's near-domination in this year's run, advertising, as expected, was limited; it followed the same themes as were evident in 1957, and continued to be found primarily in newspapers and industry publications. Socony-Mobil's corporate print advertising again was focused on the results and route information while also introducing the new Mobil signage with the tagline "You're Miles Ahead with Mobil." Also noted in this ad is that the Run "is sponsored annually by General Petroleum, far west affiliate of Socony-Mobil."[87] Appearing primarily in enthusiast magazines, an advertisement by Socony-Mobil featured first-time driver Nicky Griffen asking, "Who said lady drivers aren't as good as men?" The ad told the story of her performance driving a Lincoln Continental.[88] Magnolia Petroleum created a separate newspaper advertising campaign for their marketing area by noting, "29 Cars Averaged 18.7 miles per gallon in '58 Mobilgas Economy Run Here in the Southwest." The ad incorporated the official results for all entries.[89] In the fall of 1958, Socony-Mobil launched their new advertising theme, "The New Car Gasoline '59 Mobilgas Special," noting that one of their testing grounds was the Mobilgas Economy Run.

In the midst of 1958's disappointing sales, coupled with negative reports covering the quality of their 1957 products, Chrysler Corporation finally had some positive news to report. Its main newspaper advertising campaign focused on the Imperial's success with the headline, "Imperial Wins Again Tops All Cars in Mobilgas Economy Run ... For Third Year in a Row!" The ad also featured a photograph of the winning entry. A sidebar in this advertisement also told of the New Yorker's class win.[90] The New Yorker's achievement in scoring the highest miles per gallon for all entries was highlighted in both newspaper and selected magazines in ads that noted the "Mighty Chrysler averages 21.02 miles per gallon—*tops all other cars* in Mobilgas Economy Run." These ads featured a photograph of a close-to-identical New Yorker Sedan to the one which competed.[91] Later in the year, in an advertisement featuring the Saratoga Hardtop, appearing in magazines such as *Life* and *The Saturday Evening Post*, Chrysler pointed out that "its outstanding economy was established in the Mobilgas Economy run."[92]

Plymouth's second consecutive class win was advertised primarily in newspapers proclaiming, "Plymouth Wins Again In Gas Economy Tests!," and noting that a Belvedere hardtop finished first and second and topped all other entries in the low price class.[93] Local dealer associations advertised this success in a separate campaign featuring a partial photograph of one of the Plymouth entries, noting: "For the second consecutive year Plymouth beats all competition in its class."[94]

After a large advertising program in 1957, Auto-Lite "Power Tip" Spark Plugs relied on a single two-page ad focusing on the Chrysler Corporation victories. Featuring photographs of the class-winning models, emphasis this year was on the Imperial's performance, with the headline "Power Tip 'fires up' Sweepstakes winner for 3rd straight year in Mobilgas Economy Run."[95]

Virtually ignoring its participation, Oldsmobile only made brief mention of its Economy Run success in newspaper advertising. In one ad titled "Travel Test an Olds 88," readers were told in the text that it "won 1st place in its class in the Mobilgas Economy Run."[96]

In their 1959 showroom brochures, Chrysler introduced its new "Golden Lion V-8" by telling prospective buyers that this engine comes with "a fuel-saving pedigree, too! For Chrysler defeated all 'comers'—even the lightest 'sixes'—during the 1958 Mobilgas Economy Run."[97] In a special 1959 brochure titled "Your Next Fine Car," Imperial noted it was a "Sweepstakes Winner 3 times—Mobilgas Economy Run," accompanied by a rendering of the trophy.[98] Only a brief mention was made by Ford and Plymouth of their Economy Run performance in their 1959 brochures.

Newspaper coverage continued to be prominent, not only in the Los Angeles area, but also across the Southwest, as readers followed the progress of the drivers competing. The *Galveston Daily News* kept its readers aware of the progress along with the preparations being made for competitors at the Galvez Hotel. Naturally, these stories focused on the Alsbury family, with Mel Jr. leading in the Sweepstakes for the third straight year. Magazine articles were noticeably limited this year; this was especially evident in Floyd Clymer's *Automobile Topics*, which had the fewest pages of coverage since it began reporting on the Run in 1952. In one of the few in-depth articles, readers of *Motor Trend* were given an inside look into this year's event through the experiences of Mel Alsbury, Jr. Even with the participation of *Hot Rod Magazine*'s Ray Brock and Wally Parks, their experiences were briefly mentioned in a road-test article on the Oldsmobile 98.[99]

After the successful conclusion of the 1958 Run, both Magnolia Petroleum Company and local Galveston officials had a positive view of the city's selection as the terminus of this year's Economy Run, which also marked the first time the event entered the state of Texas. George Gregor, vice-president of Magnolia Oil Company, reported being hopeful of a return visit to Galveston:

> With officials, drivers, guests and spectators of the 1958 Mobilgas Economy Run highly pleased over finishing the event in Galveston speculation was high Thursday the 1959 run also may terminate here. George Gregor, vice-president of Magnolia Oil Co., called Galveston the "ideal finish place. It couldn't be better." Gregor emphasized, "We are pleased with every phase of the finish here. Galveston has more facilities than any city to which we have terminated the run." The oil executive, who presented three-time sweepstakes champion Mel Alsbury, Jr. and the other 1958 winners' trophies, said nothing indicates repeating the finish in Galveston in 1959, but pointed out other cities have been ending spots more than once. "There is no real reason to feel the 1959 run will be repeated here," Gregor said, "and no one will know for sure that next year's route will be until within three months of the running, but we are immensely pleased with Galveston, the reception here has been wonderful." Gregor [also] pointed out that if Galveston were the 1959 finish line, the route could be changed almost entirely. "There are many cities over the country, however, that are attempting to make reservations for the 1959 run."[100]

Even with this enthusiasm, a return to Galveston would not occur, although various Texas cities would host overnight stops in future years.

6

Kansas City and Minneapolis

1959

Los Angeles to Kansas City Via Cedar City — Grand Junction — Denver — Lincoln

April 5–9, 1959—1898.4 Miles

After the conclusion of the 1958 Economy Run, it became evident to officials of General Petroleum and USAC that changes were needed in determining winners. With the "Eisenhower Recession" beginning to ease, coupled with the growth in sales of domestic and imported economy models, the public was once again focusing on miles-per-gallon performance. For a number of years there was a desire by the officials to have an overall winner named. As a result of this thinking they continued to defend the use of a ton-miles-per-gallon formula, stressing that this "score is fair to all."[1] Criticism of the ton-mile formula only increased after the Imperial's recent third consecutive Sweepstakes win, since the only "lightweight" car to take the top place was a Ford Mainline 6 in 1953. Also, confusion in the public's mind as to the validity of the results continued to be evident where advertising campaigns by some manufacturers stressed the actual miles-per-gallon performance of their model whether or not it actually took the top prize in its class. These concerns were addressed with the announcement in the fall of 1958 of a simple change. The ton-mile formula, along with the Sweepstakes Award, was to be eliminated, with winners now being determined by their actual miles-per-gallon performance. While the USAC noted in the rules that "there will be no over-all champion," in reality the car which scored the highest miles per gallon would heavily advertise this achievement.[2] In announcing this change, Robert Minckler explained, "It is felt the straight mileage formula is more meaningful to the average motorist, for whose benefit the run is conducted."[3]

The hope by Galveston officials for a return was dashed when Robert L. Minckler, General Petroleum's president, announced on March 8 that the 1959 Economy Run would finish at the Muehlebach Hotel in Kansas City, Missouri. This would mark the first visit to a Midwestern city by the Economy Run. Again, this year's event would be the longest in history, and although only 15.2 miles longer than last year's record total, it promised to be the most rugged in several years. USAC officials, pleased with last year's procedure of withholding details of the route until hours before the start, elected to continue this practice on a permanent basis.

General Petroleum and Socony-Mobil engineers developed specially designed 21-gallon gas tanks with the goal "to eliminate all chance of fueling inaccuracy." USAC officials explained that this step had been taken "because regular gasoline tanks are often oddly shaped to fit around parts of the car. The odd shapes tend to form air pockets and make it difficult to determine precise capacity."[4] Bright red in color, these tanks were installed in the trunks of the competing cars; hooked up to the regular gas line, each included both leveling and temperature monitoring devices. They also eliminated the need for the specially constructed refueling platforms used since 1950 and opened up numerous new possibilities for starting and terminus locations.

In a continued refinement to the rules regarding the eligibility of women participating as drivers or relief drivers, new provisions were made for husband-and-wife teams to compete, as well as for "a relief driver of the opposite sex who is a member of the driver's immediate family."[5] Taking full advantage of the husband-and-wife rule was the Alsbury family with Mel, Sr., and Mildred teaming up in a Chrysler Saratoga, along with Mel, Jr., and Rae-Nell in an Imperial. The first family to compete under this rule was Visalia (California) Buick Dealer Claude L. Allred, who entered an Invicta hardtop and handled relief driver duties for his daughter Judy; at 17 years of age, she became the youngest driver in the history of the Run. Thirteen teams of women drivers and relief drivers competed; eight had previous Economy Run experience, including Mary Davis, shooting for her second Class victory. After skipping the 1958 Run, Patricia Jones and Betty Skelton returned to competition.

Eligibility was open to any stock 1959 two-door or four-door metal-top model with no limits on minimum engine size, as had been the rule in recent years. Competitive classifications were expanded to six. With the introduction of compact models from American Motors' Rambler and Stude-

At the Los Angeles Civic Center, USAC's J.C. Agajanian sends Jim Parkinson and his Cadillac Sixty-Two on his way to Cedar City, Utah. This was Cadillac's first appearance in the Economy Run since 1951 and was a private entry by Jim's import dealership, Italiano Motors. Jim would place first in the high-price class, ending the Imperial's three-year winning streak.

baker, a new compact car class was designed for those having a wheelbase of 110 inches or less. In addition, the low-price class was split, with one category for six-cylinder models and the other for eight-cylinder models. Power steering and power brakes were required on all full-sized eight-cylinder models.[6]

The elimination of the ton-miles-per-gallon formula was a major factor in attracting a total of 47 cars to compete. Not only was this the largest field to date, but it marked the first time that every make was represented. General Motors, who for years had an indifferent attitude to the Economy Run, had the largest representation with 20 cars entered, followed by Ford Motor Company with 13, Chrysler Corporation with 9, American Motors with 3 and Studebaker with 2. Dealers would continue to sponsor the majority of entries, with the Ford Motor Company and Rambler entries sponsored by their respective dealer associations. Two dealers from outside California participated for the first time, including Andy Klein Pontiac in Overland Park, Kansas, who would continue participating into the mid–1960s. Colonial Buick in Glendale, a Southern California dealer making its first appearance, would become familiar, participating each year through 1967. In an interesting partnership, Cadillac returned for the first time since 1951 with Jim Parkinson's Italiano Motors, an imported car dealership in Burbank, entering a Sixty-Two Hardtop. Jim was noted for taking top mileage honors driving an Austin Healey Sprint in the 1958 Mobil Mileage Rally.

Also of note are the eight dealers who not only sponsored entries, but handled driving duties as well. Included in this group were Mel Alsbury, Sr. (Chrysler Saratoga), Hart Fullerton (DeSoto Firedome), Dwight Hinton (Oldsmobile Super 88), Andy Klein (Pontiac Star Chief), Tom Neal (Oldsmobile Dynamic 88), Gus Retzlaff (Chevrolet Biscayne 6), Jim Retzlaff (Pontiac Catalina), and James Rush (Chevrolet Bel-Air 8). In addition to this group, Claude L. Allred (Buick Invicta) handled the co-driver duties.

With the largest field in history, the door was opened for a new crop of drivers competing for the first time. One newcomer, tapped to drive a Ford Thunderbird, was Art Chrisman, who was the current holder of the ¼ Mile Dray Racing Record at 181.81 miles per hour. Four first-time drivers, Byron Froelich, Mary Hauser, Stan Raymond and Marta Retzlaff, would become regular participants, and over the next ten years would take home a first-place trophy on at least one occasion. Five had proven their skills the previous October driving in the inaugural Mobil Mileage Rally, with two, Jim Parkinson and Lillian Warner, placing first in their respective classes. All four class-winning drivers from 1958 returned, along with several previous participants after an absence of a year or more; these included Danny Eames, Dick Griffith, Ak Miller, Vince Piggins, and Les Viland.

Following the established pattern from recent years, random car selection began in February with the usual inspection for stock status, conducted by USAC officials, followed by placement of seals on 64 major parts. No changes were deemed necessary this year in the rules covering servicing and break-in of competing cars as the teams began their preparations. For the first time since 1953, entrants were given the option of using either Mobilgas R (Regular) or Mobilgas Special. All entries elected to use Mobilgas Special.

Interesting developments occurred with the announcement that both Rambler and Studebaker would be competing in the new compact car class. Each was thought to have a good shot at achieving the best overall miles-per-gallon result. American Motors executives, who had an ongoing tense relationship with USAC, initially were not at all interested in participating in this year's or any future Economy Run. They held the belief that USAC was "rigging the rules to benefit the products of the Big Three," and additionally noted that their recent "independent fuel economy demonstrations (under the supervision of USAC's rival NASCAR) played their part in helping Rambler to achieve a spectacular financial recovery. "Who needs the Mobilgas Run?" they argued. After learning of the elimination of the ton-miles formula, it was apparent to American Motors that "the Rambler American stood an excellent chance of coming out the effective overall winner and the public could understand mpg figures well enough to recognize the best mpg of all."[7] Three Rambler models were ultimately entered, with American Motors Engineer Les Viland heading up their efforts.

For Studebaker's executives, coming off several years of serious financial issues, emphasis was on the success of its new Lark models and in meeting the demand by their dealers for additional production. Having eliminated factory support for the Economy Runs after 1955, Studebaker had given no thought to resuming this support. Prior to the closing date for entries, two Los Angeles area dealers elected to enter Lark six-cylinder and eight-cylinder models. Bill Corey, driver of the Lark Deluxe 6, only had limited time to prepare his car

Mary Hauser, driving a Chevrolet Biscayne 6, had members of her fan club on hand to help send her on the way to Cedar City. Standing near the driver's door are two members of the USAC timing crew, who were responsible for notifying each driver it was time to depart. In the first year the low-price models were divided into six-cylinder and eight-cylinder classes, Mary would place first in her class, marking Chevrolet's first win since the 1937 Gilmore Yosemite Run.

for the Run. He soon discovered that since the Lark was so new on the market, factory service instructions were not yet available. A resulting call to the factory for this information brought total surprise, and immediately, "top Studebaker engineers began commuting between the factory and Los Angeles" with "Don Hosford, chief watchdog of Studebaker Run teams for years, [coming] out of retirement."[8]

It wasn't long before problems faced Bill Corey during a break-in run to test carburetor jet needles, which were one size smaller than stock as permitted by the rules. On the return trip to the impound from Palm Springs, the engine began to ping on a long desert grade, then froze up. Finally able to restart the engine, Bill made his way back to the impound, although he encountered the same problem each time there was a grade. Studebaker engineers and mechanics worked to find out what was happening to the engine, but they were unable to remove the cylinder head, according to the rules, for a thorough inspection. "Their final conclusion was that the Lark six engine, designed specifically to operate with a stick-shift gearbox, just reacted that way with automatic transmission and the lean carb needles. There was only one thing left to do: sweat out the Run anyway. Crawl up hills with steadily decreasing throttle knowing that at any time the engine might lock up tight. Fortunately, this happened to be a perfectly good economy-driving technique, perhaps as good as any of the several used by the experts." The trouble was kept secret from the American Motors team.[9]

While USAC officials were involved in the final details of the route, advance publicity efforts began under the direction of General Petroleum's Frank Meunier. One of the main tasks for this group was to target press representatives in the Midwest, where the competitors were expected to travel through for the first time. Press releases this year placed emphasis on the new miles-per-gallon formula, along with announcing the competition of a record field "that encompass the entire American passenger car output, from the little Ramblers and Larks up to Cadillac."[10] Building up enthusiasm prior to the start, the random drawing for starting positions was held on the morning of March 30. In an effort to get the maximum amount of pre-start publicity possible, it was announced that this would be the only time prior to the start where "all drivers, co-drivers and cars (in 'war paint') will be gathered in one spot for pictures and interviews."[11] Each driver selected a rolled pennant marked with a starting position, with Bob Russo, associate editor of *Motor Life Magazine* and driving a Pontiac Bonneville, selecting No. 1. Drawing the last starting spot was Mary Davis, back behind the wheel of a Plymouth Belvedere 8. At the conclusion of the drawings, USAC officials were on hand demonstrating to the press and participants the design and function of the trunk-mounted gas tanks.

Robert Flanagan, driving a Buick Invicta, is sent on his way from the Cedar City impound by USAC's J.C. Agajanian. Bound for Grand Junction, the drivers would have sunny and warm conditions during the day on the 403.9-mile trip.

With the announcement of the route, veteran competitors noticed there was some similarity to the 1956 Run's trek over the Rocky Mountains. With overnight stops in Cedar City, Grand Junction, Denver and Lincoln, the drivers would cover 1898.4 miles over a five-day period before reaching the finish line at the Muehlebach Hotel in Kansas City. Traveling through nine states, the route had an altitude range from a low of 262 feet at the Los Angeles Civic Center Start to 11,988 feet at Loveland Pass on the Continental Divide in Colorado. With snow and ice conditions expected, each entry had chains stowed in the trunk as an added precaution.[12] Adding to the mountain, desert and prairie terrain, competitors would have to travel through 141 cities and towns, contending with 162 traffic signals. The design of a new-style route book, introduced in 1958, was termed as a success, and was repeated this year. These route books would continue to be used, with slight modifications, in all future Runs. The drivers would average 40.9335 miles per hour to stay within the 46-hour, 57-minute time limit set by USAC.

In an interesting twist, it was announced that results would not be announced daily, as "fuel economy computations will be made for three 'legs' representing different kinds of driving." The first leg would cover mountain and desert terrain over three days from Los Angeles to Denver, the second leg would cover flat prairie country to Lincoln, followed by the last leg of heavy urban traffic to Kansas City.[13] This change was never explained, but it is thought that USAC officials and refueling staff desired additional time to work out any bugs that may occur with these new tanks, although no problems were reported. This decision, however, created some unintended consequences. While the first official results would not be released until the Denver overnight stop, unofficial results were developed and reported by members of the press, including Bill Carroll of *Automotive News* and Bill Dredge of the *Los Angeles Times*, whose reports were based solely on observed refueling amounts.[14] As a result of publicizing these unofficial calculations, USAC officials would revert to reporting daily results in the future.

The use of these new fuel tanks opened up new possibilities for starting locations, since only a small gasoline tanker truck would be needed to top off each car after being driven from the Pasadena impound. The Los Angeles Civic Center parking lot was selected this year for the April 5 start, with the cars lined up in twelve rows, four abreast. Dispensing with the post-midnight starting times, a smaller crowd of spectators than had usually been seen were on hand for the early Sunday morning start. Among the industry representatives on hand at the start was Harold Churchill, president of Studebaker-Packard Corporation, in Los Angeles to attend meetings with local dealers.[15] In cool and foggy conditions, Bob Russo, driving a Pontiac Bonneville, was first to take the green flag from J.C. Agajanian at 6:00 a.m. following an unfortunate incident with press photographers. When an official from USAC attempted to clear the way, a shoving match followed involving photographers from the *Los Angeles Times* and *Los Angeles Examiner*, resulting in one photographer's being pushed to the ground along with damage to several pieces of camera equipment.[16] This is the only recorded problem of this nature during the entire history of the Economy Run.

Following Bob Russo's Pontiac at one-minute intervals, the drivers began the 462.2-mile trip to Cedar City. A driving time was set at 11 hours and 10 minutes, which was extended 7 minutes by USAC. With clear going in early Sunday morning traffic, they followed a direct route to San Bernardino, and then over Cajon Pass to Las Vegas, where the competitors arrived at the Thunderbird Hotel for their first brunch and refueling stop of the Run. As the drivers arrived around noon, with temperatures exceeding 100 degrees, there was a great deal of interest by all concerned in the refueling operation with the new gas tanks. Everything appeared to go smoothly.[17] From Las Vegas, the drivers continued to follow U.S. Route 91 into Utah bound for Cedar City, climbing to an altitude of about 6000 feet. Crossing into the Mountain Time Zone, the competitors first stopped for refueling at Green & Higbee Mobil Station, then wheeled into the impound at the Union Pacific Depot Yard. The first to arrive, just after 6:30 p.m., was Myra Buchanan in a DeSoto Firesweep. Tired after nearly 12 hours on the road, the crews were ready for a well-earned rest, but for one team there was some unexpected business to attend to. Woody Bell, driving a Rambler American Deluxe 6, discovered an error in the odometer correction formula before reaching Las Vegas. Arriving in Cedar City with only a couple of seconds to spare, Woody and his co-driver, Ken Fougner, spent the next several hours "recalculating every step of their next day's progress in the light of the real odometer correction they had come by the hard way."[18]

While there were no official results released by USAC, informal calculations showed that Woody Bell in his little Rambler American Deluxe 6 appeared to be leading all competitors. "He took on 16.6 gallons to cover the distance averaging just a hair under 28 miles a gallon. A Studebaker Lark V-8, driven by Dick Griffith, took 17.3 gallons for the distance and another Rambler, driven by Les Viland, ran the trip with 18.15 gallons." Other unofficial class leaders included Mary Hauser in a Chevrolet Biscayne 6 and Mary Davis in a Plymouth Belvedere 8, although "drivers were cautioned that these fuel consumption figures are not official, since the uniform-size tanks were not perfectly topped off today [April 5]."[19]

Prior to the competitors' heading for Grand Junction, USAC Chief Steward A.C. Pillsbury announced that for one of the few times in the history of the Run it was necessary to impose penalties on several contestants who violated the rules. In a statement from USAC:

> The exact nature of the penalties to be applied is not given in the book of Official rules. Instead, the Board of Stewards chooses the penalty which seems fair and proper in the light of the circumstances with the intent that it be sufficiently severe to discourage subsequent similar violations and to offset completely any advantage the competitor might enjoy.
>
> The following cars committed violations on the Los Angeles to Cedar City leg of the Mobilgas Economy Run, and on Monday Morning, April 6, the drivers were advised of the penalties imposed by the Board of Stewards.
>
> Car No. 4, Chevrolet (Impala), passed a truck on the right shoulder of the highway. (Vince Piggins, Driver)
>
> Car No. 11, Oldsmobile 88, followed other competing cars too closely, crossed the double lines several times. (Lillian Warner)
>
> Car No. 18, Cadillac (Sixty-Two), arrived at Cedar City Impound one minute 32 seconds late. (Jim Parkinson)
>
> Car No. 20, Pontiac (Catalina), cut corners and went through red light. (Jim Retzlaff)
>
> Car No. 40, Rambler (Deluxe), arrived at Cedar City Impound one minute 28 seconds late. (Les Viland)
>
> Car No. 41, Thunderbird, crossed the double line of the highway. (Loretta Colange)
>
> The exacted penalty was identical in each instance. The driver was instructed to proceed to the portion of the highway immediately beyond Cedar City on the route to Grand Junction, attain a maximum speed limit, (60 mph) then slow down to a complete stop, and turn off the ignition. The car then would be restarted and the procedure repeated until a total of four starts and stops had been completed.[20]

Bill Stroppe's Edsel Corsair travels through Utah's Fish Lake National Forest en route to Grand Junction. Bill could only coax 17.8975 miles per gallon out of his Corsair and placed 14th in the low-medium-price class. This would be the Edsel's second and last appearance in the Economy Run.

Later it was reported that on succeeding days, there were fewer violations reported and those were minor in nature.[21]

The drivers left Cedar City on April 6, a day that would see the competitors traveling through the towns of Beaver, Salina and Green River, which had been visited during the 1956 Run. Mary Davis, in a Plymouth Belvedere 8, was the first driver on her way at 7:00 a.m. for the 403.9 mile trip to Grand Junction with a 9-hour, 50-minute time limit set by USAC. This turned out to be the only day when a time extension was not provided. Working their way north under clear skies, the drivers headed through the mountains, passing through Fish Lake National Forest before arriving in Price. Their first stop was for brunch, served by the students and teachers at Notre Dame School's cafeteria, and immediately followed by refueling at the Hilltop Mobil Station. Leaving Price, they had to contend with a 4½-mile construction zone, gravel and all, before continuing eastward and crossing into Colorado.[22] Sunny and warm conditions prevailed as the drivers traveled across the plateau country of Utah and West Central Colorado. Following the same procedures for the previous day, the drivers first refueled at the Mann Durbin Mobil

Station in Grand Junction before arriving at the Lincoln Park Municipal Stadium. This was the second time the Lincoln Park location was selected as the overnight impound, having been used in 1956. Don Bridges, driving a Buick Electra, was the first to arrive, pulling in just after 5:30 p.m., with all competitors arriving within the time limit. The drivers were starting to show fatigue from two long days on the road and their only concern up to this point was some head winds that had been encountered from time to time.

Again no official results were released by USAC, although informal calculations continued to be reported, which showed that Woody Bell in his little Rambler American Deluxe 6 appeared to be leading all competitors with an estimated 30.7 miles per gallon. Other unofficial class leaders were Mary Hauser (Chevrolet Biscayne 6), Mary Davis (Plymouth Belvedere 8), Patricia Jones (Dodge Coronet), Art Chrisman (Ford Thunderbird) and Jim Parkinson (Cadillac Sixty-Two). Everyone was once again cautioned that "these 'homemade' calculations ... were made necessary by decision of sponsors to hold back mileage data until the end of the third day or 'desert-mountain leg' of the Run (April 5)."[23]

The toughest part of the Run was to come, as the drivers faced crossing the Continental Divide with the forecast of deteriorating weather conditions during the day.

An early start faced the drivers on April 7 as they headed eastward on the shortest leg of the Run with a 258.4-mile, 6-hour and 45-minute trip to Denver. During the day USAC officials would extend this limit by 30 minutes. Myra Buchanan, driving a DeSoto Firesweep, was the first to be flagged away at 4:00 a.m., with the rest of the field leaving at 2-minute intervals. As the drivers made their way out of Grand Junction, their first obstacle was navigating in the dark "some incredibly bad road under construction."[24] After they made their way through the scenic areas of Glenwood Canyon and the While River National Forest, highway construction again slowed their progress to Eagle, where they stopped for refueling at Palmer's Mobil Service. The temperatures were dropping as they followed U.S. Route 6 crossing Vail Pass, at an elevation of 10,603 feet, where soon afterward snow conditions were first encountered, along with head winds. With the temperature dropping to below freezing and snow beginning to fall, the crews arrived at the Moon Valley Lodge in Dillon for a brunch stop. Following the advice of USAC officials, the crews wasted no time in getting back on the road to avoid possible freezing highways as conditions were reported as getting worse.[25] As they made their way towards Loveland Pass, the highest point on the Run, a snowstorm turned the road to slush and ice, making this "the most harrowing day of the run for the drivers."[26] Descending Loveland Pass and heading to Denver, they found road conditions improved somewhat, although the temperatures remained near freezing. Fortunately no mishaps were reported as a weary group of drivers made their way to the impound at the Denver Coliseum. Bob Russo, driving a Pontiac Bonneville, was the first to arrive just after 11:00 a.m. While the drivers were interviewed

After refueling at Palmer's Mobil Station in Eagle, Colorado, Mel Alsbury, Sr. in his Chrysler Saratoga heads to Loveland Pass while USAC officials continue refueling Jim Fosdick's Ford Fairlane 500 8. The drivers would soon face falling temperatures and overcast skies.

by members of the press, including those from radio and television, all entries were topped off by USAC officials as snow began to fall.

For the first time since leaving Los Angeles, official results were announced that showed Woody Bell, driving a Rambler American Deluxe 6, achieving the best miles per gallon figure for all entries with 26.7476. The biggest development was the overall performance of the fourteen women drivers, who as a group were averaging 20.5048 miles per gallon, compared to 20.5673 for the men. Three were leading their classes — Mary Hauser (Chevrolet Biscayne 6), Mary Davis (Plymouth Belvedere 8) and Patricia Jones (Dodge Coronet) — with Loretta Colange (Ford Thunderbird) holding a second-place spot. Competition was close in three classes where less than a half a mile per gallon separated the top two spots, while in the upper-medium-price class, a mere .3759 miles per gallon separated the first five places. In the high-price class, a surprise was finding three-time Sweepstakes winner Mel Alsbury, Jr., in second place behind Jim Parkinson's Cadillac Sixty-Two. The top three performers in each class were:

Compact Car	Rambler American 6	26.7476 Miles Per Gallon
	Rambler Deluxe 6	24.4382 Miles Per Gallon
	Studebaker Lark Deluxe 6	24.0280 Miles Per Gallon
Low Price Six Cylinder	Chevrolet Biscayne	23.8835 Miles Per Gallon
	Chevrolet Biscayne	23.3611 Miles Per Gallon
	Ford Custom 300	21.5730 Miles Per Gallon
Low Price Eight Cylinder	Plymouth Belvedere	22.4168 Miles Per Gallon
	Plymouth Belvedere	22.0333 Miles Per Gallon
	Ford Custom 300	20.9825 Miles Per Gallon
Low Medium Price	Dodge Coronet	23.1096 Miles Per Gallon
	Dodge Coronet	22.1801 Miles Per Gallon
	Chrysler Windsor	20.7282 Miles Per Gallon
Upper Medium Price	Ford Thunderbird	20.4266 Miles Per Gallon
	Ford Thunderbird	20.2823 Miles Per Gallon
	Mercury Montclair	20.2608 Miles Per Gallon
High Price	Cadillac Sixty-Two	20.0182 Miles Per Gallon
	Imperial Custom	19.0695 Miles Per Gallon
	Lincoln	18.9582 Miles Per Gallon

Ready for some well-earned rest after their trek over the Rocky Mountains, the crews realized that more rough weather was in store as the snow continued to fall. The majority of the drivers were from California and virtually none had experience driving in snow conditions. They would be thoroughly tested in the coming hours.

With freezing temperatures and snow continuing to fall, the crews made their way from their hotel to the Denver Coliseum impound for the 480.7-mile trip to Lincoln. A time limit of 10 hours and 40 minutes was announced by the USAC, which was extended by 6 minutes due to weather conditions. The first car on its way at 1:00 a.m. on April 8 was a Buick Electra, driven by Don Bridges, with the remaining field leaving at 1-minute intervals. With their cars covered in snow and ice, the drivers were faced with battling snow conditions in the nighttime darkness as they headed across eastern Colorado and into Nebraska on U.S. Route 6. During these first hours, the drivers had to slow their speed to about 30 miles per hour while contending with uncertain road and traction conditions.[27] As they began to arrive at the Otto Mobil Station in McCook for refueling, the snow had already let up and road conditions were beginning to improve. The first order of business for the crews was to knock off as much snow and ice as possible that had accumulated around and under the cars during the 293-mile trip from Denver to reduce extra pounds of gas-consuming weight. As more and more of the cars arrived, a block-long backlog developed, with half the field waiting 15 to 20 minutes to refuel. After refueling, their next stop was for brunch at the First Methodist Church in Arapahoe, served by members of the community. From here the drivers had to push tight against the posted speed limits to make up for the time lost as they crossed the flat plains and were greeted by traffic as they made their way through downtown Lincoln to the impound at the Nebraska State Fair Grounds. Don Bridges, the first driver to leave Denver, led the group of tired drivers, arriving just after 1:10 p.m. While most of the competitors arrived in Lincoln within the time limit, 11 were late, ranging from as little as 10 seconds to just over 10 minutes. All late arrivals were assessed penalties by USAC. As the cars were topped off, most agreed that they had experienced the roughest weather conditions in the history of the Run, knowing full well their averages would take a drop.[28] And drop they did.

When the results were announced, it confirmed what was expected, with every driver showing a drop in miles per gallon. The field showed an average 19.6414, which was a decline of .9073 mile per gallon from the previous day's calculations. This was clear evidence of the severity of the conditions the drivers had faced. Impressive was the continuing performance of the women drivers, who as a group were averaging 19.6097 miles per gallon, compared to 19.6549 for the men. Woody Bell continued to lead all entries with 25.5070 miles per gallon in his Rambler American Deluxe 6. This represented a drop of 1¼ miles per gallon from the results computed in Denver. There were no changes in individual class leaders with most increasing their lead over second-place entries. The only change of note was in the upper-medium-price class, which remained tight with .9071 miles per gallon separating the top five places. Mel Alsbury, Sr., moved his Chrysler Saratoga into second place, dropping Loretta Colange's Ford Thunderbird into third. The top three performers in each class were:

Compact Car	Rambler American 6	25.5070 Miles Per Gallon
	Rambler Deluxe 6	23.0603 Miles Per Gallon
	Studebaker Lark Deluxe 6	22.6847 Miles Per Gallon
Low Price Six Cylinder	Chevrolet Biscayne	22.6251 Miles Per Gallon
	Chevrolet Biscayne	22.1499 Miles Per Gallon
	Ford Custom 300	20.3907 Miles Per Gallon
Low Price Eight Cylinder	Plymouth Belvedere	21.5578 Miles Per Gallon
	Plymouth Belvedere	21.2884 Miles Per Gallon
	Ford Custom 300	19.8763 Miles Per Gallon
Low Medium Price	Dodge Coronet	22.0108 Miles Per Gallon
	Dodge Coronet	21.2052 Miles Per Gallon
	Chrysler Windsor	19.9041 Miles Per Gallon
Upper Medium Price	Ford Thunderbird	19.4424 Miles Per Gallon
	Chrysler Saratoga	19.3053 Miles Per Gallon
	Ford Thunderbird	19.2688 Miles Per Gallon
High Price	Cadillac Sixty-Two	19.2797 Miles Per Gallon
	Imperial Custom	18.3594 Miles Per Gallon
	Lincoln	17.4237 Miles Per Gallon

As the tired crews gathered at the Cornhusker Hotel for announcements of the results, "they groaned at a loud-speaker announcement that snow was falling in Kansas City. If the storm spreads north to Lincoln tonight, that final short leg may mean more 60–65 m.p.h. running to make up for time lost. Tomorrow's trip will tell the final story. For almost all of the battlers for top honors, good mileage tomorrow could turn defeat into victory."[29] Fortunately, weather conditions would show improvement.

In the early morning hours of April 9, no one knew for sure who was more eager to see the Kansas City finish line, the crews or USAC officials, as the competitors faced a fairly short 293.2-mile trip. A time limit was set at 7 hours and 45 minutes, which was extended by 4 minutes due to traffic conditions. Bob Russo and his Pontiac Bonneville once again was first car on its way, leaving at 3:00 a.m., with the remaining entries leaving at 2-minute intervals. What was thought to be an easy leg turned out the opposite as the route "was over rolling country with poor passing conditions and a lot of traffic often boxing in the competitive cars mercilessly."[30] Heading north out of Lincoln, the drivers passed through Omaha and Council Bluffs before turning in a southerly direction towards the first stop of the day just south of Savannah, Missouri, for brunch at the New Del Café. After Savannah, their next stop was at Brown's Mobil Service in St. Joseph for the final refueling stop of the Run. After crossing the border into Kansas, it was on to Kansas City and the finish line at the Muehlebach Hotel, where a large crowd was on hand to welcome the competitors. Among those on hand was a group of top-performing Mobil dealers, who accompanied the competitors from Los Angeles on a specially chartered bus. Just after 11:00 a.m. Bob Russo in his Pontiac Bonneville was the first to arrive, leading a parade of dusty, snow-streaked cars to take the checkered flag from J.C. Agajanian. As the cars arrived, USAC officials began their final inspection and topping-off for the final time, with the Awards Ceremony scheduled for later in the afternoon. While waiting for the announcement of the results, many of the drivers commented that traffic had been reasonably heavy during the day.

One of the biggest "hard luck" stories in the

A snowstorm over 11,988-foot Loveland Pass resulted in slush and ice-covered roads in below-freezing temperatures. Byron Froelich carefully maneuvers his Mercury Monterey through this area, slowly making his way towards the Denver impound.

history of the Economy Runs occurred some 40 miles north of the finish line in Atchison, Kansas. Mary Davis had led the low-price eight-cylinder class since leaving Los Angeles, but her hope for a second class win was all but wiped out when a wheel on her Plymouth Belvedere 8 dropped into a deep hole in the pavement, which resulted in a broken tie-rod. Sitting on the side of the road stranded, Mary quickly located a Plymouth dealership six blocks away and was able to persuade them to provide a replacement tie-rod, which she then replaced by herself as the rules required. With her car repaired, Mary was on her way within 38 minutes. Driving hard and keeping within legal limits to make up for lost time, she crossed the finish line, although 28 minutes and 21 seconds late. After her arrival, the impound area was described as a "sea of confusion," with members of the press interviewing Mary about what had transpired, while USAC officials were busy determining the penalty for her late arrival. Then came the announcement. There would be a 1.6464 gallon penalty assessed, which resulted in a second-place finish behind Pierce Venable, driving another Plymouth Belvedere 8.[31] "I didn't beat her — the officials did," commented Venable after the final results were announced.[32] Two other drivers were also late in arriving, including class-leading driver Mary Hauser, and received gallonage penalties which did not affect the outcome.

After leaving Denver, the drivers continued to battle snow and ice as they made their way across eastern Colorado bound for McCook, Nebraska. While their cars were being refueled, the drivers worked hard to knock as much snow and ice as possible off their cars to reduce those extra gas-consuming pounds. Teenager Judy Allred relied on an oil can to knock the ice off the bumper of her Buick Invicta.

With the Sweepstakes Award now history, interest was focused on the highest mileage figure among all entries, which was set by Woody Bell and his Rambler American Deluxe 6 with 25.2878 miles per gallon. Although placing fourth (and last) in the compact car class, the best performance by an eight-cylinder model was set by a Studebaker Lark Regal with 22.2837 miles per gallon. Among the full-size models, a Chevrolet Biscayne, driven by Mary Hauser, recorded best miles-per-gallon figure for six-cylinder models with 22.3788, while a Dodge Coronet, driven by Patricia Jones, achieved the best figure for eight-cylinder models with 21.7454 miles per gallon. Dwight Hinton, a first-time participant in an Oldsmobile Super 88, recorded the worst mileage for all entries with 16.7708 miles per gallon. Other than the change in first place in the low-price eight-cylinder class, class leaders remained constant throughout the Run.

Class winners were:

CLASS	MAKE and MODEL	MILES PER GALLON
A — Compact Car	Rambler American Deluxe 6	25.2878
B — Low Price Six Cylinder	Chevrolet Biscayne	22.3788
C — Low Price Eight Cylinder	Plymouth Belvedere	21.1548
D — Low Medium Price	Dodge Coronet	21.7454
E — Upper Medium Price	Ford Thunderbird	19.1301
F — High Price	Cadillac Sixty-Two	19.0352

Detailed results for all competing cars can be found in the Appendix.

In reporting the results, USAC Officials noted that the average for the entire field was 19.4396 miles per gallon. While this was a noticeable improvement over the average attained in 1958, it was noted that this due largely to the performance of the new compact cars, along with an increase in the number of six-cylinder models competing. The competition between the men and women drivers for overall mileage averages remained close, with the men scoring 19.4595 miles per gallon compared to the women with 19.3925 ... a difference of only .6700 miles per gallon. This comparison would continue in future years. For the first time in several years, first-place victories were spread among most of the manufacturers, with Chrysler Corporation and General Motors each recording two first-place finishes, while American Motors and Ford Motor Company had one each. Only Studebaker-Packard was left out of the running. USAC also noted that all entries completed the Run, and other than the incident with Mary Davis on the last day, no other problems were reported. Overall the weather conditions could best be described as challenging! Of the six class-winning drivers, only Woody Bell and Pierce Venable had scored victories previously. Art Chrisman, Mary Hauser, Patricia Jones and Jim Parkinson were first-time winners. Although only competing in four classes, women drivers took home first-place trophies in two classes, while Mary Davis placed second in her class.

Competition between the Rambler and Studebaker Lark entries in the new compact car class was closely watched. While it was expected that the Rambler American Deluxe 6 would capture the top prize, interest centered on the duel between the six-cylinder Rambler Deluxe and Studebaker Lark Deluxe for second-place honors. Les Viland drove his Rambler into second ahead of the Studebaker by over a half-mile per gallon. Hampered by an engine problem throughout the Run, Bill Corey hung on and managed a third-place finish. Following the daily progress of their entries, Studebaker-Packard President Harold Churchill elected to be in Kansas City to greet both Lark crews after crossing the finish line.[33]

The closest competition this year was in the upper-medium-price class, with only .1352 miles per gallon separating the first four place entries. First-place winner Art Chrisman and his Ford Thunderbird held off the challenge

Art Christman crosses the finish line at Kansas City's Muehlebach Hotel in his dusty, snow-streaked Ford Thunderbird with a large group of spectators on hand to welcome the drivers. In its first Economy Run appearance, the Thunderbird took top honors in the upper-medium-price class by a slim .0600 mile per gallon over a Chrysler Saratoga.

of Mel Alsbury, Sr., and his Chrysler Saratoga by a mere .0600 miles per gallon. The margin was even closer for second and third between Mel Alsbury, Sr., and Hart Fullerton, driving a DeSoto Firedome, with a slim .0069 miles per gallon separating the two. Art Chrisman, better known for being a "leadfoot" from his recent dragster performance, explained how he prepared for economy driving by "mounting a carburetor where he could observe it under various conditions as a laboratory specimen. 'What especially struck me,' he said later, 'was how that pump squirts out the gas when you tromp on her. So I figured the thing to do was feed evenly at all times and avoid spurts. I [also] came to the conclusion you should even feed a little doing downhill if that would help you roll up the next hill without too much consumption. So that was how I drove.'"[34]

The youngest driver competing, 17-year-old Judy Allred in a Buick Invicta, placed seventh in the upper-medium-price class with 18.2003 miles per gallon. Her performance placed her ahead of Robert M. Flanagan, driving an identical Buick Invicta, in eighth with 17.8735 miles per gallon, and well ahead of the Oldsmobile Ninety-Eight and Pontiac Bonneville entries.

Cadillac, returning to the Economy Runs for the first time since 1951, was responsible for ending Chrysler Corporation's three-year string of victories in the high-price class. In the three years Cadillac had competed, it had placed first in its price class in each time.

Most drivers made the trip with windows rolled up tightly to reduce wind resistance and refrained from using power-consuming accessories in an effort to increase their mileage. Class-winning driver Mary Hauser didn't believe in such tactics, explaining that she "opened car windows when desert heat was oppressive, operated the heater during blizzards in the Rocky Mountains, and turned on the radio when she wanted news."[35] Another class winner, Jim Parkinson, added that "he didn't spare any comforts, though things got a little tedious when his radio just up and quit on him toward the last."[36]

Disappointing to many observers was the performance of the Alsbury family with their Chrysler and Imperial models. Equipped with a new, larger 413-cubic-inch engine, the best performance Mel Alsbury, Jr., could obtain was 18.2027 miles per gallon in his Imperial Custom, placing second behind Jim Parkinson's Cadillac Sixty-Two. Both the Chrysler Windsor and Saratoga models were equipped with the new 383-cubic-inch engine, and both delivered performance in the 19-miles-per-gallon range. George, driving a Chrysler Windsor, placed third in the low-medium-price class with 19.6454 miles per gallon, while Mel, Sr., driving a Chrysler Saratoga, placed second in the upper-medium-price class with 19.0701.

In general, advertising saw an increase this year, which was a result of the elimination of the ton-miles-per-gallon formula, along with a spread of several manufacturers scoring class victories. With Socony-Mobil's "The New Car Gasoline" advertising campaign, results of the 1959 Run were heavily promoted during the year. As was expected, the first to appear in newspapers and enthusiast publications covered the official results, along with a photograph of the class-winning Plymouth Belvedere 8 at the Los Angeles start. Under the headline "1959 Mobilgas Economy Run Tests Mileage of All U.S. Cars," the class winners, along with the best mileage by make, were discussed. Also a brief story of the Run by 1958 Indianapolis 500 Winner Jim Bryan was included.[37] As the year progressed, Socony-Mobil continued stressing the theme, "You're Miles Ahead with Mobil!" in advertisements for their Mobilgas Special blend that was seen in both general interest and enthusiast publications. These advertisements noted, "Only *one* gasoline gives you this proved mileage economy," which featured photos of various competing models. Available at local Mobil stations, a four-page folder was produced showing the official results for all entries. Also offered during the year was the booklet "Here's How *You* Can Get Better Gas Mileage from Your Car!," which outlined economical driving techniques to increase mileage.

American Motors was quick to advertise "Rambler Wins First Two Places" in newspapers and selected industry and enthusiast publications. Featuring photographs of both the Rambler American Deluxe and larger Rambler Deluxe, the copy announced that "Rambler has done it again!" noting too that it was "the all-time post war Mobilgas Economy Run record holder."[38] Appearing in general interest publications, including *Life* and *The Saturday Evening Post*, these ads made only a brief mention of the Economy Run performance, instead focusing mainly on the larger Rambler Custom sedan and station wagon models. Economy Run articles and pictures were found in *American Motors News Illustrated*, a newsletter designed for dealers and customers. Potential customers were told of the American's performance and its being recognized "Economy King" in its 1960 brochure. Additional details of Rambler's Economy Run performance were found in "X-Ray" comparison catalogs, also available at dealers. Not content with merely noting Rambler's Economy Run performance, American Motors continued to promote their private, NASCAR-supervised economy tests primarily using overdrive-equipped models.

Plymouth heavily advertised its third straight first-place victory competing against low-price class models. Using the theme "Others Talk Economy — Plymouth Delivers It!," newspaper advertising stressed this performance over Ford and Chevrolet. This theme continued with a two-page, black and white advertisement in *Life* magazine. Featuring a replica of the winning Belvedere hardtop, this ad noted, "Never before has a single manufacturer's V-8 won the low price class three

1959 MOBILGAS ECONOMY RUN TESTS MILEAGE OF ALL U.S. CARS

"I Saw Gasoline History Made"... Jim Bryan

JIM BRYAN, winner of the 1958 Indianapolis "500" and performance steward of the 1959 Mobilgas Economy Run, saw gasoline history in the making.

KANSAS CITY, Mo., April 9—Just four hours ago, 47 men and women, drivers in the 1959 Mobilgas Economy Run, wearily crossed the finish line in front of the Muehlebach Hotel. They were dog-tired after 4½ days and 1899 miles of some of the toughest driving I've ever seen.

Powered by Mobilgas Special, stock cars representing every U.S. make started from Los Angeles' Civic Center at dawn on April 5. First across the blistering desert, then up, up to 11,988 feet and over the jagged spine of the Rockies, then over the prairies. And by the end of the Run they had bucked 342 miles of traffic in 147 towns and cities. Deserts, mountains, prairies and cities—running at maximum legal speeds all the time!

One of the drivers summed it up for me: "Toughest, most demanding grind I've ever been through. But when I needed it, Mobilgas Special came through with plenty of smooth, knock-free power. And it came through with plenty of economy, too!"

Official Results 1959 Mobilgas Economy Run

	COMPETITIVE CLASS	OFFICIAL WINNER	MILES PER GALLON
Class A	Compact Car Class	Rambler American Deluxe	25.2
Class B	Low Price Class, Six Cylinder	Chevrolet Biscayne	22.3
Class C	Low Price Class, Eight Cylinder	Plymouth Belvedere	21.1
Class D	Low Medium Price Class	Dodge Coronet	21.7
Class E	Upper Medium Price Class	Ford Thunderbird	19.1
Class F	High Price Class	Cadillac Sixty-Two	19.0

Average of all cars entered 19.4 m.p.g.

BEST MILEAGE BY MAKE

Make	MPG	Make	MPG	Make	MPG
Buick	18.8	Edsel "6"	19.6	Oldsmobile	18.2
Cadillac	19.0	Edsel "8"	17.8	Plymouth "8"	21.1
Chevrolet "6"	22.3	Ford "6"	20.3	Pontiac	18.6
Chevrolet "8"	19.2	Ford "8"	19.6	Rambler	25.2
Chrysler	19.6	Imperial	18.2	Studebaker	22.4
De Soto	19.0	Lincoln	17.2	Thunderbird	19.1
Dodge	21.7	Mercury	18.9		

2-2448

Route of 1959 Mobilgas Economy Run

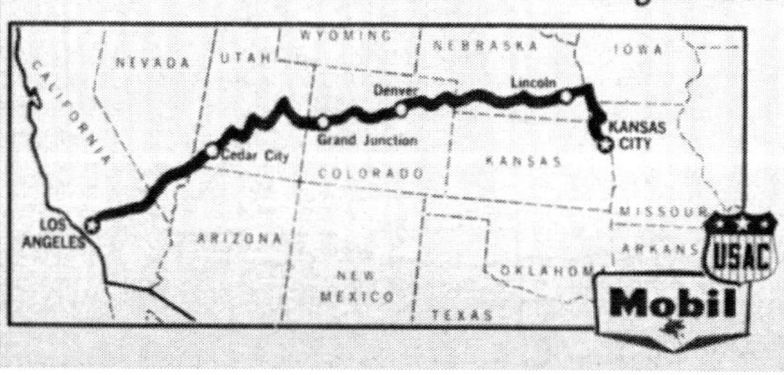

The Mobilgas Economy Run—an official test of cars competing with others in the same price class—is sponsored annually by General Petroleum, far-west affiliate of Socony Mobil, as a public service to motorists. Throughout the entire Run, cars, drivers and gasoline are carefully supervised by United States Auto Club officials.

In announcing the results of the 1959 Economy Run, the low-price class-winning Plymouth Belvedere was shown at the Los Angeles start. The story of the Run, told by 1958 Indianapolis 500 winner Jim Bryan, also featured the class winners and best mileage achieved by each make.

With the Sweepstakes Award now history, interest would now center on who would achieve the highest miles per gallon figure for all entries. This year it was Woody Bell who coaxed 25.2878 miles per gallon out of his Rambler American Deluxe, placing first in the compact car class.

straight years."[39] This same theme continued during the model year in various magazine advertisements noting its Economy Run performance. Plymouth also issued a small mailer folder, for use by its dealers, announcing "one car just proved it can give better gas mileage."[40]

Dodge quickly advertised in newspapers of its "1–2 Sweep" in the low-medium-price class with the tag-line "It Pays to Own a Dodge," listing all of the eight-cylinder competing cars in miles-per-gallon order, highlighting Dodge's top performance. This same theme was found in a magazine advertisement appearing in *Life* and *The Saturday Evening Post*, also noting that Dodge "topped all other cars from the low price V-8 field on up."[41] In a separate advertisement found in industry-related publications, it was announced that these same two Coronet models continued, under USAC supervision, to New York City, where on this segment they delivered an average of 22.1 miles per gallon.

After three years, Studebaker finally had something to celebrate with the performance of its Lark V-8. In newspapers and magazines, "The Lark By Studebaker Scores Most Miles Per Gallon Of All V-8's In All Classes In Mobilgas Economy Run" headlined the 22.28-miles-per-gallon achievement which, Studebaker noted, "proves outstanding economy of Studebaker Lark."[42] Additional advertisements, appearing in publications such as *Life* and *The Saturday Evening Post*, continued to highlight this achievement through the early months of the 1960 model year. In a separate newspaper advertisement aimed at his local community, Pasadena (California) Studebaker dealer Lloyd Pearson noted the performance of the Lark Deluxe "6" he had entered. Featuring a photograph of the Lark crossing the Kansas City finish line, the ad invited readers to stop by the dealership to see the actual car that had scored 22.4 miles per gallon. While there was no mention of the Lark's performance in the regular showroom brochures, Economy Run performance was noted in the 1960 Econ-O-Miler Taxi and Fleet Model brochures, which targeted specific customers.

Chevrolet made brief mention of the Biscayne's class-winning performance. In newspapers, a simple ad proclaimed that "Chevy 6 Wins Its Class in Mobilgas Economy Run!," noting that a Biscayne was tops in Class B. In a two-page spread appearing in *Life* and *The Saturday Evening Post* that featured an Impala convertible, readers were told that "a pair of Chevy sixes — equipped with Powerglide transmission — came in first and second in their class in this year's Mobilgas Economy Run, ahead of any other full-size car."[43] Both Cadillac and Ford chose to ignore their class-winning performances.

Explaining that tires play a role in overall economical operation, B.F. Goodrich was quick to note in press releases that four of the six winning cars (Rambler American Deluxe 6, Chevrolet Biscayne 6, Ford Thunderbird and Cadillac Sixty-Two) were equipped with their tires. B.F. Goodrich also noted that 31 of the 47 cars competing were equipped with their tires.[44]

Even though they were not class winners, both Mel Alsbury, Sr., and Hart Fullerton, both longtime Economy Run participants, promoted the performance of their respective entries. Designed by the same advertising agency, glossy heavyweight cards showing their respecting entries crossing the Kansas City finish line noted that these models scored "an amazing 19.0 miles per gallon average." A similar six-page folder was produced telling readers, "Mel Alsbury Shows You How To Save $ $ $ When You Drive..." It not only highlighted the Saratoga's performance but provided tips on how to improve gas mileage through both driving and maintenance tips.[45]

Although reporters from across the country covered the Run, newspaper articles were not as extensive as in previous years. With the official results not being released until the competitors reached Denver, after 3 days on the road, some of the interest was apparently lost after the initial pre-start publicity. Daily coverage was provided in a few newspapers, including the *Los Angeles Times*, which also reported unofficial results compiled by several automotive editors. Feature stories appearing after the conclusion of the Run, usually found in the Sunday automotive sections, were far less detailed than in previous years. Even with the pending introduction of new compact cars from the Big Three, magazine coverage was sparse and in some non-existent. *Sports Cars Illustrated*, in an issue appearing prior to the start of the 1960 Run, covered in detail the experiences of both the Rambler and Studebaker teams as they competed head-to-head for compact car class honors. After covering the Run with in-depth articles throughout the decade, *Motor Trend* only provided a capsule look at the run, publishing the results of the first three places only.[46] Coverage of the Lark's performance was featured in *Studebaker News*, a magazine designed for Studebaker dealers and staff.

This year also marked the final publication of the Mobilgas Economy Run book by well-known automotive publisher Floyd Clymer, a big supporter of the Runs. These pictorial books had been published every year except 1956, covering the events from beginning to end. These books were, and still are, the best historical reference sources available.

Nineteen fifty-nine would mark the last year that General Petroleum's name would be associated with the Mobilgas Economy Run. On January 1, 1960, General Petroleum Corporation and Magnolia Petroleum Company were fully merged into Socony-Mobil and from that date began operating under the name of Mobil Oil Company. Robert L. Minckler was named senior vice-president of Mobil Oil and elected to retire rather than relocate his home to New York.[47] The personnel associated with General Petroleum in the promotion and publicity of the Economy Run, including Advertising Manager Frank Meunier, would continue their efforts in the new merged company.

1960

Los Angeles to Minneapolis Via Flagstaff— Tucumcari — Wichita — Des Moines

April 2–6, 1960—2061.4 Miles

After three years of significant changes, Mobil Oil and USAC officials felt it was time to take a breather and only improve on the rules and procedures where necessary. At the same time, the Economy Run was ready to welcome a new batch of recently introduced compact models by Chevrolet, Ford and Chrysler set to do battle with Rambler and Studebaker. When the rules were published in the fall of 1959, the emphasis was placed on the random selection process "in order to absolutely assure the American public what cars in the Run are typical of customer cars."[48] This would be the most closely supervised Run in its history.[49]

In a continuing focus on the Midwest region, Minneapolis was announced as this year's terminus. Leaving Los Angeles on April 2, the route would cover distance of over 2000 miles in a 5-day period and would begin a trend by organizers towards longer, tougher routes. Details of the route, including overnight stops, would not be divulged until March 31.

At the same time preparations were being made, new competition surfaced in the form of the Pure Oil Economy Trials (renamed the Pure Oil Performance Trials in 1963), which were held in early February at Daytona International Speedway. Sponsored by Chicago-based Pure Oil Company and supervised by NASCAR, these trials covered economy, acceleration and braking tests and were designed to rival the

Mobilgas Economy Runs. Being held on a closed course, and not over public highways, these trials never gained the full attention of the public, although there was some interest by members of the press.[50]

Eligibility was open to any stock 1960 two-door or four-door metal-top sedan, with competitive classifications remaining identical to those used in 1959. Power steering and power brakes remained a requirement for all full-sized eight-cylinder models. In addition, the rules covering the eligibility of women participating as drivers or relief drivers continued from 1959 without change.[51] Fifteen teams of women drivers and relief drivers competed and were joined this year by four husband-and-wife teams. Women piloted all of these entries, with Marta and Jim Retzlaff the only team with previous Economy Run experience. Pat Sawyer, with her husband Bob handling relief driver duties, made her first appearance in a Chevrolet Corvair 500 and would become a regular, successful participant. Another husband-and-wife team set to compete was Mickey and Judy Thompson. Each would drive a separate entry, with Mickey in a Pontiac Catalina and Judy in a Pontiac Bonneville. Teenager Judy Allred returned for a second year as a driver, with her father, Visalia (California) Buick dealer Claude L. Allred handling relief driver duties. Another newcomer from the racing world was Vicki Wood, who had topped all drivers in the 1959 Daytona Speed Weeks with a top speed of 150.375 miles per hour.[52]

With the largest field to date, a total of 56 cars competed, almost doubling the number of entries just two years ago. Every U.S. make was represented with the exception of Lincoln. The new Mercury Comet did not participate due to the fact it was not "on the market long enough to make the run's entry deadline," according to USAC Officials.[53] General Motors again had the largest representation with 24 entries, followed by Chrysler Corporation with 17, and Ford Motor Company with 9. American Motors and Studebaker rounded out the field with 3 entries each. Individual dealers sponsoring entries, a backbone of support in previous Economy Runs, were barely in the majority as the manufacturers and their dealers' associations were beginning to have a much larger presence. In addition to those from California, dealer sponsors this year hailed from Illinois, Kansas, Minnesota, Missouri,

With his traditional leap into the air, USAC's J.C. Agajanian sends Ak Miller and his Ford Fairlane 8 (right) on their way from Los Angeles' Memorial Coliseum. Jim Parkinson in his Studebaker Hawk (left) would be sent on his way one minute later.

Ohio, Texas and Virginia, reflecting a growing awareness of the Economy Run across the country. Included among the newcomers this year were Barnett Buick in Burbank, California; Nickey Chevrolet in Chicago, Illinois; and Melvin Hilliard Chevrolet in Kansas City, Missouri, who would become regular participants through the 1960s. The lone private entry this year was the Cadillac Sixty-Two by Los Angeles businessman Leonard DeBell. This year would mark the final time DeBell, owner of DeBell Auto Sales and Southern California Plating Company, would be involved with the Economy Run, an association dating back to 1950.

With the increased number of entries, a continuing group of first-time drivers made their appearance, including several with prominent racing backgrounds who were tapped to drive either Chevrolet or Pontiac entries. These included Chuck Daigh, John Fitch, Mickey Thompson, and one of the foremost names in stock car racing, Smokey Yunick. Two other first-time drivers, Don Francisco, a writer for *Hot Rod Magazine*, and John Rich, an imported car dealer, would continue their involvement through 1968. All six class-winning drivers from 1959 returned for another shot at a first-place trophy and were joined by four previous Sweepstakes-winning drivers — Mel Alsbury, Jr., Dick Griffith, Bill Stroppe, and Les Viland. Fourteen drivers had participated the previous November in the Mobil Mileage Rally, with two, Jim Parkinson and Lillian Warner, placing first in their respective classes for the second year in a row. Random car selection commenced in February and followed procedures refined by USAC over the past several years. A more thorough inspection procedure was inaugurated, which included tear-down of selected parts to make sure that entries were "stock in every detail" before placing seals on over 50 parts, including the hood and trunk lid.[54] In an unusual move, the Run's Chief Steward A.C. Pillsbury traveled to Detroit, where he selected Chevrolet Corvair, Ford Falcon and Chrysler's Valiant entries directly from the assembly line.[55] In one particular situation, he "selected stock auto components — engine, transmission, rear, axle, etc. — then followed them down the assembly line as they were joined into a complete car."[56]

To help maintain a strictly stock status for all entries, the time allowed for servicing the competing cars was reduced to 10 man-hours, down from 12 in previous years. Additionally, new rules now limited the time a competing car could be away from the impound for break-in runs to 5 hours, along with a reduction of the maximum number of break-in miles to 1500. In previous years this mileage limit was 2000 miles.[57] Entrants

With members of USAC's timing crew standing by, J.C. Agajanian gives the green flag to Betty Skelton in a Chevrolet Biscayne 6. Standing next to the right-side windshield pillar in a light-colored jacket is this book's author, witnessing the start of the Economy Run at age 14.

were again given the option of using either Mobilgas R (Regular) or Mobilgas Special gasoline. The two Pontiac Catalina entries were the only ones who elected to use Mobilgas R.

Soon after leaving Los Angeles, the drivers had to battle strong winds as they made their way towards a refueling stop in Blythe. As Mickey Thompson in a Pontiac Catalina heads east on the San Bernardino Freeway, the bending trees in the background show the strength of these winds.

During the impound period, two changes in the rules and procedures during the Run were announced. First, only one observer would be assigned to each competing car who would ride alongside the driver to ensure all rules and traffic regulations were strictly observed. This resulted in reducing the passenger load factor to 600 pounds, compared to 750 pounds in previous Runs. Having served as USAC riding observers since 1951, senior engineering students from the California Institute of Technology (Caltech) were replaced by members from business and industry who volunteered their time and talents. Caltech students would continue to be utilized during the impound period as riding observers for break-in runs. With the successful operation of the new trunk-mounted gas tanks, each car would now be topped off at one of the refueling stops during the day with the results being announced as the competitors arrived at the impound each night. This change virtually eliminated the post-midnight starts that were so common in past years, with the competitors now heading out usually between 5:00 and 6:00 a.m. each morning.[58]

While the impound remained a hub of activity, USAC officials were putting the finishing touches on the route to Minneapolis and planning for alternate routes should conditions warrant. Mobil Oil's Advertising Manager Frank Meunier headed a team of advertising and public relations experts who worked with members of the press from across the country in making sure coverage of the Economy Run was thorough and professional. This team, which included Roger Mahey, Ken Downs and John Parkinson, would ensure that all areas were thoroughly covered by the use of press releases and personal contact. One of the main pre-start publicity events was the drawing of starting positions. Selecting from a group of rolled pennants containing a starting position, Ak Miller drew No. 1 for his Ford Fairlane 8. Judy Allred, driving a Buick Invicta Custom, drew the final spot with No. 56.

Once it was known that this year's Run would terminate in Minneapolis, interest now focused on the announcement of the course, scheduled for the afternoon of March 31. After the winter weather conditions encountered last year, USAC went for a more southerly route, covering 2061.4 miles over a five-day period. Overnight stops would be made in Flagstaff, Tucumcari, Wichita and Des Moines before reaching the finish line at the Hotel Leamington and Hotel Curtis in downtown Minneapolis.

Although the route was not as mountainous as last year, the competitors would still cover all kinds of terrain as they traveled through nine states with an altitude range from 22 feet below sea level at Indio, California, to 7270 feet at the Continental Divide in New Mexico. Additional obstacles along the route would include travel through 142 cities and towns and contending with 174 traffic signals. In order to stay within the 50-hour, 41-minute time limit set by the USAC, drivers would average a speed of 41.0980 miles per hour.

Los Angeles' Exposition Park, just south of downtown, was this year's starting location, with the cars being lined up two rows abreast curbside along Exposition Park Drive. With a small crowd of spectators on hand, and the Los Angeles Memorial Coliseum in the background, the drivers were ready on April 2 for their Saturday morning departure, aware that they would be facing the threat of severe weather and flooding conditions during the later stages of the Run.[59] Prior to the start, Marta Retzlaff's co-driver and husband Jim became too ill to compete and was replaced by Bob Newill. At 6:00 a.m.

the first driver to take the green flag was Ak Miller in a Ford Fairlane 8 with the remaining competitors leaving at one-minute intervals. Heading east on the San Bernardino Freeway, with little traffic to contend with, the drivers began a 485.3-mile trip to Flagstaff within a time limit of 12 hours and 10 minutes. As in previous years, time for refueling and 40-minute (increased from 30 minutes) brunch stops were not included in this limit. Almost immediately the drivers were "buffeted with strong headwinds all across the flat desert areas of San Bernardino County" as they made their way to Blythe.[60] Along the way they passed the lowest elevation on the Run at 22 feet below sea level in Indio before arriving in Blythe for refueling at John's Mobil Service. Immediately following this stop, the crews were ready for a break at Palo Verde Unified School, where brunch was served by a group of local volunteers. With temperatures hovering around 100 degrees, the competitors crossed the Colorado River into Arizona, then headed northward towards Prescott and Williams. After battling head winds, the drivers found their progress now hampered by slow-moving trucks in Arizona's mountainous terrain.[61] Arriving at Williams Mobil Service, all entries were topped off before heading to Flagstaff and the impound at the Flamingo Motel. Beginning this year, overnight impounds would for the most part be located at a hotel or motel where the cars, crews and USAC officials would be housed in a single location. After battling head winds and slow traffic most of the day, Ak Miller in his Ford Fairlane 8 was the first to arrive at 7:50 p.m. Here everyone was greeted by "the most exciting scene in recent economy run history. Spotlights, mounted posse riders, a western band, police and a troop of Boy Scouts waved at the widely spaced stream of cars into overnight headquarters."[62] Dick Griffith, driving a Studebaker Lark Deluxe 8, was 52 seconds late and was penalized.

To no one's surprise, Les Viland and his Rambler American Custom topped all entries, while leading the compact car class with the best miles-per-gallon figure of 27.6091. The 56-car field turned in an average of 20.5587 miles per gallon. In the battle for second place in the compact car class, Vince Piggins in his Chevrolet Corvair 500 was barely ahead of Ginny Sims and her Valiant V-100, with a mere .0155 miles per gallon separating the two. Disappointing to observers was the performance of the Ford Falcon entries, sitting in sixth and seventh places. A surprise was the strong showing of Jim Peterson and his Studebaker Hawk, leading the low-medium-price class by more than 1½ miles per gallon over the second-place entry. Chrysler Corporation's entries were leading in four classes — Mary Hauser (Plymouth Savoy 6), Mary Davis (Plymouth Belvedere 8), Mel Alsbury, Jr. (Chrysler New Yorker) and George Alsbury (Imperial Custom) — all veteran class-winning drivers. In the two low-price classes, close competition between first and second place was developing. For six-cylinder models, Mary Hauser's Plymouth Savoy was leading Pierce Venable, in another Savoy, by .1339 miles per gallon. Mary Davis's Plymouth Belvedere led Jim Fosdick, in another

A couple of residents were on hand to welcome Lute Eldridge and his Dodge Dart Seneca 8 as he arrived in Gallup, New Mexico. Over the years, groups of well-wishers along the route greeting the competitors would become common.

Belvedere, by .3873 miles per gallon among the low-price eight-cylinder models. Dodge's new Dart Seneca 6 and 8 models followed close behind in third place. The top three performers in each class were:

Compact Car	Rambler American Custom	27.6091 Miles Per Gallon
	Chevrolet Corvair 500	26.2424 Miles Per Gallon
	Valiant V-100	26.2269 Miles Per Gallon
Low Price Six Cylinder	Plymouth Savoy	24.0584 Miles Per Gallon
	Plymouth Savoy	23.9245 Miles Per Gallon
	Dodge Dart Seneca	23.8909 Miles Per Gallon
Low Price Eight Cylinder	Plymouth Belvedere	22.2197 Miles Per Gallon
	Plymouth Belvedere	21.8324 Miles Per Gallon
	Dodge Dart Seneca	21.3867 Miles Per Gallon
Low Medium Price	Studebaker Hawk	22.0860 Miles Per Gallon
	DeSoto Fireflite	20.5054 Miles Per Gallon
	Dodge Matador	20.4222 Miles Per Gallon
Upper Medium Price	Chrysler New Yorker	20.1436 Miles Per Gallon
	DeSoto Adventurer	19.6366 Miles Per Gallon
	Ford Thunderbird	19.6291 Miles Per Gallon
High Price	Imperial Custom	19.6558 Miles Per Gallon
	Cadillac Sixty-Two	17.9872 Miles Per Gallon

Leaving Flagstaff on April 3, the competitors would follow a fairly easy 495.5 miles on famed Route 66 across Arizona and New Mexico. In New Mexico, the drivers would be on the "newly-built New Mexico Freeway" which was described as "smooth as glass along almost every mile."[63] Chuck Daigh, driving a Chevrolet Bel-Air 8, was the first driver to leave at 5:00 a.m. to cover the distance within a 12-hour and 5-minute time limit with near-perfect weather conditions, including mild temperatures and winds, predicted for the entire trip. The rest of the field followed at 1-minute intervals. After crossing into New Mexico, their first stop was in Gallup for a brunch stop at the El Rancho Hotel, where the competitors were greeted by a few local residents on horseback. Heading east out of Gallup, the competitors crossed over the Continental Divide, the highest elevation on the Run at 7270 feet, before arriving in Grants for refueling at the Farmer Parish Mobil Service. From Grants, the competitors traveled through Albuquerque before arriving in Santa Rosa, where all entries were topped off at either Ira W. Smith Mobil Service or John B. Salvo Mobil Service. Prior to leaving Flagstaff, the competitors were told which station they had been assigned. From there it was a straight shot to Tucumcari and the impound at the Town House Motel, with Chuck Daigh in his Chevrolet Bel-Air 8 the first to arrive just after 5:45 p.m. All competitors arrived within the time limit, although problems plagued one driver when Verne Houle's Ford Thunderbird developed carburetor trouble 1000 feet from the day's finish. Verne was able to nurse his car into the impound; however, according to the rules he could not make any diagnosis or repairs until after his assigned starting time the next morning.[64]

The near-ideal weather conditions enabled the drivers to post an increase in mileage figures compared to the first day, with drivers reported to be "happier as they boosted their averages more than a mile per gallon."[65] With an excellent figure of 29.7052 miles per gallon, Les Viland and his Rambler American Custom continued to lead all entries. The entire field increased their average to 21.9838 miles per gallon, an improvement of 1.4251 miles per gallon. All of the class-leading drivers, with one exception, increased their leads over the second- and third-place drivers as close competition continued. With both driving a Plymouth Savoy 6, Mary Hauser's lead over Pierce Venable was trimmed to a slim .0938 miles per gallon.

In the low-medium-price class, Link Paola in a Dodge Matador moved into second place, and Byron Froelich in a Mercury Monterey moved into third. Both dropped Al Cottle in a DeSoto Fireflite into fourth place, with only .0962 miles per gallon separating the three. The top three performers in each class were:

Compact Car	Rambler American Custom	29.7052 Miles Per Gallon
	Chevrolet Corvair 500	28.0479 Miles Per Gallon
	Valiant V-100	28.0217 Miles Per Gallon
Low Price Six Cylinder	Plymouth Savoy	25.8098 Miles Per Gallon
	Plymouth Savoy	25.7160 Miles Per Gallon
	Dodge Dart Seneca	25.2995 Miles Per Gallon
Low Price Eight Cylinder	Plymouth Belvedere	23.9612 Miles Per Gallon
	Plymouth Belvedere	23.5427 Miles Per Gallon
	Dodge Dart Seneca	23.0655 Miles Per Gallon
Low Medium Price	Studebaker Hawk	23.7145 Miles Per Gallon
	Dodge Matador	21.9388 Miles Per Gallon
	Mercury Monterey	21.8823 Miles Per Gallon
Upper Medium Price	Chrysler New Yorker	21.6671 Miles Per Gallon
	DeSoto Adventurer	21.0170 Miles Per Gallon
	Ford Thunderbird	20.8339 Miles Per Gallon
High Price	Imperial Custom	21.2679 Miles Per Gallon
	Cadillac Sixty-Two	19.3357 Miles Per Gallon

On April 4, a shorter run of 416.4 miles faced the drivers as they headed in a northeast direction to Wichita. Mickey Thompson, driving a Pontiac Catalina, was the first driver on his way at 5:30 a.m. The drivers faced a 10-hour time limit on a trip over a more difficult set of highways than they encountered the previous day, which would include blustery winds for over half of the route. Verne Houle, whose Ford Thunderbird developed carburetor trouble arriving at the impound, was able to diagnose the problem as a ruptured diaphragm in the power valve and replaced the part on his own time, completing the job in 12 minutes 34 seconds. Verne was then on his way and made up this time on the road, arriving

Traveling along U.S. Route 54, Vince Piggins in a Chevrolet Corvair 500 passes a group of school students on hand to greet the drivers passing through town. After a third-place finish in the compact car class, this Corvair would next go to Colorado's Pike's Peak to complete the season's earliest climb of the 14,110-foot mountain.

in Wichita within the time limit.[66] Following U.S. 54, they passed through Texas, crossing into the Central Time Zone, and then into Oklahoma, arriving in Guymon for a brunch stop at Denny's Dine A Teria, followed by refueling at Henry Hodges Mobil Service in Hooker. Pete Novotny, driving a Cadillac Sixty-Two, took a wrong turn early in the day, going out of his way for about a mile before quickly discovering and correcting his mistake. This brought back stark memories of another wrong turn in 1953, when Pete, after getting back on course, ran out of fuel before reaching the refueling station. Crossing into Kansas, the drivers passed through Dodge City before arriving in Pratt, where entries were topped off at Leroy & Lavern Young Mobil Service. Just after 5:00 p.m., Mickey Thompson in his Pontiac Catalina was the first to arrive at the impound at the Hotel Lassen in downtown Wichita. After arriving, the drivers were anxious to check on their results that were being posted on a large scorecard in the hotel's parking lot. George Alsbury, driving an Imperial Custom, became ill with a gastric upset and was unable to drive this leg, so his co-driver and father, Mel, Sr., took over the driving duties. Upon arrival, George received medical attention and was able to resume driving the next day.[67]

Even with today's driving challenges, the overall mileage figure only dropped slightly, with the entire field averaging 21.9230 miles per gallon. Continuing to lead all entries was Les Viland and his Rambler American Custom with another excellent mark of 29.5117 miles per gallon. Close competition continued in several classes, with all the first-place drivers holding onto their leads. Ginny Sims, driving a Valiant V-100, moved into second place in the compact car class, dropping Vince Piggins and his Chevrolet Corvair 500 to third. In the low-price six-cylinder class, Mary Hauser was able to stretch her lead over Pierce Venable to .1616 miles per gallon. Both were driving a Plymouth Savoy. Jim Fosdick continued to close in on Mary Davis in the low-price eight-cylinder class. With both driving a Plymouth Belvedere, the margin separating the two was now down to .3591 miles per gallon. Filling in for his son, co-driver Mel Alsbury, Sr., continued to maintain a nearly two-miles-per-gallon margin over Pete Novotny in a Cadillac Sixty-Two. The top three performers in each class were:

Compact Car	Rambler American Custom	29.5117 Miles Per Gallon
	Valiant V-100	28.0192 Miles Per Gallon
	Chevrolet Corvair 500	27.8612 Miles Per Gallon
Low Price Six Cylinder	Plymouth Savoy	25.7505 Miles Per Gallon
	Plymouth Savoy	25.5889 Miles Per Gallon
	Dodge Dart Seneca	25.4016 Miles Per Gallon
Low Price Eight Cylinder	Plymouth Belvedere	23.7587 Miles Per Gallon
	Plymouth Belvedere	23.3996 Miles Per Gallon
	Dodge Dart Seneca	23.1274 Miles Per Gallon
Low Medium Price	Studebaker Hawk	23.6795 Miles Per Gallon
	Mercury Monterey	22.1180 Miles Per Gallon
	Dodge Matador	21.8826 Miles Per Gallon
Upper Medium Price	Chrysler New Yorker	21.6646 Miles Per Gallon
	DeSoto Adventurer	21.0185 Miles Per Gallon
	Ford Thunderbird	20.5850 Miles Per Gallon
High Price	Imperial Custom	21.2525 Miles Per Gallon
	Cadillac Sixty-Two	19.2659 Miles Per Gallon

As the drivers relaxed at the hotel, USAC officials announced that drivers would face uncertain weather and road conditions on the next leg to Des Moines. The original route, southwest of Des Moines, was reported to be under more than 4 feet of water due to the flooding of the Raccoon River, making it necessary to change the route slightly.[68]

After three days on the road, the most challenging leg faced the drivers on April 5 as they headed to Des Moines. Judy Allred, driving a Buick Invicta Custom, was the first to leave at 6:00 a.m. for the 405.7-mile trip with a 9-hour and 55-minute time limit set by USAC. The first part of the day's drive was perhaps "the easiest kind of motoring over the 200-mile stretch of the Kansas turnpike, with a 75 m.p.h. speed limit and a 40 m.p.h. minimum posted for all cars."[69] On the Kansas Turnpike, the first stop of the day was at the Topeka Service Area's Mobil Service. Leaving the Kansas Turnpike, the competitors passed through Kansas City before crossing into Missouri, and then headed north on U.S. Route 69 to their next stop in Cameron for brunch at the Rambler Café. Continuing north, they crossed into Iowa, where the route on U.S. 69 was described as tortuous with "rough two-lane roads, narrow bridges, and tractors towing farm wagon loads of hay." The drivers proceeded over miles of rural roads to their next stop in Leon, where all entries were topped off at Redman's Mobil Service. The obstacles were not over yet. As expected, a change in the route to the Des Moines impound resulted in an eight-mile detour, and the drivers encountered heavy evening traffic chiefly caused by side roads' being mired in mud as they approached the downtown area.[70] Judy Allred in her Buick Invicta Custom was the first to arrive just before 4:30 p.m. at the impound located at the Des Moines Veterans Memorial Auditorium Parking Lot. This would be the only time this year that an overnight impound was not located at a hotel or motel where the competitors were staying.

Arriving in Wichita, Pierce Venable carefully threads his Plymouth Savoy 6 through the crowded parking lot at the impound. In the background, a USAC official records the day's results after being finalized by USAC's calculating crew. Pierce would finish third in the low-price six-cylinder class in what would be his last appearance in the Economy Run.

With road and weather challenges, it was expected that the overall mileage figure would drop for a third consecutive day. This was the case, as the entire field averaged 21.6319 miles per gallon, a drop of just under three-tenths from the previous day. Les Viland and his Rambler American Custom continued to lead all entries with 29.0032 miles per gallon. With close competition continuing in several classes, especially in the low-price six- and eight-cylinder classes, there were no changes in first-place standings. In the low-price six-cylinder class, Mary Hauser hung onto her lead by a slim .0369 miles per gallon over Pierce Venable, both driving a Plymouth Savoy. Not far behind in third place was Woody Bell in a Dodge Dart Seneca. With both driving Plymouth Belvederes, Jim Fosdick continued to close in on Mary Davis in the low-price eight-cylinder class. The margin was now down to .3252 miles per gallon. In the low-medium-price class, Jim Parkinson's Studebaker Hawk maintained a comfortable lead, with the real competition now between second through fourth place. While Byron Froelich in a Mercury Monterey hung onto second place, Al Cottle in a DeSoto Fireflite swapped places with Link Paola, dropping his Dodge Matador to fourth by the slimmest of margins—.0003 miles per gallon. The top three performers in each class were:

Compact Car	Rambler American Custom	29.0032 Miles Per Gallon
	Valiant V-100	27.8084 Miles Per Gallon
	Chevrolet Corvair 500	27.5349 Miles Per Gallon
Low Price Six Cylinder	Plymouth Savoy	25.3538 Miles Per Gallon
	Plymouth Savoy	25.3169 Miles Per Gallon
	Dodge Dart Seneca	25.1944 Miles Per Gallon
Low Price Eight Cylinder	Plymouth Belvedere	23.3557 Miles Per Gallon
	Plymouth Belvedere	23.0305 Miles Per Gallon
	Dodge Dart Seneca	22.7937 Miles Per Gallon
Low Medium Price	Studebaker Hawk	23.3846 Miles Per Gallon
	Mercury Monterey	21.5927 Miles Per Gallon
	DeSoto Fireflite	21.5206 Miles Per Gallon
Upper Medium Price	Chrysler New Yorker	21.3158 Miles Per Gallon
	DeSoto Adventurer	20.7764 Miles Per Gallon
	Ford Thunderbird	20.3344 Miles Per Gallon
High Price	Imperial Custom	20.9444 Miles Per Gallon
	Cadillac Sixty-Two	19.1226 Miles Per Gallon

With storm conditions still in the forecast, the drivers knew the pressure was on to maintain or improve their positions as the final leg to Minneapolis was at hand. This fact was not lost on Mary Davis, who had been edged out on the last day by Pierce Venable in the past two years.

Ready for the final leg of the Run, the drivers faced a fairly short 258.5-mile, 6-hour and 25-minute trek to Minneapolis. Today would be the only time the Run would visit Minnesota. Mickey Thompson in his Pontiac Catalina was the first to depart at 5:00 a.m., continuing north on U.S. Route 69. Strong head winds, up to 40 miles per hour, would hamper the drivers and affect their mileage figures. USAC officials announced that an emergency refueling station would be available in Forest City, Iowa, but this would not be a compulsory stop. Eleven competitors stopped to add fuel at this point, all of whom were experiencing poor mileage and running near the bottom of their respective classes. The only stop of the day for everyone was at the Salem Lutheran Church in Lake Mills, Iowa, for brunch served by members of the congregation. Picking up U.S. Route 65, they were on a straight shot into Minneapolis, where they began to encounter heavy traffic on the outskirts of town. A large group of spectators was on hand as the competitors arrived at the finish line in front of the Hotel Leamington and Hotel Curtis. Running eighth in the low-medium-price class, Mickey Thompson decided to "speed it up" and be the first to take the checkered flag from J.C. Agajanian just before noon. Later he told Bill Dredge of the *Los Angeles Times* that he was "almost foiled by John Rich in a Buick Invicta. 'I thought I was all alone out front,' Thompson said, 'then Rich just dropped out of the sky and nearly beat me. But neither of us broke any speed laws. We just scooted through the heavy traffic into Minneapolis."[71] Three of the competitors—John Fitch (Chevrolet Biscayne 6), Patricia Jones (Dodge Dart Seneca 8) and Tommy Thomas (Rambler Deluxe 6)—arrived late, ranging from as little as 18 seconds up to 4 minutes and 34 seconds, and received penalties. After arrival, all entries were topped off for the final time as the Awards Ceremony was scheduled for the afternoon.

For the second year in a row, the highest mileage figure

Mel Alsbury, Jr., travels through farm country between Wichita and Des Moines in his Chrysler New Yorker on his way to another class victory. Mel placed first in the upper-medium-price class with 20.8715 miles per gallon in his fourth win.

for all entries was set by a Rambler American Custom driven by Les Viland with 28.3533 miles per gallon. This was a 3.0655 miles-per-gallon improvement over the mark set by the American Deluxe in 1959. Although well out of contention in the compact car class, Dick Griffith in a Studebaker Lark Deluxe 8 recorded the highest mileage figure for eight-cylinder models for the second year in a row with 23.2838 miles per gallon. Among full-size models, the best performance for by a six-cylinder model was the Plymouth Savoy's figure of 24.8265 miles per gallon, and for eight-cylinder models the Studebaker Hawk was the best with 22.9899 miles per gallon. At the other end of the spectrum, Don Francisco driving a Pontiac Bonneville returned the lowest mileage of all competitors with 16.7715. Although competition remained close in several classes to the very end, class leaders did not change from the first day's stop in Flagstaff.

Class winners were:

CLASS	MAKE and MODEL	MILES PER GALLON
A—Compact Car	Rambler American Custom 6	28.3533
B—Low Price Six Cylinder	Plymouth Savoy	24.8265
C—Low Price Eight Cylinder	Plymouth Belvedere	22.8899
D—Low Medium Price	Studebaker Hawk	22.9899
E—Upper Medium Price	Chrysler New Yorker	20.8715
F—High Price	Imperial Custom	20.5036

Detailed results for all competing cars can be found in the Appendix.

In reporting the results, USAC officials noted that the average for the entire field was 21.2259 miles per gallon. The expanded participation of compact cars resulted in a 1.7863 miles-per-gallon improvement from last year's average. In the "battle of the sexes," the men averaged 21.2882 miles per gallon, with the women close behind at 21.1137. USAC Chief Steward A.C. Pillsbury noted, "This year's higher mileages reflect progress in the automotive industry in meeting public demand for more economical driving."[72] Chrysler Corporation again dominated, placing first in four of the six classes, with American Motors and Studebaker taking the other two classes. Ford Motor Company had two third-place trophies, and General Motors took a second and a third place. All entries completed the Run and no major problems were reported, although a minor repair was made on the Ford Thunderbird entry in Tucumcari. Weather conditions were generally good, although strong gusty winds were encountered on the first and last days. USAC officials also noted, "There were less infractions of the rules this year than any previous run and none were of a serious nature."[73]

Veterans dominated the winner's circle, with Les Viland, George Alsbury and Mel Alsbury, Jr., scoring first place victories after mixed results in 1959. The two victorious women drivers, Mary Davis and Mary Hauser, each collected their second first-place trophies. Jim Peterson's victory was in his first time as a driver.

Perhaps the most closely watched competition was among the new crop of compacts. Since it was understood that the Rambler American Custom would walk away with a first-place finish, interest was on how the new Chevrolet Corvair, Ford Falcon and Chrysler Corporation's Valiant would perform. The Valiant came out on top, placing both second and fourth, with the Corvair close behind in third and fifth. Disappointing was the performance of the Falcon entries, which placed sixth and seventh. One reason for this was Ford's change in January from an axle ratio of 3.31.-to-1 to a more performance-oriented ratio of 3.56-to-1. During the impound period, USAC officials would not allow the use of the lower ratio, since it was not considered stock according to the published rules. For comparison, the American had an axle-ratio of 2.87-to-1, resulting in increased economy figures.[74] The poorest performance was the six-cylinder Studebaker Lark Deluxe in tenth (and last) place, nearly 1½ miles per gallon behind the performance of its eight-cylinder counterpart.

In the low-price classes, Plymouth and Dodge were taking all of the honors. Among the six-cylinder cars, the Plymouth Savoy took first- and third-place honors, with Dodge's new Dart Seneca moving into second place on the last leg. Competition between the first three places was close from the start, but Mary Hauser took first place for the second year in a row. Mary once again "let comfort be her guide [as] she opened the windows, played the radio, and worked the heater in the cold."[75] Woody Bell, driving a Dodge Dart Seneca, moved into second place by a mere .0069 miles per gallon over Pierce Venable in another Plymouth Savoy. Finishing well back in the standings was a Ford Fairlane and a pair of Chevrolet Biscaynes.

After two years of enduring last-day disappointments, Mary Davis hung on to take first place among the eight-cylinder cars in a Plymouth Belvedere by a "fairly comfortable" .3643 miles per gallon. This marked the fourth consecutive year that a Plymouth Belvedere 8 captured low-price class honors. Jim Fosdick, driving another Plymouth Belvedere, placed second, and Lute Eldridge placed third in a Dodge Dart Seneca. Last year's low-medium-price class winner, Patricia Jones, placed fourth in another Dart Seneca. As was the case with the six-cylinder models, a Ford (Fairlane and Starliner) and Chevrolet (Bel-Air and Impala) finished well back in the standings. Noting this performance, *Motor Life Magazine* observed: "The Plymouth-Dodge showing wasn't unpredicted but many observers thought that of the two, the Dart might produce the better figures for reasons of weight and design. Plymouth's better standings can be traced to a factor that

Jim Peterson brings his Studebaker Hawk across the finish line in Minneapolis greeted by a large group of spectators led by USAC's J.C. Agajanian. A strong first-place finish in the low-medium-price class, this would be Studebaker's eighteenth and last class win.

can't be forgotten in events of this kind, the more experienced drivers."[76]

While there was no question after the first day that Jim Peterson's Studebaker Hawk would take first place in the low-medium-price class, the battle for second through fourth place would be tight. Despite having suffered a broken speedometer en route, Link Paola managed to move his Dodge Matador into second place, dropping Byron Froelich in a Mercury Monterey to third, with Al Cottle in a DeSoto Fireflite ending up fourth. Only .1075 miles per gallon separated these three entries at the finish.[77]

Another three-way battle took place in the upper-medium-price class. Leading from the start, the Alsburys were back as Mel, Jr., maintained his lead in a Chrysler New Yorker by a margin of .5088 miles per gallon over Hart Fullerton in a DeSoto Adventurer. Despite his Ford Thunderbird's carburetor trouble in Tucumcari, Verne Houle managed to comfortably hold onto third place over Eleanor Edwards in a Chrysler Saratoga. As was the case in the low-medium-price class, the Buick, Oldsmobile and Pontiac entries finished well off the pace.

This year would be the last appearance by a DeSoto in the Economy Run after participating each year since the inaugural 1936 Gilmore Yosemite Run. During this time, DeSoto placed first only once, in 1952 with a Firedome sedan, in addition to placing second on ten occasions.

So, how did those who are prominent in racing circles fare in the Run? Not well, and all placed far down in the standings. In the low-price six-cylinder class, Betty Skelton placed sixth and John Fitch placed seventh, both driving Chevrolet Biscaynes, while in the low-price eight-cylinder class, Art Chrisman in a Ford Starliner placed seventh, and Chuck Daigh in a Chevrolet Bel-Air placed eighth. In the low-medium-price class, Mickey Thompson in a Pontiac Catalina placed eighth, while Smokey Yunick in a Pontiac Star Chief placed eleventh. With the exception of Mickey Thompson and Smokey Yunick, none would return.

Two drivers with rich Economy Run history, Dick Griffith and Bill Stroppe, made their final appearances in 1960. Bill captured the Sweepstakes Award both in 1950 and 1952 with a Mercury, while Dick captured the 1954 Sweepstakes Award in a Studebaker Land Cruiser. Both veterans competed

For the second consecutive year, Mary Hauser topped all other entries in the low-price six-cylinder class. This year driving a Plymouth Savoy, Mary averaged 24.8265 miles per gallon over the 2061.4-mile route from Los Angeles to Minneapolis. With Mary Davis also placing first in the low-price eight-cylinder class in a Plymouth Belvedere, this marked Plymouth's fourth consecutive class win.

in the Compact Car Class with disappointing results, although Dick's Studebaker Lark Deluxe recorded the best mileage of all eight-cylinder cars competing.

Following the trend of recent years, Mobil Oil advertised the results in newspapers and magazines with the headline "Mobilgas Economy Run Cars Averaged 21.23 Miles Per Gallon," continuing with their "You're miles ahead with the New Car Gasoline" slogan.[78] A separate advertisement highlighted the performance of the compacts, noting that "Compact Cars Averaged 25.50 Miles Per Gallon on the Mobilgas Economy Run!"[79] In enthusiast magazines, the "miles ahead" theme continued, focusing not only on the Economy Run mileage average, but on the performance of Mobil-powered cars in the Indianapolis 500 Mile Race and Pikes Peak Hill Climb. Available again at Mobil stations was a four-page folder showing the official results for all entries, which followed the same design and theme used in 1959.

American Motors advertised that "Rambler American Custom Wins Over All Compacts!" in newspapers and selected industry publications. With a photograph of the winning American Custom sedan, this advertisement covered aspects of the Run and included a history of Rambler's previous victories.[80] A second advertisement, again in newspapers and selected industry publications, reinforced the American's performance. With the headline "In spite of all the confusing claims about the Mobilgas Economy Run ... Rambler American Beat *All Cars* In Miles Per Gallon," the advertising copy went on to note that "almost every major car maker claims a victory of some kind."[81] This appeared to be aimed mainly at advertisements from Chevrolet about the Corvair's "Economy Proved in Mobilgas Run!" and Chrysler Corporation's ad touting the Valiant's 2nd-place showing that "beat out both of the other compact cars."[82] As the year progressed, there was only brief mention made of the American's performance in advertisements appearing in *Life* and *The Saturday Evening Post*, along with other general interest publications. The American's Economy Run performance continued to be noted in showroom brochures for the 1961 model year referring to it as "America's Recognized Economy King," while in the "X-Ray" comparison catalogs, NASCAR-supervised economy tests were also highlighted. Interestingly the NASCAR-supervised Pure Oil Economy Trials were not directly mentioned in these brochures and catalogs.

Chevrolet produced limited advertisements plugging the economy performance of the Corvair. In newspapers, a photograph of one of the competing coupes was featured with the headline "Economy Proved in Mobilgas Run!" The ad went on to detail the final mileage figures for both competing models.[83] This was followed by a single magazine advertisement, appearing after the Corvair was named *Motor Trend*'s Car of the Year, which also noted the Corvair's Economy Run performance.

As would be expected, Chrysler Corporation heavily advertised its dominating performance in both newspapers and magazines. Newspaper advertising stressed the theme "talk about economy" and provided an introduction to the individual newspaper advertising campaigns for Chrysler, Imperial and Plymouth.[84] Treated to a two-page black and white photographic layout, readers of *Life* and *The Saturday Evening Post* were told of Chrysler's overall success. The ad pointed out that Chrysler Corporation cars "won 9 out of 12 possible first and second places — 4 out of 6 firsts, 5 out of 6 seconds."[85] Other advertisements highlighted the achievement of those finishing in first, second and third place, and most included the final mileage figures.

Chrysler and Imperial's advertising was placed mainly in newspapers in the days immediately following the Run. "Chrysler Wins!" was the headline announcing the New Yorker's class win in an ad which included a portion of a telegram sent to Chrysler Division by Hollywood Chrysler dealer Mel Alsbury, Sr.[86] The Imperial's newspaper advertisement was fairly restrained: with the headline "Four-Time Winner," it announced the "Imperial completed the 2000-mile Los Angeles-to-Minneapolis in the Imperial tradition ... as a winner!"[87]

Plymouth's advertising campaign stressed the fourth consecutive win in its class in newspaper and magazine spreads. With renderings of the previous four years of Plymouth models, one newspaper advertisement headlined that it is "4 Straight Wins for Plymouth ... 4 Straight Facts for You ..." The emphasis was placed more on the eight-cylinder model's record over the six-cylinder.[88] In showroom brochures issued for the 1961 models, the four-year consecutive winning streak theme continued.

Autolite Spark Plugs capitalized on Plymouth's success in advertisements found mainly in enthusiast magazines. With a photograph of the class-winning Savoy 6 crossing the finish line, a headline announced, "Autolite Spark Plugs Sparked More Than Twice As Many Mobilgas Economy Run Winners As All Other Spark Plugs Combined."[89] Later in the year, another Autolite ad in *Stock Car Racing* combined Plymouth's Economy Run victory with that of Chevrolet and Ford victories during May.

In what would be its last successful Economy Run performance, Studebaker's newspaper advertising stressed the results for all three of their entries. "Studebaker Economy Proved Again in Mobilgas Economy Run!" was the headline accompanied by renderings of the Lark and Hawk models that competed. The ad stated, "The Lark V-8 Repeats for Second Straight Year," along with "The Hawk V-8 Sweeps Its Class." Even with the poor performance of the six-cylinder Lark, this noted that "gas mileage is only ¼ of The Lark Total Economy Story!"[90] Magazine advertising was virtually nonexistent, although in a single black-and-white advertisement for the Hawk, appearing in *Newsweek*, its Economy Run performance was noted in the tagline: "The HAWK ... 1960 Mobilgas Economy Run Winner with 22.9 mpg."[91]

Newspaper coverage was more noticeable this year, with interest centered on the battle between the compact cars competing, along with the changing weather conditions. With the focus on the Rambler American Custom leading all entries with the best overall mileage, plus Chrysler Corporation's strong performance, readers were also kept abreast of the close contests occurring in most classes, along with the flooding conditions drivers faced approaching Des Moines. An article appearing in *Time Magazine* highlighted the performance of the two class-winning women drivers, while other articles focused more on the compact car competition. Articles appearing in enthusiast magazines were more detailed than in 1959.[92] One of the more in-depth articles appeared in *Motor Trend Magazine*'s June 1960 issue, written by Bob Russo, a regular participant who sat out this year's Run. In a comprehensive overview

Two-time Sweepstakes champion and well-known race car builder Bill Stroppe stands by his Ford Falcon holding an honorable mention plaque after the awards ceremony. Bill placed seventh in the compact car class in what would be his last year behind the wheel.

that presented not only the challenges the competitors faced on the road, but also stressed USAC's efforts to keep everything above board, he concluded the Economy Run "is one of the most exacting and honest contests of its kind anywhere."[93] Coverage of the Hawk's winning performance was emphasized in *Studebaker News*' June 1960 issue for Studebaker dealers and staff. There was a photograph of the winning Hawk on the cover, and additional photographs covered the progress not only of the Hawk, but the Lark models as well, from start to finish.

After completion of the Economy Run, arrangements were made for both Chevrolet Corvair entries to travel, with all major parts remaining sealed and under the supervision of USAC Official Chad Johnson, to Pikes Peak, Colorado. On April 15, a silver Corvair 500 Coupe that had placed third in the compact car class completed the season's earliest climb, arriving at the 14,110-foot summit at 1:50 p.m. A special black-and-white folder was developed for use by Chevrolet dealers which documented the Corvair's Economy Run and Pikes Peak Hill Climb performances.[94]

7
Chicago

1961
Los Angeles to Chicago Via Tucson — Roswell — Dallas — Little Rock — St. Louis
MARCH 11–16, 1961—2560.7 MILES

Expansion was in the forefront when plans for the 1961 Economy Run were announced. With increasing emphasis on compact models, along with the arrival of additional models from the "Big Three," an opportunity was readily available for an expanded focus on the gasoline economy. This was highlighted with the reintroduction of cars equipped with a manual transmission into competition, for the first time since 1954, although restricted to compact models only. The introduction of these new competing models was only one of the new changes announced by Mobil Oil and USAC.

When the rules were announced, the biggest change involved eligibility for those named to compete as drivers. In an effort to dampen criticism that the previous mileage figures could be attained only by professionals and not by the average motorist, new provisions were established that "two cars of the same make and model may be entered provided one is driven by a person who has never before driven in the Mobilgas Economy Run. Male and female drivers and relief drivers are equally acceptable."[1] Four years after women made their first appearance in the Run, the playing field was now leveled and would remain so in the future. To further enhance the impartiality of the USAC riding observers, these assignments would now change twice each day, at the start and again at the brunch stops.

When the formal announcement of this year's Run was made by Vern A. Bellman, Mobil Oil's marketing vice-president, on February 6, a surprise was in store for the competitors. In an effort to prevent competitors from gaining detailed knowledge of the entire route, from now on the next day's actual route and running time would not be provided until after arrival at each day's overnight stop. As much as possible, this year's Run "would be a no-advantage contest."[2] Selecting Soldier Field in Chicago as the terminus, Bellman added that "the more than 2000-mile course, longest in the 25-year history of the event, will take six days [and] overnight stops will be made in Tucson, Ariz.; Roswell, N.M.; Dallas, Tex.; Little Rock, Ark.; and St. Louis, Mo."[3]

With four-door models now outselling their two-door counterparts (the only exception for 1960 models was the Ford Falcon 2-door sedan outselling the 4-door), eligibility was now limited to four-door sedan or hardtop models, reflecting those body styles being purchased by the American public. Increased emphasis was on the compact cars (those with a wheelbase of 114" or less) with three separate classes being established, expanding the total number of classes to seven. As outlined in the rules, "Class A will be for compacts having three speed manual transmissions. Class B will be for four and six cylinder compacts with automatic transmissions, and Class C for eight cylinder compacts with automatic transmissions. Entry in Class A is contingent upon a comparable model competing in either Class B or C." Reflecting on the new realities in the marketplace, the medium-price field was now consolidated into a single class. As was the case in previous years, power steering and power brakes were required on all full-sized eight-cylinder models.[4]

A record-setting field of 67 cars competed, representing every U.S. make, although American Motors declined to enter their larger Rambler Classic and Ambassador models. With the change in body style eligibility, two successful models from prior years were not allowed to compete — the Ford Thunderbird and the Studebaker Hawk. General Motors had the largest representation with 29 entries, followed by Ford Motor Company with 19, and Chrysler Corporation with 14. Rounding out the field was Studebaker with 3 entries and American Motors with 2. Compact cars made up just over half of the entries with 34 competing. Two interesting additions were made to both low-price classes with the Mercury Meteor 600 and Pontiac Catalina set to do battle with the traditional Chevrolet, Dodge Dart, Ford and Plymouth models. Both six- and eight-cylinder Mercury Meteor 600s competed.

Increased factory support and influence were in evidence, which in turn was reflected in the continued decline of individual dealers' participation, a group which had been a main-

With the Los Angeles Memorial Coliseum in the background, USAC's J.C. Agajanian sends the first car in the 1961 Economy Run on its way. Competing in the new compact car with manual transmission class, the eight-cylinder Oldsmobile F-85, driven by Stan Raymond, would be no match for the smaller six-cylinder Chevrolet Corvair, Ford Falcon and Mercury Comet entries.

stay of support since 1936. By a slim margin they still sponsored the majority of entries and included representatives from each state visited on this year's run. Two Midwestern dealers—Doane Chevrolet in Dundee, Illinois, and Ray Nolting Oldsmobile in Kirkwood, Missouri—participated for the first time and would continue their involvement over the next few years. Dealer associations were the main sponsors for 30 entries, representing those from Dodge, Ford, Lincoln, Mercury, Plymouth and Pontiac. Former Los Angeles area Hudson dealer Lee Hamer, now an imported car dealer, privately entered and drove a Cadillac Sedan de Ville. This continued to reflect the absence of support by General Motors' Cadillac Motor Car Division. Lee, participating as a driver for the first time since 1956, was the only Californian to travel east to New York the previous October to drive in the Mobil Mileage Rally. He placed third in his class, driving a Volvo.

With the change in eligibility rules, 28 teams with novice drivers participated. A number of these novice drivers would become familiar names in future years and included Bob Cahill (Plymouth Savoy), Ronnie Kaplan (Chevrolet Corvair), Bill Levy (Mercury Comet), Bob McDaniel (Plymouth Savoy), Jim Nelson (Pontiac Tempest), and Edward J. Wilder (Oldsmobile F-85). Women piloted 16 entries, with prior class winners Mary Davis and Patricia Jones returning, looking to add additional trophies to their collections. All of last year's class-winning drivers returned, hoping for a repeat performance, with Mary Hauser this year acting as co-driver for Mary Davis.

Early in 1961, the first battle in what had become over the past few years a "love-hate relationship" between American Motors and USAC officials surfaced. As the first entries were being received, USAC barred a Rambler American Custom from competing in Class A (compacts with manual transmission) "on the grounds it was not AMC's lowest-priced model." As reported in *Automotive News*, "A.C. Pillsbury, chief steward, said the barred model had a bigger engine, was a higher-priced model and was not sold in sufficient number of justify its entry. The American Deluxe and the Super have lower price tags than the Custom." Roy Abernethy, American Motors executive vice-president, countered that "no such limitation is made in the rules covering entries in the compact-car class and such a ruling prevents this year's Mobilgas Economy Run from being a test of the economy of all cars readily available to the general public."[5] An examination of the rules for the 1961 Economy Run tends to back up Roy Abernethy's argument. The following sentence appears in Section 4, Classifications For The Run: "Entry in Class A is contingent upon a comparable model competing in either Class B or C."[6] American Motors followed this rule by entering two Rambler American Customs in Class B (compact four- and six-cylinder with automatic transmission), although the rules did provide USAC officials with some leeway in rejecting a particular entry. Although no documentation has surfaced, the *Los Angeles Times* reported that not allowing this particular model to participate was on the grounds "that the model selected was not the lowest priced model in the Rambler line—a contingency involved in that class of cars."[7] This statement turned out not to be entirely accurate. Further examination of models entered in Class A reveals that three entries competing (Chevrolet Corvair 700, Oldsmobile F-85 Deluxe, and Pontiac Tempest Custom) were not the base, or lowest priced, models. To further complicate this issue, *Automobile Topics* added the American Custom, equipped with an overhead-valve engine, was not permitted in Class A "for lowest price compacts ... because there [was] another American, the Deluxe, with the old L-Head engine that costs less."[8]

Beginning in 1962, a guide to the competitive divisions which would include the make and model, number of cylinders and engine displacement for each class, would be provided to all entrants along with the rules in an effort to prevent future misunderstanding and confusion.

With random car selection commencing in late January, USAC continued to emphasize this task as one of the more important elements to ensure the stock status of all competing cars. After each entry was selected, detailed inspections commenced, following established procedures from earlier years, verifying that all parts met factory specifications and adhered to the rules that these parts were produced "in sufficient quantity and percentage of production to be considered as a representative product."[9] At this same time, "it became clear that with newer, tighter regulations, there was no longer room for quibbling or compromise with" Chief Steward A.C. Pillsbury or other USAC officials.[10] For example, in previous years, one-size leaner carburetor jets were permitted, but not this year: carburetors now had to be standard equipment as listed in each model's factory specifications.[11] Other rule changes were made that affected servicing and drivers' conduct during break-in runs. A minor change was in regards to the time allowed for servicing each competing car, being reduced to 8 man-hours. This was down from 10-man hours allowed in 1960. Break-in mileage remained at 1500. What was considered a major change involved the tires on each competing car. Most cars at this time were equipped with "synthetic rubber bias-belted tires that were not particularly good for fuel economy. It was known that identical tires that were made with natural rubber were almost 10 percent better at lower speeds, but all the cars came with the synthetics."[12] This rule change came about after it was noted that, in prior years, some participants were intentionally abusing their tires, counting on the fact that the replacement tires would be of the better natural rubber compound. Mickey Thompson was one of the more flamboyant drivers employing this tactic, and USAC had a reply.[13] Beginning this year, "the original tires and tubes, or tubeless tires that were on the car when selected, whether black or white, must be used on the Run, providing that they are regularly catalogued as standard equipment for the make and model entered."[14]

The passenger load factor of 600 pounds was eliminated beginning this year with adjustments no longer being made for weight differentials. The practice of equalizing this load factor by placing lead ballast in the trunk was discontinued. This lead ballast was thought to be a possible safety hazard in the event of an accident.[15]

As in the previous two years, entrants were given the option of using either Mobilgas R (Regular) or Mobilgas Special gasoline, with 24 entries electing to use Regular. Those selecting Mobilgas R were the Buick Special, Chevrolet Corvair (manual transmission) and Biscayne entries, as well as the entire Ford and Mercury team.

During the impound period, teams were busy servicing their entries and establishing driving techniques for the trek to Chicago. With the limited amount of time each competing car could be away from the impound, teams would extensively practice in "duplicate" models equipped with gauges and instruments to record performance. Other changes were becoming evident. After their previous successful Economy Run endeavors and the continued publicity this created, Chrysler Corporation began to slowly increase the involvement of factory engineers in pre-run preparation in both driver and co-driver positions. After an assignment to assist the Plymouth team in 1960, Chrysler Corporation engineer Bob Cahill began to realize that a better job could be accomplished by using factory personnel rather than local enthusiasts.[16] This long-term effort would start on a small scale with Bob Cahill piloting a Plymouth Savoy 6 along with another Chrysler engineer, Bob McDaniel, piloting a Plymouth Savoy 8.

Buick was another participant whose factory involvement became apparent this year but with a different approach compared to that of Chrysler Corporation. By the late 1950s, Buick's sales had stalled, and it had also developed a reputation as being a gas-guzzler. Although a Roadmaster surprised everyone by winning the high-price class quite handily in 1955, its results after that were disappointing. Buick engineer John W. Burnside was assigned to manage their efforts, and working with its engineering department assembled a team of Southern Californians as drivers and co-drivers.[17] Three drivers, Marta Retzlaff, John Rich, and Bob Russo, had previous Run experience and were joined by Barney Navarro. Los Angeles area Buick dealers joined in this effort as sponsors, and this group would continue their involvement through 1967. While other teams had a good deal of factory backing and support, none had the overall success that the Buick and Chrysler Corporation would exhibit.

While Mobil Oil's publicity machine was going full blast in promoting this year's Run, another battle between American Motors and USAC began to erupt. Already fuming at USAC's refusal to allow the American Custom to compete in Class A, American Motors reported a running change in carburetors installed as standard equipment on all overhead-valve six cylinder engines. They noted that this change had been made earlier in the year and prior to the January 23, 1961, deadline specified in the rules. By this time, both American Custom Sedans had been selected and were impounded under USAC supervision. Both were equipped with the older style carburetor no longer in production. American Motors then made a request to USAC officials to have the new carburetors installed and this was denied. In making this announcement, A.C. Pillsbury stated, "Like all cars in the run, the Ramblers were selected at random by USAC officials, transferred to our Pasadena impound and kept under guard. Subsequent checks confirmed that they were typical of the cars of that model being sold to the public." Later, additional details substantiating this decision were released by USAC's Technical Committee and Board of Stewards, which noted that although American Motors "made the claim that cars with the new carburetor were available in the Los Angeles area [and] as part of the comprehensive and country-wide [USAC survey], dealers were queried concerning any coming engineering changes. At no time did any of the dealers interviewed mention plans for a carburetor change, [and] test cars used by [Les] Viland's crew were not equipped with the new carburetor."[18] AMC continued to press its case, and neither the company nor USAC officials would back down from a looming showdown.

The dispute between American Motors and USAC over a requested change in carburetors resulted in the last-minute withdrawal of both Rambler American Custom entries. After everyone had left for Chicago, a disappointed Les Viland (left) and Les Scott, driver and co-driver of the Rambler American, stand alongside their car before making a leisurely trip home to Detroit.

While all this was going on, pre-run preparations continued, with the main publicity event being the drawings for starting positions. Once again selecting from a group of rolled pennants each designating a starting position, Stan Raymond drew No. 1 for his manual transmission–equipped Oldsmobile F-85. Another Oldsmobile F-85 would be the last car to depart since first-time driver Carol Paola drew No. 67. Interest now focused on how the drivers and co-drivers would adjust to the daily announcement of the route's next leg with the realization that they would have only a very limited time to study the route and maps. With each overnight stop known, potential routes had been explored by some of the factory teams. They sent out advance teams, one of whose tasks at hand was looking for those telltale signs that something important about to happen, like a Mobil Station freshly painted and spruced up. As always, USAC officials had a trick or two up their sleeve to make each day's route interesting and challenging.

On Friday afternoon, March 10, just before the cars were driven from the impound to the starting location at Los Angeles' Exposition Park, the detailed route for the first day's trip was released. Here the drivers learned that on the first day they would have a 524-mile trip to Tucson to be completed within a 12-hour time limit with time out for refueling and a 40-minute brunch stop along the way. Also facing the drivers were long-range weather forecasts indicating rain and cold weather could threaten much of the route.[19]

Repeating the scene from last year, the competing cars were lined up on Exposition Park Drive ready for the start on Saturday morning, March 11. Minutes before the first car was to be flagged on its way, American Motors dropped the other shoe in its dispute with USAC officials with the sudden withdrawal of both Rambler American Custom entries over the carburetor issue. Claiming that there was some prejudice against their cars, an American Motors spokesman "called the withdrawal of the two cars the 'last straw' in a controversy that started several months ago when an American Custom was barred from Class A on the grounds it was not AMC's lowest-priced model."[20] In addition, the spokesman said the "final decision to quit was made late Friday night by George Romney, president, and Roy Abernethy, executive vice-president. Almost a week [prior] some USAC officials said American Motors had threatened to withdraw, then had reconsidered after offers were made by non-factory sources to enter Rambler cars without factory sponsorship. The last-minute withdrawal was a successful attempt to keep unauthorized Rambler cars from competing."[21] A.C. Pillsbury, already in El Centro for the first brunch stop of the Run, was notified of the withdrawal as the American drivers, Les Viland and Gene Marx, prepared to head home to Detroit after the specially designed gas tanks were removed. In an attempt to put this incident to rest, A.C. Pillsbury stated that he "regretted that Roy Abernethy was 'not truly aware of the intent of the USAC ruling. Our first obligation in supervising the run is to assure the motoring public that the competing automobiles are typical production cars, the same as anyone might buy.'" Later, Mobil Oil released a statement saying that they are "merely the sponsor of the run, while USAC handles organizational matters. 'The reason we have a body like USAC handling the event is just because of the possibility that such events may happen.'"[22]

Despite this distraction, in cool and overcast weather, Stan Raymond in an Oldsmobile F-85 Deluxe was the first driver to take the green flag promptly at 6:30 a.m. Following a well-established routine, two USAC officials stood alongside each competing car with watches in hand. As their departure time approached, each driver was then given a countdown: "20 seconds, 10 seconds, 5, 4, 3, 2, 1, Go!" As USAC's starter J.C. Agajanian leaped into the air waving the green flag, the engines started (in some instances on the second or third try) and the driver was on his or her way. All of the remaining competitors followed at one-minute intervals, those starting down the line having to maneuver around the two Rambler American Customs still parked alongside the curb. Leaving Los Angeles, the drivers headed east on the San Bernardino Freeway towards Palm Springs and Indio. As they drove along the Salton Sea they passed an elevation of 205 feet below sea level, the lowest elevation on this year's Run, and then it was on to El Centro for brunch. Everyone was ready for a break after nearly 5 hours on the road. Crossing into Arizona, and into Mountain Standard Time, they made a refueling stop in Yuma. Just over a hundred miles down the road they arrived in Casa Grande, where all entries were topped off, then it was on to Tucson at the impound at the Ramada Inn. Arriving just before 8:15 p.m., Stan Raymond in his Oldsmobile F-85 Deluxe was the first to arrive, with everyone anxious to see the results from the first day. Four drivers, Lute Eldridge (Oldsmobile Dynamic 88), Dee Hildebrand (Pontiac Tempest), John Rich (Buick Special), and Ginny Sims (Plymouth Valiant), arrived late and were penalized. The penalty assessed in each case by USAC would require the driver to make up this time in the next leg, whether it was only a few minutes or a few seconds. This ruling would apply to all competitors arriving late through the St. Louis stop. A string of late arrivals would occur on the first four days of the Run.[23] Although it was a long day on the road, the terrain and weather conditions were good, as the first day's results would reflect.

With the withdrawal of the Rambler American entries, interest now focused on who would take the lead in overall mileage figures. Driving a Ford Falcon with a manual transmission, Dan "Termite" Jones set an all-time miles-per-gallon mark of 33.8834, eclipsing the mark of 33.21 set by a Willys 77 "4" in 1936. The remaining 65-car field showed an average of 23.4800 miles per gallon. Economy Run veterans leading were in all classes, with novices sitting in second place in four of the seven classes. Two classes were close. In the compact car four- and six-cylinder class, Pat Sawyer, in a Chevrolet Corvair Monza, was leading her teammate Bill Horton by a mere .0175 miles per gallon, while in the high-price class, Lee Hamer, in a Cadillac Sedan de Ville, had a .0781 lead over George Alsbury in an Imperial Custom. General Motors topped the "Big 3,"

Byron Froelich, driving a Ford Fairlane 8, drives by the lowest elevation on the 1961 Economy Run near the Salton Sea in the Southern California desert. Five days later in Chicago, Byron would place first in the low-price eight-cylinder class, ending Plymouth's four-year winning streak in this class.

placing first in three classes, with John Rich, in a Buick Special 8, also leading the compact car eight-cylinder class. In the low-price eight-cylinder class, a surprise was a pair of Ford Fairlanes and a Chevrolet Bel-Air leading the previous champion Mary Davis's Plymouth Savoy sitting in fourth place. As expected, two veteran drivers, Art Rene in a Plymouth Savoy 6 and Mel Alsbury, Jr., in a Chrysler Newport, led their respective classes by fairly comfortable margins.

The top three performers in each class were:

Compact Car — Manual Transmission	Ford Falcon	33.8834 Miles Per Gallon
	Ford Falcon	32.9634 Miles Per Gallon
	Chevrolet Corvair 500	31.9373 Miles Per Gallon
Compact Car Four and Six Cylinder	Chevrolet Corvair Monza	30.9309 Miles Per Gallon
	Chevrolet Corvair Monza	30.9134 Miles Per Gallon
	Pontiac Tempest 4	28.4970 Miles Per Gallon
Compact Car Eight Cylinder	Buick Special	26.1439 Miles Per Gallon
	Buick Special Deluxe	25.5899 Miles Per Gallon
	Pontiac Tempest Custom	25.0737 Miles Per Gallon
Standard Size Six Cylinder	Plymouth Savoy	24.0952 Miles Per Gallon
	Dodge Dart Seneca	23.6577 Miles Per Gallon
	Mercury Meteor 600	23.6062 Miles Per Gallon
Low Price Eight Cylinder	Ford Fairlane	22.4417 Miles Per Gallon
	Ford Fairlane	22.3821 Miles Per Gallon
	Chevrolet Bel-Air	22.2992 Miles Per Gallon
Medium Price	Chrysler Newport	20.5595 Miles Per Gallon
	Pontiac Star Chief	20.3797 Miles Per Gallon
	Pontiac Star Chief	20.2326 Miles Per Gallon
High Price	Cadillac Sedan deVille	19.5969 Miles Per Gallon
	Imperial Custom	19.5188 Miles Per Gallon
	Lincoln Continental	17.7250 Miles Per Gallon

On March 12, the drivers faced a 457.5-mile trip from Tucson to Roswell. With an earlier start than normal, at 4:00 a.m., Link Paola in an Oldsmobile F-85 was the first driver on the road to cover today's distance within an 11-hour and 21-minute time limit. Heading east on U.S. Route 80, the first stop, after crossing into New Mexico, was in Deming for refueling. It was here that the first mechanical failure since 1957 occurred when the transmission in Vince Piggins's Chevrolet Corvair 700 froze in third gear. Unable to continue, the Corvair was withdrawn from competition in what would be Vince's last appearance as a driver in the Run. For the remaining 64 drivers, it was on to Las Cruces for brunch at the Palms Motor Hotel Restaurant, where the competitors were ready for a break after being on the road for over 6½ hours. Leaving Las Cruces, the drivers picked up U.S. Route 70, and heading in a northeast direction passed through the southern section of the White Sands National Monument and Missile Range bound for Ruidoso Downs. After crossing the 7540-foot Apache Summit, the highest elevation on the Run, the competitors arrived in Ruidoso Downs, where the entries were topped off. From here it was a short drive to Roswell, finishing off what was the toughest day so far, with the drivers facing "a combination of blistering hill climbing, gusty side winds and narrow, rough, two-lane New Mexico highways."[24] Carol Paola, driving an Oldsmobile F-85 Deluxe, was the first to arrive, just after 4:00 p.m., at the Holiday Inn impound. Three drivers Mary Davis (Plymouth Savoy), Mel Hilliard (Chevrolet Bel-Air) and Jim Nelson (Pontiac Tempest) arrived late and were penalized.

With rougher driving conditions, most drivers were not too surprised to see their averages drop, some by close to a half-mile per gallon. Continuing to lead all entries was Dan "Termite" Jones and his Ford Falcon with 33.4043 miles per gallon, as he continued to eclipse Willys's record. The entire field turned in an average of 23.7737 miles per gallon. Two classes saw new leaders with novice drivers displacing the veterans. In the compact car four- and six-cylinder class, Bill Horton in a Chevrolet Corvair Monza took over first place from Pat Sawyer by a very slim .0112 miles per gallon. In the low-price eight-cylinder class, Robert Rice in a Ford Fairlane led teammate Byron Froelich by .0704 miles per gallon. The four other class-leading drivers held onto their leads for a second day, with the slimmest of margins found in the high-price class, where Lee Hamer's Cadillac Sedan de Ville held a .0921 miles-per-gallon lead over George Alsbury's Imperial Custom. The only other battle of note was in the medium-price class, where Mel Chastain moved into second place, dropping Mickey Thompson to third by only .0297 miles per gallon. Both were driving Pontiac Star Chiefs.

The top three performers in each class were:

Compact Car — Manual Transmission	Ford Falcon	33.4043 Miles Per Gallon
	Ford Falcon	32.4585 Miles Per Gallon
	Chevrolet Corvair 500	31.5585 Miles Per Gallon
Compact Car Four and Six Cylinder	Chevrolet Corvair Monza	30.1360 Miles Per Gallon
	Chevrolet Corvair Monza	30.1248 Miles Per Gallon
	Pontiac Tempest 4	28.0998 Miles Per Gallon
Compact Car Eight Cylinder	Buick Special	25.7762 Miles Per Gallon
	Buick Special Deluxe	25.3896 Miles Per Gallon
	Pontiac Tempest Custom	24.6646 Miles Per Gallon
Standard Size Six Cylinder	Plymouth Savoy	23.5363 Miles Per Gallon
	Dodge Dart Seneca	23.3290 Miles Per Gallon
	Mercury Meteor 600	23.2212 Miles Per Gallon
Low Price Eight Cylinder	Ford Fairlane	22.0660 Miles Per Gallon
	Ford Fairlane	21.9956 Miles Per Gallon
	Chevrolet Bel-Air	21.7487 Miles Per Gallon
Medium Price	Chrysler Newport	20.3222 Miles Per Gallon
	Pontiac Star Chief	20.0608 Miles Per Gallon
	Pontiac Star Chief	20.0311 Miles Per Gallon
High Price	Cadillac Sedan deVille	19.1603 Miles Per Gallon
	Imperial Custom	19.0682 Miles Per Gallon
	Lincoln Continental	17.4712 Miles Per Gallon

Dee Hildebrand, driving a Pontiac Tempest Custom 8, arrives at the Palms Motor Hotel Restaurant in Las Cruces, joining other competitors for a 40-minute brunch stop. With their long days of driving concentration and closely watching their speeds, the drivers found these stops to be welcome breaks each day.

What would be one of the more frustrating days for a number of drivers began at 3:45 a.m. on March 13, when Stan Raymond in an Oldsmobile F-85 Deluxe was the first driver to begin the 495.9-mile trek to Dallas. Offering a route to be covered within an 11-hour and 51-minute time limit, today was initially described as "a breeze ... with all roads flat, smooth and level."[25] Leaving Roswell, the competitors headed east on U.S. 380, and crossing into Texas entered the Central Time Zone bound for Snyder for both a brunch and refueling stop. Continuing across the Texas Plains, their next stop was in Weatherford, where all entries were topped off. As the day progressed, a number of drivers were starting to show some fatigue from three long days on the road, along with two predawn departures. Weather conditions on this day were close to being perfect, with bright sunshine and no wind to hamper their progress, but for a few drivers there would be another obstacle to overcome. Heading into Dallas on the Ft. Worth-Dallas Freeway, the competitors discovered that the directions to the impound, by skirting around the downtown area, were not too specific. Eight drivers, including Mickey Thompson, made a wrong turn onto the new Dallas Freeway, ending up 12 miles off course in the center of Dallas before discovering their error. This group "battled their way back to the route in thick traffic. Mickey Thompson, roaring mad at the mistake, scooped a pedestrian off the street, used him as a guide and paid him $20 for his services."[26] Bob McDaniel, driving a Plymouth Savoy 8, was the first driver to arrive at the Marriot Motor Hotel impound just before 5:15 p.m., where "Texas beauties, Texas cowboys, and a Texas brass band gave a roaring Texas welcome."[27] In addition to the eight "off-course" drivers, six others, including Loretta Colange (Ford Falcon), Stan Raymond (Oldsmobile F-85), Marta Retzlaff (Buick LeSabre), Paul Russo (Pontiac Catalina), Nicky Ward (Mercury Meteor 600), and Edward Wilder (Oldsmobile F-85), arrived late and received penalties. Two additional drivers, who were not identified by USAC officials, received a five-minute time penalty for failure to make a complete stop at posted intersections.[28]

By late evening, USAC officials continued to debate whether to make time allowances for the eight drivers who drove off-course. In addition to Mickey Thompson, these included Eugene Beck (Dodge Dart Seneca), Mel Chastain (Pontiac Star Chief), Mel Hilliard (Chevrolet Bel-Air), Jack Hills (Lincoln Continental), Patricia Jones (Mercury Comet), Link Paola (Oldsmobile F-85), and Bob Russo (Buick Special). Finally, "USAC stewards said they would not be charged with the additional time spent off-course. However, the drivers were forced to suffer whatever loss of gasoline mileage accrued through fuel used while they were lost."[29] Later Mickey Thompson commented that this "may have knocked me out of the running."[30]

Favorable weather and driving conditions enabled the field to increase their performance with an average of 24.0162 miles per gallon. Still maintaining a strong lead over all entries was Dan "Termite" Jones, posting a figure of 33.6917 miles per gallon with his Ford Falcon and continuing to eclipse the all-time mark by just under a half-mile per gallon. With Pat Sawyer in a Chevrolet Corvair Monza and Byron Froelich in a Ford Fairlane moving back into first place, veteran drivers were again leading in all classes. The class-leading drivers maintained their leads in what the *Los Angeles Times* called "the hottest competition in the history of this event."[31] The closest contest was between both Ford Fairlane 8 entries: a mere .0309 miles per gallon separated the two. Novice drivers held onto second-place positions in five of the seven classes. Mary Davis in her Plymouth Savoy finally was able to move into third place in the low-price eight-cylinder class, dropping Mel Hilliard in a Chevrolet Bel-Air to fourth. *Automotive News* noted that Mary, "who had won her classification or placed in four previous runs, was disappointed in her performance this year. 'I don't know what I'm doing wrong this year, but I guess four out of five will have to stand.'"[32]

The top three performers in each class were:

Compact Car—Manual Transmission	Ford Falcon	33.6917 Miles Per Gallon
	Ford Falcon	32.7001 Miles Per Gallon
	Chevrolet Corvair 500	31.8909 Miles Per Gallon
Compact Car Four and Six Cylinder	Chevrolet Corvair Monza	30.4519 Miles Per Gallon
	Chevrolet Corvair Monza	30.2937 Miles Per Gallon
	Pontiac Tempest 4	28.3863 Miles Per Gallon
Compact Eight Cylinder	Buick Special	25.9826 Miles Per Gallon
	Buick Special Deluxe	25.5516 Miles Per Gallon
	Pontiac Tempest Custom	24.8205 Miles Per Gallon
Standard Size Six Cylinder	Plymouth Savoy	23.8598 Miles Per Gallon
	Dodge Dart Seneca	23.6228 Miles Per Gallon
	Mercury Meteor 600	23.4440 Miles Per Gallon
Low Price Eight Cylinder	Ford Fairlane	22.1400 Miles Per Gallon
	Ford Fairlane	22.1091 Miles Per Gallon
	Plymouth Savoy	21.9563 Miles Per Gallon
Medium Price	Chrysler Newport	20.5605 Miles Per Gallon
	Pontiac Star Chief	20.2950 Miles Per Gallon
	Pontiac Star Chief	20.2200 Miles Per Gallon
High Price	Cadillac Sedan deVille	19.4515 Miles Per Gallon
	Imperial Custom	19.3218 Miles Per Gallon
	Lincoln Continental	17.7069 Miles Per Gallon

Although storms, with a possibility of some flooding, had been predicted earlier, clear weather prevailed as the drivers began their 421.8-mile, 10-hour and 35-minute trip to Little Rock. Today would mark the first time the Run traveled through Louisiana and Arkansas. Link Paola was the first driver to depart at 5:00 a.m. on March 14, pointing his Oldsmobile F-85 east across Texas and into Louisiana with a stop in Shreveport for refueling. The drivers then headed north into Arkansas, which brought the first "mutterings from drivers about 'Pillsbury's Road' as the route wound into back country, two-lane farm highways."[33] Always with a trick up his sleeve, Chief Steward A.C. Pillsbury managed to locate these rural roads where each driver's expertise would be tested. As they crossed back into Texas, the next stop was for brunch in Texarkana, where the first fender-bender of the Run happened to Nicky Ward's Mercury Meteor 600 6. "While she was poking through Texarkana traffic, she was struck by another car in an intersection and a strip of chrome was peeled from the left rear quarter of her new car."[34] Nicky was able to continue and no penalties were assessed by either local law enforcement or USAC officials. After this stop for brunch, it was back into Arkansas following U.S. Route 67 to Benton, where entries were topped off. The spectators on hand were here especially to welcome one of the competing Oldsmobile Dynamic 88s, entered by local Benton dealer Newbill Motor Company. From here it was on to Little Rock, where more challenges awaited. As the competitors made their way to the impound at Little Rock's Marion Hotel, heavy traffic congestion was encountered. Bill Corey in his Studebaker Lark Deluxe 6 was the first to arrive just past 4:15 p.m. After everyone had arrived, USAC officials announced that 28 were late, ranging in time from 5 seconds (Bill Krause in a Studebaker Lark Deluxe 8) to 3 minutes and 5 seconds (Bill Levy in a Mercury Comet); these received fuel penalties.[35]

The stress of the competition over the past four days was beginning to show on the drivers, many of them with limited Economy Run experience. Of the 64 drivers still in competition, only 15 had experience as drivers for five years or more. Many were grumbling over the "secrecy" surrounding the stewards' deliberations and announcement of penalties. J. Gordon Betz (USAC Referee), speaking for the stewards, told *Automotive News*:

> "The method of computing gasoline penalties given the drivers for being late is never given out. Nor are the amounts of fuel penalty posted." He continued, "The reason for this is to reduce the possibility of drivers determining whether it's to their advantage to take the penalty. We want them all to be on time. Many drivers have commented on the need for driver-steward meetings. I'd welcome such meetings. We'll do everything we can to make the run easier for them."[36]

Whether or not driver-steward meetings were held is not known.

Rural roads and traffic congestion contributed to the field's showing a small drop in their average to 23.8954 miles per gallon. Still showing a strong performance, Dan "Termite"

Another form of horsepower greeted the drivers as they arrived at the St. Louis impound after a challenging 357.5-mile trip from Little Rock. Art Rene, driving a Plymouth Savoy, would go on to finish first in the low-price six-cylinder class with 23.1540 miles per gallon. This would be Plymouth's second consecutive win in this class, and its fifth consecutive class win.

Jones led all competitors in his Ford Falcon with 33.6247 miles per gallon. Observers continued to watch Dan's performance as he continued to eclipse the all-time mileage mark. In the compact car four- and six-cylinder class, Bill Horton and Pat Sawyer, both driving Chevrolet Corvair Monzas, continued to swap places, with Bill holding a slim .0376 miles-per-gallon lead. The remaining class-leading drivers continued to maintain their leads and were able to slightly increase the margin over second-place drivers. In the low-price eight-cylinder class, a battle for second place continued, with Robert Rice's Ford Fairlane maintaining a slim .0693 miles-per-gallon lead over Mary Davis's Plymouth Savoy. General Motors continued to lead in three classes, with Chrysler Corporation and Ford Motor Company models leading in two each. One of General Motors' first-place positions was Lee Hamer's privately entered Cadillac Sedan de Ville, "entered without Cadillac's consent, and probably with some official objections."[37]

The top three performers in each class were:

Compact Car — Manual Transmission	Ford Falcon	33.6247 Miles Per Gallon
	Ford Falcon	32.6394 Miles Per Gallon
	Chevrolet Corvair 500	31.6784 Miles Per Gallon
Compact Car Four and Six Cylinder	Chevrolet Corvair Monza	30.2623 Miles Per Gallon
	Chevrolet Corvair Monza	30.2247 Miles Per Gallon
	Pontiac Tempest 4	28.2831 Miles Per Gallon
Compact Car Eight Cylinder	Buick Special	25.8253 Miles Per Gallon
	Buick Special Deluxe	25.4167 Miles Per Gallon
	Pontiac Tempest Custom	24.7821 Miles Per Gallon
Standard Size Six Cylinder	Plymouth Savoy	23.7775 Miles Per Gallon
	Dodge Dart Seneca	23.5434 Miles Per Gallon
	Mercury Meteor 600	23.2514 Miles Per Gallon
Low Price Eight Cylinder	Ford Fairlane	22.0432 Miles Per Gallon
	Ford Fairlane	21.8741 Miles Per Gallon
	Plymouth Savoy	21.8048 Miles Per Gallon
Medium Price	Chrysler Newport	20.4923 Miles Per Gallon
	Pontiac Star Chief	20.1989 Miles Per Gallon
	Pontiac Star Chief	20.0655 Miles Per Gallon
High Price	Cadillac Sedan deVille	19.4145 Miles Per Gallon
	Imperial Custom	19.2342 Miles Per Gallon
	Lincoln Continental	17.6172 Miles Per Gallon

On March 15, the drivers faced a 357.5-mile, 8-hour and 56-minute trip to St. Louis that, while it appeared on paper to be simple enough for the drivers, would turn out to be the roughest day experienced on this year's Run.[38] Promptly at 5:00 a.m., James Peterson and his Studebaker Lark Deluxe 8 was the first on the way, heading northeast on U.S. Route 67. In addition to rainstorms in the Arkansas hill country, "the road turned into a veritable roller-coaster in which upgrades were tougher than the downhill stretches, and nothing seemed to quite fall into place for any competitor."[39] While they were still in Arkansas, the second fender-bender of the Run occurred in the town of Walnut Ridge. Bob Russo, driving a Buick Special 8, "attempted to pass a farm truck on the wrong side. The truck driver swerved and creased the left side of Russo's car. Although [he] was in violation of the law, the police chief of Walnut Ridge permitted him to continue."[40] No one was injured and the damage to his left front fender was considered minor. After "an official pried out the fender and bumper with a tire iron," Bob was able to continue.[41] Well out of contention in his class, he was given a time allowance by USAC officials to compensate for this lost time. Right after crossing into Missouri, the drivers encountered both rain showers and strong head winds. The first stop was in Neelyville for refueling; then they continued for a short distance to Poplar Bluff for a brunch stop. After leaving Poplar Bluff, the drivers continued north, still on U.S. Route 67, towards Arnold-Mehville, where all entries were topped off before arriving at the impound in St. Louis. Bob McDaniel in a Plymouth Savoy 8 was the first to arrive, just after 2:30 p.m., where the competitors were greeted by music from a local high school band along with riders in medieval costumes on horseback. After a rash of late arrivals in previous days, only one driver missed the deadline, with Ak Miller in a Mercury Monterey arriving 31 seconds late.

Everyone knew from today's driving and weather conditions that their mileage figures would show a drop from the previous day, although for one driver there was a small bright spot. Dan "Termite" Jones continued to lead all competitors in his Ford Falcon with 33.3412 miles per gallon and continued to eclipse the all-time mileage mark, although his margin dropped to a slim .1312 miles per gallon. The entire field saw their overall average drop by .2142 miles per gallon to 23.6812. Continuing to swap the lead daily, Pat Sawyer's Chevrolet Corvair Monza moved back into first place in the compact car four- and six-cylinder class by a slim .1138 mile per gallon over Bill Horton. All of the other class-leading drivers maintained their positions, although three contests remained tight. In the low-price eight-cylinder class, Mary Davis finally overtook Robert Rice, moving her Plymouth Savoy into second place by a mere .0457 miles per gallon and still setting her sights on first place. So tight was the competition in this class that just over a half-mile-per gallon separated the first 7 places.

The top three performers in each class were:

Compact Car—Manual Transmission	Ford Falcon	33.3412 Miles Per Gallon
	Ford Falcon	32.2503 Miles Per Gallon
	Chevrolet Corvair 500	31.4297 Miles Per Gallon
Compact Car Four and Six Cylinder	Chevrolet Corvair Monza	29.9743 Miles Per Gallon
	Chevrolet Corvair Monza	29.8605 Miles Per Gallon
	Pontiac Tempest 4	27.9963 Miles Per Gallon
Compact Car Eight Cylinder	Buick Special	25.5855 Miles Per Gallon
	Buick Special Deluxe	25.1431 Miles Per Gallon
	Pontiac Tempest Custom	24.4936 Miles Per Gallon
Standard Size Six Cylinder	Plymouth Savoy	23.5758 Miles Per Gallon
	Dodge Dart Seneca	23.3761 Miles Per Gallon
	Mercury Meteor 600	23.0302 Miles Per Gallon
Low Price Eight Cylinder	Ford Fairlane	21.7652 Miles Per Gallon
	Plymouth Savoy	21.6177 Miles Per Gallon
	Ford Fairlane	21.5720 Miles Per Gallon
Medium Price	Chrysler Newport	20.3255 Miles Per Gallon
	Pontiac Star Chief	19.9673 Miles Per Gallon
	Pontiac Star Chief	19.9107 Miles Per Gallon
High Price	Cadillac Sedan deVille	19.2444 Miles Per Gallon
	Imperial Custom	19.0369 Miles Per Gallon
	Lincoln Continental	17.5035 Miles Per Gallon

For some of the drivers, "the sixth and last day was a blessing."[42] Below-freezing temperatures greeted the drivers as they faced the final leg into Chicago, and these temperatures took a good bite out of everyone's mileage. In addition to needing to cover the 304 miles within a 7-hour and 9-minute time limit, a full day of finish-line activities awaited everyone as Link Paola's Oldsmobile F-85 8 left at 3:30 a.m. with the rest following at one-minute intervals. Following U.S. Route 66 in Illinois, they made the first stop in Springfield for refueling, and then it was on to Lincoln for a brunch stop. Traveling on smooth but ice-cold highways, the drivers battled head winds along the way as they made their way to Chicago on freeways that led directly to South Lake Shore Drive and Soldier Field.[43] After navigating Chicago's thick but fast-moving traffic, Carol Paola in an Oldsmobile F-85 Deluxe 8 was the first to take the checkered flag from J.C. Agajanian just before 11:00 a.m. Following close behind Carol was her husband, Link, driving another Oldsmobile F-85. With a large crowd of spectators on hand and bundled up in windy 31-degree weather, the competitors arrived one-by-one, and by 12:30 p.m. everyone had arrived. Only one driver, Edward J. Wilder in an Oldsmobile F-85 8, exceeded the time limit and was penalized. As each car was inspected and topped off for the final time, activities moved to Chicago's Conrad Hilton Hotel, where post-finish activities were scheduled, including

Crossing the finish line at Chicago's Soldier Field, Danny "Termite" Jones scored an impressive 32.6800 miles per gallon in his Ford Falcon, with a 144-cubic-inch engine and three-speed manual transmission. This is the highest miles per gallon achieved by any six-cylinder car in the history of the Economy Runs. Danny came close to the all-time mark of 33.21 miles per gallon set by a four-cylinder Willys "77" 4 in the 1936 Gilmore Yosemite Economy Run.

the Awards Ceremony and Banquet. Although some drivers expected their averages to drop after today's driving conditions, the big question remained — did Dan "Termite" Jones and his Ford Falcon set an all-time mileage mark?

When the final results were announced, the highest mileage figure for all competitors was indeed Dan "Termite" Jones and his Ford Falcon with 32.6800 miles per gallon. The head winds and cold temperatures on the last leg had a significant effect on his mileage and the all-time record of 33.21 miles per gallon set in 1936 by a four-cylinder Willys 77 remained intact. The Falcon did set a record for the postwar Mobilgas Economy Runs, eclipsing the 31.053 miles-per-gallon mark set by a Nash Rambler in 1951. The lowest mileage figure was 16.6684 miles per gallon by Jack Hills in a Lincoln Continental. General Motors, who in previous years exhibited an indifferent attitude to the Economy Runs, placed first in 3 classes, with Chrysler Corporation and Ford Motor Company each placing first in two.

Class winners were:

CLASS	MAKE and MODEL	MILES PER GALLON
A — Compact Car — Manual Transmission	Ford Falcon	32.6800
B — Compact Car Four and Six Cylinder	Chevrolet Corvair Monza	29.3539
C — Compact Car Eight Cylinder	Buick Special	25.0929
D — Standard Size Six Cylinder	Plymouth Savoy	23.1540
E — Low Price Eight Cylinder	Ford Fairlane	21.3314
F — Medium Price	Chrysler Newport	19.9954
G — High Price	Cadillac Sedan deVille	18.9253

Detailed results for all competing cars can be found in the Appendix.

In reporting the results, USAC officials noted that the average for the entire field was 23.2475 miles per gallon, the second highest since 1951's mark of 23.923. In near-perfect weather conditions for most of the run, temperatures ranged from a low of 25 degrees to a high of 90. The women drivers outpaced the men this year by averaging 23.9927 miles per gallon compared to 23.0195 for the men. Even though only 16 women participated, part of the reason for their overall performance was an increased presence piloting entries in the compact car classes versus full-sized models. In covering the 2560.7-mile route, while staying within the 61-hour and 52-minute total time limit, drivers averaged 41.6314 miles per hour. USAC officials also noted there were only a few infractions of the rules; however, 34 of the entries, not counting the 8 who went astray in Dallas, were late arriving at overnight impounds at one time or another.

Veterans once again dominated, with long-time participants Mel Alsbury, Jr., Lee Hamer and Art Rene each scoring first-place finishes. This was the fifth first-place finish for three-time Sweepstakes Champion Mel Alsbury, Jr., while Dan "Termite" Jones and Pat Sawyer placed first in only their second year as drivers. For Pat, this was a small bright spot in what had been a horrific three-month period. Pat's husband (and last year's co-driver) Bob was the pilot of the ill-fated United Airlines DC-8 jetliner that collided with a TWA Constellation over Brooklyn the previous December.

Interest centered on the increased participation of compacts, especially those with manual transmission. Size mattered in this class, with the smaller models of the Chevrolet Corvair, Ford Falcon and Mercury Comet scoring the best mileage figures. The larger compacts, including the Buick Special, Oldsmobile F-85, Pontiac Tempest and Studebaker Lark, placed well down in the standings. A surprise was the poor showing of the new 4-cylinder engine in the Pontiac Tempest, its best performance being ninth with 25.2353 miles per gallon. Jim Whipple, writing in *Popular Mechanics*, observed, "Pontiac entries proved that four cylinders aren't the answer if there are V-8s around in lighter cars."[44] The Tempest outweighed both of its eight-cylinder General Motors cousins by over 200 pounds.

Compacts equipped with automatic transmission generally achieved the same results as their manual transmission counterparts with some exceptions. With the withdrawal of last year's champion, the Rambler American, the focus was now on those models from the "Big 3," which returned mixed results compared to 1960's performance. The first-place Chevrolet Corvair Monza, in recording 29.3539 miles per gallon, showed an increase of 2.3224 miles per gallon. After a disappointing performance in 1960, the Ford Falcon placed fourth with a slight increase of .0746 miles per gallon. Plymouth Valiant's performance dropped, placing sixth by recording 26.1339 miles per gallon, which was a drop of 1.1653 from 1960's result.

The new compacts from General Motors fared much better when equipped with automatic transmissions. Buick's active involvement by factory personnel was a factor in the Special's placing first and second among eight-cylinder models. Pontiac's Tempest 4-cylinder and 8-cylinder models each placed third in their respective classes, with Oldsmobile's F-85 placing near the bottom.

For Studebaker, their performance can only be described as a disappointment. Competing against other manual transmission equipped models, Bill Corey's Lark Deluxe 6 managed to place sixth in a field of eleven with 26.6707 miles per gallon. This was 6.0093 miles per gallon behind the class-winning Ford Falcon. The story was pretty much the same for the automatic transmission models. Bill Krause, participating in his first Economy Run in a Lark Deluxe 6, placed thirteenth (and last) in the compact four- and six-cylinder class, over 5½ miles per gallon behind the class-winning Chevrolet Corvair Monza. Last year's class-winning driver, Jim Peterson in a Lark Deluxe "8," placed sixth in the compact eight-cylinder class, just over 2 miles per gallon behind the class-winning Buick Special. This year would mark Studebaker's last full factory participation in the Economy Run, as these resources were more urgently needed in other areas of the corporation.

In the standard size six-cylinder class, Art Rene's Plymouth Savoy maintained its lead over Woody Bell's Dodge Dart Seneca and enabled Plymouth to record back-to-back wins over the six-cylinder competitors. Both were Economy Run veterans with experience dating back to 1950 and 1951, and Woody managed to come within .1384 miles per gallon of the first-place Plymouth. A surprise to most observers was the third-place performance of novice driver Frank Hainley in a Mercury Meteor 600. With Mercury's first six-cylinder engine (for full-size models) in its history, both Meteor 600s were equipped with the same engine that beat out both Ford entries by just under a half mile per gallon.

Ford returned to the winner's circle for the first time since 1956 as Byron Froelich drove his dark blue Ford Fairlane into first place and broke Plymouth's four-year win streak among low-price eight-cylinder cars. Mary Davis, last year's winner, couldn't overcome Byron's lead and finished second. Only .1185 miles per gallon separated the two entries. In what was considered a close contest from the start, less than a mile per gallon separated the first nine places. In a reversal of fortunes compared to their six-cylinder counterparts, the Mercury Meteor 600s could only manage fifth- and sixth-place showings, although they were only .3569 miles per gallon being the first-place Fairlane. Both the Ford and Mercury entries were equipped with the same engine, with driving skill and experience the difference between the two. Two Pontiac Catalinas, competing against the lower-priced models for the first time, placed well down, finishing in seventh and ninth.

Commenting on Ford's success this year, Dave Evans, from Ford Division's engineering office, told *Automotive News*, "The success we've had to date is due to nothing more than planning and hard work. Our people were all trained, as teams, in good driving techniques leading to good fuel economy."[45] Of the Ford Falcon's driver, Dan "Termite" Jones, he added, "His winning formula is to slow down and drive easy."[46]

In the medium-price class, Mel Alsbury, Jr., and his Chrysler Newport just missed topping the 20-mile-per-gallon mark with a figure of 19.9954. Mel's victory was the second consecutive for Chrysler and the fourth in the last five years. Pontiac Star Chiefs came in second and third. Mickey Thompson, who placed third, was unable to regain the mileage he lost after going astray arriving in Dallas. After their strong participation in recent years, only eight medium-priced makes competed, and five were lower-priced models such as the Buick LeSabre, Oldsmobile Dynamic 88 and Mercury Monterey.

Another close contest between first and second occurred in the high-price class. Separated by only .1893 miles per gallon, Lee Hamer's Cadillac Sedan de Ville hung on to beat George Alsbury's Imperial Custom, recording Cadillac's second win in three years. Both recent Cadillac victories were achieved by privately sponsored entries in the absence of factory backing. Both Lincoln Continentals placed third and fourth, well out of the running. With forthcoming eligibility changes involving this class, this would be the last year of true competition solely between Cadillac, Imperial and Lincoln.

One topic of conversation was the performance of novice drivers compared to their more experienced counterparts. Of the 28 teams with novice drivers competing, 6 placed ahead in the standings over the veterans, although one of these was "by default" after the mechanical failure of a Chevrolet Corvair. Although none of the novice drivers placed first, four placed second and five placed third. The first year for Chrysler Corporation's active factory participation as drivers resulted in fair results, although this would change over the coming years.

A few drivers with a wealth of experience would end their participation this year for various reasons, including Ak Miller, James Peterson, Vince Piggins, and Smokey Yunick. Although no longer participating as a driver, Nicky Ward would continue her affiliation with the Run over the next few years as a member of Mobil Oil's special events and publicity team.

Mobil Oil advertised the results in newspapers and industry publications using the title "USAC Certified Results" and highlighting the class winners along with the best mileage recorded by each model. At this time, Mobil was also promoting that their Regular and Special blends were now improved with Preptane, which was their "New long range mileage additive," with a new tagline: "You, Too, Can Get More Miles Between Fill-ups With Mobil."[47] In late March, Mobil Oil created a new twist by announcing the "$100,000 Mobilgas Economy Run Sweepstakes," a promotion that would continue through May 15. Published both in newspapers and general-interest publications such as *Life* and *The Saturday Evening Post*, this advertisement included an entry blank that was to be deposited by customers (and potential customers) at participating Mobil Dealers for a chance to win one of the 26 prizes. In addition to a $25,000 Grand Prize, 25 new cars were offered as prizes, and these were "duplicates of Class Winners of 1961 Mobilgas Economy Run"—3 Cadillacs, 3 Chrysler Newports, 3 Ford Fairlane 8s, 4 Buick Specials, 4 Plymouth Savoy 6s, 4 Ford Falcons and 4 Chevrolet Corvair Monzas.[48] This turned out to be a one-time event and was never promoted after the winner's names were selected. A four-page folder showing the official results for all entries, following the same format used since 1959, was once available at Mobil dealers. These folders would continue to be used in future years, although the format and design would change yearly.

As would be expected, Ford went all-out advertising Falcon's success. Appearing in newspapers and general-interest publications, a photograph of one of the competing Falcons appeared under the headline "Economy Is Another Reason Why Falcon Has Won The Battle Of The Compacts." The ad went on to detail the results of both manual transmission–equipped entries. The Peanuts crowd also appeared.[49] A second advertisement followed, this one with a two-panel design. The top panel featured a crowd surrounding a car above the headline "1st and 2nd place Mobilgas Economy Run Drivers ... reveal their secrets," with the lower panel showing one of the competing Ford Falcons as the crowd disperses. The Peanuts characters Lucy and Linus also appeared, with Lucy stating, "I wish I'd seen the other manufacturers' faces when two Falcons beat every other car competing in the Mobilgas Economy Run!" Linus added "Not me ... it's never very pleasant to see grown men cry!"[50] Later in the year a newspaper advertisement appeared featuring a rendering of a Falcon Sedan driving in a deserted area, approaching the "Last Chance For Gas!" station—the headline read "forget it!" The text went on to talk about the Falcon's achievement in the Economy Run.[51] No separate advertisements were created for the class-winning Ford Fairlane 8, although its achievement was a footnote in selected Falcon advertisements. Ford would continue to advertise Falcon's achievements in newspapers and magazines into the early months of the 1962 model year. Readers of *Ford Times* were told of the Falcon's and Fairlane's performance, along with how the women scored better in overall performance than their male counterparts.[52] Ford would continue to note Falcon's Economy Run mileage record in their showroom brochures through 1964.

Patricia Sawyer, driving a Chevrolet Corvair Monza, crosses the finish line, placing first in the compact car four- and six-cylinder class. With 29.3539 miles per gallon, this is the highest figure for an automatic transmission equipped six-cylinder car in Economy Run history. Those on hand to welcome the drivers were bundled up to protect them from the windy 31-degree weather.

Buick's advertising program centered on newspapers and dealership publications. Above a rendering of a competing Special was a headline that read "Buick Special Sweeps Class C With 25.09 MPG."[53] A second newspaper advertisement, designed for use by Buick dealers, invited readers to "*See it ... Drive it ..!* The Car That Swept Its Class In The Mobilgas Economy Run," once again using the same Special sedan rendering.[54] Buick's dealership personnel were told of the Special's achievements in a Buick Dealer Flash Sheet with the headline "Specials Run 1–2 In Economy Test." The copy also featured photographs, stories and results of the Run.[55] Additional reference to this Economy Run success was found in Buick's showroom brochures, along with a special four-page folder titled "Some High Points Of 1962 V6 & V8 Buick Economy," which featured a picture of the Special crossing the Chicago finish line.[56]

In what is probably the only magazine advertisement of its kind to date announcing a second-place finish, Dodge announced that a "Leadfoot Loses Mobilgas Economy Run Takes 2nd (Gets Only 23.0156 Miles Per Gallon)," along with a rendering of the competing six-cylinder Dart Seneca. Created by Art Director Frank Vogler at Dodge's advertising agency, Batten, Barton, Durstine & Osborne (BBDO), the copy told readers of both Woody Bell's and Dodge's accomplishments in this and in previous Economy Runs.[57]

"Plymouth Does It Again!" was the headline in a newspaper advertisement covering the achievements of its models, which began using the "5th straight year" theme of Economy Run victories. Not only did this advertisement note the six-cylinder Savoy's class win, it noted the eight-cylinder Savoy's placing second and the Valiant V-100s score of 23.16 M.P.G.[58] In magazine advertising, Plymouth stressed its "unmatched Mobilgas Economy Run record over the past five years" during the remaining months of the 1961 model year.[59] When the downsized 1962 models were announced, readers were told that "a Plymouth 6 scored higher than any other full-size car in the 1961 Mobilgas Economy Run."[60] This accomplishment would continue to be noted through early 1962. Valiant's 1962 model year introductory magazine advertising briefly noted its recent 26.13 miles-per-gallon mark. Showroom brochures covering the 1962 Full-Size Plymouth and Valiant models noted their Economy Run achievements.

Chrysler announced its success in newspapers with the headline "Newport Wins Class 'F' Mobilgas Economy Run!" with a rendering of the class-winning Newport sedan.[61] In the 1962 Chrysler showroom brochures, the Newport's mileage achievement was also noted. As expected, Cadillac completely ignored its class win.

Newspapers continued to be the prime source of coverage, with representatives from across the country reporting on the Run. After American Motors' withdrawal, daily interest was now centered on the Ford Falcon's quest to establish an all-time mileage record. Interest also focused on the obstacles drivers faced, especially on arriving in Dallas and Little Rock, where numerous teams were late. With all of this daily coverage, follow-up stories in the Sunday automotive sections were less than seen in previous years.

A Mobilgas Economy Run Fact Book, published by Mobil Oil, switched formats this year, becoming a pictorial report rather than the press information format used since 1957. Covering the highlights and personalities from this year's Run, this Fact Book also featured a day-by-day review written by Bill Dredge, automotive editor of the *Los Angeles Times*.

One of the most unusual advertisements regarding the Economy Run was created by Dodge to announce its second-place finish behind a Plymouth Savoy. Naturally no mention was made regarding the Plymouth in the text, a general comparison was made between Chevrolet and Ford models (courtesy of Chrysler Group LLC).

Several articles appeared in enthusiast magazines, with the most prominent being found in *Motor Life* and *Motor Trend*.[62] Bob Russo, writing in *Motor Life*, provided yet another one of his detailed insights on preparations and over-the-road experiences of an Economy Run driver. Because he placed well down in the standings, Bob's commentary was noticeable in highlighting the efforts of the non-winning competitors while still providing a guide to the results of those more successful drivers. Honest in his assessment of this year's event, Bob noted that he didn't think "that the greater length of the run this year — the longest ever — gives any *more* proof that good economy can be had with a well tuned automobile and careful driving habits. And I don't feel that it's necessary to make this an endurance run. It would be better to make it under less strenuous conditions and less time on the road in one day."[63] His comments were at odds with the views of the event's organizers, as future Runs would be tougher, and in some years longer, ultimately stretching from coast-to-coast.

In one of their most comprehensive looks at the Economy Run in years, a three-part article in the April 1961 issue of *Motor Trend* previewed upcoming events. In answering the question, "What does the run prove about gas mileage of 1961 cars?," the article attempted to answer "an element of confusion that always attends the economy run. What many casual observers don't understand—and it is quite understandable that they do not—are the results. When the average number of miles-per-gallon achieved by each car entry is announced, it generally is a much better average than most drivers ever realize from their own vehicles."[64] The article continued to discuss the in some detail the skill of the drivers, along with the selection and preparation of the cars in the Economy Run, to arrive at the answer to the question: "What Does The Economy Run Prove?" The answer was fairly simple. "What the man-on-the-street wants from the economy run is some kind of index to the gas mileage he can expect from his car or the one he thinks he might buy. He can get some information with a little study of the results and some applied background."[65] This same basic question was discussed in the June 1961 issue, where the results of the 1961 Economy Run were analyzed and the following observations and opinions were made:

> It is obvious that the Economy Run results are too good to be accepted by the public at large as a guide to the fuel economy capabilities of the cars involved. Yet, the Economy run does have a very worthwhile purpose. And there is no question about the integrity of the event. Mobilgas, the sponsor, and the United States Auto Club, the supervising organization, draft exceedingly strict rules and apply them relentlessly and ruthlessly. They are determined to make the Economy Run a true and accurate test. And therein probably likes the key to the dilemma. The Economy Run is not only a test, but even more so, it is a contest.
>
> In a test of machines, the operators of the machines must regulate their own efforts to arrive at impartial results. In a contest, those who operate or drive the machines are motivated by the desire to win. So then the factors of human skill and clever strategy become important. The Economy Run drivers, therefore, drive in such a manner as to exact the best possible gasoline mileage from their cars. This is natural—but it does contribute heavily to the unrealistic results.
>
> The cars in the Run are absolutely stock, in every sense of the word. The rules are very specific on this point and they are scrupulously applied. The average speeds are not fast, but fast enough to compare reasonably well with the pace of a cautious driver. In these respects the management of the Economy Run is most successful and the final mph averages are kept from truly fantastic levels. Cases of flagrant cheating by competitors are quite rare. Minor circumventions of the rules are probably common—and may influence positions within classes to some degree—but the total effect on mpg figures is relatively negligible.[66]

Over fifty years later, using different testing techniques under controlled conditions, conducted mainly by government agencies, the same basic question remains—how realistic are these mileage figures that are published as it relates to research by today's new-car buyers?

In reviewing the events of the 1961 Economy Run, it became evident to both officials from Mobil Oil and USAC that changes were in order. The public was becoming confused in understanding the results. With multiple models of a particular make competing in each class, questions arose as to why one could score so well and its teammate not so well. USAC also noted that maintaining supervision, plus calculating mileage figures, for such a large field of competitors proved to be a daunting task. These concerns would be addressed in short order.

8
Detroit

1962
Los Angeles to Detroit Via Flagstaff—Tucumcari — Tulsa — St. Louis — Chicago
March 31–April 5, 1962—2497.6 Miles

It became apparent after the conclusion of the 1961 Economy Run that changes were in order to simplify the competition in the face of an increase in new compact and intermediate-sized models, as well as engines, that were being developed by the U.S. automakers. When the rules were published in the fall of 1961, a change in the basic eligibility criteria of both cars and drivers was made that would remain in effect throughout the Sixties. Simply stated: "Only one car of each make and model will be eligible to compete in 1962. There will be no distinction between the eligibility of experienced and novice drivers, nor between men and women."[1]

A simple name change was made beginning with this year's Run, as it would now be known as the Mobil Economy Run, stressing both the corporate name and highlighting the fact that Mobil's products are more than just gasoline. However, with this name change an unintended consequence surfaced, as some newspapers and magazines, and even individuals, began to call this the "Mobile Economy Run." Despite repeated efforts to note that "there is no 'e' in Mobil," this misspelling continues in some aspects to this day.[2]

After years of speculation, Vern A. Bellman, marketing vice-president of Mobil Oil, made it official that Detroit's new Cobo Hall would be the terminus of this year's Run, marking the first time the Run would end in the Motor City. In making this announcement, Mr. Bellman reiterated Mobil's intention to "spotlight the good gasoline economy America's new cars can deliver when driven carefully through the complete variety of traffic and terrain any motorist would encounter between Los Angeles and Detroit."[3] In late February, USAC officials announced that this year's route would exceed 2000 miles with "overnight stops in Flagstaff, Ariz.; Tucumcari, N.M.; Tulsa, Okla.; St. Louis, Mo.; and Chicago, Ill."[4] Pleased with last year's results of withholding the details of each day's actual route, USAC officials announced that this practice would continue.

Additional emphasis was placed on competitive classifications and individual models' eligibility as a result of the introduction of additional models and engine options in the compact and intermediate-size field. Four classes were established for cars in these categories, with engine size being the major determining factor. As outlined in a bulletin titled "Competitive Divisions for the 1962 Mobil Economy Run," these were Class A for "Compact Cars with Manual Shift (3-speed) Transmission [and] engine displacement under 200 cubic inches"; Class B for "Small Engine Compact Cars with Automatic Transmission [and] engine displacement under 160 cubic inches"; Class C for "Large Engine Compact Cars with Automatic Transmission [and] engine displacement from 161 to 200 cubic inches)"; and Class D for "Luxury Compact Cars with Automatic Transmission [and] engine displacement from 201 to 260 cubic inches."[5] Although four-door sedan and hardtop models remained the basic body style for eligibility, two-door models could be substituted at the discretion of USAC in the compact-car classes. In classes A, B, and C, two-door models accounted for all of the entries. Another change occurred with what was formerly called the high-price class, now referred to as luxury cars, and eligibility was opened to high-line series models from Buick, Chrysler and Oldsmobile. A new requirement this year was for front and rear seat belts in all cars. Power steering and power brakes continued to be required on all full-sized eight-cylinder models, and this requirement would continue in future years.

A field of 46 cars, representing most U.S. makes, competed. Again this year Cadillac declined to participate and was joined in this category by the new intermediate-size Mercury Meteor. General Motors had the largest representation with 20 entries, followed by Chrysler Corporation and Ford Motor Company with 11 each. Rounding out the field was American Motors, having patched up their differences with the USAC officials, with 3 entries, along with a lone Studebaker Lark. Compact and intermediate-size models accounted for 24 entries and included the new Chevrolet Chevy II and Ford Fairlane models. Compacts with multiple engine options were evident this year as both four-cylinder and six-cylinder Chevrolet Chevy II models competed, along with Ford Falcons equipped with 144-cubic-inch and 170-cubic-inch engines. In the expanded luxury car class, a Buick Electra 225, Chrysler New Yorker and Oldsmobile Ninety-Eight were entered to compete with the Imperial and Lincoln Continental.

West Los Angeles Chrysler-Plymouth dealer Hart Fullerton, in a Chrysler 300, is the first driver to be flagged on his way by USAC's J.C. Agajanian. With the start of the Economy Run now located at the Pasadena impound, a smaller crowd than usual was on hand to see the competitors off to Detroit.

Factory support and participation continued to increase across the board. Chrysler Corporation engineers were active in preparation of the Dodge and Plymouth entries, and these individuals, again headed by Bob Cahill, were tapped to pilot the four Plymouth entries. Another engineer involved for the first time was Joe Strum, head of vehicle performance testing at Chrysler's Chelsea Proving Grounds, named to pilot Plymouth Valiant V-100 in the luxury compact class. Joe, who had a personal interest in the Economy Runs, brought both information and instrumentation available to provide data on fuel economy, especially when driving in hilly or mountainous terrain. This was especially helpful in calculating the "optimum speed to drive each car on any of the grades that were in common use on U.S. highways," which is an important element in economical driving techniques.[6] This type of engineering expertise by Chrysler Corporation personnel would continue to be used in the coming years. Although all three divisions received technical help from this group of engineers, for driving duties both Dodge and Chrysler took a different tactic. Dodge turned to veteran Economy Run drivers with prior class victories to their credit in tapping Woody Bell, Danny Eames and Art Rene to pilot their entries. Chrysler continued to be sponsored and driven by longtime participants Mel Alsbury and Hart Fullerton. This would be the first year for Hart participating in a Chrysler, having recently been awarded this franchise after DeSoto was discontinued in late 1960.

Buick continued its factory support to the same group of Southern California dealers and drivers who were involved in 1961, and this support would continue in future years. The remaining General Motors Divisions provided informal support, with most of their drivers having had previous Economy Run experiences. One of these drivers was Byron Froelich, who placed first last year in the low-price eight-cylinder class driving a Ford Fairlane, now named to drive the Chevrolet Impala. Ford established a "Ford Task Force, which for the second year boarded Ford drivers, navigators and mechanics in a Pasadena hotel. Here they were briefed on economy driving techniques and, as a group, made evaluation runs in Southern California."[7] American Motors' factory team was again headed by Les Viland and assisted by Les Scott, an American Motors engineer and race car driver. Studebaker was the only entry without factory support, with Bill Corey returning as both driver and co-entrant with Danyluk Motors in Glendale, California. This factory support also had a large effect on both dealers and dealer associations sponsoring various entries. Manufacturers, who increasingly became active in entering individual models, would begin to name the individual entrants. Chrysler Corporation would balance these entrants between dealers' associations, mostly located in areas where the Run would travel, and individual dealers to acknowledge those who excelled in both sales and service. One of these dealers was Dick Green Chrysler-Plymouth in Detroit, whose name would be associated with the Run though 1968. Ford would follow this same format by selecting dealers and dealer associations from each of this year's overnight stops with the exception of Flagstaff. With Detroit as the terminus city, General Motors tapped two of its well-known local dealers as entrants, Kelley Chevrolet and Packer Pontiac. This year also marked the return of Inglewood, California, Lincoln-Mercury dealer Bob Estes after an absence of six years. A successful participant with 2 Sweepstakes Awards to his credit, Bob would continue his participation through 1968.

With the change to "one car of each make and model" being eligible to compete, a number of familiar faces were absent this year. These included Mary Davis, Dick Griffith, Patricia Jones and James Peterson, all prior class winners. Another trend that would continue in future years were husband-and-wife teams as drivers and co-drivers. These teams included Woody and Lorraine Bell (Dodge Dart 6), Hart and Diana Fullerton (Chrysler 300), Stan and Lois Raymond (Oldsmobile Dynamic 88), and Tommy and Ila Thomas (Rambler American 400). Mickey and Judy Thompson also competed, but driving separate Pontiac entries, with Mickey in a Tempest Custom 4 and Judy in a Star Chief. All of last year's class-winning drivers returned, hoping for a repeat performance.

The final Mobil Mileage Rally, for imported cars, had been held the previous November. Three of this year's drivers placed first in their respective classes, Mary Hauser, Bill Levy and first-time participant Paula Murphy. Other Economy Run regulars who competed included Bill Corey, Danny Eames, Lee Hamer, and John Rich.

Random car selection commenced in early February with the rules governing preparation and break-in of competing cars remaining much as they were in 1961. To underscore the fact that these were not specially prepared cars, William Carroll of *Automotive News* accompanied USAC officials to a dealer where one of the cars was selected. To ensure that there were no chances that the entry could be tampered with, William noted, "Though we offered to drive it from the dealers to the impound, we weren't even allowed to touch it. Had the unit won, the fact that other than a USAC technician had driven the car might have been grounds for an official protest."[8] Developed and modified over the years, there were virtually no changes in this year's rules other than entry eligibility. Entrants continued to be given the option of using either Mobil Regular or Mobil Premium gasoline. Those selecting Regular included the Buick Special, along with Chevrolet's manual transmission–equipped Corvair, Chevy II, Biscayne and Bel-Air entries.

Compared to last year's pre-start developments with American Motors, the impound period remained pretty calm as teams were busy preparing for their forthcoming trek to Detroit. USAC officials were busy putting the finishing touches on their procedures for supervising and made final assignments for various duties during the actual Run. It takes a small army of volunteers to handle various duties, and these included local residents along the route to assist by installing and removing special direction signs and to act as flagmen where necessary. Mobil Oil's publicity machine was working full-time promoting this year's event while making preparations to host "nearly 100 newsmen and photographers [who] will cover this year's run for newspapers, wire services, magazines, radio and television from coast to coast."[9] Some of these press representatives would cover the run driving identical models to those competing, while others traveled by chartered aircraft between

After refueling in Blythe, Lee Hamer in his Lincoln Continental heads out the driveway bound for Arizona and the Flagstaff impound. This year would see another disappointing performance for Lincoln, as it only recorded a figure of 16.3640 miles per gallon. This would be Lincoln's final appearance in the Economy Run.

overnight stops. Again this year a private train, the "Mobil Economy Run Special," was provided for those wishing to travel directly from Los Angeles to Detroit.

With all of the drivers together one last time before the start, the traditional drawing was held for starting positions. Longtime participant Hart Fullerton drew No. 1 for his Chrysler 300; Roy Swanson in a Pontiac Catalina was the last to depart, since he drew No. 46. At this time, USAC officials announced that the start would be from the Pasadena impound's parking lot. In making this announcement, USAC said the reason for the change "was that cars have been repeatedly damaged in overnight transfer from the impound area in previous years. This new departure plan further eliminates any remote possibility of tampering with cars en route from the Pasadena [impound] to the Coliseum."[10] This was the first public indication that problems had been encountered in previous years with damage or tampering at the various Los Angeles starting locations.

On Friday afternoon, March 30, the detailed route and itinerary for the first day's trip was released. Traveling from Pasadena to Flagstaff, the drivers would need to cover a distance of 484.6 miles within an 11-hour and 40-minute time limit. This time limit, as usual, did not include time for refueling and a 40-minute brunch stop. With a limited number of available routes crossing the California desert regions into Arizona, the first day would take the competitors over some of the same roads and terrain that were used in 1960. Drivers were also advised that "some detours of [the] route are expected because of spring thaw flood conditions through Midwest states."[11] At the last minute, for reasons not disclosed, Mel Alsbury, Sr., stepped aside as driver of the Chrysler New Yorker, turning the wheel over to his youngest son, George.

On March 31, in a scene that was less spectacular than had been witnessed in prior years, all of the competing cars were arranged in the parking lot of the Pasadena impound, a large warehouse used chiefly for construction of Rose Parade floats. Far fewer spectators were on hand this year, and the vast majority of these were family members of the competitors or members of the press. West Los Angeles Chrysler-Plymouth dealer Hart Fullerton, driving a Chrysler 300, was the first to take the green flag from J.C. Agajanian at 6:30 a.m., with the remaining field leaving at one-minute intervals. Leaving Pasadena in clear sunny conditions, the drivers made their way through the suburban communities in the San Gabriel Valley, and then picked up the San Bernardino Freeway towards Palm Springs and Indio. Before leaving Indio they dropped to an elevation of 47 feet below sea level, the lowest point on this year's Run, before continuing eastward on U.S. Route 60/70 to Blythe. Arriving in Blythe after 5 hours on the road, they made their first stop at John's Friendly Mobil Service, where all entries were topped off, immediately followed by a brunch stop at the Palo Verde Union High School cafeteria. Repeating a scene from 1960, the participants were once again served by a group of local volunteers. Crossing into Arizona, and into the Mountain Time Zone, they were quickly cleared by Arizona's inspection officials and continued on U.S. Route 89 to Prescott. Climbing through the mountains of western Arizona, several drivers encountered slow-moving trucks on the steep mountain grades, which not only delayed their progress but hampered their mileage figures as well. Danny Eames, driving a Dodge Lancer 770, faced one of these obstacles, finding himself stuck not only behind a large tractor-trailer rig but also an Arizona Highway Patrol cruiser. Heading north out of Prescott, they made their way towards Williams for refueling at Gus' Mobil Service, then it was a short 32-mile trip on U.S. Route 66 to Flagstaff and the impound at the Ramada Inn. On this last portion of the trip the drivers climbed to an elevation of 7326 feet in the Coconino National Forest, the highest elevation on this year's Run. Arriving just after 8:00 p.m. in snow-covered terrain, Mary Faulkner in a Mercury Comet was first to reach the impound after twelve and a half hours on the road. As each competitor arrived, he or she was greeted by USAC officials with a packet containing the today's results along with tomorrow's route and detailed itinerary. This scene would be repeated at each night's impound. One driver, Barney Navarro in a Buick Electra 225, was over 2 minutes late and was penalized.

The question on everyone's mind centered on which "stick-shift compact" would turn in the best mileage figure ... last year's champion Ford Falcon or the Rambler American. Picking up where he left off, Dan "Termite" Jones and his Falcon recorded the best mile-per-gallon figure for all entries with 32.9733. Following close behind was Les Viland's Rambler American with a mark of 32.5381. The 46-car field turned in an average of 24.2266 miles per gallon. Pat Sawyer also picked up where she left off last year, achieving the highest miles-per-gallon figure for automatic transmission cars with 30.1922 in her Chevrolet Corvair Monza. In two classes the competition between first and second was extremely close. In the low-price eight-cylinder class, Bob McDaniel and his Plymouth Savoy had a slim .0294 miles-per-gallon lead over Bill Levy in a Ford Galaxie. In the medium-price class, Judy Thompson in her Pontiac Star Chief also had a slim .0535 lead over Stan Raymond in an Oldsmobile Dynamic 88. Finally, in the revamped luxury car class, George Alsbury in a Chrysler New Yorker had a small lead over the Oldsmobile Ninety-Eight and Buick Electra entries. Chrysler Corporation led in four classes, with General Motors leading in two. Ford Motor Company and American Motors rounded out the first-place positions, each leading in one class. Joe Strum, driving a Plymouth Valiant V-100 in the luxury compact car class, was the only novice driver in first place.

The top three performers in each class were:

Compact Cars with Manual Transmission	Ford Falcon (144 Cu. In.)	32.9733 Miles Per Gallon
	Rambler American 400	32.5381 Miles Per Gallon
	Chevrolet Chevy II 300 4	31.9906 Miles Per Gallon
Small Engine Compact Cars	Chevrolet Corvair Monza	30.1922 Miles Per Gallon
	Chevrolet Chevy II 300 4	29.0174 Miles Per Gallon
	Ford Falcon Deluxe (144)	27.9348 Miles Per Gallon
Large Engine Compact Cars	Rambler American 400	28.9678 Miles Per Gallon
	Pontiac Tempest Custom 4	28.5129 Miles Per Gallon
	Plymouth Valiant V-100	26.6671 Miles Per Gallon
Luxury Compact Cars	Plymouth Valiant V-100	26.5140 Miles Per Gallon
	Buick Special Deluxe 8	25.9049 Miles Per Gallon
	Dodge Lancer 770	25.4522 Miles Per Gallon
Standard Size Six Cylinder	Plymouth Savoy	26.6956 Miles Per Gallon
	Dodge Dart Seneca	24.3654 Miles Per Gallon
	Chevrolet Biscayne	23.0469 Miles Per Gallon
Low Price Eight Cylinder	Plymouth Savoy	22.1591 Miles Per Gallon
	Ford Galaxie	22.1297 Miles Per Gallon
	Chevrolet Impala	21.9240 Miles Per Gallon
Medium Price	Pontiac Star Chief	20.6187 Miles Per Gallon
	Oldsmobile Dynamic 88	20.5652 Miles Per Gallon
	Chrysler Newport	20.1732 Miles Per Gallon
Luxury Cars	Chrysler New Yorker	19.3708 Miles Per Gallon
	Oldsmobile Ninety-Eight	18.9094 Miles Per Gallon
	Buick Electra 225	18.7853 Miles Per Gallon

On April 1, the drivers headed for Tucumcari on a 539.8-mile, 12-hour and 30-minute trip that would follow U.S. Route 66 for most of the day. Sunshine and little wind would be helping the drivers on this trek, which would be the longest both in terms of miles and driving time in this year's Run. For the drivers, today's route would not be an April Fool's joke as the "46 cars struggled across the Continental Divide and into New Mexico."[12] Roy Swanson in a Pontiac Catalina was the first driver on his way at 6:30 a.m. After crossing into New Mexico, the drivers made the first stop in Milan at Parish Mobil Service, where all entries were topped off, followed by a brunch stop at Dougherty's Cafeteria in Grants. After spending close to 6 hours on the road, everyone was ready for a short break. Prior to reaching Albuquerque, USAC officials found a way to challenge the drivers as the route veered south for 191 miles on U.S. Routes 60 and 54, passing through several small towns and villages along the way. This route also avoided having to navigate about 10 miles of Albuquerque city traffic.[13] Rejoining U.S. Route 66 in Santa Rosa, the drivers stopped for refueling at Ellis Mobil Service before the final leg to Tucumcari. Edward Wilder, driving an Oldsmobile F-85, was the first driver to arrive around 7:45 p.m. at the Town House Motel impound, where the competitors were presented with beaded Indian ties by local residents as a memento of their visit.[14] Two drivers, Jim Wesson in a Pontiac Tempest Custom 4 and Bill Levy in a Ford Galaxie 8, arrived late and were penalized.

Traveling along U.S. Route 89 in the mountains of western Arizona created a few obstacles for the drivers as they made their way towards Prescott. Danny Eames's Dodge Lancer 770 was hampered not only by a slow-moving truck, but by an Arizona Highway Patrol cruiser as well. In the end, Danny would come out OK with a second-place finish in the luxury compact cars class.

During this leg, Mary Hauser's Chevrolet Chevy II 300 4 had problems after suffering a collapse of a vent line in the fuel system. "She stopped, disconnected the fuel line and cleared it" in just over 35 minutes and was back on her way.[15] Mary was later advised by USAC officials she would not receive a penalty for this delay.[16]

As expected, today's rugged driving conditions saw a decline in miles per gallon figures for all entries except for one. Les Viland and his Rambler American 400 not only moved into first place in his class by recording an impressive 33.2927 miles per gallon, he also achieved the best mileage figure for all entries. This mileage performance was due in part to "the long-stroke engine on the Rambler, which espouses an engineering principle opposite to those of competing cars with short piston strokes, [and] picked up mileage on the rugged hill run."[17] The entire field saw their average drop to 23.6491 miles per gallon. In the low-price eight-cylinder class, a real battle was heating up with Byron Froelich's Chevrolet Impala moving into first place. His lead over both the second-place Plymouth Savoy and third-place Ford Galaxie was a mere .0680 miles per gallon. The competition in this class was going to get even tighter. There were no changes in the remaining classes. General Motors and Chrysler Corporation were each leading in three classes, with American Motors leading in two. After two days on the road, no Ford Motor Company product was sitting in first place. The only other change of note was in the luxury compact class as Danny Eames and his Dodge Lancer 770 moved into second place, dropping John Rich and his Buick Special Deluxe 8 to third. Both first and second place entries in this class were equipped with six-cylinder engines, with the remaining entries equipped with eight-cylinder engines.

The top three performers in each class were:

Compact Cars with Manual Transmission	Rambler American 400	33.2927 Miles Per Gallon
	Ford Falcon (144 Cu. In.)	32.1843 Miles Per Gallon
	Chevrolet Chevy II 300 4	31.7508 Miles Per Gallon
Small Engine Compact Cars	Chevrolet Corvair Monza	28.9189 Miles Per Gallon
	Chevrolet Chevy II 300 4	27.5339 Miles Per Gallon
	Ford Falcon Deluxe (144)	26.7909 Miles Per Gallon
Large Engine Compact Cars	Rambler American 400	28.6142 Miles Per Gallon
	Pontiac Tempest Custom 4	28.1485 Miles Per Gallon
	Plymouth Valiant V-100	25.6342 Miles Per Gallon
Luxury Compact Cars	Plymouth Valiant V-100	25.7657 Miles Per Gallon
	Dodge Lancer 770	25.1037 Miles Per Gallon
	Buick Special Deluxe 8	24.6295 Miles Per Gallon
Standard Size Six Cylinder	Plymouth Savoy	25.7917 Miles Per Gallon
	Dodge Dart Seneca	23.8243 Miles Per Gallon
	Chevrolet Biscayne	22.7268 Miles Per Gallon
Low Price Eight Cylinder	Chevrolet Impala	21.8347 Miles Per Gallon
	Plymouth Savoy	21.7926 Miles Per Gallon
	Ford Galaxie	21.7667 Miles Per Gallon
Medium Price	Pontiac Star Chief	20.3702 Miles Per Gallon
	Oldsmobile Dynamic 88	19.8738 Miles Per Gallon
	Chrysler Newport	19.7697 Miles Per Gallon
Luxury Cars	Chrysler New Yorker	18.8408 Miles Per Gallon
	Oldsmobile Ninety-Eight	18.3053 Miles Per Gallon
	Buick Electra 225	17.9341 Miles Per Gallon

At 6:30 a.m. on April 2, Mary Hauser in a Chevrolet Chevy II 300 4 was the first driver to depart for a 469-mile trek to Tulsa. Today's trip, to be covered within an 11-hour time limit, would travel through three states. One competitor was sidelined in Tucumcari when the distributor shaft froze on Sunnie Baker's Ford Fairlane 8. She was running last in the luxury compact car class, and it was determined that there was not sufficient time to make repairs and catch up with the other competitors.[18] Heading east on U.S. Route 66, they crossed into Texas, entering the Central Time Zone, where everyone was greeted by a welcome banner announcing, "At Last You Are in Texas!"[19] Passing through Amarillo, the drivers continued to Shamrock, where all entries were topped off at Oscar's Mobil Service. Crossing into Oklahoma, they made the next stop in Elk City for brunch at the Round-up Café after being on the road for 6 hours. Then it was on to Oklahoma City for a refueling stop at Jack Reed Mobil Service before entering the Turner Turnpike. Here the drivers had a surprise in store for them. On the 85-mile stretch of the Turner Turnpike between Oklahoma City and Tulsa, a 55-mile-per-hour average had to be maintained. USAC timers and recorders were stationed at the entrance and exit, and "if any car fails to maintain at least a 55 M.P.H. average, it will receive a gasoline penalty."[20] After exiting the Turner Turnpike in Tulsa, the drivers made their way to the impound at the Trade Winds Motor Hotel. Mary Faulkner, driving a Mercury Comet, was the first to arrive just after 7:00 p.m., and the competitors were greeted by a Tulsa High School band.[21] Seven drivers including Woody Bell (Dodge Dart), Fran Foster (Ford Falcon), Bill Levy (Ford Galaxie), Barney Navarro (Buick Electra 225), Marta Retzlaff (Buick LeSabre), John Rich (Buick Special), and Pat Sawyer (Chevrolet Corvair Monza) arrived late and were penalized by "percentage changes in their use of fuel." In addition, "two drivers who failed to stop for red lights were forced to idle their engines [for] four minutes under the watchful eyes of USAC officials."[22]

Once again the entire field saw a decline in their mileage performance with the entire field averaging 23.4350 miles per gallon. Les Viland and his Rambler American 400 continued to record the highest result for all competitors with 32.5164 miles per gallon. Dan "Termite" Jones and his Ford Falcon began to close in with a difference of .8778 miles per gallon separating the two. All of the class-leading drivers continued to maintain their leads, including Pat Sawyer, even after receiving a penalty for arriving in Tulsa over 15 minutes late. In the large-engine compact car class, Tommy Thomas, driving a Rambler American 400, also faced a tight contest, as Mickey Thompson, driving a Pontiac Tempest Custom 4 in second place, closed to within .0472 miles per gallon.

The main interest in today's results centered on the com-

Where possible, USAC officials would provide direction signs or flagmen to assist the drivers along the way. With a unique style arrow, one of the flagmen sends Les Viland and his Rambler American 400 on his way towards St. Louis.

petition in the low-price eight-cylinder class, where there was a virtual tie between first and second. Byron Froelich and his Chevrolet Impala managed to hold on to first place by "about six drops of gas in a day's drive" over Bob McDaniel and his Plymouth Savoy.[23] The margin between the two was .0001 miles per gallon and would become the closest that any competitors would be in the history of the Economy Runs. Bill Levy in his Ford Galaxie was close on their heels in third. Only .0657 miles per gallon separated the first three places.

The top three performers in each class were:

Cars with Manual Transmission	Rambler American 400	32.5164 Miles Per Gallon
	Ford Falcon (144 Cu. In.)	31.6386 Miles Per Gallon
	Chevrolet Chevy II 300 4	31.2423 Miles Per Gallon
Small Engine Compact Cars	Chevrolet Corvair Monza	28.6333 Miles Per Gallon
	Chevrolet Chevy II 300 4	26.4590 Miles Per Gallon
	Ford Falcon Deluxe (144)	26.3280 Miles Per Gallon
Large Engine Compact Cars	Rambler American 400	28.1632 Miles Per Gallon
	Pontiac Tempest Custom 4	28.1160 Miles Per Gallon
	Plymouth Valiant V-100	25.4176 Miles Per Gallon
Low Price Eight Cylinder	Chevrolet Impala	21.4990 Miles Per Gallon
	Plymouth Savoy	21.4989 Miles Per Gallon
	Ford Galaxie	21.4333 Miles Per Gallon
Medium Price	Pontiac Star Chief	20.1680 Miles Per Gallon
	Oldsmobile Dynamic 88	19.6230 Miles Per Gallon
	Chrysler Newport	19.5844 Miles Per Gallon
Luxury Cars	Chrysler New Yorker	18.7333 Miles Per Gallon
	Oldsmobile Ninety-Eight	18.1743 Miles Per Gallon
	Buick Electra 225	17.8227 Miles Per Gallon

On April 3, a shorter day faced the drivers as they wound their way through a portion of the Ozarks on a 402.1-mile trek to St. Louis to be completed within a 9-hour and 20-minute time limit. Link Paola in an Oldsmobile Ninety-Eight was the first to depart, heading out at 7:00 a.m. for the Will Rogers Turnpike. Once again a 55-mile-per-hour average had to be maintained on an 87-mile stretch between Tulsa and the Missouri state line. Once again USAC timers and recorders were stationed at the entrance and exit and "if any car fails to maintain at least a 55 M.P.H. average, it will receive a gasoline penalty."[24] These timed sections would become a feature in future Economy Runs. Head winds hampered the drivers in Oklahoma as they made their way to the first stop in Joplin, where all entries were topped off at Jerry Perry Mobil Service. Following U.S. Route 166, the drivers made their way through several small towns before arriving in Springfield for brunch at The Grove, ready for a break after 4 hours on the road. Leaving Springfield they headed in a northeasterly direction on U.S. Route 66, bound for the St. Louis suburb of Bridgeton for refueling at Joe Evans Mobil Service. Encountering evening rush-hour traffic in St. Louis, the drivers made their way to the impound located at the

Holiday Inn North.[25] Edward Wilder, driving an Oldsmobile F-85, was the first to arrive, pulling in just before 5:00 p.m. For the first time since leaving Los Angeles, every driver reached the impound within the time limit set by USAC.

Head winds and traffic conditions took their toll on mileage as the entire field saw their average drop to 22.8695 miles per gallon. While Les Viland continued to maintain his overall lead among the 45 competing cars, there was a drop in his miles-per-gallon average to 31.4149. Dan "Termite" Jones continued to close in on Les with only .6643 miles per gallon separating both entries. One change occurred in class leaders as Mickey Thompson in a Pontiac Tempest Custom 4 managed to overtake Tommy Thomas's Rambler American 400 in the large-engine compact car class. Mickey, who was more noted for his speed performance than mileage, joined his wife Judy (driving a Pontiac Star Chief) as a class leader. This marked the first time a husband and wife, driving separate entries, would lead their respective classes. In the hard-fought low-price eight-cylinder class, Byron Froelich managed to open up a 0.345 mile-per-gallon lead over Bob McDaniel. General Motors now had leaders in four classes, followed by Chrysler leading in three classes, with American Motors in one. Ford Motor Company continued to be shut out of a class-leading position.

The top three performers in each class were:

Compact Cars with Manual Transmission	Rambler American 400	31.4149 Miles Per Gallon
	Ford Falcon (144 Cu. In.)	30.7506 Miles Per Gallon
	Chevrolet Chevy II 300 4	30.2938 Miles Per Gallon
Small Engine Compact Cars	Chevrolet Corvair Monza	27.6711 Miles Per Gallon
	Chevrolet Chevy II 300 4	25.7435 Miles Per Gallon
	Ford Falcon Deluxe (144)	25.6441 Miles Per Gallon
Large Engine Compact Cars	Pontiac Tempest Custom 4	27.4937 Miles Per Gallon
	Rambler American 400	27.2946 Miles Per Gallon
	Plymouth Valiant V-100	24.9755 Miles Per Gallon
Luxury Compact Cars	Plymouth Valiant V-100	25.1151 Miles Per Gallon
	Dodge Lancer 770	24.5016 Miles Per Gallon
	Buick Special Deluxe 8	23.8245 Miles Per Gallon
Standard Size Six Cylinder	Plymouth Savoy	24.9265 Miles Per Gallon
	Dodge Dart Seneca	23.0055 Miles Per Gallon
	Chevrolet Biscayne	21.9444 Miles Per Gallon
Low Price Eight Cylinder	Chevrolet Impala	21.0778 Miles Per Gallon
	Plymouth Savoy	21.0433 Miles Per Gallon
	Ford Galaxie	20.9526 Miles Per Gallon
Medium Price	Pontiac Star Chief	19.7042 Miles Per Gallon
	Oldsmobile Dynamic 88	19.2447 Miles Per Gallon
	Chrysler Newport	19.1775 Miles Per Gallon
Luxury Cars	Chrysler New Yorker	18.3672 Miles Per Gallon
	Oldsmobile Ninety-Eight	17.7626 Miles Per Gallon
	Buick Electra 225	17.4829 Miles Per Gallon

On April 4, in what would be one of the more challenging days of this year's Economy Run, the drivers faced a 306.7-mile, 7-hour and 45-minute trip to Chicago. A few miles of today's route "were on turnpikes, where driving skill is unnecessary to achieve economy [although] traffic-laden back roads and small towns" with stop-and-go driving would test each driver's skill.[26] The first driver to depart at 6:30 a.m. was Mary Hauser, pointing her Chevrolet Chevy II 300 4 in a northerly direction and crossing into Illinois. A mechanical failure would sideline another competitor when St. Louis resident Edward Wilder's Oldsmobile F-85, entered by Ray Nolting Oldsmobile in nearby Kirkwood, had problems caused by improper operation of the transmission. This became the second entry to withdraw from the Run, with both competing in the luxury compact car class. After encountering thick traffic through St. Louis mixed with rain for almost an hour, the drivers made their way to Champaign via secondary roads for a brunch stop at the Holiday Inn, followed by a stop in Paxton, where all entries were topped off at Robert Bixby Mobil Service. Near Paxton, a driving infraction was reported to USAC officials "when Bob McDaniel eased his Plymouth Savoy out to pass a truck, [after] his observer didn't notice the no-passing sign."[27] Two competing drivers and riding observers in two cars noticed the "no-passing sign" and made a complaint. Although this infraction was denied by both Bob and his USAC observer, he was given a penalty of idling his engine for two minutes.[28] The stop in Paxton would be the only refueling stop on this leg as drivers continued north towards Chicago, passing through numerous small towns along the way. Continuing to make their way into the city's South Side, the drivers headed for the impound at the 50th on the Lake Motel, located on Lake Shore Drive adjacent to Lake Michigan. Just before 3:00 p.m., Ronnie Kaplan in a Chevrolet Corvair 700, sponsored by Chicago's Nickey Chevrolet, was the first to arrive. Two drivers, Pat Sawyer (Chevrolet Corvair Monza) and Jim Wesson (Pontiac Tempest Custom 4), arrived late. Only Jim received a penalty. Twenty miles outside of Chicago, Pat became ill and stopped at, of all places, a Texaco Service Station, unable to continue. After receiving brief medical attention, she managed to continue and arrived 17 minutes late. Due to these circumstances, USAC officials did not penalize her for this late arrival. This illness, later reported to be a result of food poisoning, would continue to plague her on the final leg.[29]

Today's terrain and driving conditions contributed a little to change as averages took a slight drop, with the field recording an average of 22.8804 miles per gallon. Les Viland and his Rambler American 400 continued to record the best mileage figure with 31.3688 miles per gallon as Dan "Termite" Jones continued to close in on this lead, trailing by .6273 miles per gallon. Time for Dan to overtake Les was beginning to run out. All of the class-leading drivers maintained their positions, with everyone continuing to follow the competition in the low-price eight-cylinder class as Byron Froelich continued to lead by a margin of .0462 miles per gallon. Due to a disputed penalty from earlier in the day, an unexpected change took place as Bill Levy's Ford Galaxie moved into second place, dropping Bob McDaniel's Plymouth Savoy to third. With one day remaining in the Run, Bob would not be able to recover from this penalty.

The top three performers in each class were:

Compact Cars with Manual Transmission	Rambler American 400	31.3688 Miles Per Gallon
	Ford Falcon (144 Cu. In.)	30.7415 Miles Per Gallon
	Chevrolet Chevy II 300 4	30.3104 Miles Per Gallon
Small Engine Compact Cars	Chevrolet Corvair Monza	27.7134 Miles Per Gallon
	Chevrolet Chevy II 300 4	25.8452 Miles Per Gallon
	Ford Falcon Deluxe (144)	25.7298 Miles Per Gallon
Large Engine Compact Cars	Pontiac Tempest Custom 4	27.4987 Miles Per Gallon
	Rambler American 400	27.2386 Miles Per Gallon
	Plymouth Valiant V-100	25.0027 Miles Per Gallon
Luxury Compact Cars	Plymouth Valiant V-100	24.9994 Miles Per Gallon
	Dodge Lancer 770	24.5700 Miles Per Gallon
	Buick Special Deluxe 8	23.8282 Miles Per Gallon
Standard Size Six Cylinder	Plymouth Savoy	24.7801 Miles Per Gallon
	Dodge Dart Seneca	23.0002 Miles Per Gallon
	Chevrolet Biscayne	21.8300 Miles Per Gallon
Low Price Eight Cylinder	Chevrolet Impala	20.9519 Miles Per Gallon
	Ford Galaxie	20.9057 Miles Per Gallon
	Plymouth Savoy	20.8757 Miles Per Gallon
Medium Price	Pontiac Star Chief	19.6448 Miles Per Gallon
	Oldsmobile Dynamic88	19.2030 Miles Per Gallon
	Chrysler Newport	19.1023 Miles Per Gallon
Luxury Cars	Chrysler New Yorker	18.2679 Miles Per Gallon
	Oldsmobile Ninety-Eight	17.7025 Miles Per Gallon
	Buick Electra 225	17.3759 Miles Per Gallon

All of the competitors appreciated a mid-afternoon arrival time in Chicago. Pat Sawyer was not the only driver battling some medical issues as colds, sore throats and influenza kept the USAC medical staff busy treating drivers and other crew members.[30] A full day was on the agenda for tomorrow as drivers would have nearly 300 miles of driving, followed by various finish line and awards ceremonies planned in Detroit.

After covering 2202.2 miles since leaving Los Angeles, everyone was ready for the final 295.4-mile leg to Detroit, to be completed within a 6-hour and 30-minute time limit. It was an early call for everyone as Pat Sawyer, driving a Chevrolet Corvair Monza, was the first to leave at 4:00 a.m. Leaving Chicago via the Calumet Skyway, they quickly crossed into Indiana and entered the Indiana Tollway, where they would encounter a heavy rainstorm for the first hour of the day. Although this section would not be timed, competitors were advised to maintain a 52-mile-per-hour average on this nearly 100-mile stretch "to insure [*sic*] arriving at [the] finish on time."[31] Crossing into the Eastern Time Zone, the competitors left the Tollway at Middlebury and headed north into Michigan, bound for Kalamazoo. After nearly three and a half hours on the road, their first stop in Kalamazoo was for refueling at Boers Mobil Service, quickly followed by brunch at Chicken Charlie's Restaurant. Leaving Kalamazoo, the competitors picked up Interstate 94, heading east toward Detroit's new Cobo Hall. Pat Sawyer continued to battle illness and at one point almost called it quits. Not one to give up, she pushed herself all the way into Detroit and crossed the finish line just before 11:30 a.m., well ahead of other competitors.[32] So early was her arrival that J.C. Agajanian, the photographers and most spectators were not yet on hand. Arriving just prior to noon, Link Paola and his Oldsmobile Ninety-Eight led the remaining 43 drivers across the finish line. One of the largest crowds in recent years, most of whom were from Detroit's automotive industries, were on hand to welcome the competitors. Everyone arrived in Detroit within the time limits and each entry was impounded and topped off for the final time. Interest continued to center on the competition between Chevrolet, Ford and Plymouth in the low-price eight-cylinder class.

When the final results were announced, Les Viland and his Rambler American 400 held on and recorded the highest mileage figure for all entries with 31.1105 miles per gallon. This marked the third time in four years that a Rambler American took top honors. Following close behind was Dan "Termite" Jones's Ford Falcon with 30.4941 miles per gallon. The lowest mileage figure was 16.3640 miles per gallon by Lee Hamer in a Lincoln Continental, which was almost identical to its results in 1961. Women took the top prize in two classes, with General Motors topping all manufacturers placing first in four classes. Chrysler Corporation placed first in three and American Motors in one. Although none of its entries placed first, Ford Motor Company took home three second-place trophies.

Class winners were:

CLASS	MAKE and MODEL	MILES PER GALLON
A — Compact Cars with Manual Transmission	Rambler American 400	31.1105
B — Small Engine Compact Cars	Chevrolet Corvair Monza	27.0178
C — Large Engine Compact Cars	Pontiac Tempest Custom 4	27.2950
D — Luxury Compact Cars	Plymouth Valiant V-100	24.8525
E — Standard Size Six Cylinder	Plymouth Savoy	24.6149
F — Low Price Eight Cylinder	Chevrolet Impala	20.7788
G — Medium Price	Pontiac Star Chief	19.4797
H — Luxury Cars	Chrysler New Yorker	18.1139

Detailed results for all competing cars can be found in the Appendix.

At Detroit's Cobo Hall, Bob Cahill crosses the finish line in his Plymouth Savoy 6 to an enthusiastic greeting from USAC's J.C. Agajanian. At the Economy Run's first finish in Detroit, a large crowd of automotive industry-related personnel were on hand to welcome the participants. Placing first in the standard-size six-cylinder class, Plymouth scored a third consecutive win in this class and its sixth consecutive class win.

In reporting the results, USAC officials noted that the average for the entire field was 22.6980 miles per gallon. While this was a drop from last year's average of 23.2475 miles per gallon, a smaller percentage of compact cars participated this year compared to full-sized models. The 11 women participating averaged 22.8781 miles per gallon, compared to the men with 22.3509. A factor in the women's performance was their primarily driving in the compact car classes, as there were only 4 driving full-sized models. Covering a distance of 2497.6 miles, the route to Detroit encompassed 1394.2 miles of country driving, 830.8 miles of expressway (or freeway) driving, and 272.6 miles of city driving. Driving through 183 cities and towns, the drivers had to contend with 256 traffic signals along the way. Concerns that the competitors would encounter spring thaw flood conditions did not materialize, as good weather prevailed with temperatures ranging from a low of 20 to a high of 80 degrees. Keeping within the time limit of 58 hours and 55 minutes set by USAC, the field averaged a speed of 42.7019 miles per hour. This miles-per-hour average was the highest in four years. USAC officials noted that there were only a few infractions of the rules, and drivers showed an improvement with their "on-time" performance compared to last year. Only 11 drivers arrived at an overnight impound late, and of these, one was for a mechanical issue and the other due to illness.

Drivers with previous driving experience captured first-place trophies in all classes except for the luxury compact car class, where first-time driver Joe Strum in a Plymouth Valiant V-100 took the top prize. Longtime participant Les Viland scored his fifth victory, while another regular, George Alsbury, scored his fourth. With their individual class victories, Mickey and Judy Thompson became the only husband-and-wife team in the history of the Economy Runs to capture first-place trophies while driving separate entries. Both Pat Sawyer and Byron Froelich notched "back-to-back" wins, with Bob Cahill placing first in only his second year participating.

As noted in *Car Life Magazine*, "Oldtimers on the Mobil Economy Run are inclined to check the first day's results and then forget the contest until the final awards are presented. They figure they already know the winners. This year they got fooled — but good!"[33] This was true in two of the four compact car classes. After finding himself in second place in the class for compact cars with manual transmission, Les Viland in his Rambler American 400 passed Dan "Termite" Jones's Ford Falcon on the second day and never looked back. Although Dan slowly narrowed the gap between the two, the margin between the Rambler American and Ford Falcon at the end was .6164 miles per gallon. Bill Horton, driving Chevrolet's new 4-cylinder Chevy II 300, placed third, slightly over a mile per gallon off the pace. Although a direct comparison to last year's results is not possible, all of the compacts in this class turned in respectable miles-per-gallon figures, although lower than achieved in 1961. Of the four entries who had participated last year, only the Pontiac Tempest Custom 4 showed a year-to-year increase, while the remaining three (Chevrolet Corvair, Ford Falcon and Studebaker Lark) recorded declines ranging from 1.8920 to 2.4818 miles per gallon.

Another close contest was in the large-engine compact car class between Mickey Thompson's Pontiac Tempest Custom 4 and Tommy Thomas's Rambler American 400. Maintaining a slim lead for the first three days, Tommy was unable to hold off the challenge from Mickey, who finished in first place with a margin of .2023 miles per gallon. First-time driver William McNulty in a Plymouth Valiant V-100 was solidly in third place, although he finished nearly two and a half miles per gallon off the pace.

The remaining two compact car classes, small-engine compact and luxury compact, were pretty much settled from the first day's results. Pat Sawyer, despite becoming ill on the last two days of the run, maintained a comfortable lead over Leon Funderburk's Ford Falcon and Mary Hauser's Chevrolet Chevy II 300. Despite a problem with her Chevy II's fuel line on the second day, Mary maintained second place for the first five days, only to be nosed out on the last leg by Leon by a slim .0071 miles per gallon. Another first-time driver, Joe Strum in a Plymouth Valiant V-100, also maintained a comfortable lead over veterans Danny Eames in a Dodge Lancer 770 and John Rich in a Buick Special Deluxe 8. Competing against two 6-cylinder entries, John placed a respectable third while recording the best miles-per-gallon figure for all 8-cylinder models. Two cars competing in this class suffered mechanical problems which sidelined the Ford Fairlane and Oldsmobile F-85 entries.

The compacts equipped with automatic transmissions overall also turned in respectable results, although none of the 7 who had participated in 1961 showed a year-to-year increase. These declines ranged from .0767 miles per gallon for the Ford Falcon (equipped with the 144-cubic-inch engine) to a decline of 2.3361 for the class-winning Chevrolet Corvair Monza.

In the standard size six-cylinder class, the competition between the Plymouth Savoy and Dodge Dart was cemented from the start and ended with the same first- and second-place finish that occurred in 1961. Bob Cahill, competing as a driver for only the second time, surprised observers as he recorded a 1.7799 miles per gallon margin over veteran driver Woody Bell. The remaining two entries, a Chevrolet Biscayne and Ford Galaxie, placed well back, nearly three to four miles per gallon off the pace.

Some of the closest competition in the history of the Economy Runs occurred in the class for low-price eight-cylinder cars, featuring the traditional rivals of Chevrolet, Ford and Plymouth. From the time the first day's results were released, with Bob McDaniel's Plymouth Savoy holding a slim 0.294 miles-per-gallon lead over Bill Levy's Ford Galaxie, everyone knew this would be a close-fought battle for first place. On the second day, Byron Froelich's Chevrolet Impala jumped from third to first by a margin of .0421 miles per gallon. The margin after the third day was a record-setting .0001 miles per gallon in favor of the

Another Plymouth scored a class win in the 1962 Economy Run when this Valiant V-100, driven by Joe Strum, took top honors in the luxury compact car class. In a two-year time span, the compacts from Detroit's "Big Three" would each capture a first-place trophy in their respective classes.

Chevrolet Impala, and stretched on the fourth day to .0345 miles per gallon. At this point disaster struck Bob McDaniel's chances to retake first place when a disputed penalty saw him drop into third place behind Bill Levy's Ford Galaxie. At the finish line, the margin between Byron Froelich's Chevrolet Impala in first and Bob McDaniel's Plymouth Savoy in third was .0647 miles per gallon. The other four competitors in this class could only watch from the sidelines as they finished well back in the standings. *Car Life* summed it up best: "Froelich, Levy and McDaniel had managed to turn the snail-paced Economy Run into something really exciting."[34] So true.

Some of the closest competition in the history of the Economy Runs took place in the low-price eight-cylinder class, with Byron Froelich bringing his Chevrolet Impala home in first place. In the end, a slim .0647 mile per gallon separated the first three places, with a Ford Galaxie in second and a Plymouth Savoy in third. This was Byron's second consecutive win in the class after placing first in a Ford Fairlane in the 1961 Economy Run.

The surprise in the medium-price class was the strong performance of the Pontiac Star Chief and Oldsmobile Dynamic 88 models, which placed first and second. Although competition was keen throughout the Run, Judy Thompson and her Pontiac Star Chief managed to maintain a nearly half-mile-per-gallon lead over both second- and third-place entries from the beginning. The third-place finish by veteran Mel Alsbury, Jr., and his Chrysler Newport was a surprise to most observers as the veteran of seven previous Economy Runs struggled against the competition of both General Motors products being driven by lesser-experienced drivers.

The results were reversed in the luxury car class as George Alsbury's Chrysler New Yorker finished first with a comfortable margin over Link Paola's Oldsmobile Ninety-Eight and Barney Navarro's Buick Electra 225. The more traditional luxury models, Imperial Crown and Lincoln Continental, placed fourth and fifth. Less than a mile per gallon separated the first four entries.

In addition to the previously mentioned string of victories by the Rambler American, other highlights regarding class winners include:

- Chevrolet scored its first class win since 1959 with the Impala 8 holding on to place first in the low-price eight-cylinder class. This also marked the first time a V-8 powered Chevrolet placed first.
- Chrysler recorded its fifth class win in 6 years when the New Yorker placed first in the luxury cars class. Each of Chrysler's class victories since 1957 had been driven by a member of the Alsbury family, either George (this year's driver) or Mel, Jr.
- Plymouth notched its sixth consecutive class win with the Savoy 6 taking first place in the class for standard size six-cylinder cars. This was also the third year in a row Plymouth topped all standard size six-cylinder models.
- Pontiac scored its first class wins since 1956 when the Star Chief took first place in the medium-price class, along with the Tempest Custom 4 placing first in the large-engine compact car class.

After two years of disappointing performances, this year would mark the final time Lincoln would be a participant. Going back to the days of the Gilmore Economy Runs, Lincoln captured the Sweepstakes Award twice: in 1941 with a twelve-cylinder Custom, and again in 1951, when it set an all-time ton miles per gallon record. Additionally, Lincoln took home nine first place trophies, the final time being in 1954. Throughout the latter half of the fifties, Lincoln's performance and participation was spotty at best, although it placed second in 1955 and 1958.

Two drivers with first-place trophies in their display cases, Danny Eames and Dan "Termite" Jones, would end their participation this year. Both drivers have the distinction of setting mileage records during their involvement with the Economy Runs. Danny Eames scored the highest mileage for all 8-cylinder cars competing in 1953 with 23.4189 miles per gallon in a Dodge Coronet. Dan "Termite" Jones did him one better in 1961 when he posted the highest mileage for all cars competing in the 1961 Run with 32.6800 miles per gallon in a Ford Falcon. This mark is also the second highest in the history of the Economy Runs, dating back to the record set by a four-cylinder Willys 77 in 1936.

The majority of advertising was seen in the days and weeks immediately following the conclusion of the Run, with very little occurring during the remainder of the year. With the exception of some newspaper advertising campaigns, a trend started of reduced advertising that would become more evident in the coming years. As had been their practice, Mobil Oil advertised the results in both newspapers and industry publications, highlighting only the individual class winners. The focus of their advertising this year centered on a new rating system for gasoline quality called the "Megatane Rating," which enabled the driver "to judge not just Octane qualities, but the total performance of your gasoline."[35]

After last year's absence, Rambler was quick to advertise its achievement in both newspapers and industry publications. In an all-text layout, advertising announced, "Rambler American '400' Wins Class-A 31.11 M.P.G. Tops All Compact Cars, Manual Shift, in Mobil Economy Run — Los Angeles to Detroit," while introducing a new tagline: "Top Economy and Top Performance Get a Rambler."[36] With a photograph of the winning American crossing the finish line, its mileage achievements were highlighted in the economy compact edition of the 1963 "X-Ray" comparison catalogs available at Rambler dealers. Featuring a somewhat blurry photograph of the Rambler American, Champion Spark Plugs noted its products were used in the Rambler's success delivering the best gas mileage.[37]

Although placing second to the Rambler American for the overall best mileage, Ford still promoted that "a Ford Falcon scored 30.5 miles per gallon in the 1962 Mobil Economy Run," while continuing to proclaim, "A Falcon still holds the record for highest gas mileage for a 6 or 8 in the 26-year history of the run!" Featuring *Peanuts* characters Charlie Brown and Lucy, this advertisement was found in both newspapers and industry publications.[38] In an edition of *News Flash* for Ford salesmen, the April 1962 issue was not devoted only to the Falcon's second-place showing, but also highlighted the Falcon's manual transmission and the 8-cylinder Galaxie's results compared to the competition from the Chevrolet entries. In noting the close competition between the Galaxie and Impala, where the difference was .03 miles per gallon, Ford noted that the difference in cost of operation between the two "over 10,000 miles of driving [is] just .22," which was based on a .32-per-gallon gas price.[39]

Chevrolet advertised its Economy Run performance in newspapers, covering the range of the nine models that competed by showing their final miles-per-gallon figures with the headline "Chevrolets Demonstrate Outstanding Economy in Mobil Run!"[40] What makes this advertisement unusual is Chevrolet's decision not to note the individual class wins for both the Corvair Monza or Impala entries.

Pontiac's double win was featured in newspaper ads with a Star Chief announcing, "Hey, little brother, I guess we showed 'em in the 1962 Mobil Economy Run from Los Angeles to Detroit. I won Class G, medium-priced, standard size cars." To which the Tempest replied, "And don't forget my win over all the compacts with large engine and automatic shifts."[41] This same theme was featured in a small 4-page brochure available at Pontiac dealers. This folder featured photographs of the winning models along with their drivers and included tips on getting better mileage.[42]

Even though Buick placed third in the luxury compact car class, it was quick to advertise its performance in various newspapers. Under the headline "V-8 Buick Special Racks Up 23.68 M.P.G. Gives Class D 6's a scare in Mobil Economy Run," readers were told that the "Buick Special Aluminum V-8 beats all Compact and Full Size V-8's." The ad also highlighted the winning mileage in that class.[43] Buick's 1963 showroom brochures also noted this achievement. As the Economy Run progressed, Buick's public relations staff was busy getting the word out on their product's overall progress. Through the use of "Flash Bulletins" or "Special Bulletins," results were published for the four classes in which Buick models were competing, stressing the third-place positions, by narrow margins, of the Special Deluxe "8" and Electra 225. Although they were shown, no mention was made of the poor performance of the Special "6" and LeSabre entries.

Plymouth advertised the 6-cylinder Savoy's class win with a simple newspaper layout featuring 6 large marks noting, "*Plymouth Makes It Six Straight*— Wins Class "E" in 1962 Mobil Economy Run!" The ad went on to highlight Savoy's prior success.[44] This "winning streak" was also noted in Plymouth's 1963 showroom brochures. In somewhat of a surprise, there was no mention of Valiant's class-winning achievement in general publications, although advertisements noted its performance in specialized publications aimed at fleet customers.

Chrysler noted the New Yorker's success in several ways. In newspapers and industry publications, an advertisement featured a rendering of the New Yorker and a headline that announced, "Chrysler New Yorker Wins Class 'H' in Mobil Economy Run! (Tops Luxury Car Class)." The ad noted, "This 1962 victory makes Chrysler a *5-time winner* in its class in the past 6 years!"[45] Chrysler's success in the Economy Runs was noted throughout the model year and continued with the introduction of the 1963 models.

Chrysler had another trick up their sleeve, scoring what was termed "an advertising first" in announcing their victory. A huge outdoor sign, directly across from Cobo Hall, read: "Welcome Mobil Economy Run Competitors!" Three other canvas signs were ready for display. Minutes after the final results were announced, confirming that the New Yorker placed first, a sign was lowered reading, "Welcome Champ! Chrysler New Yorker wins Class H Mobil Economy Run with 18.11 m.p.g. average." As the sign was lowered, a painter entered the final figure in red numerals. Two other signs were ready in the event the Newport captured the medium-price class, or if both won.[46]

With a send-off from USAC's J.C. Agajanian, Mildred Alsbury drives her Chrysler New Yorker under the start banner and heads for Las Vegas. With her husband, Hollywood Chrysler-Plymouth dealer Mel Alsbury, Sr. as her co-driver, Mildred would place first in the luxury car class, scoring back-to-back wins for both Chrysler and the Alsburys.

With close to one hundred representatives from newspapers and magazines on hand to cover the Run, the public was kept fairly well informed on events. Interest appeared to be heaviest in the Southern California area, where the bulk of the participants called home. Newspapers in several communities, including Long Beach and Pasadena, focused their reporting on their local driver's progress. Prominent in newspaper coverage was the continuing battle between Chevrolet, Ford and Plymouth in the low-price eight-cylinder class, along with the success of Pontiac drivers Mickey and Judy Thompson. In addition to reporting the days' results, human interest stories told of mechanical problems and issues facing several drivers, along with one fighting off an illness toward the end. Fewer follow-up stories were seen this year in Sunday automotive sections, beginning a trend that would ultimately see these features disappear in a few years' time. *Automotive News* continued its tradition of providing a day-by-day review to its readers, most of whom were automotive industry–related. Enthusiast magazines continued to report on the Run, although articles were noticeably less extensive than in the past.[47] Some of the more comprehensive articles appearing in *Car Life*, *Car and Driver* and *Popular Science* followed the same theme, mainly reporting on the final results while highlighting the close contest between Chevrolet, Ford and Plymouth in the low-price class.

Several reporters drove over the route in a manual transmission–equipped Ford Fairlane 8 identical to Ginny Sims's entry. Over the course of the Run they learned first-hand the expertise of, and stress levels experienced by, Economy Run drivers. In summing up both this year's Run and his experiences as one of those reporters driving, William Carroll of *Automotive News* noted, "Careful driving gains good economy. Beyond this, there's not much anyone can do to influence or handicap results of the Mobil Economy Run. There's economy in every car when it leaves the factory. All you have to do is drive like a miser to get it."[48] Most reporters shared the same conclusion when the results were analyzed. Fifty years later, William's comment remains true.

1963

Los Angeles to Detroit Via Las Vegas — Salt Lake City — Cheyenne — Omaha — Chicago

March 30–April 4, 1963—2373.2 Miles

Generally pleased with the outcome of the 1962 Economy Run, Mobil Oil and USAC officials decided basically to leave well enough alone. Rumors, however, had surfaced that Mobil would discontinue the Economy Run as the trend in new cars was pointing more toward performance rather than economy. As preparations began for the 1963 Run, Roger Mahey, Mobil Oil's manager of special events and publicity, put this rumor to rest for the time being saying: "We have had a number of expensive surveys taken, the most recent late last year. The findings were so positive, in terms of public understanding, that we've been encouraged to put even more money into it than ever before." Additionally he noted that the 1963 Economy Run "will be Mobil's print advertising and promotion event of the year." It was estimated that the cost to put on this year's Run would be around $400,000.[49]

For the second year in a row, Detroit's Cobo Hall was selected as the terminus in what would be "the longest and toughest test in the 27-year history of the economy run." Taking a more northern route than those used in recent years, the Run "will make overnight stops in Las Vegas, Nev.; Salt Lake City, Utah; Cheyenne, Wyo.; Omaha, Neb.; and Chicago, Ill., during the 2500-mile, 6-day trip." USAC officials also stated, "The course is layed [*sic*] out so that cars will encounter every driving condition possible on a long trip. Competing cars will have to reach cruising speeds of 65 miles-per-hour on highways, climb roads over 8500 feet high, and battle through rush-hour traffic in cities."[50]

In an effort to have competing models reflect those models being purchased by the majority of motorists, a revision was made to the competitive classifications. Compact and intermediate-sized models comprised four of the eight classes with engine size now being the main factor in determining their eligibility. When the "Competitive Divisions for the 1963 Mobil Economy Run" was published, breakdown was Class A for "Small Engine Compact Cars [with an] engine displacement from 144 through 153 cubic inches"; Class B for "Medium Engine Compact and Intermediate-size Cars [with an] engine displacement of 169 through 170 cubic inches"; Class C for "Large Engine Compact and Intermediate-size Cars four and six cylinders [with an] engine displacement of 194 through 225 cubic inches"; and Class D for "Intermediate Size Cars with V-8 Engines."[51] The special class for compacts equipped with a manual transmission was eliminated, and all cars were required to be equipped with automatic transmission. It was shown that "manual shifts are being sold largely as sporting accessories to motorists who *like* to shift. Four-speeds with floor control are the ones that sell while the popularity of the old three-speeds-on-the-column is on the wane." *Car Life Magazine* observed that these class breakdowns "will stimulate some rough competition."[52] No changes were deemed necessary in the remaining classes for full-sized models. Four-door sedan and hardtop models remained the basic body style for eligibility although USAC officials may substitute two-door models in the compact car and intermediate size classes. Of the ten cars competing in the compact car classes, two-door models accounted for all the entries with the exception of a four-cylinder Chevrolet Chevy II 300.

When the rules were published, it was determined that only a few technical changes were needed. While Chief Steward A.C. Pillsbury remained "in charge of all technical and mechanical details of the Run," Chad Johnson was named to the new position of assistant chief steward after serving as a steward for a number of years. This was done in part to relieve Art, now in his seventies, of some of the day-to-day responsibilities both before and during the Run. Although no problems had been previously made public, all competitors were now notified that any unnecessary stops or delays, whether on the road or at refueling stops, would result in a penalty. Another area where the rules were tightened involved advertising, and the need for clarity, which "may require the use of bold face, underlining or other appropriate techniques of emphasis for class designation, model, mpg and name of event."[53]

A field of 36 cars competed, which represented most U.S. makes. While this number was fewer than the 46 cars that competed in 1962, 9 of those earlier entries were equipped with manual transmissions which were no longer eligible. A few makes declined to compete, including Cadillac; Lincoln, along with Mercury's intermediate-sized Meteor; and American Motors' Rambler Classic and Ambassador models. The reason for the Rambler Classic's absence was that it would have competed directly against the smaller Rambler American in the same compact car class. General Motors had the largest representation with 16 entries, followed by Chrysler Corporation with 11 and Ford Motor Company with 7. Rounding out the field were single entries from American Motors and Studebaker. Compact and intermediate-size models accounted for 15 entries. Two were from private individuals and competed without any factory support. Don Bridges and his Santa Monica Nickelodeon Restaurant entered a Dodge Custom 880, and Bill Corey entered a Studebaker Lark Regal 8.

Factory support continued, and in some instances expanded, this year reflecting to a degree the importance placed by the manufacturers on their individual models' performance. This was especially evident in Chrysler Corporation's involvement of their engineers not only in the preparation of the Dodge and Plymouth entries, but in the selection of the drivers and co-drivers as well. Moving into the driver's seat this year were two of Chrysler's engineers, Bob Checkley and Scott Harvey, and these two would enjoy a measure of success. At this time, Scott was leading the point standings in the Sports Car Club of America sports car rally competition.[54] While attaining the best possible gasoline mileage was the prime goal, the Run also had elements of a "time and distance" rally, so experience in events of this type would be beneficial. Jim

Whipple, auto editor of *Popular Mechanics Magazine*, was selected to pilot the Dodge 330 6 with Chrysler engineer Paul Mallwitz as co-driver. This selection would begin a tradition over the next several years of having special "non-engineering types" behind the wheel of one or more of the Dodge entries. This would also provide an opportunity for Dodge public relations personnel to promote additional publicity for both Dodge and the Economy Run. Again this year, Chrysler's entries would continue to be handled by Mel Alsbury and Hart Fullerton, joined by Culver City's Danny McGroo Chrysler-Plymouth entering an Imperial Custom.

Ford Motor Company did not provide factory support this year although two of their dealers, Tom Frost in Warrenton, Virginia, and Bob Estes in Inglewood, California, stepped in to assist. Both had been active in the USAC for a number of years and currently were serving as vice-presidents, Tom for the Eastern and Bob for the Western United States.[55] Art Rene moved over to head a team in preparing the Ford Motor Company entries after a three-year involvement as a driver for Plymouth and Dodge entries. Art would drive the Ford 300 8, hoping to add another first-place trophy to his display case. American Motors' and General Motors' divisions continued with their involvement as had been the practice in prior years.

Dealers were in the majority this year as sponsors, although manufacturers remained active in naming individual entrants. Only four dealer associations were named this year, and these represented the Dodge and Rambler American entries. Some of the new names to appear this year as entrants included Tom Frost Ford in Warrenton, Virginia. A Ford dealer since 1937, Tom for years was involved with the AAA Contest Board and USAC, serving as honorary steward of the Economy Run. His dealership would continue as a sponsor through 1968. Another Ford dealer being recognized was William L. Hughson Ford in San Francisco. Awarded a Ford franchise in 1903, he had the distinction of being the first Ford dealer appointed by Ford Motor Company.[56] This was noted on his Falcon's entry with the slogan "*The First Ford Dealer in the World*," which marked the first time any slogans or other achievements were allowed on competing cars. This slogan was again used in 1964, and after that, slogans of this type would not be seen again on Economy Run entrants. In a sponsorship combination for the Oldsmobile Dynamic 88 with Paola Oldsmobile in La Crescenta was Irving Air Chute Company, known in part for their manufacture of automobile seat belts.

The days of total dealer involvement, both as an entrant and a driver, were slowly coming to a close. Only three dealers in this category competed: Hart Fullerton (Chrysler 300) and Melvin Hilliard (Chevrolet Impala 8) were joined this year by Mandy Williams (Oldsmobile Ninety-Eight). Two dealers, Mel Alsbury, Sr. (Chrysler New Yorker), and Jim Retzlaff (Buick LeSabre), acted as co-drivers for their wives as part of the four husband-and-wife teams competing. The other two husband-and-wife teams were Hart and Diana Fullerton (Chrysler 300) and Stan and Lois Raymond (Oldsmobile Dynamic 88). Mickey and Judy Thompson returned driving separate entries, although switching models from last year with Mickey in a Star Chief and Judy in a Tempest Custom 4. Women were named as drivers for eight entries, including all of the cars competing in the small engine compact class. Five of last year's winning drivers returned, with George Alsbury, Byron Froelich and Joe Strum sitting out this year's Run. Both George and Byron would return in future years.

Random car selection commenced in mid–February, and USAC officials did not limit their scope to Southern California in locating competing cars. An example was the Dodge 330 6, selected from a dealer's stock in Missouri, where it was inspected for stock status before being transported back to the Pasadena impound under USAC supervision.[57] For the next six weeks the impound was a hub of activity with serving and break-in runs being accomplished under the close supervision of USAC officials. To dispel the public's thinking that these cars were specially built and prepared, *Road & Track* magazine was invited to observe the activities during the impound period. The magazine's representatives noted that nothing could be done to a competing car without a USAC official present and most of this was limited to standard dealer service. Entrants were no longer permitted to work on the carburetor; only limited adjustments were permitted, and were done by USAC personnel as entrants observed from behind a barrier. *Road & Track* concluded that "because of the elaborate safeguards practiced by USAC, the chance for hanky-panky in the Economy Run is, for all practical purposes, nonexistent."[58] Assisting the USAC, the Automobile Club of Southern California furnished one of their Ford F-100 Service Patrol vehicles and its driver to provide assistance as needed should a competing car encounter trouble during the Run. Ready for any emergency, the truck carried a wide supply of items including batteries, motor oil, transmission and brake fluid, plus a complete set of keys for each car. The Auto Club's presence would continue to be seen in future years.[59] As in prior years, competitors also spent this time in honing their driving skills in replicas of the cars selected for the Run which were equipped with special fuel meters and vacuum gauges to measure the engine's performance under different driving conditions. Entrants again were given the option of using either Mobil Regular or Mobil Premium gasoline.

While USAC officials finalized plans and personnel assignments for the Run, Mobil Oil's publicity personnel continued to get the word out to newspapers, radio and television stations along the route. Frank Meunier continued active in all of Mobil Oil's publicity efforts while working more behind the scenes as other members of his team had more visible roles.

Mickey Thompson makes a hard right turn with his Pontiac Star Chief off U.S. Route 99, heading towards California's Antelope Valley. After winning the medium-price class in 1962, the Star Chief could do no better than fourth this year with 19.27 miles per gallon.

Press representatives from across the country, hosted by Mobil Oil's Roger Mahey, traveled over the route on a chartered United Airlines DC-7, stopping at each overnight's impound to file stories on each day's events. Others, including *Los Angeles Times*' Bob Thomas and *Automotive News*' William Carroll, covered the route driving replicas of competing cars to experience first-hand the skills needed to achieve the best possible miles per gallon. This group included *Car Life Magazine*'s Dennis Shattuck and Gene Booth, who covered the route in a Chevrolet Chevy II 300 6 that had been used as one of the practice cars for the Chevrolet team.[60]

On Monday, March 25, the traditional drawing for starting positions was held at the Pasadena impound, with drivers again selecting their number from a group of rolled pennants. Kansas City (Missouri) Chevrolet dealer Melvin Hilliard drew No. 1 for his Chevrolet Impala 8; Bob Checkley in a Plymouth Valiant V-100 brought up the rear when he drew No. 36. USAC officials announced that again this year the start would be from the Pasadena impound.

On newsstands in March, *Car Life Magazine* made a fairly detailed study of the field and, in one of the only articles of its type, made predictions for each class winner and runner-up. Taking into account prior performance, engine size and driver's expertise, the magazine's Tom Swain saw these entrants placing one-two:

Class A — Small Engine Compacts	Chevrolet Chevy II 300 4 Chevrolet Corvair Monza
Class B — Medium Engine Compacts	Plymouth Valiant V-100 Dodge Dart 170
Class C — Large Engine Compacts	Pontiac Tempest Custom 4 Rambler American 440
Class D — Intermediate Size Eight Cylinder	Ford Fairlane Buick Special Deluxe
Class E — Standard Size Six Cylinder	Plymouth Savoy Chevrolet Biscayne
Class F — Low Price Eight Cylinder	Plymouth Savoy Chevrolet Impala
Class G — Medium Price	Chrysler Newport Dodge Custom 880
Class H — Luxury Cars	Chrysler New Yorker

Tom, who did not select a runner-up to the Chrysler New Yorker, added, "Because the Mobil Economy Run is a great test of both driver and car, any of our favorites could be nosed out by a man (or woman). Traffic conditions and slight variations within a specific car model can influence the results."[61] As the Run progresses, we will see how close these predictions came true.

On Friday, March 29, the first day's detailed route and itinerary was released. At this point, everyone knew that USAC officials would ensure that this year's Run would present several challenges. Following a new and out-of-the-way route from Pasadena to Las Vegas, the drivers would cover a 384.7-mile route within an 8-hour and 53-minute time limit over a track that was described as "a devious course which carried the

After refueling in Yermo, California, Paula Murphy heads out in her Ford 300 6 for the first overnight stop in Las Vegas. Another Ford, a 300 8 driven by Art Rene, is being refueled by USAC officials. No match in economy against the Chevrolet, Dodge and Plymouth entries, both Fords would finish last in their respective classes.

lineup more than 100 miles out of its way."[62] As had been the custom, the daily time limits did not include refueling and 40-minute brunch (or lunch) stops. An added element this year was that the long-range weather forecasts were calling for snow and high winds on the Salt Lake City-to-Cheyenne leg.[63]

On March 30, in a scene reminiscent of last year, all of the competing cars were arranged in the parking lot of the Pasadena impound, which, at 960 feet, was the lowest elevation on this year's Run. Bright sunshine greeted the small crowd of spectators, mainly comprising participants' family members and members of the press, who were on hand at 6:00 a.m. when Mel Hilliard, driving a Chevrolet Impala 8, was the first to take the green flag from J.C. Agajanian. Leaving at one-minute intervals, the drivers made their in a northwest direction picking up U.S. Route 99 to Gorman before heading east on California 138 over two-lane roads towards the Antelope Valley in a route that on paper somewhat resembled a hook. Passing through Lancaster, they made their first stop of the day in Hesperia for lunch at the Hesperia Inn after being on the road for just over 4 hours. Then it was on to Lucerne Valley, from which they headed north on local roads to Barstow. There they picked up U.S. Route 91 towards Yermo, where entries were topped off at Yermo Mobil Service. Continuing to follow U.S. Route 91, crossing the 4731-foot Mountain Pass, the drivers continued on to their destination was Las Vegas. The first stop was for refueling at Joe Montoya Mobil Service, then to the overnight impound at the Las Vegas Convention Center, where they arrived just after 4:00 p.m. after nearly 10 hours on the road. Among those greeting the drivers was Kathee Francis, known locally as Miss Thunderbird Hotel.[64] USAC officials were also on hand with today's results and tomorrow's route and detailed itinerary. This stop would be the only impound this year which was not located at the hotel or motel where participants were lodged overnight.

Pat Sawyer's Chevrolet Corvair Monza encountered problems on this opening leg, caused by a faulty vacuum advance line. This was repaired after her arrival in Las Vegas and Pat was able to continue.[65]

To anyone who had followed the Economy Run for the past several years, it was not a surprise that Les Viland and his Rambler American 440 turned in the best mileage for all entrants with a figure of 28.43 miles per gallon. Close on his heels for overall best mileage was Mary Hauser in her Chevrolet Chevy II 300 4 with 27.87 miles per gallon. Both Les and Mary were also leading their respective classes. Three classes saw competition that was quite close. In the class for medium-engine compacts, Bill Levy's Ford Falcon Deluxe had a .54 miles-per-gallon lead over Bob Checkley's Plymouth Valiant V-100. In the low-price eight-cylinder class, Bob Cahill's Dodge 330 had a slim .03 miles-per-gallon lead over Bob McDaniel's Plymouth Savoy. Close behind in third was Don Edmond's Chevrolet Bel-Air, with .14 miles per gallon separating the first three places. Those following the Run started to wonder if a repeat of last year's close contest could be developing. In the medium-price class, Stan Raymond's Oldsmobile Dynamic 88 had a slim .59 miles-per-gallon lead over Mel Alsbury, Jr.'s

Chrysler Newport. John Rich (Buick Special Deluxe 8), Scott Harvey (Plymouth Savoy 6) and Mildred Alsbury (Chrysler New Yorker) all opened up a fairly comfortable lead in their respective classes. General Motors and Chrysler Corporation each held first place in three classes, with Ford Motor Company and American Motors each leading in one.

The top three performers in each class were:

Small Engine Compacts	Chevrolet Chevy II 300 4	27.87 Miles Per Gallon
	Chevrolet Corvair Monza	26.13 Miles Per Gallon
	Ford Falcon Deluxe (144)	25.91 Miles Per Gallon
Medium Engine Compacts	Ford Falcon Deluxe (170)	26.06 Miles Per Gallon
	Plymouth Valiant V-100	25.52 Miles Per Gallon
	Mercury Comet Custom	25.44 Miles Per Gallon
Large Engine Compacts	Rambler American 440	28.43 Miles Per Gallon
	Pontiac Tempest Custom 4	26.31 Miles Per Gallon
	Chevrolet Chevy II 300	23.62 Miles Per Gallon
Intermediate Size Eight Cylinder	Buick Special Deluxe	23.90 Miles Per Gallon
	Ford Fairlane	21.09 Miles Per Gallon
	Pontiac Tempest	20.88 Miles Per Gallon
Standard Size Six Cylinder	Plymouth Savoy	24.04 Miles Per Gallon
	Dodge 330	22.89 Miles Per Gallon
	Chevrolet Biscayne	21.72 Miles Per Gallon
Low Price Eight Cylinder	Dodge 330	20.65 Miles Per Gallon
	Plymouth Savoy	20.62 Miles Per Gallon
	Chevrolet Bel-Air	20.51 Miles Per Gallon
Medium Price	Oldsmobile Dynamic 88	19.77 Miles Per Gallon
	Chrysler Newport	19.18 Miles Per Gallon
	Pontiac Catalina	19.12 Miles Per Gallon
Luxury Cars	Chrysler New Yorker	18.12 Miles Per Gallon
	Oldsmobile Ninety-Eight	17.21 Miles Per Gallon
	Imperial Custom	17.15 Miles Per Gallon

On March 31, the drivers headed for Salt Lake City for a 467.4-mile trip to be covered within a 100-hour and 49-minute time limit. Drivers would encounter "warm spring skies but [would be] buffeted by strong desert winds" throughout the day.[66] Heading north out of Las Vegas, they would travel on Interstate 15 and U.S. Route 91 for the first 300 miles before another detour waited. Crossing into Utah and entering the Mountain Time Zone, they made the first stop of the day in Cedar City, where each entry was topped off at Green & Higbee Mobil Service. This is the same station where competitors made a refueling stop in 1959. Further along U.S. Route 91, the next stop was in Beaver, a city visited several times in previous Runs, for lunch at Anderson's Ponderosa Café, where the participants were ready for a break after nearly six hours on the road. Leaving Beaver, they made their north to Holden, where they veered off onto various U.S. and Utah routes covering 165 miles towards Salt Lake City, climbing to an elevation of 6020 feet at one point. As they arrived in the late afternoon, the first stop was for refueling at Anderson's Mobil Service before heading to the impound at the World Motor Hotel. The competitors were ready for a well-earned break after close to twelve hours on the road. They would need this time as more challenges were on tap for tomorrow.

Driving along South Las Vegas Avenue, Al Cottle in his Mercury Monterey Custom passes the iconic "Welcome to Fabulous Las Vegas" sign as he makes his way to the Las Vegas Convention Center impound. Today, although the background is completely built up, the sign is still in the same location.

Despite the wind and mountainous terrain, most entries slightly improved their performance from the first day's results by recording an overall average of 21.61 miles per gallon. Les Viland's Rambler American 440 continued to lead all entries with the best miles-per-gallon average of 28.59, a slight increase from yesterday's results. Mary Hauser's Chevrolet Chevy II 300 4 continued close on his heels for overall mileage honors with 28.15 miles per gallon. While Mary continued to lead the small-engine compacts class, Pat Sawyer in a Chevrolet Corvair Monza began to close the gap after making repairs to her malfunctioning vacuum line. No changes were noted in the first three positions for each class. Competition became even closer in the medium-engine compacts class, with Bill Levy's Ford Falcon continuing his lead over Bob Checkley's Plymouth Valiant V-100 and Mary Faulkner's Mercury Comet Custom, though the margin between the three was a slim .47 miles per gallon. This appeared to be the contest to watch as Bob Cahill's Dodge 330 and Stan Raymond's Oldsmobile Dynamic 88 slightly opened their lead over second-place competitors. A close contest was developing between second- and third-place entries in the medium-price class, where a mere .03 miles per gallon separated Mel Alsbury, Jr.'s Chrysler Newport and Jim Wasson's Pontiac Catalina.

The top three performers in each class were:

Small Engine Compacts	Chevrolet Chevy II 300 4	28.15 Miles Per Gallon
	Chevrolet Corvair Monza	26.82 Miles Per Gallon
	Ford Falcon Deluxe (144)	26.06 Miles Per Gallon
Medium Engine Compacts	Ford Falcon Deluxe (170)	26.22 Miles Per Gallon
	Plymouth Valiant V-100	25.77 Miles Per Gallon
	Mercury Comet Custom	25.75 Miles Per Gallon
Large Engine Compacts	Rambler American 440	28.59 Miles Per Gallon
	Pontiac Tempest Custom 4	26.72 Miles Per Gallon
	Chevrolet Chevy II 300	24.00 Miles Per Gallon
Intermediate Size Eight Cylinder	Buick Special Deluxe	23.74 Miles Per Gallon
	Ford Fairlane	21.31 Miles Per Gallon
	Pontiac Tempest	20.82 Miles Per Gallon
Standard Size Six Cylinder	Plymouth Savoy	24.05 Miles Per Gallon
	Dodge 330	23.18 Miles Per Gallon
	Chevrolet Biscayne	21.90 Miles Per Gallon
Low Price Eight Cylinder	Dodge 330	21.02 Miles Per Gallon
	Plymouth Savoy	20.82 Miles Per Gallon
	Chevrolet Bel-Air	20.58 Miles Per Gallon
Medium Price	Oldsmobile Dynamic 88	20.10 Miles Per Gallon
	Chrysler Newport	19.29 Miles Per Gallon
	Pontiac Catalina	19.26 Miles Per Gallon
Luxury Cars	Chrysler New Yorker	18.03 Miles Per Gallon
	Oldsmobile Ninety-Eight	17.08 Miles Per Gallon
	Imperial Custom	17.01 Miles Per Gallon

Although not the longest day in terms of miles, the 483.6-mile trip from Salt Lake City to Cheyenne on April 1 would be the most demanding and challenging the drivers would face. After studying the route through Wyoming, including a climb over the Rocky Mountains, everyone knew this 11-hour and 17-minute trip would be no April Fool's joke as once again Chief Steward A.C. Pillsbury had a few tricks up his sleeve. Heading northeast out of Salt Lake City, the drivers crossed into Wyoming following Interstate 80 / U.S. Route 30, also known as the Lincoln Highway, passing through several mountain communities including Evanston and Green River. This year would mark the only time the Run traveled through Wyoming. Just west of Rock Springs, the first stop of the day was at the Log Inn for brunch after being on the road for just over four hours. Following brunch, it was a short drive into Rock Springs for a stop at John's Super Mobil Service, where all entries were topped off. Continuing to follow Interstate 80 / U.S. Route 30 across the Great Divide Basin, the drivers encountered gravel road surfaces where crews were constructing portions of Interstate 80. Each year, Chief Steward Art Pillsbury tried to "plot out a path to confound drivers and engineers. This year his 'special' treat was a detour from U.S. Route 30 just outside of Rawlins, Wyo., that would send the caravan to Medicine Bow National Forest in Colorado [*sic*]."[67] In late afternoon, while traveling through this area, the competitors reached an elevation of 9010 feet, the highest elevation on this year's Run. There had been earlier forecasts for snow and ice in this region; however, the weather turned out to be ideal. A few hours later, a snowstorm did develop.[68] The drivers continued on to Laramie, arriving in late afternoon, and after navigating the downtown streets they arrived at Tatham's Mobil Service for refueling and were greeted by a good-sized group of spectators. Back on Interstate 80 / U.S. Route 30, the drivers still had a 48-mile trek to Cheyenne and the impound at the Hitching Post Motel. Pulling into the motel's parking lot was a welcome end of a long day's drive, even though they had been "spared the snow and storm conditions which they feared were in store for them."[69] Even so, the competitors still looked to the skies as "the weatherman predicts heavy snows across the plains [where] the cars will be traveling."[70]

After three days on the road, comments began to surface from drivers and factory team representatives "expressing unhappiness over the route." In answering these comments, Chief Steward Art Pillsbury said, "We want to make it typical. Tourists don't always stick to the freeways when they go across country. That's why we've got the cars on back country highways, on some secondary gravel roads and in the high mountains. This thing should be realistic," he said, "and side routes put a premium on driver skill, too."[71] The next leg would present more challenges not only for the drivers, but for USAC officials as well.

In spite of the uphill climb over the Rockies, the drivers managed to post a gain of nearly a half-mile per gallon by recording an overall average of 22.18 miles per gallon. Some of this increase was "attributed to tail winds ... and the absence of expected bad weather."[72] Les Viland's Rambler American 440 continued to lead all entries in miles per gallon with a figure of 29.34. At this stage it looks like Les's lead was secure, barring any trouble on the remaining legs. Mary Hauser's Chevrolet Chevy II 300 4 continued close on his heels with 28.95 miles per gallon. While all class leaders maintained their positions, competition became even tighter in two classes. In the medium-engine compact car class, Bill Levy's Ford Falcon continued to lead Bob Checkley's Plymouth Valiant V-100 by a slim margin of .29 miles per gallon. Another contest suddenly tightened up. In the low-price eight-cylinder class, Bob Cahill's Dodge 330 saw its lead over Bob McDaniel's Plymouth Savoy shrink to a mere .16 miles per gallon. Another close contest continued in the medium-price class, where a mere .05 miles per gallon separated second-place Mel Alsbury, Jr.'s Chrysler Newport from Jim Wasson's Pontiac Catalina. In the luxury car class, Marty Gordon moved his Imperial Custom into second place, dropping Mandy William's Oldsmobile Ninety-Eight to third.

The top three performers in each class were:

Small Engine Compacts	Chevrolet Chevy II 300 4	28.95 Miles Per Gallon
	Chevrolet Corvair Monza	27.94 Miles Per Gallon
	Ford Falcon Deluxe (144)	26.51 Miles Per Gallon
Medium Engine Compacts	Ford Falcon Deluxe (170)	26.87 Miles Per Gallon
	Plymouth Valiant V-100	26.58 Miles Per Gallon
	Mercury Comet Custom	26.39 Miles Per Gallon
Large Engine Compact	Rambler American 440	29.34 Miles Per Gallon
	Pontiac Tempest Custom 4	27.57 Miles Per Gallon
	Chevrolet Chevy II 300	24.63 Miles Per Gallon
Intermediate Size Eight Cylinder	Buick Special Deluxe	24.25 Miles Per Gallon
	Ford Fairlane	21.93 Miles Per Gallon
	Oldsmobile F-85	21.50 Miles Per Gallon
Standard Size Six Cylinder	Plymouth Savoy	24.60 Miles Per Gallon
	Dodge 330	23.89 Miles Per Gallon
	Chevrolet Biscayne	22.68 Miles Per Gallon
Low Price Eight Cylinder	Dodge 330	21.48 Miles Per Gallon
	Plymouth Savoy	21.32 Miles Per Gallon
	Chevrolet Bel-Air	21.12 Miles Per Gallon
Medium Price	Oldsmobile Dynamic 88	20.59 Miles Per Gallon
	Chrysler Newport	19.73 Miles Per Gallon
	Pontiac Catalina	19.68 Miles Per Gallon
Luxury Cars	Chrysler New Yorker	18.56 Miles Per Gallon
	Imperial Custom	17.43 Miles Per Gallon
	Oldsmobile Ninety-Eight	17.31 Miles Per Gallon

On April 2, the drivers pointed their cars east towards Omaha for the 501.5-mile trek with a time limit of 11 hours and 16 minutes. Today would be the longest in terms of miles covered and would feature open highway driving most of the day while staying ahead of a large storm system that had been predicted for today. Before leaving Cheyenne, the competitors were warned to "make sure to check average speeds on freeways or tollways over the entire distance so that the Run schedule can be maintained. Depending upon using these sections of highways to make up time at the end of the day may prove to be disappointing. Also, loitering or impeding the normal flow of traffic will be reported."[73] Following U.S. Route 30, the drivers crossed into Nebraska, later entering the Central Time Zone, before arriving in North Platte, where all entries were topped off at Tomahawk Mobil Service. Just under an hour later, they arrived in Gothenburg for brunch at Our Lady of Good Counsel Catholic Church. Here volunteers from the church prepared and served the participants, who were ready for a short break after being on the road for close to six hours. Continuing to follow U.S. Route 30, the drivers picked up U.S. Route 34 in Grand Island, heading for Lincoln for a refueling stop. For most of the day the drivers had been contending with strong head winds and crosswinds, with gusts as high as 60 miles per hour, "which pitched the cars crazily over what has been considered a 'straight as a string' course through Nebraska."[74] As the drivers arrived in Lincoln, human error resulted in a third of the fleet's being directed onto a new freeway section, which was not on the official route. A local volunteer, assigned as a flagman on the outskirts of Lincoln, thought he was doing the drivers a favor by sending them on this nicer, smoother route rather than on the official route, which was bumpy while undergoing construction. Before his actions could be corrected, he had sent the drivers 3.6 miles out of their way and in turn bypassed some city traffic.[75] After regrouping and navigating Lincoln's city streets, these drivers finally made their way to Jess Quigley North East Mobil Service. After refueling, they picked up Interstate 80 bound for Omaha, tired and frustrated, making their way to the impound at Schimmel's Indian Hills Inn.

That evening, once tempers cooled, USAC officials met to discuss what was called "an unprecedented mixup" in Lincoln.[76] There was only one answer, as announced by USAC: "A 281.5 mile segment of the Official Course between North Platte and Omaha, Nebraska was ruled out from fuel consumption computation for all cars by the Board of Stewards after 12 competing cars were misdirected by an improperly placed flagman."[77] This would be the only time in the Economy Run's history where a segment was invalidated for all competitors. While "it is doubtful that [this] ruling had any serious effect in the final results," there would always be a feeling by some second- and third-place finishers that they perhaps could have pulled off a win had this flagman has not send a third of the competitors on the wrong route.[78] In another development, Fran Foster, driver of the Ford Falcon Deluxe competing in the small-engine compacts class, dropped out and returned to Los Angeles after being notified her husband, Lloyd, had suffered a stroke. Fran's co-driver, Irma Payne, took over the driving duties.[79]

Despite the obstacles the drivers faced today, they managed to maintain their overall average with a figure of 22.19 miles per gallon. Les Viland in a Rambler American 440 continued his strong lead over all entries in miles per gallon with a figure of 29.26. Mary Hauser in her Chevrolet Chevy II 300 4 continued close behind with 28.71 miles per gallon, although she was facing some additional competition in the small-engine compacts class. Recovering from mechanical problems on the first day, Pat Sawyer in a Chevrolet Corvair Monza continued to chip away at Mary's lead, now trailing her by a half-mile per gallon. While all class leaders maintained their positions, it was really getting even tighter in the medium-engine compacts class. Bill Levy's Ford Falcon held on to a mere .01 miles per gallon lead over Bob Checkley's Plymouth Valiant V-100 in a contest everyone continued to watch closely. In the low-price eight-cylinder class, Bob Cahill's Dodge 330 was able to slightly stretch its lead, now by a slim .24 miles per gallon, over Bob McDaniel's Plymouth Savoy. In the only other close contest of note, Mel Alsbury, Jr.'s Chrysler Newport was able to maintain second place in the medium-price class over Jim Wasson's Pontiac Catalina by stretching his margin to .13 miles per gallon.

The top three performers in each class were:

Small Engine Compacts	Chevrolet Chevy II 300 4	28.71 Miles Per Gallon
	Chevrolet Corvair Monza	28.21 Miles Per Gallon
	Ford Falcon Deluxe (144)	26.39 Miles Per Gallon
Medium Engine Compacts	Ford Falcon Deluxe (170)	26.74 Miles Per Gallon
	Plymouth Valiant V-100	26.73 Miles Per Gallon
	Mercury Comet Custom	26.33 Miles Per Gallon
Large Engine Compacts	Rambler American 440	29.26 Miles Per Gallon
	Pontiac Tempest Custom 4	27.67 Miles Per Gallon
	Chevrolet Chevy II 300	24.56 Miles Per Gallon
Intermediate Size Eight Cylinder	Buick Special Deluxe	24.13 Miles Per Gallon
	Ford Fairlane	21.87 Miles Per Gallon
	Pontiac Tempest	21.46 Miles Per Gallon
Standard Size Six Cylinder	Plymouth Savoy	24.61 Miles Per Gallon
	Dodge 330	23.98 Miles Per Gallon
	Chevrolet Biscayne	22.80 Miles Per Gallon
Low Price Eight Cylinder	Dodge 330	21.53 Miles Per Gallon
	Plymouth Savoy	21.29 Miles Per Gallon
	Chevrolet Bel-Air	21.15 Miles Per Gallon
Medium Price	Oldsmobile Dynamic 88	20.59 Miles Per Gallon
	Chrysler Newport	19.88 Miles Per Gallon
	Pontiac Catalina	19.75 Miles Per Gallon
Luxury Cars	Chrysler New Yorker	18.68 Miles Per Gallon
	Imperial Custom	17.53 Miles Per Gallon
	Oldsmobile Ninety-Eight	17.23 Miles Per Gallon

It was an early call on April 3, as the drivers began to leave Omaha before sunrise for a 499.2-mile trek to Chicago. This would be the longest day on the road, although not in actual miles, with a time limit set by USAC of 12 hours and 9 minutes. Today's route would feature open highway driving on U.S. Route 6 across Iowa and Illinois, and would include stops in the Iowa cities of Newton for top-off and Amana for brunch at the Ox Yoke Inn. Drivers continuing to be plagued with strong head and side winds while they also kept a close watch on the skies. "Outside of Davenport, tornado warnings were issued, but twisters failed to develop."[80] Crossing into Illinois, the drivers made their way to Joliet for a refueling stop before continuing on to the impound at the Holiday Inn in Lansing, a suburb south of Chicago near the Indiana border. Arriving after another long day put the competitors "within a gas tank of their destination of Detroit."[81] With fatigue setting in on all of the competitors, one missed instruction could change everything. And it did, on this leg where a mistake would hamper one class leader's performance. Mary Hauser, driving a Chevrolet Chevy II 300 4, made a wrong turn in Joliet and ended up traveling 21 miles out of her way in rush-hour traffic before getting back on course.[82] In a close contest these extra miles could spell the difference between first and second place, although this would not be known until reaching Detroit.

After five days of intense driving, covering 2336.4 miles since leaving Los Angeles, one competitor commented, "You just don't have time for yourself. Complete, constant concentration is required: the throttle must be held just so, the speedometer watched and the navigator listened to. Miss one instruction and you can 'blow the whole bit.'"[83] Without a doubt, this was on the mind of the 36 drivers and co-drivers who were more than ready to see the finish line in Detroit tomorrow.

After another demanding leg, the drivers continued to maintain an overall average of 22.17 miles per gallon, only .02 less than the previous day. Les Viland and his Rambler American 440 continued to maintain a strong lead over all entries with a figure of 29.26 miles per gallon. Mary Hauser's Chevrolet Chevy II 300 4 continued close behind in overall mileage performance with 28.77 miles per gallon while at the same time maintaining a small lead over Pat Sawyer's Chevrolet Corvair Monza. The difference between these two ladies, both competing in the small-engine compacts class, was .61 miles per gallon. There was one change on the leader board. In the medium-engine compacts class, Bob Checkley eased his Plymouth Valiant V-100 ahead of Bill Levy's Ford Falcon, leading by a slim .07 miles per gallon. This contest would continue to be watched closely as the final leg to Detroit would determine the winner. The remaining seven class leaders maintained their positions and with the exception of the low-price eight-cylinder class, all of the drivers were able to pad their leads. In that class Bob Cahill's Dodge 330 continued ahead of Bob McDaniel's Plymouth Savoy, although the lead slimmed to .20 miles per gallon. This contest too would not be settled until tomorrow. The only other close contest of note was in the medium-price class involving second- and third-place positions. Although unable to close in on first place, Mel Alsbury, Jr.'s Chrysler Newport was able to stretch his small lead the over third-place Jim Wasson by a slim .19 miles per gallon.

Top: **It was a busy scene in Newton, Iowa, as USAC officials refueled Barney Navarro's Buick Electra 225, Roy Swanson's Chevrolet Biscayne 6 and Stan Raymond's Oldsmobile Dynamic 88. With open highway driving all day, the drivers had to contend with strong head and side winds, although severe weather which had been predicted did not develop.** ***Bottom:*** **Arriving at Detroit's Cobo Hall well ahead of the other competitors, Hart Fullerton crosses the finish line in his Chrysler 300 with a "wave of the hat" from a USAC official. Running last in the medium-price class, Hart instead pressed on to be the first to cross the finish line, and arrived before USAC's J.C. Agajanian was ready with his checkered flag.**

The top three performers in each class were:

Small Engine Compacts	Chevrolet Chevy II 300 4	28.77 Miles Per Gallon
	Chevrolet Corvair Monza	28.16 Miles Per Gallon
	Ford Falcon Deluxe (144)	26.29 Miles Per Gallon
Medium Engine Compacts	Plymouth Valiant V-100	26.75 Miles Per Gallon
	Ford Falcon Deluxe (170)	26.68 Miles Per Gallon
	Mercury Comet Custom	26.23 Miles Per Gallon
Large Engine Compacts	Rambler American 440	29.26 Miles Per Gallon
	Pontiac Tempest Custom 4	27.62 Miles Per Gallon
	Chevrolet Chevy II 300	24.50 Miles Per Gallon
Intermediate Size Eight Cylinder	Buick Special Deluxe	24.09 Miles Per Gallon
	Ford Fairlane	21.81 Miles Per Gallon
	Pontiac Tempest	21.45 Miles Per Gallon
Standard Size Six Cylinder	Plymouth Savoy	24.63 Miles Per Gallon
	Dodge 330	23.95 Miles Per Gallon
	Chevrolet Biscayne	22.81 Miles Per Gallon
Low Price Eight Cylinder	Dodge 330	21.49 Miles Per Gallon
	Plymouth Savoy	21.29 Miles Per Gallon
	Chevrolet Bel-Air	21.14 Miles Per Gallon
Medium Price	Oldsmobile Dynamic 88	20.54 Miles Per Gallon
	Chrysler Newport	19.88 Miles Per Gallon
	Pontiac Catalina	19.69 Miles Per Gallon
Luxury Cars	Chrysler New Yorker	18.68 Miles Per Gallon
	Imperial Custom	17.52 Miles Per Gallon
	Oldsmobile Ninety-Eight	17.26 Miles Per Gallon

With an early start, a fairly comfortable 318.3-mile, 6-hour and 15-minute drive faced the drivers as they headed for the finish line at Detroit's Cobo Hall, where a full schedule of activities was on tap. Hampered by head winds for most of the day, the drivers crossed into Indiana traveling on Interstates 80/90 and the Indiana Tollway past Gary, where they headed north on Indiana highways to Michigan City and an early brunch stop at the Golden Sands Café. Continuing on U.S. Route 12, they briefly crossed through Michigan, entering the Eastern Time Zone, before heading south on U.S. Route 31 and back into Indiana. Arriving in Roseland, they stopped for refueling at Gen Mock Mobil Service. Then it was back to the Indiana Tollway, heading east, and after crossing into Ohio, they picked up the Ohio Tollway bound for Toledo. This would mark the Run's first visit to Ohio, but not the last.

For the first time since 1958, an Oldsmobile topped all entries in the medium-price class. With the husband-and-wife team of Stan and Lois Raymond as driver and co-driver, the Dynamic 88 recorded a figure of 20.01 miles per gallon for a comfortable lead over Mel Alsbury, Jr. and his Chrysler Newport in second.

From Toledo the drivers headed north on Interstate 75 towards Detroit, where they picked up Interstate 94 and followed last year's route to Cobo Hall. Arriving ahead of the other competitors, Hart Fullerton in a Chrysler 300 was first to cross the finish line, receiving a "wave of the hat" from a USAC official as J.C. Agajanian was not yet in place with the traditional checkered flag. Around noontime, the remaining drivers arrived and were greeted by a crowd of spectators, most of whom represented the auto industry, that was somewhat smaller than last year. As the entries were topped off for the final time, participants enjoyed lunch while waiting for the Awards Ceremony scheduled for 2:45 p.m. Everyone's interest was centered on the competition in three classes: small-engine compact cars, medium-engine compact cars, and low-price eight-cylinder. That evening the annual Mobil Mileage Dinner was held at the Sheraton Cadillac Hotel.

When the results were announced, once again Les Viland and his Rambler American 440 recorded the highest mileage figure for all entries with 28.61 miles per gallon. This marked the fourth time in five years a Rambler American recorded the top mileage figure. Barney Navarro, driving a Buick Electra 225, had the distinction of recording the lowest mileage figure with 16.60 miles per gallon. A change in first place occurred in two classes, small-engine compacts and medium-engine compacts. Overall, General Motors and Chrysler Corporation each placed first in three classes, and Ford Motor Company and American Motors in one each. Women took the top prize in two classes.

Class winners were:

CLASS	MAKE and MODEL	MILES PER GALLON
A — Small Engine Compacts	Chevrolet Corvair Monza	27.90
B — Medium Engine Compacts	Ford Falcon Deluxe	26.29
C — Large Engine Compacts	Rambler American 440	28.61
D — Intermediate Size Eight Cylinder	Buick Special Deluxe	23.81
E — Standard Size Six Cylinder	Plymouth Savoy	24.33
F — Low Price Eight Cylinder	Dodge 330	21.20
G — Medium Price	Oldsmobile Dynamic 88	20.01
H — Luxury Cars	Chrysler New Yorker	18.43

Detailed results for all competing cars can be found in the Appendix.

USAC officials reported the average for the entire field was 21.84 miles per gallon. While this was a drop from 1962's average of 22.6980 miles per gallon, two factors came into play. This year's course and weather conditions were more rugged, and all entries were equipped with automatic transmission. Nine compacts competing in 1962 were equipped with manual transmission, which skewed 1962's average. The 8 women drivers showed the men a thing or two, averaging 23.49 miles per gallon compared to the men's average of 21.36. A factor in the women's performance involved the cars they were selected to drive, with 5 of the 8 piloting compact-sized models. Covering 2373.2 miles, this year's route included 525 miles of expressway driving along with 356 miles in city streets while contending with 340 traffic signals. This official distance was shorter than the 2654.7-mile route actually driven, as a result of the 281.5-mile segment in Nebraska that was ruled out due to human error. To keep within the adjusted 54-hour and 8-minute time limit, the drivers averaged 43.77 miles per hour, which was a mile an hour faster than 1962's average. These averages had crept up steadily over the past five years, but came nowhere close to 1958's mark of 45.2224 miles per hour. There were no major mechanical issues and all entries completed the Run. The Automobile Club of Southern California's Service Patrol provided some limited assistance, including replacing headlights broken by flying gravel and replacing a tire which had a faulty tread.[84] One change in drivers was made when Irma Payne took over for Fran Foster in a Ford Falcon Deluxe competing in the small-engine compacts class.

Looking at the eight drivers standing in the winner's circle, half were first-time champs. The "first-timers" were Mildred Alsbury (Chrysler New Yorker), Scott Harvey (Plymouth Savoy), Bill Levy (Ford Falcon Deluxe) and Stan Raymond (Oldsmobile Dynamic 88). Bob Cahill and Pat Sawyer notched back-to-back wins, with Pat scoring her third consecutive victory. Les Viland scored his sixth win, and John Rich his second. Two husband-and-wife teams, the Alsburys and Raymonds, were successful in their respective classes.

Pat Sawyer, in her Chevrolet Corvair Monza, managed to move into first place in the small-engine compacts class by a margin of .38 miles per gallon over Mary Hauser in a Chevrolet Chevy II 300 4. After a mechanical problem on the opening day, Pat began to shrink Mary's lead each day, and her win was virtually assured when Mary made her wrong turn coming into Joliet, driving 21 miles out of her way. This marked the third year in a row a Chevrolet Corvair Monza captured a first-place trophy.

The medium-engine compacts class was settled after final top-off in Detroit. In a contest closely watched since leaving Los Angeles, Bill Levy edged his Ford Falcon Deluxe back into first place over Bob Checkley's Plymouth Valiant V-100 by a slim margin of .03 miles per gallon. "The closeness of the figures, 26.29 for the Ford and 26.26 for the Valiant, was explained by one USAC official as 'temperature correction on the final top-off.'"[85] In his third year competing as a driver, this marked Bill's first class win. With a margin of .70 miles per gallon separating the four competing cars in this class, the last-place performance of the Dodge Dart 170 was a disappointment for those associated with Chrysler Corporation.

For the second year in a row, and the fourth in five years, a Rambler American achieved the best mileage for all entries. Driving a Rambler American 440, Les Viland recorded a figure of 28.61 miles per gallon, placing first in the large-engine compact cars class (courtesy of Chrysler Group LLC).

Les Viland's performance in the large-engine compacts class was never questioned as he maintained well over a mile-per-gallon lead over Judy Thompson driving a Pontiac Tempest Custom 4 in what would be her last year of participation. The main reason for this performance was American Motors' overall theme of economy over performance in the regular production specifications for the Rambler American. As explained in *Car Life Magazine*, the American 440 hardtop competing this year was "listed with 2.87:1 axle ratio for automatic-transmission models with the 195.6-cu. In., 125-bhp, ohv aluminum-block engine, this lightweight stroker is just made for economical running." Added to this is that Les Viland was "perhaps one of the country's leading carburetion experts [and] has been perfecting his technique on a multitude of Mobil Runs and ranks as one of the best drivers in the business."[86]

A surprise to most observers was the performance of John Rich and his Buick Special Deluxe in the intermediate-size eight-cylinder class. Opening up a strong lead from the start, John's Buick Special finished with a margin of 2.34 miles per gallon over pre–Run favorite Jim Fosdick and his Ford Fairlane in second. This was John's second first-place finish in the past three years, which helped to validate Buick's recent commitment to Economy Run performance.

First-time Economy Run driver Scott Harvey put his rally-driving skills to good use as he placed his Plymouth Savoy first in the standard-size six-cylinder class. This marked the fourth year in a row a Plymouth captured a win in this class, and in addition was the seventh year in a row a Plymouth model took first place. Placing second was another first-timer, *Popular Mechanics* auto editor Jim Whipple driving a Dodge 330. Only .64 miles per gallon separated these two competitors, with a Chevrolet Biscayne and Ford 300 finishing well off the pace.

Another close contest from the beginning took place in the low-price eight-cylinder class. Bob Cahill, driving a Dodge 330, edged out Bob McDaniel, driving a Plymouth Savoy, by a slim margin of .16 miles per gallon. Without the daily drama that was seen in last year's event, a victory by the Dodge was not assured until the final top-off. This was the first time a Dodge was in the winner's circle since 1959. Placing third, only .23 miles per gallon behind the leader, was Don Edmunds driving a Chevrolet Bel-Air. Last year's champion, the Chevrolet Impala, finished nearly a mile-and-half gallon off the winner's pace, with the Ford 300 a distant fifth.

The winning performance of the Oldsmobile Dynamic 88 in the medium-price class was another surprise. This was Oldsmobile's third class win in the past ten years. Scoring his first win in four years as a driver, Stan Raymond maintained a fairly comfortable margin from the start over a Chrysler Newport driven by Economy Run veteran Mel Alsbury, Jr. Competition in this class remained close with .74 miles per gallon separating the first four places, and a Pontiac Catalina and Star Chief placing third and fourth. With a fourth-place finish, this would be Mickey Thompson's last year participating.

Maintaining a comfortable lead from the start, Mildred Alsbury piloted her Chrysler New Yorker to win over three other luxury models, although experiencing a couple of bumps along the way. A cold engine that stalled twice plagued her at the start, followed later by a report she drove "some 14 miles with her emergency brake on."[87] A grandmother of twelve, and wife of Hollywood Chrysler-Plymouth dealer Mel Alsbury, Mildred was enjoying her second time behind the wheel, having last driven in 1957. This was the fourth consecutive class win for Chrysler and its sixth in seven years. Another Chrysler product, an Imperial Custom driven by Marty Gordon, placed second, followed by General Motors' Oldsmobile Ninety-Eight and Buick Electra 225.

Car Life Magazine's predictions for class winners were accurate in two classes, with the Plymouth Savoy 6 and Chrysler New Yorker placing first. Until the final top-off, Tom Swain was right on the mark in the small-engine compacts and medium-engine compacts classes, where the predicted winners placed second. Two of his second-place picks, the Rambler American 440 and Buick Special Deluxe, took top honors, while the reverse occurred with the Plymouth Savoy 8 and Chrysler Newport, which placed second.[88]

After enjoying considerable success in the Economy Runs dating back to 1936, a Studebaker crossed the finish line for the last time. With factory support ending after 1961, Bill Corey continued his efforts in both preparing and driving a Lark Regal and struggled to a fifth-place finish in the intermediate-size eight-cylinder class. Bill's Lark "apparently was suffering from distributor trouble and thus certainly not functioning to maximum economy," and finished 3.26 miles per gallon off the pace.[89] This was a disappointing end to nineteen years of competition (the only year Studebaker did not compete was 1937) that saw two consecutive back-to-back Sweepstakes Awards (1939–1940 and 1954–1955), along with eighteen first-place performances. After its last Sweepstakes Award in 1955, Studebaker's performance began to decline, although a Hawk placed first in the 1960 Economy Run.

Advertising and promotion continued to be seen in the days and weeks immediately following the Run, but were far less prominent than seen previously, especially in magazines. Mobil Oil continued announcing the results in both newspapers and industry publications, although two different formats were used. Newspaper advertising featured the headline, "Results 1963 Mobil Economy Run ... demonstrate long mileage performance of Mobil gasoline in new cars," with the class winners prominently shown along with mileage results for each competing car. A separate format was used for industry publications, such as *Automotive News*, where individual performance was stressed with no mention of class winners. A new tagline was introduced: "Mobil — the only gasoline that proves its mileage performance in public under actual road conditions."[90] This was in direct reference to the Pure Oil Performance Trials held at Daytona International Speedway. A small 14-page fan-fold-style folder with the results and overview of this year's Run was available at Mobil service stations.

As expected, Rambler's achievement was noted in newspaper advertisements with the headline "Rambler American '440' Wins Class C 28.61 Miles Per Gallon." While the text emphasizes the Rambler's title "America's Economy King," no mention was made that the American achieved the highest mileage figure for all entries.[91] Limited mention of the Economy Run was made in magazine advertising which stressed that Rambler Americans "have won in every economy run entered with highest miles per gallon," in reference to both the Mobil Economy Runs and Pure Oil Performance Trials, along with a photograph of trophies from both events.[92] This same theme continued in their showroom brochures for the 1964 models, including the "X-Ray" comparison catalogs.

In a limited campaign, Ford tied in the Falcon's Economy Run win to its recent success in the Monte Carlo Rallye under a headline that read "Total Performance." Appearing in newspapers and industry publications, photographs of both winning cars were accompanied with comments from *Peanuts* characters Charlie Brown and Linus.[93] In a brochure titled "The cars we build for two kinds of Sunday driving," devoted to Ford's competition success, one section covered the Falcon's Economy Run win and included a photograph of the winning car.[94]

Although it had placed second in a close contest with a Ford Falcon, newspaper advertising stressed the Valiant's mileage, noting that "Valiant's a great *economy* car!"[95] In general interest publications, Valiant's mileage performance was advertised in a somewhat unorthodox way. Featuring a picture of a camel and a Valiant V-100 under the headlines, "Camel Gets 13.33 M.P.G. in Sahara Economy Run" and "Valiant Gets 26.26 M.P.G. in Mobil Economy Run," the text told readers, "Camels suffer by comparison to Valiants. So do most other compact cars. A Valiant V-100 just delivered 26.26 miles per gallon averaging 44 mph in Mobil Economy Run."[96]

Appearing in general interest magazines, such as *Life* and *The Saturday Evening Post*, Buick announced, "Special 4-dr. sedan tops all Class D middle size cars with V-8 engines in Mobil Economy Run," along with a photograph of the winning sedan, while emphasizing its top mileage performance among all 8-cylinder entries.[97] Buick's dealership personnel were told of the Special's achievements in a 4-page folder with the headline "Flash! Buick Special V-8 Tops *All* V-8's With 23.81 MPG." It also featured photographs, background and daily results of the Run.[98]

Advertising for Plymouth's class win was limited to newspapers. With the headline "Terrific combination!" the text noted the Savoy's class win "makes *7 years in a row* that Plymouth has been a class winner."[99] Only a brief mention was made of Economy Run performance in Plymouth's showroom brochure for the 1964 full-size models.

Oldsmobile advertised the Dynamic 88's win only in general interest magazines, asking potential buyers to "Try This One for Savings!" while noting that a husband-and-wife team was victorious over medium priced models.[100] This was the first and only national magazine advertising effort for Oldsmobile in their three Economy Run victories dating back to 1954.

Dodge did not advertise its Economy Run success in either newspapers or magazines, although its class win was mentioned in its 1964 showroom brochure for full-size models. No mention was made in any form of advertising noting class wins for the Chevrolet Corvair Monza or Chrysler New Yorker.

With press representatives covering the Run, traveling to each overnight stop via air or on the ground, newspapers remained the main source of information on daily events. Traditionally the largest readership was in Southern California, the home of most participants, as a result of which local drivers' progress was of interest. For the most part, the first day's results told the story, although this year competition was close in two of the compact car classes which kept the reader's attention. Aside from the flagman issue in Lincoln, events this year were pretty quiet from a press point of view. Writing in *Auto Sports Magazine*, W.W. Jackson observed:

> Most writers that accompany the Run are constantly on the lookout for unusual happenings that can vary the daily stories. Through the strict handling of the Run by USAC and Mobil, the safety record is excellent and little or no near tragedy can be reported. The fact that the competition closely resembles a six-day rally with drivers holding to time and cheating difficult to effect, keeps the news items somewhat restricted to the start and finish. This year was no exception.[101]

After filing their final stories from Detroit, most newspaper auto editors returned to focus on other automotive-related events in their area while waiting for a hoped-for invitation to return for the 1964 Run.

Articles appearing in enthusiast magazines were more limited in scope than in previous years, in a trend that would be a cause for concern.[102] The only comprehensive articles found appeared in *Car Life* and *Popular Mechanics*, and while reporting from different perspectives, both were on the same theme: preparation of the entries and individual driving techniques. *Car Life Magazine*'s Dennis Shattuck and Gene Booth reported on events of the Run by following the route in one of Chevrolet's test cars while comparing their mileage to the competing 6-cylinder Chevy II 300.[103] *Popular Mechanics*' Jim Whipple provided an in-depth look at what it takes to drive in the Economy Run, which cumulated in his second-place finish driving a 6-cylinder Dodge 330. In answering two questions—"Do drivers or gimmicks win them?" and "What does the Mobil Run prove?"—he observed, "Basically it shows that any standard American production car can make phenomenally good mileage at reasonably fast average speeds if fantastic amounts of concentration and a fair amount of brain power are expended on the part of two people. Mobil is simply supporting the manufacturer's claim that such a car, properly broken in and well tuned, can do it." Jim concluded his experiences by answering the question, "Was the Mobil Run fun, despite the nerve strain and fatigue? You bet! Next year, I'll be back at it shooting for first in my class!"[104] Mobil Oil published their annual Fact Book featuring results, stories and pictures in what would be the last issue. With competition for attention coming from other competitive events, *Motor Trend* and *Popular Science* generally ignored the Run this year. Although *Motor Trend* reported the final results, both publications devoted their editorial efforts to the Pure Oil Performance Trials conducted the previous February at Daytona International Speedway.[105]

While basically satisfied with this year's Run, there was something missing, and a need arose to spike the interest of the general public while keeping the interest high in the minds of writers for various national publications. With the cost involved in staging the Economy Run each year, keeping the Mobil name in the forefront was paramount as competition in the economy arena was increasing and a cause for concern. One area of competition for attention was the Pure Oil Performance Trials, covering both acceleration and braking performance in addition to fuel economy, that was sanctioned by USAC's rival NASCAR. These concerns would be addressed as plans were made for next year's Run.

9
New York

1964
Los Angeles to New York Via Phoenix — El Paso — Fort Worth — Memphis — Indianapolis — Harrisburg
April 3–9, 1964—3243.8 Miles

The opening of the New York World's Fair, set for April 22, 1964, would be one of the factors in determining the focus of this year's Economy Run. By the fall of 1963, Mobil Oil was already building its pavilion at the Flushing Meadows site "where visitors will be able to test their driving skills on an electronic simulation of [the] Economy Run. Electronic sensing equipment will evaluate each person's driving skills throughout the test and present each contestant with his miles per gallon score. After demonstrating their driving skills, winners of the economy run will receive awards certifying their driving skill."[1] This sparked speculation that the World's Fair site would be the terminus of the 1964 Economy Run, marking the first time it would travel coast-to-coast.

While speculation continued on the route for the Run, Chevrolet created additional interest by announcing the formation of the Chevy Teen Team. Preparations for this program were already well underway by late November when, with the country in shock over the assassination of President Kennedy, a low-keyed announcement was made. "Declaring that Chevrolet was anxious to spotlight the driving capabilities of America's teenagers," it "would abandon the customary use of veteran adult drivers for its entries." The members of the team would be selected from "Junior Achievement organizations in ten major metropolitan areas."[2] Chevrolet Motor Division's General Manager Semon E. (Bunkie) Knudsen stated that, in spotlighting the capabilities of teenagers, "we believe that the 6.5 million boys and girls under 20 who hold drivers' licenses are capable of being the best and safest drivers of any segment of our population."[3]

On February 25, Chief Steward A.C. Pillsbury announced "the longest and toughest course in the history of the mileage-minded classic," which would indeed terminate at the World's Fair's Mobil Pavilion on April 9, covering a course "more than ten times the distance of the original Run from Los Angeles to Yosemite in 1936." Over a seven-day period the Run "will make overnight stops in Phoenix, Ariz.; El Paso and Fort Worth, Tex.; Memphis, Tenn., Indianapolis, Ind., and Harrisburg, Pa. [over a] course [that] will include driving, weather and road conditions typical of those encountered by the average motorist on a cross-country trip." Always one for some surprises along the way, Pillsbury added with a grin that the "sixth leg should shake-up the drivers. The Run from Indianapolis to Harrisburg will be one of the longest day's travel in the 28-year history of the Mobil Economy Run. To complete the segment within the prescribed time limit, drivers must maintain a healthy average of miles per hour. And, over that portion of the Run they can anticipate such terrain problems as hills, city congestion and turnpike traffic." In another unusual feature, USAC officials planned a visit to the Indianapolis Motor Speedway, where each competitor would make one lap around the famed Brickyard. Pillsbury also added, "With the drivers accountable for every drop of gas, I think we can call this the slowest competitive lap anyone will ever see at Indianapolis."[4]

With continued development of new model offerings by Detroit's automakers and the growing popularity of intermediate-sized models, changes in competitive classifications were again called for. These new classifications, announced with the release of the "Competitive Divisions for the 1964 Mobil Economy Run," dealt with regards to compact and intermediate-size cars. Class A was set for "Small Engine Compact Cars [with an engine displacement from] 160 through 170 cubic inches"; Class B for "Large Engine Compact Cars [with an engine displacement from] 190 through 327 cubic inches"; Class C for "Intermediate Size Cars (6 Cylinders); and Class D for "Intermediate Size Cars (8 Cylinders)." Two models were specifically excluded this year, the 4-cylinder Chevy II and the Falcon with the 144-cubic-inch engine, since "these cars are not available in sufficient quantity to meet eligibility requirements."[5] Class B for large-engine compact cars would provide some interesting competition since both six-

and eight-cylinder models would be competing. Included in this group were Chevrolet Chevy II, Dodge Dart and Plymouth Valiant models with their recently introduced V8 engines. No changes were necessary in the four classes designated for full-sized models. Four-door sedan and hardtop models continued to be eligible, with two-door models allowed in the compact and intermediate size classes at the discretion of USAC. Only four-door models were selected by USAC officials.

After fourteen years of continual refinement, aside from the new competitive classifications, the only noticeable change in the rules involved required accessories on the competing cars. In the interest of safety, outside rear view (driver's side) mirrors were required, although it appears USAC officials missed one. Photographs taken of George Alsbury's Chrysler Newport during the run show that this had not been installed. While power steering and brakes continued to be required on full-size eight-cylinder models, power steering was now required on all intermediate-size eight-cylinder models.[6] Three cars in the luxury automobile class (Buick Electra 225, Chrysler New Yorker Salon and Imperial Crown) were equipped with cruise control, but it was not used.[7]

The Dodge 330 8, driven by New York taxi driver Herman "Hy" Plotkin, passes the lowest elevation on the 1964 Economy Run near Indio, California. In his only appearance as a driver, Hy would place second in the low-price eight-cylinder class with 21.4649 miles per gallon.

When the entries closed on March 7, a field of 45 cars was set to compete, representing virtually every U.S. make. Lincoln and Studebaker did not participate, along with models that included the Rambler Classic, Rambler Ambassador and Dodge 880. American Motors' decision not to enter either the Classic or Ambassador mirrored that of last year, as these would be competing in the same class with the smaller American. General Motors comprised the largest group with 23 entries, followed by Chrysler with 12 and Ford Motor Company with 9. American Motors rounded out the field with a single Rambler American. Reflecting a variety of models being offered by Detroit's automakers, compact and intermediate-sized models comprised half of the field. After a two-year absence, Cadillac returned, although without factory support. A private entry sponsored by Concrete Coring Company of Los Angeles, the Cadillac Sixty-Two was driven by the firm's president, Ed Dempsey.

Factory support and involvement generally mirrored that seen in 1963, although an increase was noted with Ford returning, although its support was limited. This involvement was especially evident with Chevrolet's efforts in the training and preparation of the teenage crews who would handle its nine entries. Under the direction of Chevrolet engineer Vince Piggins, assisted by Bill Horton and Paul Prior, the 20 Teen Team members arrived in Southern California on March 1 and began a comprehensive training program covering all aspects of economy driving along with basic car-handling skills. Both Vince and Bill had prior Economy Run experience as drivers, which give the teenagers a comprehensive look at what they would experience on the road ahead. The teenagers created and adopted a "Keep Smooth, Keep Safe" slogan, which taught "that the same driving smoothness and long range judgment which produces maximum use of every drop of fuel accomplished the other basic requirement of the Mobil Run — safety through absolute obedience to traffic laws." In addition to classroom-style instruction, test runs were made throughout Southern California, and as far away as Dallas, as they accumulated a total of 7000 miles of practice during the month.[8] While not providing direct factory support, Mercury joined Chevrolet in selecting teenagers to pilot one of their entries, naming Julian "Bo" Jett as driver for the six-cylinder Mercury Comet 202. Winner of the 1963 National Teenage Safe Driving Road-E-O Winner, sponsored by the U.S. Junior Chamber of Commerce and Lincoln-Mercury, Jett would have as his co-driver the 1963 Miss Teenage America, Darla Banks, who was that year's Lincoln-Mercury's Youth Safety Spokesperson.[9] Although three of the Chevrolet drivers were 18 years old, the distinction of being youngest driver to compete remained with Judy Allred. In her first appearance in 1959, driving a Buick Invicta entered by her father's Visalia (California) Buick dealership, Judy was 17 years old.

Members from Chrysler Corporation's engineering staff were heavily involved in the preparation of both the Dodge and Plymouth entries. This year Dodge selected two individuals from differing backgrounds to pilot their full-size models. Jim Whipple, auto editor of *Popular Mechanics Magazine*, returned to pilot the six-cylinder Dodge 330, while Herman "Hy" Plotkin, a New York taxi cab driver, was tapped to pilot the eight-cylinder Dodge 330. In keeping with a New York theme, Dodge "chose a taxi driver for the run on the premise that by virtue of his job a cabbie should be eminently suited for the grueling competition." Chrysler Corporation Engineer Bob Cahill noted, "One of the basic necessities in this run is the ability of a driver to remain behind the wheel of an automobile for a 12-hour period each day for a week without becoming exhausted or rattled. Cab drivers in general spend long hours driving, with a good share of them working twelve hour days. And any person that can ply his way through New York traffic for twelve hours a day without becoming unglued certainly is well suited for the Mobil Run."[10] Years later it was revealed that the Plymouth and Dodge team had a trick up their sleeve. Writing in the October 1995 issue of *High Performance MoPar*, Bob Cahill revealed the use of a special advance crew. After their first car arrived and received the itinerary for the next day's trip, this crew "could drive quickly (not necessarily legally) over the route, getting the pertinent stoplight timings, grade readings when possible, and any other useful information. The next day we would be more prepared than the competition."[11] No doubt other teams used similar tactics, although actual driving and weather conditions faced by the drivers during daytime hours would have a definite impact on their mileage performance, even with this advance information.

Chrysler's remaining entries would continue to be handled by last year's participants, Mel Alsbury, Hart Fullerton and Danny McGroo. American Motors' preparation of its single entry continued under the direction of Les Viland, assisted by other company engineers. Joining Buick in supporting the teams, both before and during the Run, both Oldsmobile and Pontiac provided limited support, although this support would increase over the next few years.

Dealers continued to be in the majority as sponsors, with manufacturers remaining active in naming individual entrants. Dealer associations returned as sponsors for the Dodge and Rambler American entry, joined this year by Chevrolet Motor Division's "Chevy Teen Team." With New York as this year's terminus, five of that state's dealers participated for the first time. These included Airport Motors, Inc. (Chrysler-Plymouth) in Jackson Heights, Armory Garage (Chrysler-Plymouth) in Albany, Flinn Motor Corporation (Ford) in Larchmont, Rex Pontiac in Jackson Heights, and Rogers Pontiac in Brooklyn.

Only two dealers acted as both entrants and drivers this year, Hart Fullerton in a Chrysler 300 and Mandy Williams in an Oldsmobile Ninety-Eight. As a result of the Chevy Teen Team concept, Kansas City Chevrolet dealer Melvin Hilliard had to sit out this year's event. Two returning

Refueling time for the two Dodge Dart entries at Ed Malcom Mobil Service Texarkana, Arkansas. The Dodge Dart 170 6 (left), driven by Bob Checkley, would place first in the small-engine compacts class, while the Dodge Dart 270 8, driven by George Miller, would place fourth in the large-engine compacts class.

husband-and-wife teams, Hart and Diana Fullerton (Chrysler 300) and Stan and Lois Raymond (Oldsmobile F-85 Deluxe 8), were joined by first-time participants Ed and Karen Dempsey (Cadillac Sixty-Two). A participant since 1950, Pete Novotny returned once again to driving duties, piloting an Oldsmobile Jetstar 88, after three years as a co-driver. Women were named as drivers for ten entries and included prior class winners Mary Davis and Patricia Jones, both returning after an absence of two years. Only one of last year's class winners was absent, as Pat Sawyer also had to sit out this year due to the Chevy Teen Team concept. Mildred Alsbury returned to handle co-driving duties for her son Mel, Jr.

Random car selection commenced during the second half of February, and for the next five weeks, USAC officials and individual teams were busy with inspection, servicing and break-in runs. During this time, entrants were given the option of using either Mobil Regular or Mobil Premium gasoline. Eleven entries selecting Regular included the Buick LeSabre along with both Buick Specials, Chevrolet's Corvair, Chevy II, Chevelle, Biscayne and Bel-Air entries. Working on their individual driving skills, both in their actual competing cars or in specially equipped replicas, this time was critical to the seventeen drivers sitting in the driver's seat for the first time. As noted earlier, this time was important for the teenage teams competing in the Chevrolet and Mercury Comet 202 6 entries.

Seeking to increase publicity of the Run in newspapers and magazines, Mobil Oil's publicity personnel continued to be actively involved in advance preparations while at the same time bringing new reporters onboard from eastern-based newspapers. Dispensing with their Economy Run Fact Book, Mobil created a more comprehensive Press Book, which provided detailed information on this year's event, including biographies of each driver. While some press representatives traveled over the route on a chartered airliner, others would be covering the route driving replicas of competing cars. Among those covering the Run from the ground were regulars Bob Thomas of the *Los Angeles Times* and Bill Carroll of *Automotive News*, joined by some from East Coast newspapers, including Henry Ward of the *Pittsburg Press*.

Buick got off to an early start on this year's publicity campaign and in somewhat of a unique manner. Greeting competitors and visitors to the Pasadena impound was a large billboard, located on adjoining property overlooking the parking lot, which read: "You have about 3,000 miles to drive. Wouldn't you really rather drive a Buick?" It was a take on their current tagline, "Wouldn't you really rather have a Buick?"[12] This would remain in place until all of the drivers were well on their way to New York.

The traditional drawing for starting positions was held the week prior the start at the Pasadena impound. Dispensing with the rolled pennants used in prior years, drivers selected their starting position, hidden from view on circular cards hanging from a large board. Chevy Teen Team member Richard Doyle drew No. 1 for his Chevrolet Chevelle 300 6, with veteran driver Pete Novotny drawing the final position, No. 45, for his Oldsmobile Jetstar 88. USAC officials announced that the start would again be from the Pasadena impound. As the time for practice and preparation drew to a close, everyone's attention focused on the release in a few days of the first day's routing.

On Thursday, April 2, the first day's detailed route and itinerary was released, and once again A.C. Pillsbury had an interesting detour planned for the drive to Phoenix. On what was described as "a less than 'straight as an arrow' route, deviously laid out by USAC," the drivers would face a 443.4-mile route to be covered within a 9-hour and 40-minute time limit.[13] Continuing the custom of previous years, these limits would not include refueling and 40-minute brunch (or lunch) stops. This time-out from driving times would continue in future years. As a reminder that a faster pace was on tap for this year, a new feature was found in the daily itinerary booklets. On the opening page there was a note regarding today's trip that noted: "You must arrive at Phoenix Refuel Station with a 46.89 MPH average if you expect to arrive at Impound on time, as heavy traffic congestion will be encountered. By following the time schedule on the itinerary you will reach the Impound with a two and one-half (2½) minute cushion." Another new feature was the addition of an asterisk in the instructions "where particular attention to the route described must be observed."[14] These reminders would continue throughout the Run. As indicated by long-range forecasts, drivers again would face weather-related challenges over a distance of more than 3100 miles.

On the morning of April 3, competitors, officials and spectators were on hand at the Pasadena impound, where the cars were ready to set out for the cross-country trek. Cool temperatures with clear skies and bright sunshine greeted everyone when Chevy Teen Team member Richard Doyle in a Chevrolet Chevelle 300 6 was the first to take the traditional green flag from an enthusiastic J.C. Agajanian at 6:30 a.m. The remaining drivers took off at the traditional one-minute intervals. Following the same route that was used in 1962, the drivers made their out of the Los Angeles metropolitan area and headed for Palm Springs and Indio following Interstate 10. A "frustrating delay" occurred near Indio when a freight train held up a group of cars. Paula Murphy, driving a Buick Electra 225, laughingly remarked, "That freight train at least gave us an extra chance to visit the powder room."[15] Before leaving Indio they dropped to an elevation of 47 feet below sea level, the lowest point on this year's Run, as they followed U.S. Routes 60 and 70 towards Desert Center, where their first challenge awaited. Following two-lane roads, the drivers headed northeast towards Parker. As they neared Parker, some of the drivers were faced with another delay caused by a military convoy.[16] Crossing the Colorado River into Arizona, entering the Mountain Time Zone, they made their first stop in Parker at F.E. "Mac" Mobil Service, where the entries were topped off, followed by brunch at the Paradise Harbor Restaurant. Arriving at the top-off station, George Alsbury damaged the rear door of his Chrysler Newport after he "dinged a fire hydrant," later explaining, "It was that or a truck."[17] After nearly six hours on the road, everyone was ready for a short break before heading southeast on Arizona 95 and 72 towards Hope. Returning to U.S. Routes 60 and 70, the drivers headed towards Wickenburg and then on to Phoenix. They arrived in Phoenix around 5:30 p.m. for their first refueling stop at Gene Fullmer Mobil Service before continuing to the impound located at the Ramada Inn. After navigating Phoenix traffic, the first to arrive at 6:00 p.m. was Mary Hauser in a Pontiac Star Chief. Three drivers, "Red" Eason (Oldsmobile F-85 Deluxe 6), Scott Harvey (Plymouth Savoy 6) and Karole Roth (Chevrolet Biscayne 6), arrived late and were penalized. If there was any doubt in the competitors' minds that this would be a challenging seven-day trek, this doubt was erased when they read details of tomorrow's route, along with weather forecasts for Arizona and New Mexico.

For the third year in a row, perennial leader Les Viland's Rambler American 440 recorded the highest miles-per-gallon average for all competitors with a figure of 27.8779. The entire field turned in an average of 21.5557 miles per gallon. In addition to Les, two drivers were on pace for back-to-back wins, with John Rich leading in the intermediate-size eight-cylinder class and Bob Cahill leading in the low-price eight-cylinder class. Bob Checkley, driving a Dodge Dart 170 6, had a fairly comfortable lead among the small-engine compacts.

A lot of attention this year would be on the ten teenage drivers and their performance compared to the more experienced adult drivers. With an average of 21.9048 miles per gallon, the teenagers were ahead of the adults with an average of 21.4560. One of the factors in the teens' performance was their greater participation in the compact and intermediate-size classes. Described as a surprise performance, 19-year-old Bo Jett's Mercury Comet 202 led the intermediate-size six-cylinder class in a close contest in which .05387 miles per gallon separated the first three places. The best performance by a member of the Chevy Teen Team was in the large-engine compacts class, where Jean Fitzgerald's Chevrolet Chevy II 100 6 was in second place behind Les Viland's Rambler American.

Close competition was also seen in the full-size six-cylinder class, where Scott Harvey's Plymouth Savoy had a .0299 lead over Jim Whipple's Dodge 330, with Chevy Teen Team's Karole Roth well off the pace in third. Both Chrysler Corporation and General Motors were leading in 3 classes, with American Motors and Ford Motor Company with one each.

There were two other surprises with today's results. In the medium-price class, Marta Retzlaff's Buick LeSabre had strong a 1.2700 mile-per-gallon lead over Mary Hauser's Pontiac Star Chief in second place. This would be Marta's best performance to date in six years of competition. The other

To mark the Economy Run's first visit to Indianapolis, arrangements were made for the drivers to make one lap around the famous speedway before being refueled in the pit area. Driving through the gates is *Popular Mechanics*' Jim Whipple in a Dodge 330, about to make what was called "the slowest lap in Indianapolis history!" Jim would go on to place second in the full-size six-cylinder class.

surprise was in the luxury car class, where Ed Dempsey's Cadillac Sixty-Two had a small lead over veteran Mel Alsbury, Jr.'s Chrysler New Yorker Salon.

It appeared that this year's Run would set a record as the fastest to date, with today's average speed of 46.1545 miles per hour.

The top three performers in each class were:

Small Engine Compacts	Dodge Dart 170	25.9137 Miles Per Gallon
	Plymouth Valiant V-100	25.2265 Miles Per Gallon
	Ford Falcon Futura	23.5321 Miles Per Gallon
Large Engine Compacts	Rambler American 440	27.8779 Miles Per Gallon
	Chevrolet Chevy II 100 6	22.7320 Miles Per Gallon
	Plymouth Valiant V-100 8	22.5707 Miles Per Gallon
Intermediate Size Six Cylinder	Mercury Comet 202	24.8475 Miles Per Gallon
	Buick Special Deluxe	24.8088 Miles Per Gallon
	Pontiac Tempest	24.7938 Miles Per Gallon
Intermediate Size Eight Cylinder	Buick Special Deluxe	23.3925 Miles Per Gallon
	Mercury Comet Caliente	22.1205 Miles Per Gallon
	Ford Fairlane 500	21.9741 Miles Per Gallon
Full Size Six Cylinder	Plymouth Savoy	24.7230 Miles Per Gallon
	Dodge 330	24.6931 Miles Per Gallon
	Chevrolet Biscayne	22.3091 Miles Per Gallon
Low Price Eight Cylinder	Plymouth Savoy	22.0774 Miles Per Gallon
	Dodge 330	21.7223 Miles Per Gallon
	Ford Custom 500	20.7263 Miles Per Gallon
Medium Price	Buick LeSabre	21.1407 Miles Per Gallon
	Pontiac Star Chief	19.8707 Miles Per Gallon
	Chrysler Newport	19.5168 Miles Per Gallon
Luxury Cars	Cadillac Sixty-Two	18.6128 Miles Per Gallon
	Chrysler New Yorker Salon	18.4759 Miles Per Gallon
	Oldsmobile Ninety-Eight	18.2804 Miles Per Gallon

On April 4, the drivers headed for El Paso on a 500.7-mile trek to be covered within an 11-hour time limit. Once again they would tackle one of Chief Steward A.C. Pillsbury's well-known detours through the mountains of Arizona and New Mexico. Karole Roth in a Chevrolet Biscayne 6 was the first on the road, leaving at 5:00 a.m., heading southeast, where the drivers encountered heavy rain for 40 miles before reaching Tucson.[18] Leaving Tucson, they followed U.S. Route 80 through historic Tombstone before arriving in Bisbee for brunch at the Copper Queen Hotel. While they were traveling through Tombstone, a second minor incident occurred when a young girl "with a brand new driver's license clipped a rear fender" of Jim Fosdick's Ford Custom 500 8.[19] Leaving Bisbee, which at 5380 feet was the highest elevation on this year's Run, the drivers headed for Douglas. Along this 25-mile stretch they battled snow, rain and ground fog along with snow-slick roads and winds up to 30 miles per hour before arriving at Douglas Mobil Service, where all entries were topped off.[20] Making their way north on U.S. Route 80, they crossed into New Mexico, passing through Lordsburg and Las Cruces headed for El Paso. After they had crossed into Texas, their first stop in El Paso was at Mesa Mobil Service for refueling. They then navigated city streets to the impound at the Ramada Inn East. Thanks to a cold front moving through the area, the drivers continued to battle high winds which caused live television coverage of their arrival in El Paso to be canceled.[21] Arriving around 4:00 p.m., Karole Roth in a Chevrolet Biscayne 6 was the first to arrive after a second long day on the road. Despite today's challenges, all of the drivers arrived within the prescribed time limit. New York taxi driver Hy Plotkin, commenting on today's trip and adverse weather conditions, said, "I saw everything out there today," and that "it didn't bother me." Driving a Dodge 330 8, he was running second behind Bob Cahill's Plymouth Savoy 8. He added, "This thing is easy, nothing to it."[22]

After the arrival in El Paso, a switch in co-drivers was made for the two Chryslers entered by Hollywood's Mel Alsbury. Mildred Alsbury moved over to handle co-driver duties for her son George in a Chrysler Newport, while Bill Brandes joined Mel, Jr., in a Chrysler New Yorker Salon.[23] Since 1960, Mel, Jr., and Bill had teamed up, capturing class wins in both 1960 and 1961. With George's Newport well off the pace in the medium-price class, it was felt that Bill's expertise would be needed in a battle that was considered quite close for first place in the luxury car class.

Everyone's performance took a hit today as both the route and weather conditions resulted in a drop of nearly a mile per gallon on average. Continuing to record the best mileage for all entries, Les Viland's Rambler American 440 turned in a figure of 26.6751 miles per gallon. This reflected a drop of 1.2028 miles per gallon from yesterday's results. In the intermediate-size six-cylinder class, Bill Levy moved his Buick Special Deluxe into first place, dropping the first day's leader, Bo Jett and his Mercury Comet 202, into third. Bill joined three other drivers working on back-to-back class wins. In the remaining seven classes, the first day's leaders retained their positions, and for most by fairly comfortable margins. General Motors was leading in four classes, followed by Chrysler Corporation leading in three and American Motors in one. A surprise performance was Buick's leading in three classes with two Specials along with a LeSabre. The only class where competition was fairly close was in the luxury car class, with Ed Dempsey's Cadillac Sixty-Two maintaining a .1915 mile-per-gallon lead over Mel Alsbury, Jr.'s Chrysler New Yorker Salon. The entire field saw their average drop to 20.6654 miles per gallon. The teenagers stayed ahead of the adults with an average of 20.9243 miles per gallon compared to the adults' 20.5914. In the large-engine compact car class, Chevy Teen Team's Jean Fitzgerald's Chevrolet Chevy II 100 6 dropped to third. Today's average

speed of 46.1725 miles per hour continued to reflect the fast pace of this year's Run.

The top three performers in each class were:

Small Engine Compacts	Dodge Dart 170	24.6310 Miles Per Gallon
	Plymouth Valiant V-100	24.4170 Miles Per Gallon
	Ford Falcon Futura	22.5039 Miles Per Gallon
Large Engine Compacts	Rambler American 440	26.6751 Miles Per Gallon
	Plymouth Valiant V-100 8	21.8911 Miles Per Gallon
	Chevrolet Chevy II 100 6	21.8149 Miles Per Gallon
Intermediate Size Six Cylinder	Buick Special Deluxe	23.9136 Miles Per Gallon
	Pontiac Tempest	23.5268 Miles Per Gallon
	Mercury Comet 202	23.2004 Miles Per Gallon
Intermediate Size Eight Cylinder	Buick Special Deluxe	22.4586 Miles Per Gallon
	Oldsmobile F-85 Deluxe	21.1677 Miles Per Gallon
	Ford Fairlane 500	21.0756 Miles Per Gallon
Full Size Six Cylinder	Plymouth Savoy	23.9282 Miles Per Gallon
	Dodge 330	23.7251 Miles Per Gallon
	Chevrolet Biscayne	21.0858 Miles Per Gallon
Low Price Eight Cylinder	Plymouth Savoy	21.2503 Miles Per Gallon
	Dodge 330	20.6166 Miles Per Gallon
	Ford Custom 500	19.7070 Miles Per Gallon
Medium Price	Buick LeSabre	20.2916 Miles Per Gallon
	Oldsmobile Jetstar 88	19.0612 Miles Per Gallon
	Pontiac Star Chief	18.7298 Miles Per Gallon
Luxury Cars	Cadillac Sixty-Two	17.9696 Miles Per Gallon
	Chrysler New Yorker Salon	17.7781 Miles Per Gallon
	Oldsmobile Ninety-Eight	17.4282 Miles Per Gallon

Crossing the finish line in front of the Mobil Pavilion at the New York World's Fair, Les Viland in his Rambler American 440 receives the usual enthusiastic greeting from USAC's J.C. Agajanian. A smaller crowd of spectators was on hand this year since the World's Fair site was not yet open to the public. Again this year, Les would achieve the best mileage figure for all entries with 27.8336 miles per gallon.

After reviewing the route information for the next leg to Fort Worth, the competitors knew that they were in for a record-setting day. Fortunately for them, the weather forecasts were showing a marked improvement.

On April 5, the drivers faced a 592.6-mile trek to Fort Worth over a route that was pretty straightforward, stretching across western and central Texas. This would be the second-longest leg to date in the history of the Economy Runs, the record being 599.3 miles in the 1941 Gilmore Grand Canyon Run. With a time limit of 12 hours, drivers would need to average close to 49 miles per hour in order to stay on schedule. The first to leave at 4:00 a.m. was John Rich in a Buick Special 8, picking up Interstate 10 / U.S. Route 80 bound for Odessa. After the problems experienced on the first two days, the drivers would be treated to pleasant weather conditions, making for easier going, as they crossed the Texas plains.[24] Just east of Van Horn they crossed into the Central Time Zone, continuing to Odessa, where all entries were topped off at Woodrow Wilson Mobil Service. Continuing eastbound on U.S. Route 80, the competitors picked up sections of Interstate 20 as they made their way to Big Spring for a brunch stop at Cosden Country Club. After nearly 7 hours on the road, covering 337 miles since leaving El Paso, the drivers found the 40-minute stop a welcome break. At one point on this leg, "it was necessary to rope a steer to get it off the highway," as it was blocking one of the Buick entries.[25] Continuing on Interstate 20 / U.S. Route 80, the competitors headed for Weatherford for refueling at Beck Mobil Service, and then it was a short 22-mile trip to Fort Worth and the impound at the Western Hills Motel. Arriving at 5:45 p.m., Edward Wilder in an Oldsmobile Dynamic 88 was the first of the weary drivers to reach the impound, and was greeted by a mounted posse joined by local residents who were on hand to watch the cars as they arrived. Only one driver, Chevy Teen Team's James Sny in a Chevrolet Chevelle Malibu 6, arrived late and was penalized. Not a bad record for late arrivals, considering the distance the competitors covered today.

After arriving in Fort Worth, Louis Delight stepped aside as co-driver for Mandy Williams in an Oldsmobile 98 for reasons not known. Louis was replaced by Dale Smith.[26]

After covering 1536.7 miles, some of the veteran drivers were wondering how the teenagers were coping and whether they were beginning to buckle under the pressure. Vince Piggins, who headed the Chevy Teen Team, replied, "Not so. They're doing a beautiful job. They're disappointed by the standings, but certainly not discouraged." Jean Fitzgerald, driver of the Chevrolet Chevy II 100 6, added, "I feel great. I'm not kidding. It's really fun."[27]

When the results for the third day were released, each competitor showed an increase in performance, with some improving by over a mile per gallon. As expected, Les Viland's Rambler American 440 continued to turn in the best mileage for all entries with 27.6786 miles per gallon. If the current class leaders' performances continued, they would retain these first-place positions by fairly comfortable margins. Two classes continued to have competition which would keep the level of interest fairly high for those following the Run. In the full-size six-cylinder class, Scott Harvey's Plymouth Savoy had a .1826 mile per gallon lead over Jim Whipple's Dodge 330, while in the luxury car class, Ed Dempsey's Cadillac Sixty-Two saw its lead over Mel Alsbury, Jr.'s Chrysler New Yorker Salon shrink to a slim .1159 mile per gallon. Buick's strong performance enabled General Motors to continue to lead in four classes, followed by Chrysler Corporation leading in three and American Motors in one. The entire field saw their average increase to 21.5615 miles per gallon. The teenagers continued to stay ahead of the adults by recording an average of 21.8458 miles per gallon while the adults averaged 21.4802. Even maintaining this overall average, Chevy Teen Team's drivers were disappointed to be running in last place in five of the six classes in which they were competing. Jean Fitzgerald saw her Chevrolet Chevy II 100 6 drop into fourth place behind George Miller in a Dodge Dart 270 8 by only .0447 mile per gallon. With an increase in the average speed to 49.7271 miles per hour, this year's Run was on pace to become the fastest in history.

The top three performers in each class were:

Small Engine Compacts	Dodge Dart 170	25.8718 Miles Per Gallon
	Plymouth Valiant V-100	25.5837 Miles Per Gallon
	Ford Falcon Futura	23.4423 Miles Per Gallon
Large Engine Compacts	Rambler American 440	27.6786 Miles Per Gallon
	Plymouth Valiant V-100 8	22.9342 Miles Per Gallon
	Dodge Dart 270 8	22.7483 Miles Per Gallon
Intermediate Size Six Cylinder	Buick Special Deluxe	25.0914 Miles Per Gallon
	Pontiac Tempest	24.7503 Miles Per Gallon
	Mercury Comet 202	24.3342 Miles Per Gallon
Intermediate Size Eight Cylinder	Buick Special Deluxe	23.6542 Miles Per Gallon
	Oldsmobile F-85 Deluxe	22.1403 Miles Per Gallon
	Ford Fairlane 500	22.0808 Miles Per Gallon
Full Size Six Cylinder	Plymouth Savoy	25.2220 Miles Per Gallon
	Dodge 330	25.0394 Miles Per Gallon
	Chevrolet Biscayne	22.0636 Miles Per Gallon
Low Price Eight Cylinder	Plymouth Savoy	22.1331 Miles Per Gallon
	Dodge 330	21.4179 Miles Per Gallon
	Ford Custom 500	20.5578 Miles Per Gallon
Medium Price	Buick LeSabre	21.1601 Miles Per Gallon
	Oldsmobile Jetstar 88	19.9141 Miles Per Gallon
	Pontiac Star Chief	19.6365 Miles Per Gallon
Luxury Cars	Cadillac Sixty-Two	18.5680 Miles Per Gallon
	Chrysler New Yorker Salon	18.4521 Miles Per Gallon
	Oldsmobile Ninety-Eight	17.9782 Miles Per Gallon

On April 4, another long day faced the drivers as they headed on a 512.6-mile leg to Memphis within an 11-hour and 30-minute time limit. At 5:00 a.m., Pete Novotny in his Oldsmobile Jetstar 88 was the first driver on the way, heading out of Fort Worth on terrain that would be relatively flat, although today would end up as a challenge for most of the drivers. Fran Foster was 4 minutes late in leaving after receiving the starting signal as the throttle linkage on her Ford Fairlane 6 needed to be lubricated. This was done on her own time and not deducted from her driving time.[28] After passing through Dallas, the drivers continued on Interstate 30 / U.S. Route 67 across eastern Texas to Texarkana and the first stop of the day at Ed Malcolm Mobil Service, where all entries were topped-off. Now in Arkansas, the drivers continued on U.S. Route 67 to Emmet for a brunch stop at Arkla Village's Chuckwagon after nearly 6 hours on the road. Continuing through Arkansas, the drivers picked up Interstate 30 as they neared Little Rock, then picked up U.S. Route 70 towards Memphis. Before arriving in West Memphis for refueling, the competitors passed a marker which noted the Economy Run, from its beginning, had covered 1,000,000 miles.[29] After crossing the Mississippi River into Tennessee, the drivers encountered heavy evening traffic in Memphis as they made their way to the impound at the Holiday Inn. Gordon Madison, driving a Ford Custom 500 6, was the first to arrive just before 5:30

Plymouth scored double wins for full-size models for the first time since 1960, topping both the full-size six-cylinder and low-price eight-cylinder classes. Scott Harvey drove his Savoy 6 to Plymouth's fifth consecutive win in the six-cylinder class. Bob Cahill drove his Savoy 8 to capture the eight-cylinder class, putting Plymouth on top for the first time in that class since 1960.

p.m. Among those on hand to greet the competitors were performing members of a Dixieland jazz band. Fourteen drivers, including class leaders Bob Checkley (Dodge Dart 170 6) and Les Viland (Rambler American 440), along with five members of the Chevy Teen Team, arrived late, ranging from just over 2 to as much as 7½ minutes, and were penalized. Three competing cars encountered minor mechanical problems during today's drive, all of which would be repaired in the morning, as more challenges would face the drivers.

For the second day, all averages improved: Les Viland nursed his Rambler American 440 just over the 28-miles-per-gallon mark, continuing to achieve the best miles per gallon with a figure of 28.0938. There were no changes in positions, with competition remaining fairly close in two classes. In the full-size six-cylinder class, Scott Harvey's Plymouth Savoy saw its lead shrink to .1173 mile per gallon over Jim Whipple's Dodge 330. In the luxury car class, it was getting tighter with Ed Dempsey's Cadillac Sixty-Two now leading by a mere .0317 mile per gallon over a hard-charging Mel Alsbury, Jr.'s Chrysler New Yorker Salon. This would appear to be the contest to watch over the next three days. The entire field was again able to increase their average to 21.9592 miles per gallon, which was an improvement of 4/10th a mile per gallon. The teenagers maintained their lead over the adults with an average of 22.3137 miles per gallon. The adults' average was 21.8579. Chevrolet Teen Team's Jean Fitzgerald moved her Chevrolet Chevy II 100 6 back into third place, dropping George Miller and his Dodge Dart 270 8 to fourth. Only .0960 mile per gallon separated the two drivers. Today's driving pace slowed as the field averaged 44.6483 miles per hour on the drive from Dallas.

The top three performers in each class were:

Small Engine Compacts	Dodge Dart 170	26.3731 Miles Per Gallon
	Plymouth Valiant V-100	26.0515 Miles Per Gallon
	Ford Falcon Futura	23.9193 Miles Per Gallon
Large Engine Compacts	Rambler American 440	28.0938 Miles Per Gallon
	Plymouth Valiant V-100 8	23.3286 Miles Per Gallon
	Chevrolet Chevy II 100 6	23.2215 Miles Per Gallon
Intermediate Size Six Cylinder	Buick Special Deluxe	25.6602 Miles Per Gallon
	Pontiac Tempest	25.3764 Miles Per Gallon
	Mercury Comet 202	24.8080 Miles Per Gallon
Intermediate Size Eight Cylinder	Buick Special Deluxe	24.0236 Miles Per Gallon
	Oldsmobile F-85 Deluxe	22.5590 Miles Per Gallon
	Ford Fairlane 500	22.4833 Miles Per Gallon
Full Size Six Cylinder	Plymouth Savoy	25.6633 Miles Per Gallon
	Dodge 330	25.5460 Miles Per Gallon
	Chevrolet Biscayne	22.4996 Miles Per Gallon
Low Price Eight Cylinder	Plymouth Savoy	22.4288 Miles Per Gallon
	Dodge 330	21.7555 Miles Per Gallon
	Ford Custom 500	20.9671 Miles Per Gallon
Medium Price	Buick LeSabre	21.6255 Miles Per Gallon
	Oldsmobile Jetstar 88	20.3440 Miles Per Gallon
	Pontiac Star Chief	19.9842 Miles Per Gallon
Luxury Cars	Cadillac Sixty-Two	18.7756 Miles Per Gallon
	Chrysler New Yorker Salon	18.7439 Miles Per Gallon
	Oldsmobile Ninety-Eight	18.2852 Miles Per Gallon

Buick was the big winner in this year's Economy Run, placing first in three classes. Marta Retzlaff and her Buick LeSabre topped all cars in the medium-price class by a comfortable margin. Buick Specials topped all cars for intermediate-size models, with Bill Levy placing first among six-cylinder models and John Rich among eight-cylinder models. This was the first time since 1954 that one make placed first in three classes.

As the drivers headed out of Memphis on April 7, they would face several challenges on the 460.2-mile, 10-hour and 40-minute journey to Indianapolis. Hart Fullerton, driving a Chrysler 300, was the first to leave at 5:00 a.m., following U.S. Route 51 through eastern Tennessee. Three cars were delayed in leaving, by as much as 25 minutes, while minor repairs were made. These were Mary Faulkner's Mercury Comet Caliente 8 (replaced transmission vacuum valve), Fran Foster's Ford Fairlane 6 (replaced throttle linkage), and Gordon Madison's Ford Custom 500 6 (repaired hood latch). All repairs were completed on their own time and this was not deducted from the day's time limit.[30] Leaving in light rain showers, the drivers soon had to contend with two flooded stretches of the highway about 65 miles north of Memphis. Although water was from 6" to 10" deep in spots over a quarter-mile stretch, the drivers were not too hampered navigating through this obstacle.[31] Crossing into Kentucky, the drivers followed U.S. Routes 45 and 60 to Henderson, near the Indiana border, for a brunch stop at the First Christian Church, where they were hosted by members of the congregation. Tennessee and Kentucky presented everyone with challenges including "unbelievably narrow roads, trucks, tractors and rolling terrain. No highspeed freeways."[32] This would be the only time the Economy Run would travel through these two states. Following U.S. Route 41, the drivers crossed into Indiana for their next stop in Evansville, where all entries were topped off at Axton's Mobil Service. After leaving Evansville, they entered the Eastern Time Zone and headed for Vincennes, where they picked up Indiana Route 67, passing through numerous towns and cities on their way towards Indianapolis. Having dealt with various obstacles and high winds along the way, the drivers had the highlight of today's trip finally in sight. Arriving in the town of Speedway, the drivers entered through the gates of the Indianapolis Motor Speedway, where they made one lap around the famous Brickyard. After completing what was then called the slowest lap in Indianapolis history, the drivers entered the Pit Area for refueling at "Tony Hulman's Mobil Service" before continuing on to the Indianapolis Motor Speedway Motel located adjacent to the track. Hart Fullerton, driving a Chrysler 300, was the first to arrive at 5:20 p.m., with all drivers arriving within the set time limits. For Hart, today featured a memorable trip around the speedway: since 1952, he had been a sponsor of cars competing in the annual Memorial Day 500 Mile Race. After five long days over what was termed by most as a rugged course, the competitors were beginning to look forward to wrapping things up in New York. However, when they studied the routing for tomorrow's trip to Harrisburg, they knew there was still another long day ahead.

Even with today's challenges the mileage averages remained pretty much on par with yesterday's results, although there were two significant changes in the standings. Continuing his strong performance, Les Viland continued to achieve the best miles per gallon average for all entries as his Rambler American recorded a figure of 28.0144. In the large-engine compacts class, Chevrolet Teen Team's Jean Fitzgerald moved her Chevrolet Chevy II 100 6 back into second with a small margin over Bob Knoll's Plymouth Valiant V-100 8. The big news of the day occurred in the luxury car class when Mel Alsbury, Jr. moved his Chrysler New Yorker Salon into first place over Ed Dempsey in a Cadillac Sixty-Two. It remained very close: only .0884 miles per gallon separated these two entries. There were no lead changes in the remaining seven classes, with one other contest continuing to be fairly close. In the full-size six-cylinder class, Scott Harvey's Plymouth Savoy saw its lead shrink once again as Jim Whipple's Dodge 330 moved to within .0641 mile per gallon. These two classes continued to be the ones to watch over the next two days. The entire field saw their average drop ever so slightly to 21.9440 miles per gallon. Although they didn't have any chance to win in the classes in which they were competing, the teenagers continued an overall lead over the adults, posting an average of 22.3589 miles per gallon versus the adults' mark of 21.8254. Chrysler Corporation was now leading in four classes, with General Motors leading in three and American Motors one. Buick continued to be the star performer, leading in three classes. One other result of note: Chevy Teen Team's Ronnie Cherry finally moved his Chevrolet Chevy II Nova 8 out of last place in the large-engine compacts class. Today's driving pace slowed for a second day in a row, with the field averaging 43.5491 miles per hour.

The top three performers in each class were:

Small Engine Compacts	Dodge Dart 170	26.3350 Miles Per Gallon
	Plymouth Valiant V-100	26.1012 Miles Per Gallon
	Ford Falcon Futura	23.9885 Miles Per Gallon
Large Engine Compacts	Rambler American 440	28.0144 Miles Per Gallon
	Chevrolet Chevy II 100 6	23.3706 Miles Per Gallon
	Plymouth Valiant V-100 8	23.2996 Miles Per Gallon
Intermediate Size Six Cylinder	Buick Special Deluxe	25.5555 Miles Per Gallon
	Pontiac Tempest	25.4439 Miles Per Gallon
	Mercury Comet 202	24.7896 Miles Per Gallon
Intermediate Size Eight Cylinder	Buick Special Deluxe	23.9000 Miles Per Gallon
	Oldsmobile F-85 Deluxe	22.5463 Miles Per Gallon
	Ford Fairlane 500	22.4412 Miles Per Gallon
Full Size Six Cylinder	Plymouth Savoy	25.5423 Miles Per Gallon
	Dodge 330	25.4782 Miles Per Gallon
	Chevrolet Biscayne	22.4854 Miles Per Gallon
Low Price Eight Cylinder	Plymouth Savoy	22.3790 Miles Per Gallon
	Dodge 330	21.5749 Miles Per Gallon
	Ford Custom 500	20.9475 Miles Per Gallon
Medium Price	Buick LeSabre	21.6301 Miles Per Gallon
	Oldsmobile Jetstar 88	20.2884 Miles Per Gallon
	Pontiac Star Chief	19.9657 Miles Per Gallon
Luxury Cars	Chrysler New Yorker Salon	18.8483 Miles Per Gallon
	Cadillac Sixty-Two	18.7599 Miles Per Gallon
	Oldsmobile Ninety-Eight	18.3085 Miles Per Gallon

As the competitors read the sixth day's itinerary, a note at the beginning would set the tone for the trip to Harrisburg. It said in part, "This sixth day is possibly the toughest one day's leg ever encountered on a Mobil Economy Run. We have taken you farther than usual so you can enjoy the beautiful, outstanding facilities of the Holiday Motor Hotel in Harrisburg, Penn. This was done purposely so you will have a very short seventh day into New York."[33] And the competitors thought the previous five days were tough.

On April 8, the drivers faced a 550-mile, 12-hour and 20-minute trip to Harrisburg that would feature "narrow roads through numerous small towns in Ohio and West Virginia," along with some highway and Interstate driving thrown in for a good mixture.[34] With an early starting time, Les Viland in a Rambler American 440 was the first to leave Indianapolis at 4:30 a.m., following U.S. Route 40 / Interstate 70 and crossing into Ohio. After battling early rush-hour traffic in Columbus, the drivers made the first stop after nearly 6 hours on the road in Norwich for brunch at Baker's Restaurant. From there it was a short drive through New Concord, home of astronaut John Glenn, the first American to orbit the Earth, then to Cambridge, where all entries were topped off at Mark's Mobil Service. At this stop, the competitors were presented with a plaque noting "astronaut John Glenn's life in Cambridge, Ohio."[35] Presentations of this type were fairly common during the cross-country trip. With their trip half over, the drivers continued eastward on U.S. Route 40, where they crossed the Ohio River and the narrow chimney of West Virginia, the only time the Economy Run would travel through this state. Crossing into Pennsylvania, they followed Interstate 70 before entering the Pennsylvania Turnpike towards Harrisburg. This would be one of the fastest legs of the Run, as drivers were cautioned that they would need to average 54.03 miles per hour over the next 152 miles.[36] Leaving the Pennsylvania Turnpike in Carlisle, the drivers followed U.S. Route 11 into Harrisburg, where their first stop was for refueling at Samuel Hughes Mobil Service. Then it was a short 15-mile trek to the impound at the Holiday Motor Hotel. Bob Checkley, driving a Dodge Dart 170 6, was the first to arrive at 5:35 p.m. Eight drivers, George Alsbury (Chrysler Newport), Woody Bell (Pontiac Catalina), Win "Red" Eason (Oldsmobile F-85 Deluxe 6), Jim Fosdick (Ford Custom 500 8), Pete Novotny (Oldsmobile Jetstar 88), Karole Roth (Chevrolet Biscayne 6), Les Viland (Rambler American 440) and Edward Wilder (Oldsmobile Dynamic 88), arrived late, ranging in time from 1 minute to well over 7 minutes, and received penalties. Les Viland later explained that earlier in the day he missed a turn and had driven off course to an extent "that cost him 12 minutes and four or five miles of fuel."[37]

Today was tough, no doubt about it, and the averages reflected this. Only one driver, Chevy Teen Team's Richard Doyle, managed to improve the mileage on his Chevrolet Chevelle 300 6, while the others all saw declines.[38] Although hampered by driving off course and a penalty for late arrival, Les Viland continued to hold on to the best miles-per-gallon average for all entries. His Rambler American dropped below the 28-mile-per-gallon mark with an overall average of 27.9594. All of the leading drivers continued to maintain their positions. In the full-size six-cylinder class, Scott Harvey's Plymouth Savoy slightly expanded its lead by a margin of .1345 mile per gallon over Jim Whipple's Dodge 330. After six days there remained one contest that would not be settled until the final top-off in New York. In the luxury car class, Mel Alsbury, Jr. in a Chrysler New Yorker Salon maintained his lead over Ed Dempsey in a Cadillac Sixty-Two by a scant .0896 mile per gallon. Today the entire field saw their average drop slightly to 21.8664 miles per gallon. The teenagers continued their lead over the adults, posting an average of 22.2822 miles per gallon versus the adults' mark of 21.7476. In the large-engine compacts class, Chevy Teen Team's Ronnie Cherry's Chevrolet Chevy II Nova 8 ended up back in last place, being passed by Don Royer's Ford Falcon Deluxe 8. Although contests at the bottom of individual classes usually don't rate reporting, their contest started to tighten up after the third day, and now only .0188 mile per gallon separated the two. With the faster pace on the Pennsylvania Turnpike, the field covered today's leg averaging 44.7646 miles per hour.

The top three performers in each class were:

Small Engine Compacts	Dodge Dart 170	26.1742 Miles Per Gallon
	Plymouth Valiant V-100	25.9804 Miles Per Gallon
	Ford Falcon Futura	23.9609 Miles Per Gallon
Large Engine Compacts	Rambler American 440	27.9594 Miles Per Gallon
	Chevrolet Chevy II 100 6	23.3019 Miles Per Gallon
	Plymouth Valiant V-100 8	23.1561 Miles Per Gallon
Intermediate Size Six Cylinder	Buick Special Deluxe	25.4990 Miles Per Gallon
	Pontiac Tempest	25.3916 Miles Per Gallon
	Mercury Comet 202	24.6358 Miles Per Gallon
Intermediate Size Eight Cylinder	Buick Special Deluxe	23.8617 Miles Per Gallon
	Oldsmobile F-85 Deluxe	22.4714 Miles Per Gallon
	Ford Fairlane 500	22.3898 Miles Per Gallon
Full Size Six Cylinder	Plymouth Savoy	25.3820 Miles Per Gallon
	Dodge 330	25.2475 Miles Per Gallon
	Chevrolet Biscayne	22.3833 Miles Per Gallon
Low Price Eight Cylinder	Plymouth Savoy	22.2618 Miles Per Gallon
	Dodge 330	21.5067 Miles Per Gallon
	Ford Custom 500	20.8314 Miles Per Gallon
Medium Price	Buick LeSabre	21.4902 Miles Per Gallon
	Oldsmobile Jetstar 88	20.2592 Miles Per Gallon
	Pontiac Star Chief	19.9302 Miles Per Gallon
Luxury Cars	Chrysler New Yorker Salon	18.8308 Miles Per Gallon
	Cadillac Sixty-Two	18.7412 Miles Per Gallon
	Oldsmobile Ninety-Eight	18.2336 Miles Per Gallon

With the finish line in sight, it was a fairly quiet scene in Harrisburg as a group of weary competitors readied themselves for the short drive and festivities on tap for tomorrow. The members of the press following the Run on the chartered airliner bypassed this stop, heading directly for New York, where results and the highlights of today's drive were waiting for them to file their stories.

After setting records for the longest days, both in mileage and driving time, today's final leg to New York would be the shortest to date in the event's history. Leaving Harrisburg on April 9, it would be a fairly quick 184.3-mile, 3-hour and 30-minute trip to New York. West Covina Oldsmobile dealer Mandy Williams was the first to leave at 5:00 a.m., taking his Oldsmobile Ninety-Eight on both the Pennsylvania Turnpike and New Jersey Turnpike, heading towards New York. Today's weather would feature overcast skies with the drivers encountering winds up to 30 miles per hour. Arriving in Elizabeth, the drivers crossed Goethal's Bridge to Staten Island, where the entries were topped off for the final time. Just before 8:15 a.m., with very light snow falling, Edward Wilder in his Oldsmobile Dynamic 88 was the first to arrive. A long seven days' drive was over with the festivities just getting underway. After each entry was topped off, the drivers continued to the Staten Island Ferry, where they were remained until the final mileage calculations were completed. After a 20-minute crossing to Manhattan, the drivers followed F.D.R. Drive to the Queens Midtown Tunnel; then they headed for Flushing Meadows. Due to the traffic congestion in the New York City area, it was decided that the competing cars would proceed to the finish line in a caravan, led by the eight class winners, accompanied by a police escort. This would usher in a new tradition for traveling to the terminus location and thus eliminated what had been in previous years a contest to see who could cross the finish line first. Led by Class "A" winner Bob Checkley in a Dodge Dart, the drivers were flagged across the finish line, located in front of the Mobil Pavilion, by J.C. Agajanian. With two weeks remaining before the World's Fair opened to the public, the spectators on hand were basically invited guests, representing those affiliated with this year's Run and their families. Shortly after the last car received the checkered flag, a luncheon and awards ceremony was held at the Top of the Fair Restaurant, hosted by Fred H. Moore, president of Socony-Mobil Oil Co.

For the third consecutive year, Les Viland and his Rambler American 440 recorded the highest mileage for all entries with 27.8336 miles per gallon, at the same time capturing the large-engine compacts class. At the other end of the spectrum, Marty Gordon in an Imperial Crown recorded the lowest mileage with 17.2633 miles per gallon. Among the teenagers, the best mileage figure was turned in by Bo Jett and his Mercury Comet 202 6 with 24.5321 miles per gallon, placing third in the intermediate-size six-cylinder class. This year belonged to Chrysler Corporation and General Motors' Buick Division. In addition to the American Motors class win, Chrysler Corporation products placed first in four classes. Buick followed with three class wins, which validated the division's support and efforts in achieving success.

Class winners were:

CLASS	MAKE and MODEL	MILES PER GALLON
A — Small Engine Compacts	Dodge Dart 170	26.1191
B–Large Engine Compacts	Rambler American 440	27.8336
C — Intermediate Size Six Cylinder	Buick Special Deluxe	25.2986
D — Intermediate Size Eight Cylinder	Buick Special Deluxe	23.7441
E — Full Size Six Cylinder	Plymouth Savoy	25.2955
F — Low Price Eight Cylinder	Plymouth Savoy	22.1963
G — Medium Price	Buick LeSabre	21.3670
H — Luxury Cars	Chrysler New Yorker Salon	18.7737

Detailed results for all competing cars can be found in the Appendix.

USAC officials noted that the entire field averaged 21.7733 miles per gallon, which was a slight drop from last year's average of 21.84. Although their individual efforts did not result in any first-place finishes, the teenagers managed to outperform the adults by recording an average of 22.2078 miles per gallon. The adults as a whole averaged 21.6491 miles per gallon. Covering a 3243.8-mile route, the drivers averaged 46.2700 miles per hour to keep within the overall time limit of 70 hours and 40 minutes. This was the fastest pace to date, surpassing the previous average of 45.2274 miles per hour set in 1958. For the first time in a number of years, no breakdown between city and open highway driving was provided. Although three entries required minor repairs during the Run, all entries reached the finish line intact. Despite the average speed and weather conditions, with temperatures ranging from a low of 34 to a high of 87, no accidents were reported in what historically had been a very safe event.

Of the eight drivers in the winner's circle, two were first-timers, with Bob Checkley (Dodge Dart 170 6) and Marta Retzlaff (Buick LeSabre) joining the other winners. With the exception of Mel Alsbury, Jr. all of this year's winning drivers scored back-to-back wins, with Bob Cahill and Les Viland scoring their third consecutive wins.

Pretty much from the first day, the eventual winners had been determined in the compact and intermediate-size classes. After losing out on first place last year by .03 mile per gallon, Bob Checkley in a Dodge Dart 170 6 maintained a comfortable lead through the Run over the other three compacts competing in the small-engine compacts class. This gave Dodge its first class win with one of its compact models. There was never any doubt that Les Viland's Rambler American 440 would top all other entries in the large-engine compacts class. Les placed first by a wide 4.6154 miles per gallon margin over Chevy Teen Team Jean Fitzgerald's Chevrolet Chevy II 100 6 in second. This particular class had an unusual mix of entries, with four of the six equipped with eight-cylinder engines. The real competition here was between second and fourth place, where .3427 mile per gallon separated the three.

For intermediate-size models, it was all Buick after the second day, when Bill Levy's Special Deluxe moved into first place in the intermediate-size six-cylinder class while maintaining a small lead over Ginny Sims' Pontiac Tempest and Bo Jett's Mercury Comet 202. It was close at the end with Bill's margin over second place a mere .0492 mile per gallon. Although not receiving the total factory effort as members of the Chevy Teen Team, Bo was one of two teenagers to place third in their respective classes. It was John Rich all the way in the intermediate-size eight-cylinder class as he maintained over a mile per gallon lead over the other entries with his Special Deluxe.

In the full-size and low-price classes, it was Plymouth all the way. Competing in the full-size six-cylinder class, Scott Harvey kept his Plymouth Savoy one step ahead of Jim Whipple's Dodge 330, although the margin between the two at one time was a little as .0299 mile per gallon. Although Chevy Teen Team's Karole Roth placed third in her Chevrolet Biscayne, both Chevrolet and Ford finished well off the pace. In the low-price eight-cylinder class, it also was Plymouth and Dodge placing first and second, with Bob Cahill's Plymouth Savoy maintaining a comfortable lead over New York cab driver Hy Plotkin's Dodge 330. Competing against other models in the popular low-price field, this was Plymouth's eighth consecutive class win dating back to 1957. This also marked the second time since 1960 Plymouth scored a double win, with both six- and eight-cylinder models topping their respective classes. Although by a smaller margin than was seen for the six-cylinder models, Ford and Chevrolet also finished well off the pace of the leaders, with Jim Fosdick's Ford Custom 500 placing third.

Although the Plymouth and Dodge teams were piloted, for the most part, by factory engineers, this use of sometimes criticized "professional drivers" does not fully explain their overall success. One factor over the past several years was the

Capturing the luxury cars class for the third consecutive year is a Chrysler New Yorker Salon, driven this year by Mel Alsbury, Jr. This first-place finish was a fitting finale for Hollywood Chrysler-Plymouth dealer Mel Alsbury, Sr., participating in his last Economy Run. A regular supporter since 1950 and dealer entrant since 1952, Mel's entries would accumulate three consecutive Sweepstakes Awards along with twelve first-place finishes.

recognized performance of Chrysler Corporation's Torqueflite transmissions, along with the handling advantages of its torsion-bar suspension system. Both Plymouth and Dodge models, after being downsized in 1962, also enjoyed a definite weight advan-tage of between two and six hundred pounds over competing Chevrolet and Ford models. For example, the Plymouth Savoy 8 had a listed weight of 3210 pounds, with the Dodge 330 8 at 3325 pounds. This compares to the Chevrolet Bel-Air 8 at 3475 pounds and Ford's Custom 500 8 at 3659 pounds. With larger and heavier models being introduced by Plymouth and Dodge in 1965, this weight advantage would all but disappear.[39]

The real surprise this year was the performance of Marta Retzlaff, driving her Buick LeSabre to place first in the medium-price class. Marta had been competing since 1959, and her previous best finish was fifth in 1961, also driving a Buick LeSabre. From the first day, she maintained a fairly steady lead over Pete Novotny's Oldsmobile Jetstar 88 by just over a mile per gallon and never really looked back. With a Pontiac Star Chief and Catalina placing third and fourth, General Motors products produced their overall best performance in this class in the years they had participated. Buick personnel noted that one of the factors in this year's successful effort was the new Super Turbine automatic transmission, a similar version of which was also used in the winning Special models. The real disappointment was the fifth-place performance of the Chrysler Newport. Usually a strong contender in this class, the Newport finished well off the pace, trailing the winning LeSabre by 1.8601 miles per gallon.

In the luxury car class, Mel Alsbury, Jr. was able to keep his Chrysler New Yorker Salon in first place by a slim .1228-mile-per-gallon margin over Ed Dempsey's Cadillac Sixty-Two. Although Ed had a small lead from the start, Mel began to chip away at this lead, moving into first place on the fifth day and capturing his sixth first-place award. This win also marked the fifth consecutive year a Chrysler had placed first in its class. Mandy Williams piloted his Oldsmobile Ninety-Eight into third place, only .6566 mile per gallon behind the class-leading New Yorker.

One cannot dismiss the overall enthusiasm and skill displayed by the members of the Chevy Teen Team. In meeting the team at the finish line, Chevrolet General Manager Semon E. "Bunkie" Knudsen said, "We are gratified that our teenagers demonstrated that young people can be among the nation's best and safest drivers."[40] Although their individual results were disappointing, overall they "lived up to the faith expressed by Mr. Knudsen and other Chevrolet officials."[41] Two members placed well in the standings, with Jean Fitzgerald taking a second-place trophy and Karole Roth a third-place trophy, and the entire team "winning praise for their safe, responsible driving from Mobil Run officials and newsmen across the country."[42] Had she not been competing directly against Les Viland and his Rambler American 440 in the class for large-engine compacts, it is almost certain that Jean Fitzgerald and her Chevrolet Chevy II 100 would have taken the top honors. Bob Thomas, auto editor of the *Los Angeles Times*, noted, "There's definitely no substitute for experience," adding that the teenagers held up well during the seven-day trip: "The experienced drivers all had predicted that the kids would 'flake'—come apart after a few of the early morning departures (4 a.m.) plus 12 or more hours of driving in a day. They didn't, though. They got steadily better, thus proving that experience does command. Also, U.S. Auto Club observers said that the kids, if anything, were too good or cautious, failing as a result to take advantage of situations on which the experienced drivers capitalized."[43]

Although they achieved recognition from other drivers and USAC officials for their dedication and efforts, only one member of the Chevy Teen Team would return to participate in future Economy Runs. Jean Fitzgerald returned in 1965 as co-driver for the Chevrolet Chevelle Malibu 6. A few prominent names in Economy Run history would also see their involvement as drivers come to a close. Mary Davis, the first woman to capture a first-place trophy in 1957, decided to call it quits after a disappointing fifth-place finish in a Pontiac Tempest 8. Three drivers with long records of competition dating back to 1950 would also "retire": Woody Bell, Jim Fosdick and Art Rene. Art would continue his involvement as a USAC official in the coming years.

The conclusion of this year's Economy Run would marked the final appearance of one of its most prominent participants, Mel Alsbury and his Hollywood Chrysler-Plymouth dealership. Mel, who retired after selling his dealership earlier in the year, had had a continuous presence since 1950 during his days as general manager for Paul Bobst, Inc. This continued after he acquired the franchise in 1951. His rise to prominence began in 1952 with his first price class win in a Chrysler Saratoga. Over these thirteen years Mel's entries would accumulate three consecutive Sweepstakes Awards (1956–1958) along with twelve first-place and five second-place trophies. No other dealer would come close to the record attained by his entries. Mel participated only once as a driver in a Chrysler Saratoga, placing second in 1959. With Mel's departure, only two dealers with active participation dating back to 1950 remained, Southern California's Bob Estes and Hart Fullerton. Both would remain active over the next four years.

With the domination this year by both Chrysler Corporation and General Motors' Buick Division, advertising and promotion was far less prominent than in previous years. Mobil Oil continued its previous year's practice of publishing the results in newspapers and industry-related publications such as *Automotive News*. These ads featured the headline: "Here are results of the 1964 Mobil Economy Run" noting that "45 cars travel coast to coast ... 3,243 miles!" Listing both class winners and the complete results, Mobil pointed out that "mileage is the measure of power!," which both in mileage and in all-round performance was the benefit of their two blends of gasoline. To note that this performance had been proved in public, a new tagline was introduced: "Mobil—the Economy Run Gasoline."[44] Mobil service stations once again offered a small 8-page folder detailing the results of the Run, while at the same time introducing the concept of the Mobil Pavilion at the New York World's Fair.

All of Chevrolet's entries this year were handled by teenagers who formed the Chevy Teen Team. Crossing the finish line is Jean Fitzgerald driving a Chevrolet Chevy II 100 6. She turned in the best performance of all the teenagers, placing second in the Large Engine Compacts Class. Another Chevy Teen Team member, Karole Roth driving a Chevrolet Biscayne, placed third in the full-size six-cylinder class.

Rambler's achievement again was noted in newspapers announcing "Rambler American 440 Wins Class B With 27.83 M.P.G.," with a photograph of Les Viland and the winning entry. Noting it "Beats All Other Large-Engine Compacts in Mobil Economy Run," the text also discussed the American's new styling and increased popularity with the buying public, resulting in "sales up 46% over last year!"[45] One error did occur in the first release of this copy. In the *Los Angeles Herald-Examiner*, the mileage figure was mistakenly listed at 28.83 M.P.G., but was quickly changed before appearing in later editions. Featuring a photograph of the Rambler American on the open road, Champion Spark Plugs noted, "For top economy, Rambler installs Champion spark plugs right at the factory."[46] American Motors made only a brief mention of its mileage performance in their 1965 showroom brochures and "X-Ray" comparison catalogs, focusing instead on the redesigned Classic and Ambassador models using the "Sensible Spectaculars" theme.

Buick was quick to point out its three class wins with simple advertising themes. Using text only, a headline reading "We only won three classes" appeared in newspapers, noting that this feat was accomplished in three out of the four classes Buick entered. The text told readers of the first-place accomplishments of the six- and eight-cylinder Buick Specials and the Buick LeSabre, adding, "As for Class H, we just didn't win it. Oh well, you can't win 'em all," but did not mention the Electra finishing well off the pace by placing fourth.[47] Appearing in the July issue of *Motor Trend*, Buick advertised its three class victories, focusing on the performance of their new Super Turbine automatic transmissions. The text also included one of the few references to Buick's use of regular grade gasoline.[48] This Economy Run performance would be featured in ads through next several months, including a 60-second commercial seen on television. An illustrated 8-page folder, mailed to current and prospective Buick owners, featured Marta Retzlaff talking about her success competing against other medium-priced cars. The double wins for the Special were also discussed, along with driving tips for both for safety and economy.[49] Buick dealers were provided with an 8-page folder, devoted to both the Economy Run victories and the new Super Turbine Transmissions. Featuring photographs of the three winners crossing the New York Finish Line, this folder pointed out the benefit of the Super Turbine transmission for prospective customers.[50] In the 1965 model year, this achievement continued to be discussed in Buick's showroom catalogs printed for both the United States and Canada.

Not content to let Buick claim all the laurels, Chrysler Corporation advertised the "Most Wins By One Manufacturer in 1964 Mobil Economy Run" in general interest magazines.[51] This advertisement featured text outlining not only the individual achievements by Dodge and Chrysler but also highlighted Plymouth's 8-year string of victories. While only brief mention was made in their individual showroom brochures, a section on the Economy Run victories was featured in a special Plymouth performance brochure. Titled "Wherever Plymouth goes ... things happen!," this catalog also highlighted various racing events where Plymouth had excelled, noting, "All of Plymouth's eggs are not in the raw speed, sheer power basket!"[52]

Chevrolet took the opportunity in newspaper advertising to acknowledge the performance of its Teen Team Members in the Run. Featuring a photograph of the team's members standing with the second-place Chevy II 100 6 sedan, the text told the story of bringing this group of teenagers together, highlighting that it was "the first time any company had relied entirely on drivers with limited experience in this exacting competition." Although there was no mention of their individual standings, it was noted how well they performed as a group in miles-per-gallon performance, Chevrolet added, "The final results are a tribute to the high degree of driving skill displayed by the Chevy Teen Team representing the youth of America."[53] Chevrolet produced a special brochure, available at Chevrolet dealers, which highlighted the Teen Team's driving skills to achieve top mileage figures.[54] Their experiences were also highlighted in the September 1964 issue of *Friends* magazine for Chevrolet dealers and customers.[55]

Articles covering the events of this year's Economy Run were generally on the same level that was seen in 1963. Newspapers continued to remain the main source of coverage, with reporters filing stories daily whether traveling between overnight stops via chartered airliner or driving along with the competitors. The emphasis this year was placed on the overall performance of the teenagers who were competing against the more experienced adults. These stories also focused on some of the effects of the long hours behind the wheel that were needed to complete the cross-country route.

Those few articles that appeared in enthusiast magazines mainly dealt with the Chevy Teen Team.[56] Articles appearing in *Hot Rod* and *Popular Science* focused in the training involved for these teens as they prepared to compete against the more experienced adults. The article appearing in *Popular Science* was written as a group of letters to friends from Chevrolet Corvair 700 driver Earl Schilling telling of his experiences in preparing for the cross-country event. This same theme was found in Chevrolet's *Friends* magazine for its dealers and customers. Although competing as a driver for his second and final time, *Popular Mechanics*' Jim Whipple highlighted only the results of what he called a "considerably less exciting Mobilgas Economy Run" in his column "Detroit Listening Post."[57] While there were no comprehensive stories written on the Run, *Auto and Motor Sports Illustrated*'s John Hawkins wrote an overview of each day's events while focusing on the efforts by USAC officials to maintain the integrity of the Run. Certain that there had to be a gimmick, John also joined writer Ocee Ritch on the Memphis-to-Indianapolis leg and handled the navigator's duties. Joined by former driver Bill Corey, acting as an observer, he experienced the real world of Economy Run competition. Over the seven days, unable to find any irregularities, he concluded there were "no gimmicks, no chicane, the USAC supervision is purely wonderful. For the rest — just skill, patience and very great know how. And these run drivers have one other thing going for them: good navigators."[58] Again this year, *Motor Trend* focused its reporting on the Pure Oil Performance Trials.

Forty-two years later, a look back at this year's Run appeared in the *New York Times*. Written by the Plymouth Valiant V-100 8 driver Bob Knoll, the article recounted his experiences both in preparation and in driving on the Run. Explaining that friction, both inside the engine and in the running gear, is an enemy of fuel economy, Knoll said the team "set out to reduce the cars' rolling resistance. Each day when we took our cars from the impound area we drove them as close to wide-open as was prudent in the mountain roads around Los Angeles." While bringing today's fuel costs and concerns into focus, Bob thought it might be a "good time to consider a revival of the economy run." Recounting his experience in competing, and noting that there were spectators on hand at each overnight impound, Bob concluded that if the Economy Run were revived, "it would not surprise me if, just as in 1964, curious spectators lined the streets when the caravan rolled into town."[59]

Although there was an overall decline in publicity in the months following, the 1964 Economy Run could still be viewed as an overall success with its first cross-country outing. No doubt, as mentioned previously, a good degree of this success was due to the performance of the teenage drivers of the Chevy Teen Team and the publicity surrounding their participation. Over the coming months, competition from two other oil companies' events and promotions would continue to challenge Mobil's longtime efforts in publicizing mileage performance. Royal Dutch Shell joined the mileage bandwagon by announcing a new ingredient in their Super Shell brand. Called Platformate, this was designed to help increase gasoline mileage. A television commercial appearing in 1964 showed this mileage advantage in a private demonstration. Filmed in California's Mojave Desert, 10 convertibles were given the identical amount of fuel, half with Platformate and half without. After 4 miles of travel, those without this ingredient slowed to a stop while those with Platformate continued on a short distance further.[60] The impact of this campaign would not be felt immediately.

Another event at the New York World's Fair would have an impact on the 1965 Economy Run: the Ford Mustang was introduced to the public on April 17. During the next five months, thousands of visitors daily at the Mobil Pavilion were able to experience some of the skills involved in Economy Run driving. From personal experience visiting the World's Fair in the summer of 1964, this was not as easy as it looked. While the exhibits from Detroit's Big Three automakers featured examples of their products, the Mobil Pavilion did not have displays of any of the winning cars or trophies.

1965

Los Angeles to New York Via Flagstaff— Pueblo — Kansas City — Chicago — Niagara Falls — Albany

April 5–11, 1965—3266.1 Miles

With the New York World's Fair set for another season, there was never any question in the minds of Mobil Oil and USAC officials that New York would again be 1965's destination. That issue easily settled, there was a more pressing item on the agenda. Over the past several years, comments would surface regularly concerning the miles-per-gallon figures being achieved and how realistic these really are. While taking into account the careful, balanced driving style used to achieve these mileage results, a faster pace over the past five years was also a contributing factor. During this period, the average speed increased from 40.9335 miles per hour in 1959 to 46.2700 in 1964. But in an effort to make mileage figures more realistic to the average motorist, even faster driving times would be mandated by USAC officials.

When the announcement of the 1965 Run was made on March 1, a faster trip was just one of the changes announced by Chief Steward A.C. Pillsbury. Noting that the distance to New York's Times Square would be more than 3200 miles, he said overnight stops will be in Flagstaff, Pueblo, Kansas City, Chicago, Niagara Falls and Albany. Pillsbury noted, "More significant than the distance is the fact this will be the fastest Run. Required minimum average speed coast-to-coast will be above 50 miles per hour, as compared to last year's 46.27. We believe that the higher speeds will make the miles per gallon average the most realistic ever achieved." He went on to add, "This Run will be one of the toughest on the drivers. They [will] face snow-covered passes on the Continental Divide, and one day will be the longest ever attempted ... more than 600 miles. Mix this in with high-speed turnpikes, stop-and-go city traffic and assorted streets and highways in varying conditions, all at speeds comparable to those an average driver would maintain, and our drivers will know they've been cross country."[61]

Another change was the addition of a new class designed for "Detroit's newest sports-type cars" — Class S for Special Sports. Prompted by the introduction of Ford's Mustang and Plymouth's Barracuda, this class was designed for two-door hardtop or fastback models equipped with bucket seats and "powered by the lowest displacement V-8 that would be considered standard with no options permitted." They would be

Glen "Ole" Olsen, driving an Imperial Crown, receives as enthusiastic sendoff from USAC's J.C. Agajanian and cheerleaders from West Covina High School as the first to start the 1965 Economy Run. Ole would run out of fuel fourteen miles from the Toledo refueling station, the first driver to suffer this fate since 1955.

"equipped with one two-barrel carburetor and have [a] standard compression ratio."[62] Additional models that would compete in this class were Chevrolet's Chevelle Malibu SS, Dodge's Dart GT, and Mercury's Comet Caliente Sport. In announcing this new class, A.C. Pillsbury added, "Their inclusion reflects the popular acceptance by the motoring public of the new personal car models."[63]

Other than the addition of the special sports class, no significant changes were needed in the competitive classifications. When the "Competitive Divisions for the 1965 Mobil Economy Run" were announced, only a minor change was made regarding eligibility in Class B for large-engine compact cars, where the maximum engine displacement was reduced to 290 cubic inches from 327 cubic inches in 1964.[64] With the exception of the class for special sports, four-door sedan and hardtop models continued eligible to compete, with two-door models again allowed in the compact and intermediate-size classes at the discretion of USAC. As was the case last year, only four-door models were selected by USAC officials.

Aside from the provision for the new class for special sports, no changes were made in the rules or for required accessories on competing cars.[65]

After entries closed on March 6, a field of 49 cars represented every U.S. make except Lincoln. "When asked why there was no Lincoln, a United States Auto Club representative said 'because the Lincoln-Mercury people did not want to enter one.'"[66] With its production moved to Canada, Studebaker was no longer eligible to compete, as the rules stipulated that entry is "open to stock 1965 model cars of U.S.A. manufacture and design."[67] The only other "no-show" of note was the absence of the Rambler Classic, which again would be in direct competition with the smaller American. With interest being focused on the class for special sports, no eight-cylinder models were entered in the class for large-engine compacts. This was no doubt a reflection of the results in the 1964 Run, when six-cylinder models outpaced the eight-cylinder models. General Motors products again comprised the largest group of entries with 21, followed by Chrysler with 14 and Ford Motor Company with 11. American Motors rounded out the field with 3 entries. With Cadillac again declining to provide factory sponsorship or support, USAC officials reached out to the West Covina (California) Chamber of Commerce, who elected to enter a Sedan de Ville; the chamber's president, Stan Seeds, handled the driving duties. The chamber would continue with their active participation through 1968. Reflecting the growing popularity of compact and intermediate-size models, this group accounted for over half of the total entries.

With factory support continuing, nearly every manufacturer was involved in one way or another. Chrysler Corporation, Buick and American Motors continued providing engineering resources and personnel who had been involved with the Run for the last several years. Ford's involvement increased with the assignment of two of its engineering department's special test drivers, Dave Apple and Al Johnson, to lead this year's efforts. Teaming up with Southern Californians, most with performance or racing backgrounds, to handle driving duties, Dave stepped in to drive the Falcon Futura 6, with Al behind the wheel of the Fairlane 8. Continuing to rely on Inglewood dealer Bob Estes for the main support of its entries, Mercury again named teenagers to pilot one of its entries, following the successful results of last year. Neb Aberly, 1964 National Teenage Safe Driving Road-E-O Winner, was selected to drive its eight-cylinder Comet 404 with Jo-Jo Shutty, Lincoln-Mercury's Youth Safety Spokesperson, as co-driver.[68]

Teenagers were named as drivers for the two Dodge Dart entries, although one selection was not without some reservations. Marty Payne, the first and only African American to participate in the Runs, was chosen to drive the six-cylinder Dart after winning a driver education–inspired contest at Santa Monica High School. As a sign of the times, concern was expressed by some Mobil Oil officials of possible problems or incidents along the route.[69] None were reported. Dareth Rich, active with the International Car Club Association, was named to drive the Dart GT competing in the special sports class. Members of Chrysler Corporation's engineering staff returned to handle driving duties for the Chrysler New Yorker, Dodge Coronet and all Plymouth entries. Patricia Jones, who placed first in the medium-price class driving a Dodge in 1959, returned to pilot a Dodge Custom 880. Hart Fullerton returned to sponsor and drive a Chrysler Newport, while the Imperial Crown was sponsored by Beverly Hills Chrysler-Plymouth and driven by a member of the dealership's staff.

Chevrolet drastically reduced their involvement this year, with individual dealers providing personnel for the preparation of their entries. Two dealers, Kansas City's Mel Hilliard and Chicago's Nickey Chevrolet, provided their expertise from past Economy Run efforts to the newcomers. The only member of

With a roaring send-off, Al Cottle, a regular participant since 1951, drives his Mercury Monterey under the start banner for the first day's trip to Flagstaff. This was the fourth year Al had driven a Mercury Monterey entered by another regular, Inglewood dealer Bob Estes. The USAC riding observer is Smirle Morris, one of the many who volunteered their time each year to assist in various ways the supervision and conduct of the Economy Run.

the 1964 Chevy Teen Team to return was Jean Fitzgerald, who was selected as co-driver of the Chevelle Malibu 6.

Pontiac turned to freelance writer Don Francisco to head up its efforts; he also returned to driving duties for the first time since 1960. Joining the Pontiac team as drivers included two well-known names, Art Arfons and Jean Calvin. Art, a professional drag racer, was better known as the "fastest man on wheels" after setting a world's land speed record in October 1964. Driving his jet-powered "Green Monster," he recorded a top speed of 536.71 miles per hour. Jean was well-known as an associate editor of *Sports Car Graphic* magazine.

Oldsmobile increased its support by also taking over sponsoring duties, with all of its entries entered by the Oldsmobile Dealers of U.S.A. After driving Chrysler entries for their father beginning in the mid–1950s, both George and Mel Alsbury, Jr. moved over to Oldsmobile, with George driving a Dynamic 88 and Mel, Jr. in a Ninety-Eight.

Dealer associations took on a greater role this year as sponsors, representing Dodge, Ford, Oldsmobile, and Rambler. Dealers participating for the first time, located in cities along the route, included: Gotham Ford in New York City; Hunter Brothers Motors (Chrysler-Plymouth) in Pueblo; Jess Hunter Motors (Ford) in Pueblo; E. Koeppel, Inc. (Chrysler-Plymouth) in Jamaica, Long Island; Sites Brothers Motors (Chrysler-Plymouth) in Kansas City; and Wilson Motors (Chrysler-Plymouth) in Flagstaff.

Three dealers participated as both entrants and drivers: Hart Fullerton (Chrysler Newport), Mel Hilliard (Chevrolet Corvair Monza) and E.L. "Kirk" Kirkpatrick (Chevrolet Chevelle Malibu SS). This was Kirk Kirkpatrick's first participation either as a driver or entrant in the Run. Oldsmobile dealer Mandy Williams originally had planned to drive the Oldsmobile Ninety-Eight, but stepped aside to handle co-driving duties for Mel Alsbury, Jr. Four husband-and-wife teams returned: Ed and Karen Dempsey (Oldsmobile F-85 6), Hart and Diana Fullerton (Chrysler Newport), Stan and Lois Raymond (Oldsmobile F-85 Deluxe 8), Tommy and Ila Thomas (Rambler Ambassador 880 6). They were joined by first-timers Edward "Budd" and Kay Liburdi (Mercury Comet Caliente Sport), Kirk and Elaine Kirkpatrick (Chevrolet Chevelle Malibu SS), and Glen "Ole" and Helene Olsen (Imperial Crown). Another husband-and-wife team, Gordon and Scherlee Madison, each drove separate entries. Both in Chevrolets, Gordon piloted a Chevy II Nova 6 with Scherlee in a Chevelle Malibu 6. The mother-and-daughter team of Pat and Pam Sawyer returned for a second try at first place after their successful efforts in 1963 with a Chevrolet Corvair Monza. Women were named as drivers for ten entries and included prior class winners Patricia Jones, Marta Retzlaff and Pat Sawyer. All of last year's class-winning drivers returned to defend their titles.

With random car selection and inspections beginning in late February, the impound area was once again the center of activity. For the next four weeks, both USAC officials and individual teams made preparations for the trip to New York. The insistence by USAC officials of unquestionable stock status was best illustrated by Don Royer, driver of the Chevrolet Impala 8, who was responsible for preparing the two entries for Mel Hilliard Chevrolet. Writing in the April 1965 issue of *Mobil Dealer News* Royer recalled:

> Most difficulties encountered with the cars during break-in may be corrected or repaired, although USAC is rather finicky as to some of these. My Impala had, during break-in and throughout the Run, a rattle that was initially thought to be the exhaust pipe interfering with the frame. Later this was found to be a piece of linkage extending from the transmission through a frame member. This can normally be corrected by removal of a small edge of the frame member but was not permitted by USAC as this would serve to lighten the car by a fraction of an ounce. This car was also delivered without a cigarette lighter element and USAC insisted that this be installed. Due to a build-up of tolerances, the car also had a very, very slight end play in one of the rear axles, a normal condition and correctable by an available spacer. As this spacer was not a genuine Chevrolet part, it could not be installed. Unequivocally, I can say that the cars on the Mobil Economy Run are truly stock.[70]

Entrants were given the option of using either Mobil Regular or Mobil Premium gasoline. Both six- and eight-cylinder models of the Buick Special and Pontiac Tempest entries chose Regular. One cannot overstate the depth of planning by both USAC and Mobil Oil officials that was required to move the hundreds of persons involved from point-to-point. While detailed planning undertaken by USAC officials, from developing the route to calculating the daily results, has been discussed in previous chapters, unique situations occur from year to year to challenge even the most experienced. Traveling into Canada presented a unique situation that involved customs officials from both countries. For several months, and well before the overnight stops were announced, USAC officials began quiet contacts with customs and immigration officials in both countries to expedite the movement of several hundred participants across the border without revealing the true nature of the event. As detailed in the March 1966 issue of *Motor Trend*, one USAC representative took trips to Montreal and Ottawa for meetings with Deputy Ministers of Customs and Immigration, [and] six times to Niagara Falls, the actual crossing point, [where he] did a masterful selling job to the bureaucrats involved. Eventually, of course, he had to let the officials in on the secret, but not until the Run was irrevocably committed to cross the border. The upshot was that the customs inspections for both countries were carried on at the overnight hotel in Canada, and all competing cars were listed in advance at the border and allowed to pass without stopping.[71]

To ensure that the competing cars were well taken care of on the Run if a problem should occur, the Automobile Club of Southern California again provided one of their service patrol trucks. With service patrol driver Raymond Brooks following the last competing car each day, equipment was available to assist in any way when necessary.

Advance publicity efforts and accommodations for the press representatives covering the Run continued under the direction of Mobil Oil's Frank Meunier, assisted by Roger Mahey and John Parkinson. While most of the press representatives would travel between overnight stops on a chartered United Airlines DC-6B, additional coordination resulted in the loan of practice cars for press representatives to use in covering the Run from the ground. This year there was one very special press car. A completely restored 1940 Dodge Deluxe Sedan was used by selected press representatives traveling along the route, working with Jack McFarland, manager of the Dodge News Bureau in Los Angeles. It was advertised as a duplicate of the model which competed in that year's Gilmore Yosemite Economy Run, achieving 23.76 miles per gallon. Mileage comparisons would be made between it and the more modern Dodge entries.

One of the last events prior to the start was the drawing for starting positions. First-time driver Ole Olsen drew No. 1 for his Imperial Crown, with Scherlee Madison drawing the final spot, No. 49, for her Chevrolet Chevelle Malibu 6. Dispensing with the Pasadena impound at the starting location, USAC officials announced the start would be from the recently opened Mobil Repair Center located adjacent to the San Bernardino Freeway in West Covina. Given that this year's Run would be a challenge, the focus now was on the forthcoming release of the first day's routing and driving time. USAC officials earlier had predicted that the average miles-per-gallon average would hit an all-time low. As more details of the route and driving time were released, two factors confirmed this statement: "First, cars will have to average more than 50 m.p.h. for the entire distance [and] most of it will cover high-speed throughways." USAC added, "Secondly, the possibility of snow and poor weather, particularly in the northeast sections where the event reaches into Canada for the first time, could present gas-consuming problems for drivers."[72]

On Sunday, April 4, the competing cars were driven from the Pasadena impound to the Mobil Repair Center in West Covina in light rain conditions. Arriving at the Repair Center, decorated with red, white and blue bunting along with checkered-flag pennants, all entries were topped off before being parked in their starting order for tomorrow morning's start. For most drivers and factory representatives, this was their first true chance to compare the mileage performance of each entry. Several drivers were observed looking intently over the shoulders of USAC officials who handling the refueling process, noting the amount of fuel being added. One driver in particular, American Motors' Les Viland, was observed meticulously recording this information for those cars competing against the Ramblers. With the cars ready for tomorrow's start, the first day's route summary and detailed itinerary were released, a task that was given to the current

After contending with snow flurries and wet road surfaces, Scott Harvey arrives at the Flagstaff impound in his Plymouth Fury II after the 488.5-mile trip from West Covina. This is the third time since 1960 the Economy Run would make an overnight stop in Flagstaff. Scott would go on to place first in the low-price eight-cylinder class, marking Plymouth's ninth consecutive class win.

Miss West Covina. Compared to the previous two years, there were no special detours planned on this first leg by A.C. Pillsbury to challenge the drivers. Covering a distance of 488.5 miles, the trip to Flagstaff was a fairly straight shot with a time limit set at 9 hours and 46 minutes. This year would mark the third visit to Flagstaff for an overnight stop since 1960 by a different route. There was no doubt now that this year's Run would be the fastest in its history, together with weather conditions that would present challenges to the drivers. Generally the weather along the route had been unseasonably bad this spring, with snowfall throughout the Rocky Mountains the heaviest in over a decade.[73]

Early in the morning on April 5, the competitors and officials were joined by one of the largest groups of spectators in a number of years ready for their trip east. Under cloudy skies, Ole Olsen in an Imperial Crown was the first driver to receive the green flag at 6:15 a.m. from J.C. Agajanian. The West Covina High School band was joined by its cheerleading squad in giving everyone an enthusiastic send-off. With two quick turns out of the Repair Center's driveway, they were heading east on the San Bernardino Freeway (Interstate 10) in chilly temperatures towards the town of Colton, where they picked up Interstate 15 towards Barstow. Just outside of Barstow, the first minor incident occurred when Dareth Rich's Dodge Dart GT temporarily got stuck in some mud. As explained later, Dareth was attempting to pull off the road to make an emergency stop when "the car skidded onto the soft shoulder, getting stuck fast in the mud." It took "10 minutes to get the car free and back on the road." Neither Dareth nor her co-driver, Larry Rathgeb, would explain the reason for the stop.[74] From Barstow, the drivers headed east on U.S. Route 66 towards Needles for their first stop of the day at Harold Strout's Mobil Service, where each entry was topped off, quickly followed by brunch at St. Ann's Catholic Church. As in prior years, church and community organizations were readily available in preparing and serving brunch to the participants. Continuing to follow U.S. Route 66, they quickly crossed into Arizona, entering the Mountain Time Zone, headed for Flagstaff. Throughout their trip across the desert, the drivers battled crosswinds; then, with snowstorms and winter conditions earlier in the day in and around the Flagstaff area, they had to contend with snow flurries and wet roads, although they had been cleared of ice and snow.[75] Their first stop was for refueling was at E.D. Babbit Mobil Service, also the city's local Ford dealer, before continuing on to the impound at the Americana Motor Hotel. Greeted by subfreezing temperatures, the first to arrive at 5:45 p.m. was Patricia Jones, driving a Dodge Custom 880, with all drivers finishing this leg within the prescribed time limit. Patricia, who started the Run with a heavy cold and high temperature, required medical treatment after arrival.[76] Greeted by USAC officials with today's results and details of tomorrow's routing, the drivers quickly knew the trip to Pueblo would prove to be a challenge, as rough going was expected over Colorado's Wolf Creek Pass. Some 500 inches of snow had already fallen on the pass this season and more was forecast at the time the competitors would cross the summit.[77]

If the first day's results proved anything, it was that the Ford Falcon and Rambler American were going to battle it out for the highest mileage honors in this year's Run. While Les Viland's American 440 recorded the best overall miles-per-gallon figure of 27.0894, it was only a mere .0526 ahead of Byron Froelich's Ford Falcon. Since both drivers were competing in separate classes, it appeared they were well set for first-place finishes as each maintained a comfortable margin over the drivers in second place. The entire field turned in an average of 21.0127 miles per gallon, with the three teenagers averaging 22.0841. The focus this year would be on the performance of the special sports class, who turned in an average of 20.7660 miles per gallon. At this early stage it appeared that close contests were on tap for half of the classes. In the intermediate-size eight-cylinder class, Bob Cahill's Dodge Coronet had a .3202 lead over Art Hesskamp's Plymouth Belvedere II, with last year's winner, John Rick's Buick Special Deluxe, in third. In the full-size six-cylinder class, Bob Checkley's Plymouth Fury II was ahead of Pat Sawyer's Chevrolet Biscayne and Ginny Sims' Ford Custom, although only .7081 mile per gallon separated the three. By less than a mile per gallon, three drivers led their respective classes: Darrell Droke's Ford Fairlane 6 in the intermediate-size six-cylinder class; Scott Harvey's Plymouth Fury II 8 in the low-price class; and Art Bertelsen's Chrysler New Yorker in the luxury car class. Jean Calvin's Pontiac Catalina recorded a comfortable 1.0785-mile-per-gallon lead over Art Arfon's Pontiac Star Chief. In the special sports class, George Miller's Plymouth Barracuda had the lead over Dareth Rich's Dodge Dart GT and Bill Levy's Ford Mustang 2+2, with .7013 miles per gallon separating these three. Despite her trouble earlier in the day, Dareth managed a mere .0142 lead over third place. The surprise in the first day's results was the performance of last year's winning Buick models. The Special Deluxe 6 was sitting in eighth place, while the Special Deluxe 8 and LeSabre were both shown in third. Other than Dareth Rich placing second in her class, the other two teenage drivers were not faring all that well. Marty Payne's Dodge Dart 6 was in third in the small-engine compacts class, well over 2 miles per gallon off the pace. Ned Aberly's Mercury Comet 404 8 was in fifth in the intermediate-size eight-cylinder class, and although competition in this class was close, he was 1.3653 miles per gallon behind the leading Dodge Coronet. Today's average speed of 50.3559 miles per hour put this year's Run on track to set a record for the fastest in history.

The top three performers in each class were:

Small Engine Compacts	Ford Falcon (170 Cu. In.)	27.0368 Miles Per Gallon
	Plymouth Valiant 100	25.2915 Miles Per Gallon
	Dodge Dart	24.5565 Miles Per Gallon
Large Engine Compacts	Rambler American 440	27.0894 Miles Per Gallon
	Ford Falcon Futura (200)	25.5643 Miles Per Gallon
	Chevrolet Chevy II Nova	23.2643 Miles Per Gallon
Intermediate Size Six Cylinder	Ford Fairlane	25.3842 Miles Per Gallon
	Mercury Comet 202	24.6756 Miles Per Gallon
	Plymouth Belvedere I	24.2725 Miles Per Gallon
Intermediate Size Eight Cylinder	Dodge Coronet	21.6728 Miles Per Gallon
	Plymouth Belvedere II	21.3526 Miles Per Gallon
	Buick Special Deluxe	20.9862 Miles Per Gallon
Full Size Six Cylinder	Plymouth Fury II	22.5401 Miles Per Gallon
	Chevrolet Biscayne	21.9160 Miles Per Gallon
	Ford Custom	21.8320 Miles Per Gallon
Low Price Eight Cylinder	Plymouth Fury II	20.8575 Miles Per Gallon
	Ford Galaxie 500	20.0571 Miles Per Gallon
	Chevrolet Impala	19.4236 Miles Per Gallon
Medium Price	Pontiac Catalina	20.9131 Miles Per Gallon
	Pontiac Star Chief	19.8346 Miles Per Gallon
	Buick LeSabre	19.7020 Miles Per Gallon
Luxury Cars	Chrysler New Yorker	17.9256 Miles Per Gallon
	Buick Electra 225	17.1047 Miles Per Gallon
	Oldsmobile Ninety-Eight	16.6726 Miles Per Gallon
Special Sports	Plymouth Barracuda	22.0753 Miles Per Gallon
	Dodge Dart GT	21.3882 Miles Per Gallon
	Ford Mustang 2+2	21.3740 Miles Per Gallon

In a somewhat lighthearted commentary, the last sentence of the first day's route summary said: "Sleep fast—we'll have an early start in the morning."[78] This note prepared the drivers for their trek to Pueblo on April 6. Jean Calvin, driving a Pontiac Catalina, was the first driver on the way at 4:00 a.m. on a 584.8-mile trip to be completed within a time limit of 12 hours even. In Flagstaff, "the temperature was down to about 20, the cars were covered with ice, fogged up, and almost impossible to see, necessitating the use for the first time of the heaters and defrosters, run at a very low speed so as to conserve as much electricity as possible."[79] Even with some weather concerns in the forecast, USAC officials said, "When the sun rises, today's run over the Navajo Trail and the mountains of Southern Colorado promises to be one of the most scenic ever experienced on any Mobil Economy Run."[80] From Flagstaff, the drivers headed north on U.S. Route 89 to Tuba City, where they turned in a northeast direction on Arizona 64 toward the Four Corners Monument. In addition to the cold temperatures, for the first two hours the drivers also had to contend with patches of ice on the road.[81] Briefly crossing into New Mexico, the drivers crossed into Colorado following Colorado 40 and 789 towards Cortez and Mancos, where heavy fog was encountered in the Mesa Verde area.[82] Arriving in Mancos, they made their first stop of the day at Riverside Mobil Service, where all entries were topped off, then it was a quick 27-mile trip to Durango for brunch at the Grand Palace.

Their driving day half-over, there was more mountain driving in front of them as they headed for the Continental Divide. Following U.S. Route 160, they ascended over Wolf Creek Pass Summit, at 10,850 feet the highest elevation on this year's Run. It had snowed the previous night on Wolf Creek Summit, and highway maintenance crews managed to keep the highway open, although it was covered by 2–3" of ice and packed snow. All of the cars got through safely without having to use chains.[83] Snow began to fall minutes after the last car reached the summit. After they had traveled through Alamosa and crossing the 9249-foot Sangre de Cristo Pass Summit, USAC officials noted, "Now it's downhill all the way to New York!"[84] Before arriving in Walsenburg, the drivers contended with road construction before reaching Interstate 25 and heading north. Arriving in Pueblo, the drivers navigated its city streets to James W. Mathews Mobil Service for refueling, then it was back to the Interstate for a quick 4-mile trip to the impound at the Continental Pueblo Motor Hotel. Just after 4:45 p.m., Jean Calvin's Pontiac Catalina, having been the first to leave Flagstaff, was the first to arrive, pleased at the sight that no snow was anywhere to be found. Don Bridges, driving a Rambler Ambassador 880 8, arrived 4 minutes and 46 seconds late and was penalized. Patricia Jones, driver of the Dodge Custom 880, continued to battle a heavy cold and high temperature which required further medical treatment.[85]

Today's terrain resulted in everyone's averages dropping, some fairly significantly. Continuing to lead all entries with the highest miles-per-gallon figure, Les Viland saw his Rambler American 440 record a figure of 26.2098. This was a decline of .8796 mile per gallon from yesterday's results. Byron Froelich and his Ford Falcon continued to be a strong contender for the top mark with 25.8950 miles per gallon, although this was a sizeable 1.1418 drop from yesterday's mark. Two classes saw new leaders emerge. Art Hesskamp's Plymouth Belvedere was leading among the intermediate-size eight-cylinder models with Bob Cahill's Dodge Coronet in second. The margin between these two was tight—.0354 mile per gallon. Pat Sawyer moved her Chevrolet Biscayne into first in the full-size six-cylinder class by a margin of .1212 mile per gallon over Bob Checkley and his Plymouth Fury II. Competition remained tight in three other classes where just over a half-mile per gallon separated first three positions. The only other changes of note involved second-place positions. In the medium-price class, Marta Retzlaff moved her Buick LeSabre

After a long day of mountain driving, including crossing the 10,850-foot Wolf Creek Pass Summit, Darrell Droke arrives at the Pueblo refueling station in his Ford Fairlane 6. Here everyone was greeted by a large number of local residents on hand to view the refueling process. Scenes like this were typical as the drivers made their way across the country.

into second by a slim .0147 mile per gallon over Art Arfon's Pontiac Star Chief. In the special sports class, Bill Levy moved his Ford Mustang 2+2 into second by the slimmest of margins with .0027 mile per gallon separating Bill's Mustang and Dareth Rich's Dodge Dart GT now in third. The entire field saw their average drop to 20.4962 miles per gallon, which was a half-mile per gallon lower than yesterday's results. The other averages being watched this year also dropped, with the teenagers coming in with 21.5336 miles per gallon and the special sports models with a figure of 20.3346. While Marty Payne's Dodge Dart 6 remained in third place in the small-engine compacts class, Ned Aberly was able to move his Mercury Comet 404 8 into fourth place in the intermediate-size eight-cylinder class. With the terrain and driving conditions, today's average speed dropped to 48.9963 miles per hour, which was still considered to be a fast pace.

The top three performers in each class were:

Small Engine Compacts	Ford Falcon (170 Cu. In.)	25.8950 Miles Per Gallon
	Plymouth Valiant 100	24.4473 Miles Per Gallon
	Dodge Dart	23.6013 Miles Per Gallon
Large Engine Compacts	Rambler American 440	26.2098 Miles Per Gallon
	Ford Falcon Futura (200)	25.0432 Miles Per Gallon
	Chevrolet Chevy II Nova	22.8496 Miles Per Gallon
Intermediate Size Six Cylinder	Ford Fairlane	25.0162 Miles Per Gallon
	Mercury Comet 202	23.9123 Miles Per Gallon
	Dodge Coronet	23.6378 Miles Per Gallon
Intermediate Size Eight Cylinder	Plymouth Belvedere II	20.9288 Miles Per Gallon
	Dodge Coronet	20.8934 Miles Per Gallon
	Buick Special Deluxe	20.3681 Miles Per Gallon
Full Size Six Cylinder	Chevrolet Biscayne	21.6121 Miles Per Gallon
	Plymouth Fury II	21.4909 Miles Per Gallon
	Ford Custom	21.1826 Miles Per Gallon
Low Price Eight Cylinder	Plymouth Fury II	20.3921 Miles Per Gallon
	Ford Galaxie 500	19.6555 Miles Per Gallon
	Chevrolet Impala	18.8954 Miles Per Gallon
Medium Price	Pontiac Catalina	20.0910 Miles Per Gallon
	Buick LeSabre	19.6488 Miles Per Gallon
	Pontiac Star Chief	19.6341 Miles Per Gallon
Luxury Cars	Chrysler New Yorker	17.2585 Miles Per Gallon
	Buick Electra 225	16.7252 Miles Per Gallon
	Cadillac Sedan de Ville	16.4245 Miles Per Gallon
Special Sports	Plymouth Barracuda	21.8457 Miles Per Gallon
	Ford Mustang 2+2	20.8147 Miles Per Gallon
	Dodge Dart GT	20.8120 Miles Per Gallon

There was an added surprise on the Flagstaff-to-Pueblo leg for two reporters covering the Run, Bob Thomas of the *Los Angeles Times* and Bill Carroll, West Coast editor of *Automotive News*. While at the Four Corners Monument area for refueling, they spotted two Oldsmobile test drivers along with a revolutionary new car which confirmed rumors that progress on the first American front-wheel drive automobile since the ill-fated Cord of the 1930s was on schedule. Whether this was a planned or unplanned event was never determined. This car turned out to be the new Oldsmobile Toronado, scheduled for introduction as a 1966 model in the fall.[86]

When the competitors read the routing and itinerary to Kansas City, they realized that this would be, in one sense, a record breaker. Another pre-dawn start was also on tap.

On April 7, the drivers faced the trip to Kansas City, which would be the longest in terms of mileage in the history of the Run. At 607 miles, this eclipsed by 7.7 miles the previous record of 599.3 miles set in the 1941 Gilmore Grand Canyon Run. Ready for the 11-hour and 56-minute trip, "Ole" Olsen in his Imperial Crown was the first driver on their way at 4:00 a.m. After struggling with transmission issues on his Chevrolet Corvair Monza, Mel Hilliard was 12 minutes and 10 seconds late in leaving as he took time to investigate the problem, which resulted in replacing the transmission modulator valve. This was done on his own time and was not deducted from the driving time. The drivers headed east on U.S. Route 50 through eastern Colorado, then crossed into Kansas, passing through numerous small communities as they entered the Central Time Zone. Arriving in Dodge City, they made the first stop at the Silver Spur Restaurant for brunch, followed immediately by a stop at Tommy Council Mobil Service, where all entries were topped off. After driving at near-maximum speed limits since leaving Pueblo, their day was not quite half over when they faced a small challenge from A.C. Pillsbury. Known for his famous detours, he directed the drivers through the heart of Old Dodge City, while at the same time having them contend with several compulsory stops through town. Not only did the drivers have this small detour, they also began to encounter strong crosswinds of between 35 and 50 miles per hour and had to fight the steering wheel for control. They would also deal with overcast skies and the threat of rain all the way to Topeka.[87] After briefly rejoining U.S. Route 50, they then followed U.S. Route 56, heading in a northeast direction bound for Ellsworth. In Kinsley, the competitors passed a sign marking a halfway point between New York and San Francisco, 1561 miles in each direction. Just past Ellsworth they picked up U.S. Route 40 / Interstate 70 and the Kansas Turnpike bound for Topeka, where all entries were refueled at Wilson's Turnpike Mobil Service, located at the Topeka Service Area. With a little more than an hour to go, they continued on to Kansas City and the impound at the Holiday Inn. In warm, near 80-degree weather, Scott Harvey, driving a Plymouth Fury II 8, was the first to arrive just before 5:45 p.m. after nearly thirteen hours on the road. The competitors were welcomed by a good-sized turnout of local residents, which included live coverage by NBC's affiliate station WDAF-TV.[88] Nine drivers, including George Alsbury (Oldsmobile Dynamic 88), Dave Apple (Ford Falcon Futura 6), Mel Hilliard (Chevrolet Corvair Monza), Kay Kimes (Pontiac Tempest Custom 8), Bud Liburdi (Mercury Comet Caliente Sport), Ole Olsen (Imperial Crown), Stan Raymond (Oldsmobile F-85 Deluxe 8), Pat Sawyer Chevrolet Biscayne 6), and Ginny Sims (Ford Custom 6), arrived late, ranging in time from 1 minute and 6 seconds to 6 minutes and 14 seconds, and were penalized. Ginny Sims in a Ford Custom 6 drove off-course for a duration of three-and-half minutes which caused in her late arrival. Jean Calvin summed up today's trip: "To stay on time across the Midwest, we were driving at top of the speed limit (60–65) whenever possible and still worried about being on time at night."[89] After 1680.3 miles, their trek across country was half-over in terms of miles traveled. The competitors were beginning to show the strain of these first three days.

Mel Hilliard guides his Chevrolet Corvair Monza out of the Dodge City refueling station, heading to Kansas City with another large number of local residents on hand. Early in the Run, Mel struggled with transmission issues which resulted in his having to replace a modulator valve. He could do no better than placing fourth among the small-engine compacts.

With one exception, everyone's mileage showed improvement, with several classes showing competition continuing to be very close. Perennial favorite Les Viland in a Rambler American 440 continued to have the best overall mileage at 26.8306 miles per gallon for all participants. Close on his heels was Byron Froelich and his Ford Falcon with 26.8100—a difference of only .0206 mile per gallon. So tight was this contest that Bob Thomas of the *Los Angeles Times* noted, "Measuring that [difference] in feet and inches, Viland's advantage a gallon is slightly more than 105 feet."[90] Both drivers had comfortable leads in their respective classes and barring any unforeseen problems, they would capture first-place trophies in New York. Art Hesskamp's Plymouth Belvedere continued to lead the intermediate-size eight-cylinder class with Bob Cahill's Dodge Coronet in second. The margin between these two tightened up to a mere .0141 mile per gallon. Bob Checkley moved his Plymouth Fury II back into first in the full-size six-cylinder class by another tight margin of .0456 mile per gallon over Pat Sawyer and her Chevrolet Biscayne. With Scott Harvey's Plymouth Fury II continuing to lead in the low-price eight-cylinder class and George Miller's Plymouth Barracuda in the special sports class, Plymouth was shown as a leader in four classes. This was the first time any individual make "had placed so many of their cars at the top of their classes. Although there are four days of fierce competition remaining, the Plymouths will be fighting to hold their leads."[91] With the exception of the low-price eight-cylinder and medium-price classes, the remaining five class-leading drivers maintained fairly comfortable margins over their second-place rivals. The only other change of note was in the intermediate-size six-cylinder class, where Cass Szpieg moved his Dodge Coronet into second, dropping Mary Faulkner's Mercury Comet 202 to third. The entire field saw their average improve to 21.0381 miles per gallon, with the teenagers also improving their average to 22.1351. The special sports models also showed an improvement, averaging 20.9544 miles per gallon despite a significant drop in one of the competing cars, with Kirk Kirkpatrick's Chevrolet Chevelle Malibu SS mileage dropping to 17.0718. There was no reason given for this decline of a mile and a half per gallon from the previous day's results while all other drivers showed improved figures. Today's trip continued to show the fast pace of the run with the field averaging 50.9650 miles per hour over the 607-mile route.

The top three performers in each class were:

Small Engine Compacts	Ford Falcon (170 Cu. In.)	26.8100 Miles Per Gallon
	Plymouth Valiant 100	24.9285 Miles Per Gallon
	Dodge Dart	24.2369 Miles Per Gallon
Large Engine Compacts	Rambler American 440	26.8306 Miles Per Gallon
	Ford Falcon Futura (200)	25.9086 Miles Per Gallon
	Chevrolet Chevy II Nova	23.3726 Miles Per Gallon
Intermediate Size Six Cylinder	Ford Fairlane	25.7477 Miles Per Gallon
	Dodge Coronet	24.4382 Miles Per Gallon
	Mercury Comet 202	24.3796 Miles Per Gallon
Intermediate Size Eight Cylinder	Plymouth Belvedere II	21.5269 Miles Per Gallon
	Dodge Coronet	21.5128 Miles Per Gallon
	Buick Special Deluxe	20.9754 Miles Per Gallon
Full Size Six Cylinder	Plymouth Fury II	22.0662 Miles Per Gallon
	Chevrolet Biscayne	22.0206 Miles Per Gallon
	Ford Custom	21.8745Miles Per Gallon
Low Price Eight Cylinder	Plymouth Fury II	20.9592 Miles Per Gallon
	Ford Galaxie 500	20.3206 Miles Per Gallon
	Chevrolet Impala	19.2769 Miles Per Gallon
Medium Price	Pontiac Catalina	20.3718 Miles Per Gallon
	Buick LeSabre	20.2374 Miles Per Gallon
	Pontiac Star Chief	20.2005 Miles Per Gallon
Luxury Cars	Chrysler New Yorker	17.6671 Miles Per Gallon
	Buick Electra 225	17.0291 Miles Per Gallon
	Cadillac Sedan de Ville	16.8662 Miles Per Gallon
Special Sports	Plymouth Barracuda	22.3670 Miles Per Gallon
	Ford Mustang 2+2	21.6473 Miles Per Gallon
	Dodge Dart GT	21.4137 Miles Per Gallon

On April 8, another challenging day, over mostly level terrain, awaited the drivers as they prepared for the long 543.3-mile drive to Chicago. Today would be one of the fastest trips on the Run, with a time limit set at 10 hours and 10 minutes. Everyone had a chance to sleep in, since Bob Cahill in a Dodge Coronet 8 was the first driver on his way at 6:00 a.m. This was a welcome break from the previous two days' 4:00 a.m. starts, although the drivers would have to contend again with strong crosswinds of between 35 and 50 miles per hour for most of the day. After leaving the impound they immediately crossed into Missouri, following Interstate 70 / U.S. Route 40 across the state towards Wright City, where all entries were topped off at Brookfield Mobil Service. At this point, Paula Murphy experienced a recurrence of back trouble, attributed to a slipped disk, and was forced to relinquish her driving duties. With her co-driver, Jan Detrick, taking over as driver, Paula was transported at St. Mary's Hospital in East St. Louis, where she was placed in traction.[92] She remained hospitalized for three days before returning home to Los Angeles. Dan Scoville handled the co-driver duties for Jan for the rest of today's trip to Chicago. After crossing the Mississippi River into Illinois, the drivers headed north on U.S. Route 66 to Edwardsville for brunch at Schneithorst's Restaurant. In many of the communities in Missouri and Illinois, the schools would dismiss their students so they could join local residents in welcoming the drivers as they drove through area. Although their day was half over, additional challenges arose as they continued on U.S. Route 66 towards Joliet when weather conditions began to deteriorate. As described by some of the drivers, "At one time the rain on the windshield was so heavy that it was standing at least a half inch think at the bottom of the windshield and the wipers wouldn't take it off."[93] Arriving in Joliet in heavy rain, the drivers made their way through local streets to Joliet Mobil Service for refueling. From here it was a 50-mile trek on Interstate 55 to Chicago, where they picked up the Tri-State Tollway. With heavy rain continuing, along with thunderstorms and tornado warnings in the area, the drivers safely reached the impound at the O'Hare Inn.[94] Bundled up for protection from the rain and cold, J.C. Agajanian was on hand as always to greet the first to arrive, "Kirk" Kirkpatrick in his Chevrolet Chevelle Malibu SS, pulling in at 4:45 p.m. Five drivers including Jean Calvin (Pontiac Catalina), Ed Dempsey (Oldsmobile F-85 6), Fran Foster (Ford Galaxie 500 8), Gordon Madison (Chevrolet Nova 6) and Ole Olsen (Imperial Crown), arrived up to 17 minutes late and were penalized. Three of these, Ed Dempsey, Fran Foster and Ole Olsen, drove off-course at some point during the day. In Ole's case, he made a wrong turn coming into Chicago, which took him off-course for several miles. As explained by his wife and co-driver Helene, "We knew we had made the wrong turn as soon as it happened, but it still took 17 minutes to work our way back to the right on-ramp."[95] The extra miles he drove would prove problematic on tomorrow's leg.

After arriving at Chicago's O'Hare Inn, Don Scoville relinquished co-driver duties for Jan Detrick and her Buick Special Deluxe 6. Replacing Don as co-driver was George Pollin.[96]

That evening the drivers learned the details of the last two days' travels. In an unprecedented move, A.C. Pillsbury released the remaining route summary and detailed itinerary for the legs from Niagara Falls to Albany and from Albany to New York. In explaining this move, he stated: "Some material regarding the balance of the run from Chicago to New York was removed from one of the cars. This contained information normally released to the drivers the night before the next day's leg. On the possibility that certain competitors may have learned this data, we are released the same information to all competitors." He further explained "that the move was taken to protect the integrity of the run and to be fair to all entrants."[97]

Despite the speed and weather conditions encountered on this leg, most drivers were able to improve their mileage performance. There was one major exception. In the battle for achieving the best miles per gallon for all entries, Byron Froelich was able to move his Ford Falcon past Les Viland and his Rambler American 440 with a margin of .1291 miles per gallon. Byron was able to improve his average to 26.9498 miles per gallon over this leg, while Les barely stayed even with his previous results at 26.8207. For the past four years there has been a rivalry between American Motors and Ford for overall mileage honors, and this year was no exception. "A Ford spokesman had predicted that if they could stay close to Rambler through the mountainous west, they would pull away and win after the roads flattened out from Kansas City east. With one long day and two short ones ahead, this is the battle that all the experts will be watching."[98] At this point, should no major problems occur, the leaders in seven of the nine classes were assured of receiving first-place trophies. In addition to Byron Froelich and Les Viland in the two compact car classes, these were Art Bertelsen (Chrysler New Yorker), Darrell Droke (Ford Fairlane 6), Scott Harvey (Plymouth Fury II 8), and George Miller (Plymouth Barracuda). In what was becoming a seesaw battle for positions, several changed hands as competition continued to remain close. In one that had remained tight since the second day, Bob Cahill moved his Dodge Coronet back into first place in the intermediate-size eight-cylinder class over Art Hesskamp's Plymouth Belvedere II by an ever-so-slim margin of .0034 mile per gallon. In the full-size six-cylinder class, Pat Sawyer in her Chevrolet Biscayne did likewise, moving back into first by only .0513 mile per gallon over Bob Checkley's Plymouth Fury II. With these two changes, Plymouth was also knocked out of Economy Run history that had been made the day before. Other changes of note involved second- and third-place positions, with Art Arfons moving his Pontiac

With the Oneida Tower in the background, Bill Levy's Ford Mustang 2+2 crosses the Rainbow Bridge at Niagara Falls heading for Albany, New York. This year marked the first and only time the Economy Run would travel to Canada when it made an overnight stop in this popular area. Competing in the special sports class, Bill's Ford Mustang would finish second behind George Miller's Plymouth Barracuda.

Star Chief into second in the medium-price class by .0223 mile per gallon over Marta Retzlaff's Buick LeSabre. In the luxury car class, Stan Seeds and his Cadillac Sedan de Ville was able to overtake John Meyers's Buick Electra 225 by a slim margin of .0165 mile per gallon. Cadillac's performance surprised most observers with the lack of factory interest and support while competing against the factory-backed Buick and Chrysler teams. Today's average mileage again showed a slight improvement to 21.1161 miles per gallon. The three teenagers continued to show strong performance with an average of 22.1494 miles per gallon, while the special sports models came in with 21.0037. If yesterday's speed was fast, today's was even faster, with the field averaging 53.6478 miles per hour despite the windy and rainy conditions. Even faster speeds were on tap.

The top three performers in each class were:

Small Engine Compacts	Ford Falcon (170 Cu. In.)	26.9498 Miles Per Gallon
	Plymouth Valiant 100	25.1019 Miles Per Gallon
	Dodge Dart	24.3526 Miles Per Gallon
Large Engine Compacts	Rambler American 440	26.8207 Miles Per Gallon
	Ford Falcon Futura (200)	25.8961 Miles Per Gallon
	Chevrolet Chevy II Nova	23.6279 Miles Per Gallon
Intermediate Size Six Cylinder	Ford Fairlane	25.9188 Miles Per Gallon
	Dodge Coronet	24.5529 Miles Per Gallon
	Mercury Comet 202	24.4271 Miles Per Gallon
Intermediate Size Eight Cylinder	Dodge Coronet	21.5782 Miles Per Gallon
	Plymouth Belvedere II	21.5748 Miles Per Gallon
	Buick Special Deluxe	21.0461 Miles Per Gallon
Full Size Six Cylinder	Chevrolet Biscayne	22.1977 Miles Per Gallon
	Plymouth Fury II	22.1464 Miles Per Gallon
	Ford Custom	21.6900 Miles Per Gallon
Low Price Eight Cylinder	Plymouth Fury II	20.8891 Miles Per Gallon
	Ford Galaxie 500	20.2580 Miles Per Gallon
	Chevrolet Impala	19.3872 Miles Per Gallon
Medium Price	Pontiac Catalina	20.5881 Miles Per Gallon
	Pontiac Star Chief	20.3711 Miles Per Gallon
	Buick LeSabre	20.3548 Miles Per Gallon
Luxury Cars	Chrysler New Yorker	17.8163 Miles Per Gallon
	Cadillac Sedan de Ville	17.0165 Miles Per Gallon
	Buick Electra 225	17.0000 Miles Per Gallon
Special Sports	Plymouth Barracuda	22.3611 Miles Per Gallon
	Ford Mustang 2+2	21.7854 Miles Per Gallon
	Dodge Dart GT	21.4474 Miles Per Gallon

One more long day faced the drivers on April 9 as they prepared to leave Chicago for the 590-mile, 11-hour and 27-minute trip to Niagara Falls. Ed Dempsey in an Oldsmobile F-85 6 was the first driver to leave at 4:00 a.m., with heavy rain continuing, heading south towards Interstate 90 and through the heart of Chicago. With this early start the drivers were able to miss Chicago's morning commute traffic, and by sunrise they had crossed into Indiana. Now on the Indiana Toll Road (Interstates 80 and 90), they entered the Eastern Time Zone prior to reaching South Bend. After traveling nearly 160 miles across Indiana, the drivers crossed into Ohio; with the weather clearing, they picked up the Ohio Turnpike (Interstate 80 and 90) towards Toledo and their first stop of the day at Coy's Mobil Service. Here all entries were topped off. Fourteen miles short of the station, Ole Olsen experienced "the worst thing that can happen to a Mobil Economy Run driver": his Imperial Crown's engine sputtered and then he rolled to a stop, out of gas.[99] After receiving an emergency supply of fuel from the Automobile Club of Southern California's service patrol vehicle, he continued to the top-off station, where he was informed by USAC officials of his disqualification. In an explanation of this, a USAC official said, "A car must reach the designated refueling station under its own power."[100] Up to this point, he had been running last in the luxury car class while also recording the lowest miles per gallon average of all competitors. This marked the fourth time in the Run's history a competitor ran out of gas. In 1953, this also resulted from a wrong turn made by Pete Novotny in a Chevrolet Two-Ten. Pete got back on the right road, although he was unable to reach the Boise impound before running out of gas.

Leaving Toledo, they continued along the south shore of Lake Erie at a somewhat slower pace, while at the same time contending with numerous traffic signals, reaching Lorain for lunch at the Castle Restaurant. After a break from being on the road for six and a half hours, the drivers then headed on Ohio Route 2 towards Cleveland, where they rejoined Interstate 90 headed for the Austinburg Mobil Service Center for the second refueling stop of the day. In an unusual occurrence, Don Royer's co-driver, Jack Sparks, took over driving duties for the 82 miles to the next refueling station. Running third (and last) in their class, they made the change to give Jack an opportunity to actually drive in the Economy Run.[101] Continuing on Interstate 90, the drivers passed through Pennsylvania to New York, where they picked up the New York Thruway towards Buffalo and Niagara Falls. After crossing the Peace Bridge, the participants were cleared by Canadian customs in less than one hour, making this the first truly international Mobil Economy Run, then continued to the impound at the Fallsway Motor Hotel. Shortly before 5:30 p.m., Scott Harvey, driving a Plymouth Fury II 8, was the first of what was now 48 competitors to arrive. No doubt from the strain of five long, intense days on the road, ten drivers arrived late and were penalized. In addition to class-leading drivers Jean Calvin (Pontiac Catalina) and Darrell Droke (Ford Fairlane 6), this group included Dave Apple (Ford Falcon Futura 6), Don Bridges (Rambler Ambassador 880 8), Ed

With Niagara Falls in the background, Patricia Sawyer drives her Chevrolet Biscayne over the Rainbow Bridge back into the United States and heads for Albany. Placing first in the full-size six-cylinder class, Pat would end Plymouth's five-year winning streak in this class.

Dempsey (Oldsmobile F-85 6), Mary Faulkner (Mercury Comet 202 6), Fran Foster (Ford Galaxie 500 8), Hart Fullerton (Chrysler Newport), Marta Retzlaff (Buick LeSabre), and Ginny Sims (Ford Custom 6). Hart Fullerton missed an exit on the New York Thruway and drove off-course for 10 minutes, which resulted in his arriving 6 minutes and 50 seconds late. Tired as they were, most of the participants were eager to make the two-minute walk to see "the most beautiful sight ever seen on any Mobil Economy Run"—Niagara Falls.[102]

Speed and weather conditions resulted in a decline in everyone's mileage performance by more than a third. At this point two things were clearly evident. The mileage figures being recorded were becoming more realistic, and the close competition contributed to increased interest by the both press and public. The battle for the best miles per gallon for all entries continued as Les Viland moved his Rambler American 440 past Byron Froelich's Ford Falcon by an ever-so-slim margin of .0281 mile per gallon. Both drivers were the only ones with an average exceeding 26 miles per gallon, with Les recording a figure of 26.2676, compared to Byron's figure of 26.2395. With competition still very close, there were no changes among the nine class leaders. In the intermediate-size eight-cylinder class, Bob Cahill's Dodge Coronet stretched its lead over Art Hesskamp's Plymouth Belvedere II to a margin of .1125 mile per gallon. In the full-size six-cylinder class, Pat Sawyer's Chevrolet Biscayne saw her lead shrink to a slim .0440 mile per gallon over Bob Checkley's Plymouth Fury II. In the medium-price class, second-place Art Arfons's Pontiac Star Chief continued to close in on Jean Calvin's Pontiac Catalina, closing the gap to .1804 mile per gallon. There were no changes between second- and third-place positions. Today's average mileage declined to 20.7779 miles per gallon, with the three teenagers now averaging of 21.6585. The miles per gallon average for the closely watched special sports models was 20.5217. Although the vast majority of today's driving was on Interstate highways, today's average speed declined to 51.7844 miles per hour.

The top three performers in each class were:

Small Engine Compacts	Ford Falcon (170 Cu. In.)	26.2395 Miles Per Gallon
	Plymouth Valiant 100	24.5421 Miles Per Gallon
	Dodge Dart	23.9159 Miles Per Gallon
Large Engine Compacts	Rambler American 440	26.2676 Miles Per Gallon
	Ford Falcon Futura (200)	25.2810 Miles Per Gallon
	Chevrolet Chevy II Nova	23.0833 Miles Per Gallon
Intermediate Size Six Cylinder	Ford Fairlane	25.4451 Miles Per Gallon
	Dodge Coronet	23.8809 Miles Per Gallon
	Mercury Comet 202	23.8556 Miles Per Gallon
Intermediate Size Eight Cylinder	Dodge Coronet	21.1825 Miles Per Gallon
	Plymouth Belvedere II	21.0700 Miles Per Gallon
	Buick Special Deluxe	20.6210 Miles Per Gallon
Full Size Six Cylinder	Chevrolet Biscayne	21.6655 Miles Per Gallon
	Plymouth Fury II	21.6215 Miles Per Gallon
	Ford Custom	21.2847 Miles Per Gallon
Low Price Eight Cylinder	Plymouth Fury II	20.3569 Miles Per Gallon
	Ford Galaxie 500	19.7534 Miles Per Gallon
	Chevrolet Impala	19.0631 Miles Per Gallon
Medium Price	Pontiac Catalina	20.2115 Miles Per Gallon
	Pontiac Star Chief	20.0331 Miles Per Gallon
	Buick LeSabre	19.9749 Miles Per Gallon
Luxury Cars	Chrysler New Yorker	17.4793 Miles Per Gallon
	Cadillac Sedan de Ville	16.7065 Miles Per Gallon
	Buick Electra 225	16.6593 Miles Per Gallon
Special Sports	Plymouth Barracuda	21.8345 Miles Per Gallon
	Ford Mustang 2+2	21.1875 Miles Per Gallon
	Dodge Dart GT	20.9916 Miles Per Gallon

On April 10, the competitors were treated to a late morning start for a "short" 302.4-mile trip across New York State to Albany. With ideal weather conditions, the drivers were given a time limit of 5 hours and 33 minutes, although another challenge would be presented to the drivers. After a leisurely morning to "rest, relax, and enjoy the beauty" of Niagara Falls, Don Royer, driving a Chevrolet Impala 8, was the first to leave at 10:00 a.m.[103] After crossing back into the United States, the drivers made their first stop in Buffalo at Daniel Knoll Mobil Service, where all entries were topped off, then entered the New York Thruway bound for Syracuse and brunch at the Trivet House Restaurant. From Syracuse it was on to Herkimer for refueling at Fran Bruckner Mobil Service. At this point the drivers returned to the New York Thruway and soon began a timed control section, where drivers "must average a minimum of 60 miles per hour between [Gate #30] and the toll booth at the Albany Exit #24." With a distance of 72.4 miles, "the maximum net running time is *1 hour, 12 minutes, 23 seconds*. An official USAC Timing Station will be set up at the start and finish of the timed control section. Penalties will be applied if you do not comply with the above."[104] Three drivers, Don Francisco (Pontiac Tempest 6), "Kirk" Kirkpatrick (Chevrolet Chevelle Malibu SS) and George Miller (Plymouth Barracuda), ran over on this timed section from as little as 15 seconds to a full minute and were penalized. After completing this timed control section in Albany, it was a short 2-mile drive to the impound at the Thruway Motor Inn, where another large group of spectators, joined by a local high school band, greeted the participants. At 4:20 p.m., the first to arrive was Bob Cahill in his Dodge Coronet 8. One driver, George Alsbury in an Oldsmobile Dynamic 88, arrived 2 minutes 14 seconds late and was penalized.

Looking back on the Economy Run's visit to Canada, USAC and Mobil officials could not be more pleased on how smoothly everything went. As a testament to USAC's prior planning and attention to details, not only were the 48 competing cars cleared quickly, but so was "an entourage of some 75 other vehicles carrying nearly 300 USAC officials, newsmen and technical personnel necessary to the Run." A.C. Pillsbury commented, "The Canadian welcome we received was one of the most heartwarming ever. Customs, the motels that housed us, and the restaurants where our group ate and the citizens themselves — couldn't have been more gracious. We're going back again."[105] Despite his positive statement, a return to Canada would not be in the cards.

This leg would be the fastest to date in the history of the Economy Run with an average speed of 55.0340 miles per hour, thanks in part to the timed control section, which was "designed to make run mileages a realistic par for the average motorist."[106] Naturally, with this speed, the miles per gallon average would drop for a second day in a row. Les Viland and his Rambler American 440 continued to lead all entries with the best miles-per-gallon average with 25.9593. Following close on his heels by a slim .0207 mile per gallon was Byron Froelich and his Ford Falcon with a figure of 25.9386. All of yesterday's leaders maintained their positions. In the classes where competition remained very close, the margins between first and second were slightly increased. Barring any unforeseen incidents on the final leg, it appeared that in three classes first place would not be decided until the final top-off. The closest competition continued to be in the full-size six-cylinder class, where Pat Sawyer's Chevrolet Biscayne's lead stretched to a still-slim .0718 mile per gallon over Bob Checkley's Plymouth Fury II. In the luxury car class, a battle continued between second- and third-place drivers, with John Meyers's Buick Electra 225 closing to within a mere .0089 mile per gallon of Stan Seeds's Cadillac Sedan de Ville. Today's average mileage declined once again to 20.5563 miles per gallon. The teenagers continue to display their overall strong performance compared to the adults with an average of 21.4492 miles per gallon. The five special sports models were averaging 20.3146 miles per gallon.

The top three performers in each class were:

Small Engine Compacts	Ford Falcon (170 Cu. In.)	25.9386 Miles Per Gallon
	Plymouth Valiant 100	24.3452 Miles Per Gallon
	Dodge Dart	23.7508 Miles Per Gallon
Large Engine Compacts	Rambler American 440	25.9593 Miles Per Gallon
	Ford Falcon Futura (200)	24.7815 Miles Per Gallon
	Chevrolet Chevy II Nova	22.8006 Miles Per Gallon
Intermediate Size Six Cylinder	Ford Fairlane	25.0141 Miles Per Gallon
	Dodge Coronet	23.6432 Miles Per Gallon
	Mercury Comet 202	23.5125 Miles Per Gallon
Intermediate Size Eight Cylinder	Dodge Coronet	20.9839 Miles Per Gallon
	Plymouth Belvedere II	20.8379 Miles Per Gallon
	Buick Special Deluxe	20.4292 Miles Per Gallon
Full Size Six Cylinder	Chevrolet Biscayne	21.4522 Miles Per Gallon
	Plymouth Fury II	21.3804 Miles Per Gallon
	Ford Custom	21.0292 Miles Per Gallon
Low Price Eight Cylinder	Plymouth Fury II	20.0901 Miles Per Gallon
	Ford Galaxie 500	19.5304 Miles Per Gallon
	Chevrolet Impala	18.8788 Miles Per Gallon
Medium Price	Pontiac Catalina	20.0004 Miles Per Gallon
	Pontiac Star Chief	19.8056 Miles Per Gallon
	Buick LeSabre	19.7668 Miles Per Gallon
Luxury Cars	Chrysler New Yorker	17.3570 Miles Per Gallon
	Cadillac Sedan de Ville	16.5241 Miles Per Gallon
	Buick Electra 225	16.5152 Miles Per Gallon
Special Sports	Plymouth Barracuda	21.6525 Miles Per Gallon
	Ford Mustang 2+2	20.9625 Miles Per Gallon
	Dodge Dart GT	20.7805 Miles Per Gallon

Placing first in the intermediate-size eight-cylinder class by a slim margin, Bob Cahill heads his Dodge Coronet towards New York's Times Square and the finish of the 1965 Economy Run. This year marked Bob's fourth consecutive class victory, two driving a Dodge and two a Plymouth.

The finish line in Times Square was in sight, and once again the detailed planning and advance preparations of USAC officials would become clearly evident. On April 11, two days after the longest leg in the history of the Run, this would be the shortest in history with a 149.3-mile trip to New York City. Marta Retzlaff in her Buick LeSabre was the first driver on the way at 7:45 a.m. for a quick 2-hour and 37-minute trip on the New York Thruway. After crossing the Tappan Zee Bridge, at 15 feet the lowest elevation on this year's Run, their first stop of the day was at the Ardsley Service Area's Mobil Service Center, where all entries were topped off for the final time. Just after 10:15 a.m., Ed Dempsey's Oldsmobile F-85 6 was the first to arrive. Only one driver, Al Cottle, in a Mercury Monterey, arrived late, by just over a minute and half. From here, the drivers continued through Harlem and into Manhattan, where they navigated city streets to a parking area on West 54th Street to regroup. After a brief pause here, the special activities and ceremonies would commence. With the Economy Run finishing in Times Square, USAC officials worked with city officials to temporarily block Times Square to traffic. Scheduling the finish on a Sunday would result in minimum disruption for the short period of time it took for the cars to cross under the finish line banner. In honor of this, New York City Mayor Robert F. Wagner, Jr. renamed Times Square as Economy Run Square for the day.[107] With a police escort leading the competitors in parade fashion down Seventh Avenue, the largest turnout of spectators in the history of the Run lined the sidewalks of Times Square to greet the participants. Led by Class "A" winner Byron Froelich in a Ford Falcon, they were given an enthusiastic welcome, led by J.C. Agajanian waving the checkered flag. After making an illegal street maneuver onto 44th Street, the drivers headed for the Mobil Oil Building located on East 42nd Street. With the class-winning cars on display in front of the building, Socony-Mobil Oil President Fred H. Moore presented them with their trophies. What many regarded as one of the most successful Economy Runs had come to a close.

In the end, Les Viland held off the challenge from Byron Froelich, bringing his Rambler American 440 across the finish line with the highest mileage for all entries at 25.6524 miles per gallon. In what had been a close contest from the start, the margin between the American and Falcon was a slim .0360 miles per gallon as both drivers captured their respective compact car classes. Noting that close to a pint of gasoline separated these two entries, Les Viland remarked, "There were very few opportunities to 'featherfoot' this time. All I tried to do was hold a steady foot."[108] Another driver with a winning legacy crossed the finish line with the lowest mileage for all entries. Driving an Oldsmobile Ninety-Eight this year, Mel Alsbury, Jr. could only manage a miles per gallon average of 15.7911. Prior to his running out of gas on the fifth day, Ole Olsen's Imperial Crown held this distinction. Among the three teenage drivers, the best mileage figure was Marty Payne in a Dodge Dart 6 with 23.5016 miles per gallon. All manufacturers shared space this year in the winner's circle, with Chrysler Corporation placing first in four classes, followed by Ford Motor Company and General Motors each first in two classes and American Motors first in one. Five classes were decided by less than a mile per gallon.[109]

Class winners were:

CLASS	MAKE and MODEL	MILES PER GALLON
A — Small Engine Compacts	Ford Falcon (170 Cu. In.)	25.6164
B–Large Engine Compacts	Rambler American 440	25.6524
C — Intermediate Size Six Cylinder	Ford Fairlane	24.7134
D — Intermediate Size Eight Cylinder	Dodge Coronet	20.7682
E — Full Size Six Cylinder	Chevrolet Biscayne	21.1748
F — Low Price Eight Cylinder	Plymouth Fury II	19.8292
G — Medium Price	Pontiac Catalina	19.8423
H — Luxury Cars	Chrysler New Yorker	17.2609
S — Special Sports	Plymouth Barracuda	21.4231

Detailed results for all competing cars can be found in the Appendix.

With this year's faster pace, USAC officials reported that the entire field averaged 20.3472 miles per gallon. This showed a decline of 1.4261 miles per gallon over 1964's average of 21.7733 and reflected efforts of both USAC and Mobil Oil to make these mileages more realistic to the average motorist. The other two averages closely watched included the three teenagers, who once again outperformed the adults, with 21.2101 miles per gallon, and the special sports models with 20.1140. This year's route was 3266.1 miles in length, only 22 miles longer than last year's, but still the longest distance in the Run's history. The drivers faced a shorter driving time limit of 63 hours and 29 minutes. This resulted in the fastest pace in the history of the Run, with drivers averaging 51.6830 miles per hour while contending with virtually every type of weather conditions along the route. This pace was 6.4556 miles per hour faster than the previous record set in 1958. This year saw several developments, including one entry failing to finish after running out of gas along, and a driver change due to hospitalization. One entry required minor repairs and no tire problems or accidents were reported.

Four drivers in the winner's circle were first-timers, with Art Bertelsen (Chrysler New Yorker), Jean Calvin (Pontiac Catalina), Darrell Droke (Ford Fairlane 6) and George Miller (Plymouth Barracuda) joining veteran champions. Both Bob Cahill and Les Viland scored their fourth consecutive wins, with Scott Harvey scoring his third. In addition to his consecutive win streak, this also marked Les Viland's eighth first-place finish. Although not consecutive, Byron Froelich and Pat Sawyer also scored their third class victories.

From the start, it was certain that both Byron Froelich's Ford Falcon in the class for small-engine compacts and Les Viland's Rambler American 440 for the large-engine compacts would each place first by comfortable margins. As detailed earlier, the real contest here would be which of the two would achieve the best mileage for all entries, with the American again coming out on top. Many of those following this year's Run were privately rooting for the American. With their sales in a decline over the past two years, a strong finish "might considerably enhance the Rambler image as the industry's most economical car. A loss could mar the Rambler image."[110] Among the four small-engine compacts competing, teenager Marty Payne placed his Dodge Dart 6 in third, finishing 2.1148 miles per gallon behind the winning Falcon. A disappointing performance was that of the Chevrolet Corvair Monza. A previous three-time class winner (1961 to 1963), the Corvair struggled with early transmission problems and ended up recording 20.2960 miles per gallon, the lowest among the six-cylinder compacts.

Competing against eight other intermediate-size six-cylinder models, Darrell Droke's Ford Fairlane maintained a strong lead from the start, finishing with a comfortable 1.3150 miles-per-gallon margin over Cass Szpieg's Dodge Coronet. Chrysler Corporation and Ford Motor Company products dominated the first four places. General Motors' best finish was fifth. The Buick Special Deluxe, last year's winner in this class, was well down in the standings from the start, and its performance was further hampered by Paula Murphy's relinquishing driving duties in Kansas.

The intermediate-size eight-cylinder class was one of the two classes where competition was extremely close from the start. Bob Cahill managed to keep his Dodge Coronet in first despite a continuing challenge from Art Hesskamp in a Plymouth Belvedere II. With a very slim .0034 mile-per-gallon margin between the two at one point, in the end Bob was able to stretch this slim margin to .1580. Last year's winner, John Rich in a Buick Special Deluxe, placed third, although only .4611 mile per gallon behind the leader. John was faced with stronger competition this year from the intermediate-sized models from Dodge and Plymouth which were considered as full-size models for the past two years.

In a traditional matchup of the three popular full-size, low-price models from Chevrolet, Ford and Plymouth, one contest was closely watched, while the other was pretty much settled from the start. Among the six-cylinder models, Pat Sawyer's Chevrolet Biscayne was able to maintain its lead over Bob Checkley's Plymouth Fury II by finishing with a slim margin of .0991 mile per gallon. In this class, only .3989 mile per gallon separated the three entries, with Ginny Sims's Ford Custom in third. For the eight-cylinder models, it was Scott Harvey's Plymouth Fury II in front all the way, finishing with a comfortable margin of .5452 mile per gallon over Fran Foster's Ford Galaxie 500. This win by the Plymouth Fury II marked its ninth consecutive class victory. Don Royer's Chevrolet Impala placed third, 1.2147 miles per gallon off the pace. Don and Chevrolet's performance would show a marked improvement beginning next year.

In the medium-price class, General Motors continued to show strong performance with a first-place finish for the fourth year in a row. In her first time behind the wheel, Jean Calvin kept her Pontiac Catalina in front from the start. It what ultimately turned out to be a close contest, Art Arfons in a Pontiac Star Chief placed a strong second, only .1949 mile per gallon behind the Catalina. Commenting on his first time competing in the Run, Art commented that it was "tougher than Bonneville" and that he "likes every grueling minute of it."[111] Last year's winner, Marta Rezlaff in a Buick Le Sabre, was a close third and missed out on second place by a mere .0154 mile per gallon.

A Chrysler New Yorker scored its fourth consecutive victory in the luxury car class, this time without a member of the Alsbury family behind the wheel. Driven by Chrysler Corporation special car programs engineer Art Bertelsen, the New Yorker maintained a comfortable lead from the start. This victory also marked the eighth year a Chrysler had been in the winner's circle. A battle for second place continued throughout the Run, with Stan Seeds's Cadillac Sedan de Ville prevailing over John Meyer's Buick Electra 225 by a small margin of .0607 mile per gallon. Disappointment is the only word Mel Alsbury, Jr. would use to describe this year's performance. After ten years driving Chryslers for his father's dealership, his first year behind the wheel in an Oldsmobile Ninety-Eight resulted in a fourth (and last) place finish, 1.4698 miles per gallon off the pace.

In the class that was perhaps the one watched the closest, George Miller in a Plymouth Barracuda maintained his lead over Bill Levy in a Ford Mustang 2+2, although by a small margin. Both of the new

In front of the largest crowd of spectators in the history of the Economy Runs, small-engine compact-cars class winner Byron Froelich's Ford Falcon 6 leads the convoy of cars across the finish line in New York's Times Square. Right behind is the large-engine compact cars class winner Les Viland's Rambler American 440. Both drivers had a friendly battle from the start for the top mileage honors, with Les coming out on top by a slim .0360 mile per gallon.

"pony cars" did outperform the other competitors, although the battle between second and third place was close for most of the Run. Despite being stuck in the mud on the first day, teenager Dareth Rich turned in a strong performance with her third-place finish. It was close, as only .2065 mile per gallon separated second and third. Mercury's Comet Caliente Sport and Chevrolet's Chevelle Malibu SS both finished well off the pace, over 2 miles per gallon behind the Barracuda.

There were many comments during this year's Run about the higher average speed, which in turn resulted in lower, more realistic mileage figures compared to previous events. In a comparison of mileage results for selected models for 1964 and 1965, these lower averages (with one exception) become quite clear:

Make	1964 Miles Per Gallon	1965 Miles Per Gallon
Buick Special Deluxe 8	23.7441	20.3071
Ford Falcon 6 with 170 Cu. In. Engine	23.8341	25.6164
Oldsmobile F-85 6	23.2406	22.0422
Oldsmobile F-85 Deluxe 8	22.4083	19.8068
Plymouth Valiant 100 6	25.8921	24.0943
Plymouth Savoy / Belvedere I 6	25.2955	23.1156
Plymouth Savoy / Belvedere II 8	22.1963	20.6102
Rambler American 440 6	27.8336	25.6524

All of these models had the same basic body design and engine size in 1964 and 1965 as well as drivers of comparable experience and background. As pointed out earlier, the 1964 Plymouth Savoy and 1965 Belvedere models were identical in size. Engineers from Chrysler Corporation and American Motors prepared and drove the Plymouth and Rambler entries, while the Buick Special Deluxe 8 and Oldsmobile F-85 Deluxe 8 were driven by the same individuals. Different drivers piloted the Oldsmobile F-85 6, both having limited Economy Run experience. Only one model showed an increase its mileage performance, the Ford Falcon 6, equipped with the 170-cubic-inch engine. The main factor in 1965's improved mileage figure was due mainly to the involvement of engineers and support from Ford in 1965. Local personnel were responsible for preparation and driving duties in 1964. Delivering more realistic mileage figures by Economy Run drivers was on the horizon.

Although they were only three in number, the teenagers again displayed their enthusiasm and skill. Both Marty Payne and Dareth Rich, both driving Dodge Darts, took home third-place trophies, with Ned Aberly placing fifth in his Mercury Comet 404 8. The other teenager who was participating as a co-driver, Jean Fitzgerald, placed seventh in a Chevrolet Chevelle Malibu 6. None from this group would return in future years.

One driver with a long history of participation would hang up his driving gloves this year. Al Cottle, who first competed as a driver in 1951, had scored first-place victories on two occasions. His first victory was in 1951 with a Ford Deluxe 8 after serving as a co-driver in 1950. This was followed by his second class win in 1955 with a Buick Roadmaster. This win was considered by many observers to be a surprise, accomplished without any factory support in Buick's first participation in the Economy Run. Since 1962, Al had piloted a Mercury Monterey for longtime participant Bob Estes Lincoln-Mercury.

Even with each manufacturer recording class victories, advertising and promotion continued to show a decline from previous years. Mobil Oil continued its advertising efforts to publicize the results in both newspapers and industry related publications. Two different styles were utilized. Immediately following the finish, simple text-only advertising headlined "Mobil Economy Run Official Results" showed the results for each entry. Asking the question, "What does this mean to you?," this advertisement went on state, "Simply this: while other gasolines talk about good mileage, Mobil High Energy Gasoline proves its mileage potential in public."[112] Later a second advertisement appeared, featuring a photograph of the cars crossing the Times Square finish line. With the headline "America's toughest gasoline mileage test ends in Times Square," the text generally followed the same theme of proving mileage potential in public.[113] Both of these introduced a new tagline, "The gasoline that proves its performance in public," taking direct aim at those promotions by Pure Oil and Shell Oil Companies. Continuing to promote these results, Mobil produced a television commercial highlighting this year's Run, appearing on such shows as CBS' *The Jackie Gleason Show* and ABC's *Wide World of Sports*. Mobil service stations continued to offer an 8-page folder detailing this year's results. Mobil's name would continue to be in the forefront in advertising from the automotive manufacturers.

American Motors advertised the American's results by combining this feat with the introduction of the new fastback Marlin. "Shucks!" was the headline on an ad which went on to say, "Here we were set to talk all about the snappy new things Rambler had come up with," followed by, "Rambler American 440 wins Class B with 25.65 m.p.g."[114] As AMC was attempting to change its image with redesigned 1965 Rambler Classic and Ambassador models, very little mention of the American's Economy Run performance was made in advertising during the rest of the year.

The emphasis this year by Chrysler, Ford and General Motors was on performance and not economy. Other than a small folder titled "Here's Proof of Plymouth's Outstanding Gas Economy," which outlined Plymouth's Economy Run achievements, this year's results were virtually ignored by the Big Three.[115]

Due to a combination of factors, including the introduction of sports models coupled with several close contests, articles this year showed a marked increase. As before, newspapers remained the main source of coverage, with daily stories appearing in cities across the country. One embarrassing event that made the headlines was "Ole" Olsen's running out of gas in Ohio. More detailed than in previous years, a few of these stories focused on the real-world cost facing motorists in cross-country trips. As Bob Thomas, auto editor of the *Los Angeles Times*, pointed out: "If you're planning to drive across the country soon, it's going to cost you about $28 more to do it in style." Comparing the results of Les Viland's Rambler American to Mel Alsbury, Jr.'s Oldsmobile Ninety-Eight, Bob concluded, "The compact car driver is sacrificing space, size and power but he is saving $28 to go from Los Angeles to New York by not going in style."[116]

A resurgence of interest by enthusiast magazines was evident with an increase in the number of articles.[117] With several publications providing an overview of the Run, in both words and pictures, two in particular focused on the drivers. *Sports Car Graphic*'s Jean Calvin provided readers with a firsthand look at the preparation and competition as she competed in a Pontiac Catalina, finishing first in the medium-price class. *Car Craft* took a different approach by focusing the majority of its coverage on the participation of the teenage drivers. Most publications agreed that "this year's Mobil Economy Run was a honest test of car and driver over 3000 grueling miles."[118] Prior to the 1966 Economy Run, *Motor Trend* took a look back at the preparations involved in staging the Run, focusing on USAC's advance planning while at the same time keeping the details secret.[119] Forty years after the crossing of the Times Square finish line, *Motor Trend* brought back a little history to its readers by publishing a photograph of this finish with a short thumbnail sketch of the Economy Run. For a new generation of readers, this was perhaps their first look at the Economy Run in action.[120]

The participation of Marty Payne was a focus of articles in *Jet* and *Ebony*, publications aimed at the African American community. While *Jet* only announced Marty's selection to drive the Dodge Dart 6 on the "basis of [his] scholastic, school activity and potential driving ability," *Ebony* followed his progress through words and pictures of his participation which cumulating in a third place finish.[121]

Mobil Oil continued to tell its dealers the story of each year's Run in a special edition of the *Mobil Dealer News*. This year's article was through the eyes of Don Royer, driver of the Chevrolet Impala 8, and followed his experiences from preparation to crossing the finish line in Times Square.[122] Chrysler-Plymouth dealers and their staff were told of this year's successes in the April 1965 issue of *Chrysler-Plymouth Times*.

Placing first in the new special sports class was George Miller and his Plymouth Barracuda. With the cooperation of city officials, Times Square was temporarily blocked to all traffic to allow for the drivers to cross the finish line. USAC's J.C. Agajanian was on hand with his traditional checkered flag greeting.

The 1940 Dodge press car nicknamed "Old Dependable" successfully made it all the way to Times Square. Selected newspaper and magazine writers, who kept a record of its gasoline and oil consumption, also used this opportunity to do a comparison road test between the 1940 and 1965 Dodge models. The drivers were attempting to meet or exceed the 23.76 miles per gallon achieved when the Deluxe 6 placed second in the 1940 Gilmore Yosemite Run. In the end, they didn't even come close, as the Dodge averaged 16.4 miles per gallon on the trip. Its oil usage was poor due to a broken oil line discovered en route.[123]

Mobil Oil and USAC officials were genuinely pleased with the overall outcome of this year's

Economy Run. As Frank Meunier, Mobil Oil's products promotions manager, observed: "This is the most successful chapter in the history of the run. It was also the fastest — 51.68 miles per hour — and the most realistic. The mixture of terrain, altitudes, road conditions and weather were very close to what anyone could expect making a trans-continental tip this time of year."[124] The challenge now facing those involved was how to expand and improve on the last two Economy Runs to New York while keeping the interest of the public in high gear.

10

Boston

1966

Los Angeles to Boston Via Tucson — Albuquerque — Wichita — St. Louis — Cleveland — Philadelphia

March 14–20, 1966—3301.4 Miles

With the dawn of a new year, corporate changes were announced, with one resulting in a new visual look for the Economy Run. Marking its 100th anniversary, the sponsoring company adopted a name change by dropping the Socony name; the company was now known as the Mobil Oil Corporation. In line with this, a new corporate logo was unveiled which downplayed the famous "Flying Horse" insigne used for many years. The new design featured "Mobil" in blue and red on a white background. Even as corporate discussions continued to be held on the future of the Economy Run, plans were unveiled for a longer and tougher contest that was experienced in 1965.

On February 16, Chief Steward A.C. Pillsbury announced that the destination of this year's Run would be Boston's new Prudential Center. He said, "The drivers will be tackling the longest route ever (over 3300 miles) and they will be asked to maintain a new top average speed of over 51 miles per hour. From the Los Angeles starting line the drivers will point for overnight stops at Tucson, Albuquerque, Wichita, St. Louis, Cleveland and Philadelphia." Several other changes were announced, including "the requirement for California approved anti-smog devices in all cars [as] within a short period of time air pollution control equipment may be standard on all new Detroit products." Additionally, "Last year's class for [special] sports cars and fastbacks with bucket seats has been eliminated, but cars of that type will vie with their standard-bodies cousins in the eight competitive classes for '66."[1] With this announcement, a long-held desire by Massachusetts native Art Pillsbury for the Run to travel to New England was now being realized. An addition to USAC's staff this year was the appointment of Art Rene as a steward, later named assistant to the chief steward, bringing with him sixteen years of Economy Run experience as a driver and team leader.

With the elimination of the special sports class, changes were naturally required for the competitive classifications. With the release of the "Competitive Divisions for the 1966 Mobil Economy Run," provisions were made for sports models to compete in each class along with a big change to the two classes for compact cars. Reflecting the growing popularity of two-door models, eligibility in all classes was opened to these in addition to four-door models. Class A was now limited to compact six-cylinder cars (with an engine displacement of 164 through 232 cubic inches), and Class B was now for compact eight-cylinder cars (with an engine displacement of 273 through 289 cubic inches). The classes for intermediate-size and full-sized models remained identical to those used in 1965, although a modification to engine displacement was made in Class C for intermediate-size six-cylinder cars, increasing the maximum displacement to 250 cubic inches. As outlined earlier, provisions were made for sports models to compete in the compact eight-cylinder, intermediate-size eight-cylinder, low-price eight-cylinder and medium-price classes.[2]

Aside from changes in competitive classes and eligibility, the only rules change of note involved the requirement of all cars to conform to California smog control regulations.[3] Additionally, all cars competing in the luxury car class were equipped with air conditioning, although it was not operated during the Run.[4]

After entries closed on February 12, the field of 54 cars was set to compete. It was the third largest in history, representing every U.S. make except for Lincoln. Within a week, however, the field would be reduced to 52 with the withdrawal of the six-cylinder Dodge Dart and Plymouth Valiant 100 models. No reason was given, as both models had already been selected and were at the Pasadena impound undergoing preliminary inspection. General Motors continued to have the largest representation with 29 entries, followed by Ford Motor Company with 12, Chrysler Corporation with 8 and American Motors with 3. Chrysler Corporation scaled back their participation in comparison to previous years. In addition to pulling the six-cylinder Dart and Valiants, Chrysler limited their intermediate-size entries to a single Plymouth Belvedere

6 and Dodge Coronet Deluxe 8, and did not enter a Barracuda to again challenge the Ford Mustang.

Compact and intermediate-size models accounted for 31 of the 52 entries. With two-door models fully eligible to compete, interest centered on two models participating for the first time. Joining the Buick Riviera in the luxury car class was the new front-wheel drive Oldsmobile Toronado. With its 385-horsepower engine, the Toronado also had the distinction of being the most powerful entry.

There were two non-factory entries this year. With Chrysler Corporation declining to provide sponsorship or support for the Imperial, members of the Baldwin Park (California) Rotary Club stepped in by entering a Crown hardtop with its past president, Bart Parker, handling the driving duties. The West Covina Chamber of Commerce returned, entering a Cadillac Sedan de Ville with its past president, Stan Seeds, named as driver.

USAC's J.C. Agajanian gives Carl Westervelt, driving a Ford Falcon Futura 6, the traditional send-off as he leaves Los Angeles' Ambassador Hotel bound for Tucson. A smaller crowd of spectators was on hand this year to see the drivers on their way. With Mobil's new logo now prominent, a new and more subdued lettering style on the cars is clearly evident.

As one walked through the doors into the impound, the first thing that stood out was an entirely new look for the competing cars. For the first time since 1950, an entirely new lettering design was introduced, with the new Mobil logo now being prominent, and was a stark contrast to the brightly lettered cars in prior years. The Flying Red Horse insigne which had graced the hoods and front doors of the cars was gone, along with the USAC emblem, replaced by a simple statement "SUPERVISED BY UNITED STATES AUTO CLUB" on the rear fenders.

Factory support continued on par with those efforts seen in 1965, and no changes were reported in those team leaders responsible for preparing their entries. With Don Bailey stepping in as team captain, Inglewood Lincoln-Mercury dealer Bob Estes continued to handle the majority of support for the Mercury entries, again assisted by Santa Monica's Lynch Lincoln-Mercury. Paul Prior, an engineer with Chevrolet Motor Division, was on hand to provide technical assistance to individual dealers who were tasked with preparing their own entries. All of this preparation and attention to detail was due to the fact that "after the over–3000 road miles to Boston the winning margin will likely be a few thousandths of a gallon if past accomplishments on the Run are to be used as a criteria."[5]

Mercury continued to spotlight the skills of teenage drivers by selecting Harold Limbird, winner of the 1965 Cal State Safe Driving Auto Rodeo, to pilot its eight-cylinder Comet Cyclone. Linda Sullivan, this year's Lincoln-Mercury Youth Safety Spokesperson, would handle co-driving duties. Lynch Lincoln-Mercury, in cooperation with *Car Craft Magazine*, reached out to Southern California's Chaparels Car Club for a team to handle its Comet Cyclone GT entry. Frank Mowrer, president of the Chaparels, was tapped as driver, with fellow club member Bob Dunker handling co-driver duties. Two Chevrolet dealers also selected teenagers as drivers. Bernard Chevrolet-Oldsmobile in Libertyville, Illinois, named Jim Finnessy to drive its six-cylinder Chevelle 300 after his winning a safe-driving essay contest held in Lake County High Schools. Carolyn Thomas, daughter of well-known race car builder Bill Thomas, was named by Cone Chevrolet in Fullerton, California, to drive its eight-cylinder Bel-Air.

While many familiar names returned as drivers this year, a diverse group of first-time participants were named as drivers and co-drivers bringing a new look to the Run. Gail Hire, who appeared as Holly MacGregor in Paramount Motion Pictures' *Red Line 7000*, was selected by Pontiac to drive its eight-cylinder Tempest. Named as Gail's co-driver was singer Carol Connors, a rock 'n' roll favorite best known as lead vocalist in 1958's *Billboard* hit "To Know Him Is to Love Him." National Hot Rod Association (NHRA) 1966 Winternationals Drag racing champion Shirley Shahan, who raced a Plymouth as the Drag-on-Lady, was selected to drive the six-cylinder Fury I. All of the remaining Dodge and Plymouth entries were driven by Chrysler Corporation engineers.

In a partnership with American Motors, the eight-cylinder Ambassador DPL was sponsored by the City of Kenosha, Wisconsin. George Sorenson, a student at Kenosha Technical Institute, was selected as the driver, with Kenosha's Mayor Eugene R. Hammond as co-driver. During the Run, Mayor Hammond would act as Kenosha's "goodwill ambassador," as he would be visiting with various mayors and other officials during the cross-country trip.[6]

Dealer associations saw a reduced role this year as sponsors, representing American Motors, Dodge and Oldsmobile entries. Based on this year's route, Dodge selected those dealer associations from Boston, Newark and Philadelphia as sponsors. In addition to several from Southern California, dealers participating for the first time included E.D. Babbitt Motor Co. (Ford) in Flagstaff, Arizona; Bernard Chevrolet in Libertyville, Illinois; Bob Ford, Dearborn, Michigan; Harris Auto Sales (Chrysler-Plymouth) in Providence, Rhode Island; Pontiac Village in Brookline, Massachusetts; Sea Crest Pontiac-Cadillac in Lynn, Massachusetts; and Bob Tasca Ford Sales in East Providence, Rhode Island.

Only two dealers participated both as entrants and drivers. Joining longtime participant Hart Fullerton (Chrysler New Yorker) was Greg Bucilla (Oldsmobile F-85 Cutlass) of Prestige Oldsmobile in the Monterey Park community of Los Angeles. Participating for the first time, Greg's Cutlass was the only entry not affiliated with the Oldsmobile factory team. Although his dealership entered both a Chevrolet Chevy II Nova SS and an Impala, Mel Hilliard did not return as a driver. Four husband-and-wife teams returned: Hart and Diana Fullerton (Chrysler New Yorker), Gordon and Scherlee Madison (Chevrolet Chevy II 100 8), Stan and Lois Raymond (Oldsmobile Jetstar 88), and Tommy and Ila Thomas (AMC Ambassador 880 6), joined by newcomers C.K. and Betty Enoch in a Chevrolet Chevy II 100 6. Gordon and Scherlee Madison drove separate entries last year. The mother-and-daughter team of Pat and Pam Sawyer were back, defending last year's win, once again in a Chevrolet Biscayne 6. Women were named as drivers for twelve entries, including prior class winners Jean Calvin, Mary Hauser, Patricia Jones, Marta Retzlaff, and Pat Sawyer. Paula Murphy returned, again in a Buick Special 6, after having to step aside midway during last year's run due to a recurring back problem. All three entries in the standard-size six-cylinder class were driven by women. Three of last year's class-winning drivers, Art Bertelsen, Darrell Droke, and Scott Harvey, were absent this year. All three would return in 1967.

Byron Froelich heads his Ford Falcon 6 into Arizona, bound for a brunch stop in Yuma. Again this year, USAC officials made special arrangements with Arizona's Department of Agriculture to expedite movement of the competitors through its Inspection Station.

Random car selection and inspections began in early February and repeated those scenes from prior years. The impound remained a bustling scene of activity as the cars were serviced and given break-in test runs. As before, these break-in runs were limited to 1500 miles. Entrants were given the option of using either Mobil Regular or Mobil Premium gasoline, with 12 choosing Regular. This group included both Buick's six- and eight-cylinder Specials, along with Chevrolet's Biscayne 6, Bel-Air 8, Corvair Monza, Chevelle Malibu, and all of the Chevy II and Chevelle 300 six- and eight-cylinder models.

Continuing in their traditional roles were Mobil Oil's Frank Meunier, Roger Mahey and John Parkinson, coordinat-

The first collision between competing cars in the history of the Economy Run occurred two miles east of Yuma, Arizona. Bob Checkley, driving a Dodge Dart 8, was forced to brake quickly to avoid hitting another car that allegedly pulled out in front of him. Al Johnson, driving a Ford Fairlane 500 6, was unable to stop in time, rear-ending Bob Checkley's Dart. The USAC riding observer in the Ford was injured after striking the windshield and required treatment at Yuma's Park View Hospital. The extensive damage sustained to the Ford Fairlane forced it to withdraw from the Run. Using a baling-wire approach to secure the trunk lid, the Dodge Dart was able to continue and placed first in the compact eight-cylinder class.

ing all of the publicity and press arrangements. While most of the press representatives would again travel between stops on a United Airlines DC-6B, some would cover the Run from the ground in cars loaned by the various manufacturers. Among the auto editors in the group was Bob Fendell of the *New York World Telegram and Sun*, driving a Plymouth Satellite hardtop, equipped with a Hemi 426-cubic-inch engine. In addition to providing his readers with the daily events on the Run, he conducted a cross-country test report. These daily reports included his personal miles-per-gallon results, which ranged from 10.8 to 12.1. His average speed was much higher than those of the competing cars.[7]

The traditional drawing for starting positions was held several weeks prior to the start. For the third year in a row, another first-time driver, C.K. Enoch, drew No. 1 for his Chevrolet Chevy II 100 6. Bringing up the rear was Stan Seeds, drawing No. 52 for his Cadillac Sedan de Ville. At this time, USAC officials announced that Los Angeles' famed Ambassador Hotel, located on Wilshire Boulevard, would be the location of the start.

On Sunday afternoon, March 13, the competing cars were driven from the Pasadena impound to the Ambassador Hotel. At this time it was announced that Chevrolet engineer Paul Prior would replace John Alerich behind the wheel of the Chevrolet Caprice. Before being parked in proper order to tomorrow's start, all entries were topped off, then the pre-start festivities commenced. After a festive barbeque, hosted by Mobil Oil for all participants, the first day's route summary and detailed itinerary was released by USAC officials. The drivers learned they would cover the 515.9-mile trip to Tucson, following the same basic route used in 1961, with a time limit set at 9 hours and 35 minutes. As in prior years, time for refueling and brunch stops was not included in this limit. For the 12 drivers who had competed in 1961, it became very clear this would be the fastest Run in history. Covering virtually the same number of miles, this year's time limit was shorter by 2 hours and 25 minutes compared to 1961. With the Run on the road earlier than in previous years, the participants faced the possibility in the coming days of adverse weather and difficult road conditions as late season storms could be encountered. Another area which would have an impact on mileage figures was the California-required smog control devices on all entries and the effect these would have on fuel economy. This is "a factor considered significant because all new cars will be equipped with them beginning in 1968."[8] That night, Mercury's team captain and the driver of the Park Lane, Don Bailey, was stricken with the flu and was forced to withdraw. Since there was insufficient time to locate a replacement driver, the Park Lane would not compete, and the field was reduced to 51 cars.

In the pre-dawn darkness on March 14, the participants were on hand for their trip to New England. At 6:00 a.m., with a small group of spectators on hand, C.K. Enoch in a Chevrolet Chevy II 100 6 was the first driver to be sent on the way by USAC's J.C. Agajanian, with the remaining entries leaving at the traditional one-minute intervals. Today the drivers would have dry, warm weather conditions and no wind to contend with as they traveled across the California and Arizona deserts. Following a familiar route leaving Los Angeles, the drivers headed east on the San Bernardino Freeway / Interstate 10 towards Indio, where they turned onto California 86 towards El Centro. Between the towns of Coachella and Salton Sea, the route dropped to an elevation of 206 feet below sea level, marking the lowest point on this year's Run. Arriving in El Centro, the drivers made their way to Oneal & Havens Mobil Service, where each entry was topped off. Following U.S. Route 80, they crossed into Arizona, entering the Mountain Time Zone, heading for Yuma and brunch at the Stardust Motor Hotel. Their first day on the road was now half over. As they were leaving Yuma, still on U.S. Route 80, the first collision between competing cars in the history of the Run occurred. Two miles east of town, Bob Checkley, driving a Dodge Dart 8, was forced to slam on his brakes, making an emergency stop to avoid hitting another car that allegedly pulled out in front of him. Following behind was Al Johnson, driving a Ford Fairlane 500 6, who was unable to stop in time, striking the left-rear of Bob Checkley's Dart. The Ford sustained major front-end damage, resulting in its withdrawal. USAC observer Edgar T. Marshall's head hit the windshield of the Ford, suffering facial cuts that required treatment at Yuma's Park View Baptist Hospital, where he was admitted and listed in good condition. Although his car was heavily damaged, Bob Checkley was able to continue, as a baling-wire approach was used to secure the trunk lid in position.[9] While the cleanup continued, the remaining drivers continued east on U.S. Route 80 and Arizona 84 to Casa Grande for a refueling stop at Warren Don Mobil Service. Then it was a 65-mile trek to Tucson and the impound at the Hiway House. Just before 5:30 p.m., C.K. Enoch in a Chevrolet Chevy II 100 6 was the first of the 50 remaining cars to arrive. Just over an hour and a half later, Bob Checkley became the center of attention when he arrived in his damaged Dodge Dart 8. Four drivers, Art Arfons (Pontiac Star Chief Executive), John Meyers (Buick Electra 225), Bobby Spears (Ford Fairlane 500 8) and Carole Stroud (Chevrolet Chevy II Nova 6), arrived late and were given penalties.

As the events of the day were recounted, USAC officials "pointed out that today's accident marked the first injury in the 30-year history of the run, which has had 1,220,000 miles without a reportable accident."[10] While this claim was widely reported, it is not entirely true, as it ignores an incident involving Les Viland in 1956. On the first day of that year's Run,

Les' Nash Ambassador 6 partially left the road and overturned, resulting in minor injuries and lacerations to his co-driver and both USAC observers. Although their injuries were not as serious as those suffered on this day by Ed Marshall, these three were treated at a local hospital and released. This event, almost to the day ten years earlier, actually was the first time anyone had been injured in the Run's history.

Almost forgotten were the first day's results, which showed perennial leader Les Viland and his Rambler American 440 recording the best mileage for all entries with 25.4034 miles per gallon. As a result of a faster pace today, his figure was 1.6860 miles per gallon lower than last year's first-day result. In renewing their competition from last year, Byron Froelich in his Ford Falcon was close behind, only .4083 mile per gallon off the pace. The entire field turned in an average of 20.7292 miles per gallon. Bob Checkley had something to be pleased about as his damaged Dodge Dart was leading in the compact eight-cylinder class by nearly a mile per gallon over Fran Foster in a Ford Mustang 2+2. Two Chrysler Corporation products were leading in the intermediate-size cars, with Bob Binder's Plymouth Belvedere I 6 and George Miller's Dodge Coronet Deluxe 8 leading both Buick Special entries sitting in second place. For the remaining four classes it was all General Motors. Chevrolets were leading in two classes, with Pat Sawyer's Biscayne 6 out front among the full-size six-cylinder models and Don Royer's Impala 8 leading the low-price models. Art Arfons's Pontiac Star Chief Executive held a small lead over three other medium-priced General Motors models, while Mel Alsbury, Jr.'s Oldsmobile Toronado was out in front of the seven competing luxury models. A friendly rivalry between two Economy Run veterans was developing as Mel was holding a .04755 mile-per-gallon lead over Hart Fullerton's Chrysler New Yorker. Close competition was beginning to develop in seven of the eight classes. With the exception of Don Royer's 1.0806 mile per gallon lead over Carl Diehl's Dodge Polara 318, the difference between first and second place was less than a mile per gallon. In three classes this difference was less than a half-mile per gallon. While averaging 19.6531 miles per gallon, all four teenagers competing turned in individual results that were disappointing. Their best showing was fifth place for Harold Limbird's Mercury Comet Cyclone 8 and Carolyn Thomas's Chevrolet Bel-Air 8. Frank Mower's Mercury Comet Cyclone GT had the distinction of recording the lowest miles per gallon for all entries with a figure of 14.7672, nearly 8 miles per gallon behind the class-leading Dodge Coronet Deluxe 8. In addition to Frank's inexperience, his Cyclone GT was more performance-oriented, being equipped with a four-barrel 390-cubic-inch engine. Today's average speed of 53.9898 miles per hour confirmed this year's Run would be a record setter.

The top three performers in each class were:

Compact Six Cylinder	Rambler American 440	25.4034 Miles Per Gallon
	Ford Falcon	24.9951 Miles Per Gallon
	Chevrolet Corvair Monza	24.7198 Miles Per Gallon
Compact Eight Cylinder	Dodge Dart	22.6257 Miles Per Gallon
	Ford Mustang 2+2	21.6319 Miles Per Gallon
	Ford Falcon	20.5986 Miles Per Gallon
Intermediate Size Six Cylinder	Plymouth Belvedere I	24.9951 Miles Per Gallon
	Buick Special	24.1976 Miles Per Gallon
	Pontiac Tempest Custom	24.1278 Miles Per Gallon
Intermediate Size Eight Cylinder	Dodge Coronet Deluxe	22.4807 Miles Per Gallon
	Buick Special Deluxe	21.8885 Miles Per Gallon
	Ford Fairlane 500	20.6651 Miles Per Gallon
Standard Size Six Cylinder	Chevrolet Biscayne	22.2693 Miles Per Gallon
	Plymouth Fury I	21.6490 Miles Per Gallon
	Ford Custom	21.3155 Miles Per Gallon
Low Price Eight Cylinder	Chevrolet Impala	20.9257 Miles Per Gallon
	Dodge Polara 318	19.9451 Miles Per Gallon
	Plymouth Fury I	19.7837 Miles Per Gallon
Medium Price	Pontiac Star Chief	20.2175 Miles Per Gallon
	Oldsmobile Jetstar 88	19.9850 Miles Per Gallon
	Buick LeSabre	19.9641 Miles Per Gallon
Luxury Cars	Oldsmobile Toronado	18.4895 Miles Per Gallon
	Chrysler New Yorker	18.0140 Miles Per Gallon
	Buick Riviera	17.4591 Miles Per Gallon

On March 15, the drivers began a 461.3-mile trip through "vast stretches of the southwest's wide open spaces" and mountain terrain to Albuquerque, with a time limit set at 9 hours and 19 minutes.[11] The drivers would follow the same route used in 1955, and the overall road conditions showed a big improvement over those encountered eleven years earlier. Bob Checkley, driving his damaged Dodge Dart 8, was the first driver to be sent on his way at 6:30 a.m., following U.S. Route 80 / Arizona 86 towards the New Mexico border. Part of today's travel in New Mexico would be on portions of Interstate 10 as they made their way to Lordsburg for a stop at Parker's Mobil Service, where all entries were topped off. Continuing eastward, they picked up U.S. Route 85, headed for Truth or Consequences for brunch at the Sierra Country Fair Grounds. This marked only the second time, the first being in 1955, the Run had stopped in Truth or Consequences. Here a western barbeque was hosted by the Sierra County Sherriff's Posse and their Ladies Auxiliary, and a special event had been planned. As explained in today's route description, "members of the staff of the Carrie Tingley Hospital for (Disabled) Children, maybe even some of the kids," would be on hand to welcome the participants. "The whole town here has adopted these children, and because they wanted to help us by serving such an unusual lunch, we asked what we could do for Truth or Consequences. This is the reason everyone was asked to bring toys and games to the impound."[12] After this special stop, they headed north on U.S. Route 85 / Interstate 25 on a fairly direct route for the next 150 miles to Albuquerque. Throughout the day, weather conditions remained dry and warm, making for ideal driving conditions, although gusty winds were encountered from time to time. After navigating Albuquerque's local streets, they made the first stop for refueling at Globe Mobil Service, where they were greeted by a Mariachi Latin band, along with local residents who served the participants "rattlesnake steaks, Fritos and chili conqueso" while their cars were serviced. Then it was a quick three mile trek on Interstate 40 to the impound at the Holiday Inn. Here local residents, along with students from the University of New Mexico, were joined by Indian dancers who turned out to give the participants a warm Western welcome.[13] Just after 4:30 p.m., Bob Checkley in his Dodge Dart 8 was the first to arrive. Frank Mower in his Mercury Comet Cyclone GT arrived close to 3 minutes late and was penalized.

Today's "mountainous terrain and a series of small towns with reduced speed limits" resulted in averages taking a dip, as the field averaged 20.0627 miles per gallon.[14] Although still showing the highest miles-per-gallon figure for all entries with 24.3987, Les Viland in his Rambler American 440 was facing extremely close competition in the compact six-cylinder class. Byron Froelich's Ford Falcon was only .1252 mile per gallon behind in second place, with Bill Levy's Chevrolet Corvair Monza close behind in third. In addition to the competition in his class, Les was also being pressed for top mileage honors by Bob Binder in a Plymouth Belvedere I 6, who was only .2392 mile per gallon behind. With one exception, yesterday's leaders maintained their positions. That exception was in the medium-price class, as Marta Retzlaff moved her Buick LeSabre into first place, dropping Art Arfons in a Pontiac Star Chief Executive into second by a margin of .2217 mile per gallon. In this class, less than a mile per gallon separated all four entries. While Don Royer's Chevrolet Impala maintained a fairly comfortable lead in the popular low-price eight-cylinder class, the real competition was between second and fifth place. Only .1874 mile per gallon, which is about an eye-dropper of gas, separated Bob Cahill's Plymouth Fury I in second place from Carl Diehl's Dodge Polara 318, shown in fifth place with two Chevrolets sandwiched in between.[15] In the luxury car class, two veterans battled for second place,

With a group of Native Americans on hand to welcome the participants, Mel Alsbury, Jr., pulls his Oldsmobile Toronado into the Albuquerque impound. The appearance of the all-new Toronado marks the first time since 1937 a front-wheel drive car had competed in the Economy Run. Mel placed first in the luxury car class for his seventh win as a driver and his first in an Oldsmobile.

with Pete Novotny moving his Oldsmobile Ninety-Eight into second over Hart Fullerton's Chrysler New Yorker by a margin of .1318 mile per gallon. The teenagers turned in an average of 19.2258, although Carolyn Thomas managed to move her Chevrolet Bel-Air 8 into third place. With today's roads and terrain, the average speed dipped to 49.9247 miles per hour.

The top three performers in each class were:

Compact Six Cylinder	Rambler American 440	24.3987 Miles Per Gallon
	Ford Falcon	24.2735 Miles Per Gallon
	Chevrolet Corvair Monza	24.2105 Miles Per Gallon
Compact Eight Cylinder	Dodge Dart	21.9830 Miles Per Gallon
	Ford Mustang 2+2	20.9426 Miles Per Gallon
	Ford Falcon	20.2793 Miles Per Gallon
Intermediate Size Six Cylinder	Plymouth Belvedere I	24.1595 Miles Per Gallon
	Buick Special	23.4346 Miles Per Gallon
	Pontiac Tempest Custom	23.3195 Miles Per Gallon
Intermediate Size Eight Cylinder	Dodge Coronet Deluxe	21.7830 Miles Per Gallon
	Buick Special Deluxe	21.2672 Miles Per Gallon
	Ford Fairlane 500	20.0093 Miles Per Gallon
Standard Size Six Cylinder	Chevrolet Biscayne	21.7945 Miles Per Gallon
	Plymouth Fury I	21.2509 Miles Per Gallon
	Ford Custom	20.3874 Miles Per Gallon
Low Price Eight Cylinder	Chevrolet Impala	20.5358 Miles Per Gallon
	Plymouth Fury I	19.2129 Miles Per Gallon
	Chevrolet Bel-Air	19.1292 Miles Per Gallon
Medium Price	Buick LeSabre	19.6239 Miles Per Gallon
	Pontiac Star Chief Executive	19.4022 Miles Per Gallon
	Pontiac Catalina	19.1061 Miles Per Gallon
Luxury Cars	Oldsmobile Toronado	17.5219 Miles Per Gallon
	Oldsmobile Ninety-Eight	17.1811 Miles Per Gallon
	Chrysler New Yorker	17.0493 Miles Per Gallon

On March 16, everyone had an early wake-up call as they prepared for the 583.6-mile trip to Wichita. With a time limit of 11 hours and 19 minutes, today would feature, for the first time in history, three refueling stops along the way. Mike Jones, driving a Buick Riviera, was the first driver on his way at 4:30 a.m., heading east on U.S. Route 66 / Interstate 40 towards Tucumcari. Although rain and snow had been predicted earlier across the Texas and Oklahoma panhandles, drivers would have "a beautiful spring day with temperatures in the 70s," and would have to deal with some gusty winds along the way.[16] After leaving Clines Corners, they reached an altitude of 6970 feet, marking the highest elevation on this year's Run. In Tucumcari, the first stop of the day was at Crump Parks Mobil Service, where all entries were topped off. Leaving Tucumcari, the drivers headed northeast on U.S. Route 54, crossing into Texas, entering the Central Time Zone, and continuing into Oklahoma. Arriving in Goodwell, the participants stopped for brunch at Panhandle A&M College's Student Union Building. Served by student volunteers, the menu featured "beef that had been actually raised on the campus by the students," along with entertainment provided by the school's dance band.[17] Back on the road, continuing to follow U.S. Route 54, the drivers headed to Hooker for the second refueling stop at Hodges Mobil Service. Crossing into Kansas, they continued eastward on U.S. Route 54, traveling through numerous cities and towns, with their reduced speed zones, for the next 210 miles. In many of these towns, residents came out to welcome the participants, and usually were joined by local high school bands.

During the day, problems surfaced for the Plymouth Belvedere I 6 team when a slow leak developed in one of the rear tires. This leak continued and "threw off the odometer readings, which caused the navigator to miss a turn on the route, and the team lost their location. They became aware of the leaking tire as it went flat, and they had to stop and change it."[18] Arriving in Wichita, the drivers made their first stop for refueling at Weslink Mobil Service before continuing on to the impound. Arriving in the downtown area, the drivers were directed to the rooftop parking area of Union National Bank, located next door to the Hotel Lassen, the official impound location. The first to arrive at 5:40 p.m. was C.K. Enoch's Chevrolet Chevy II 100 6. After a frustrating day on the road, Bob Binder, driving a Plymouth Belvedere I 6, arrived nearly two and a half minutes late and was penalized.

Mileages slightly improved on today's trek. Les Viland continued to record the highest miles-per-gallon figure of 24.6491, although he was being pressed by several other drivers. Les managed not only to hold off his two closest competitors in the compact six-cylinder class by .1745 mile per gallon, but also a hard-charging Bob Binder in a Plymouth Belvedere I 6. The difference between Les' Rambler American and Bob's Plymouth Belvedere was a slim .0888 miles per gallon, and was reported to be "about ⅔-cup of gasoline in the 1560 miles from Los Angeles."[19] All of the class leaders maintained their positions, with Bob Checkley's damaged Dodge Dart 8 maintaining a fairly comfortable lead in the compact eight-cylinder class. Competition continued to remain close in six classes. The closest was in the medium-price class, with Marta Retzlaff's Buick LeSabre holding on to a .2317 mile-per-gallon lead over Art Arfons's Pontiac Star Chief Executive. In the low-price eight-cylinder class, it was Chevrolet placing one-two-three with Don Royer holding a comfortable lead. The real battle continued between second and fifth place. With Bob Cahill's Plymouth Fury I dropping down to fourth place, Carolyn Thomas's Chevrolet Bel-Air moved into second and Paul Prior's Chevrolet Caprice into third. Only .2425 mile per gallon separated second through fifth place. In the intermediate-size six-cylinder class, Don Francisco moved his Pontiac Tempest Custom into second, dropping Paula Murphy in her

Buick Special to third by a margin of .0814 mile per gallon. The field recorded an overall average of 20.2835 miles per gallon with the teenagers showing a slight improvement to 19.4962. Today's average speed for the 583.6-mile trip from Albuquerque was 51.9228 miles per hour.

The top three performers in each class were:

Compact Six Cylinder	Rambler American 440	24.6491 Miles Per Gallon
	Ford Falcon	24.4746 Miles Per Gallon
	Chevrolet Corvair Monza	24.2369 Miles Per Gallon
Compact Eight Cylinder	Dodge Dart	22.3275 Miles Per Gallon
	Ford Mustang 2+2	21.2775 Miles Per Gallon
	Ford Falcon	20.5657 Miles Per Gallon
Intermediate Size Six Cylinder	Plymouth Belvedere I	24.5603 Miles Per Gallon
	Pontiac Tempest Custom	23.7284 Miles Per Gallon
	Buick Special	23.6470 Miles Per Gallon
Intermediate Size Eight Cylinder	Dodge Coronet Deluxe	22.0310 Miles Per Gallon
	Buick Special Deluxe	21.6358 Miles Per Gallon
	Ford Fairlane 500	20.3119 Miles Per Gallon
Standard Size Six Cylinder	Chevrolet Biscayne	22.1219 Miles Per Gallon
	Plymouth Fury I	21.5474 Miles Per Gallon
	Ford Custom	20.6345 Miles Per Gallon
Low Price Eight Cylinder	Chevrolet Impala	20.7023 Miles Per Gallon
	Chevrolet Bel-Air	19.3213 Miles Per Gallon
	Chevrolet Caprice	19.2538 Miles Per Gallon
Medium Price	Buick LeSabre	19.8491 Miles Per Gallon
	Pontiac Star Chief Executive	19.6174 Miles Per Gallon
	Pontiac Catalina	19.2282 Miles Per Gallon
Luxury Cars	Oldsmobile Toronado	17.8354 Miles Per Gallon
	Oldsmobile Ninety-Eight	17.3935 Miles Per Gallon
	Chrysler New Yorker	17.1831 Miles Per Gallon

On March 17, the drivers would have a 447-mile trip across Kansas and Missouri almost entirely on turnpike or interstate highways with a time limit set at 8 hours and 15 minutes. For the 28 drivers who competed last year, today's scenery would look very familiar. At 6:30 a.m., Bob Binder in his Plymouth Belvedere I 6 was the first driver to leave, heading for the Kansas Turnpike, with its speed limit of 80 miles per hour, where the drivers would face the first challenge of the day. As explained in today's itinerary, "Beginning at the Toll Both, you *MUST* average a minimum of 60 miles per hour between that point and the Official TOP-OFF station on the Turnpike in Topeka, Kansas. The total miles from the Toll Booth to the TOP-OFF station is 133.5. The maximum net running time is 2 hours, 13 minutes, 30 seconds. An Official USAC Timing Station will be set up at the start and finish of the time control section. *Penalties will be applied if you do not comply with the above.*"[20] Three drivers — Fran Foster (Ford Mustang 2+2), John Meyers (Buick Electra 225), and Carl Westervelt (Ford Falcon Futura 6) — exceeded the time limit, ranging from 13 seconds to 1 minute and 9 seconds, and were penalized.

Completing this timed control section, they arrived at Wilson's Turnpike Mobil Service, where all entries were topped-off. Unseasonably warm weather would be on tap today with temperatures in the high 70s; however, blowing dust and gusty winds of up to 25 miles per hour would create a driving challenge.[21] Continuing on the Kansas Turnpike (Interstate 70), they traveled through Kansas City, crossing the Mississippi River into Missouri, and headed for Independence and a brunch stop at the Cracker Neck Country Club. Back on Interstate 70, the drivers' next stop was at Wright City Mobil Service for refueling, which marked the second return visit to a refueling location that was used last year. The other return visit was to Wilson's Turnpike Mobil Service earlier in the day. After another 44 miles of Interstate highway driving, they arrived in St. Louis in time for the evening rush hour traffic jams as they made their way to the impound at the Chase Park Plaza Hotel.[22] At 3:30 p.m., Fran Foster, driving a Ford Mustang 2+2, was the first to arrive. Two drivers, Patricia Jones (Oldsmobile F-85 8) and Pete Novotny (Oldsmobile Ninety-Eight), arrived late and received penalties. In Pete Novotny's case, he drove off course in St. Louis for less than a half-mile, which resulted in his arriving nearly 2 minutes late. After settling in at the Chase Park Plaza and reviewing the itinerary for tomorrow, the participants knew right off the trip to Cleveland was going to be one long, tough day. Tomorrow's weather forecast for heavy rain would add to the challenge. Those following the Run continued to compliment speed ace and Firestone Tire test driver Art Arfons on his handling "economy driving like a seasoned pro." Talking with Bruce Young of the *Long Beach Press-Telegram*, Art sighed, "This is a lot harder than it looks. When I drive on the [Bonneville] Salt Flats I am in the car only two minutes and only use power for 35 seconds. I use 60 gallons in just 35 seconds. Now I must stretch every gallon."[23]

Despite today's fast driving pace of 54.4510 miles per hour, most averages managed to remain fairly close to the yesterday's results. Continuing to post the highest miles-per-gallon figure for all entries at 24.7823, Les Viland and his Rambler American 440 also opened up a slight margin over his two closest competitors in the compact six-cylinder class. Continuing their friendly rivalry from last year, Byron Froelich kept his Ford Falcon close on Les' heels, running .3476 mile per gallon off the pace. Bob Binder's tire problems yesterday with his Plymouth Belvedere I 6 saw his average drop to 23.9860 miles per gallon. This pretty much took him out of contention to challenge Les for the best mileage. What was a slim .0888 miles per gallon difference yesterday grew to .7963 today. All of the class leaders continued to maintain their positions as competition remained close in five classes, with the closest in both intermediate-size classes. In the class for six-cylinder models, Don Francisco's Pontiac Tempest Custom was able to close within .1924 miles per gallon of Bob Binder's Plymouth Belvedere, while among the eight-cylinders, George Miller's Dodge Coronet Deluxe held a .1786 lead over John Rich's Buick Special Deluxe. At this point, barring any unforeseen circumstances, it appeared that Don Royer's Chevrolet Impala would capture the trophy in the low-price eight-cylinder class, with only a slim .1476 mile-per-gallon separation between second and fifth place. The only change in this hotly contested class was Paul Prior's Chevrolet Caprice moving into second by a mere .0273 mile per gallon, dropping Carolyn Thomas's Chevrolet Bel-Air into third. The field's overall average remained fairly constant from yesterday's at 20.2409 miles per gallon. The four teenagers also stayed pretty much on par, averaging 19.4394 miles per gallon, with Carolyn Thomas showing the best individual performance in third place. Frank Mowrer's Mercury Comet Cyclone GT, equipped with one of the largest engines among the eight-cylinder intermediates, barely stayed above 14 miles per gallon after four days with a figure of 14.0035.

The top three performers in each class were:

Compact Six Cylinder	Rambler American 440	24.7823 Miles Per Gallon
	Ford Falcon	24.4347 Miles Per Gallon
	Chevrolet Corvair Monza	24.3516 Miles Per Gallon
Compact Eight Cylinder	Dodge Dart	22.2739 Miles Per Gallon
	Ford Mustang 2+2	21.3115 Miles Per Gallon
	Ford Falcon	20.3669 Miles Per Gallon
Intermediate Size Six Cylinder	Plymouth Belvedere I	23.9860 Miles Per Gallon
	Pontiac Tempest Custom	23.7936 Miles Per Gallon
	Buick Special	23.7006 Miles Per Gallon
Intermediate Size Eight Cylinder	Dodge Coronet Deluxe	21.8434 Miles Per Gallon
	Buick Special Deluxe	21.6648 Miles Per Gallon
	Ford Fairlane 500	20.2091 Miles Per Gallon
Standard Size Six Cylinder	Chevrolet Biscayne	21.9617 Miles Per Gallon
	Plymouth Fury I	21.4007 Miles Per Gallon
	Ford Custom	20.5576 Miles Per Gallon
Low Price Eight Cylinder	Chevrolet Impala	20.6391 Miles Per Gallon
	Chevrolet Caprice	19.2319 Miles Per Gallon
	Chevrolet Bel-Air	19.2046 Miles Per Gallon
Medium Price	Buick LeSabre	19.8755 Miles Per Gallon
	Pontiac Star Chief Executive	19.6166 Miles Per Gallon
	Pontiac Catalina	19.2409 Miles Per Gallon
Luxury Cars	Oldsmobile Toronado	17.7550 Miles Per Gallon
	Oldsmobile Ninety-Eight	17.3337 Miles Per Gallon
	Chrysler New Yorker	17.1358 Miles Per Gallon

The competitors faced another long day on March 18 as they made the 575.4-mile trek to Cleveland in adverse weather conditions. With a routing over major U.S. routes and Interstate highways, for the drivers today's 11-hour and 53-minute trip would be most difficult faced in several years. Rain, heavy at times, would be a factor on the entire trip. With an early 4:00 a.m. start, Pete Novonty, driving an Oldsmobile Ninety-Eight, was the first driver on his way. Leaving St. Louis, they quickly crossed into Illinois, where they picked up Interstate 70 and U.S. Route 40 bound for Indiana, where they entered the Eastern Time Zone. While traveling through Terre Haute, Mary Faulkner's Mercury Comet 202 6 was involved in a minor collision with a local resident's car, with "damage limited to a couple of rumpled fenders." After a short delay, Mary was on her way once again.[24] Heavy rain with gusty winds up to 20 miles per hour would hamper the drivers as they made their way to the first stop of the day in Plainfield, where all entries were topped off at Summerlot's Mobil Service. Heading for Indianapolis, everyone was ready for a break after being on the road for over 5 hours when they arrived at Mickler's Sirloin Inn for brunch. Then, continuing east on U.S. Route 40 and into Ohio, they continued to pass through numerous towns, contending with their reduced speed limits along the way. In Springfield, Pat Sawyer's Chevrolet Biscayne 6 was involved in a rear-end collision resulting in minor damage. All of the occupants escaped injury, although Pat commented later, "We're a little sore, but otherwise OK, I think."[25] Arriving in Columbus, the drivers then headed north on Interstate 71. In Mansfield, the drivers stopped for refueling at Expressway Mobil Service, where USAC officials were seen using plastic sheeting as protection from the rain during the refueling process.

With more than an hour and a half still to go, the drivers continued to make their way to downtown Cleveland and the impound at the Hollenden House. Parking on the Hotel's rooftop parking area, Pete Novotny in his Oldsmobile Ninety-Eight was the first to arrive at 5:40 p.m. Just before 7:30 p.m., Bobby Spears and his Ford Fairlane 500 8 finally pulled in after being involved in the third collision of the day, in which he rear-ended a non-competing car. With the left front fender heavily damaged, repairs were made to ensure the headlight remained in operating condition, thus allowing Bobby to continue. The only driver to arrive late was Fran Foster in a Ford Mustang 2+2, running 3 minutes and 18 seconds over the time limit; she was penalized. "The snow predicted for Cleveland failed to materialize, actually disappointing some drivers who believed that adverse highway conditions might give them a relative advantage that would permit them better mileage than their competition."[26] Not all drivers shared this thought. Frank Mowrer, driving a Mercury Comet Cyclone GT, commented, "We were so nervous from driving in the rain on the two lane highways that we couldn't each much brunch."[27] Without a doubt, after battling wind, rain and two-lane traffic, the weary drivers were happy to see this day now behind them. It was a rough one ... and expensive, too!

Today's weather conditions saw everyone's mileage drop from the previous day, although "positions appeared to be sta-

Two prior medium-price class winners, Stan Raymond and Jean Calvin, contend with slow city traffic as they make their way to the St. Louis impound. In a close contest between second and fourth place, Stan's Oldsmobile Jetstar 88 was sitting in fourth while behind him, Jean's Pontiac Catalina was in third.

bilizing" and no changes were shown in the first three places in each class.[28] Continuing to record the best mileage for all entries with 24.4461 miles per gallon, Les Viland's Rambler American 440 also increased his margin over Byron Froelich's Ford Falcon to .3598 mile per gallon. In two classes where drivers had fairly comfortable leads, their cars looked more like survivors, both Bob Checkley's Dodge Dart 8 and Pat Sawyer's Chevrolet Biscayne 6 having being in separate collisions. Competition remained close in four classes where less than a half-mile per gallon separated first and second place. The closest competition was in the intermediate-size eight-cylinder class, where George Miller's Dodge Coronet Deluxe maintained a slim .1476 lead over John Rich's Buick Special Deluxe. Everyone's eyes continued to be on the competition in the low-price eight-cylinder class. Although there were no position changes, a mere .1217 mile per gallon separated Paul Prior's Chevrolet Caprice in second from Carl Diehl's Dodge Polara 318 sitting in fifth. Everyone associated with General Motors and Oldsmobile continued to be pleased with the performance of the new Toronado, which maintained its first-place position. Today's average made a small drop to 19.9752 miles per gallon, with the teenagers also showing a small drop to 19.1991. For the first time in history, a competitor saw his average drop below 14 miles per gallon, as Frank Mowrer's Mercury Comet Cyclone GT recorded a figure of 13.9334. Weather and road conditions contributed to a slower speed on today's drive, showing an average of 48.6602 miles per hour.

The top three performers in each class were:

Compact Six Cylinder	Rambler American 440	24.4461 Miles Per Gallon
	Ford Falcon	24.0863 Miles Per Gallon
	Chevrolet Corvair Monza	24.0121 Miles Per Gallon
Compact Eight Cylinder	Dodge Dart	21.9533 Miles Per Gallon
	Ford Mustang 2+2	20.9824 Miles Per Gallon
	Ford Falcon	20.0474 Miles Per Gallon
Intermediate Size Six Cylinder	Plymouth Belvedere I	23.6109 Miles Per Gallon
	Pontiac Tempest Custom	23.4075 Miles Per Gallon
	Buick Special	23.3176 Miles Per Gallon
Intermediate Size Eight Cylinder	Dodge Coronet Deluxe	21.4824 Miles Per Gallon
	Buick Special Deluxe	21.3348 Miles Per Gallon
	Ford Fairlane 500	19.9207 Miles Per Gallon
Standard Size Six Cylinder	Chevrolet Biscayne	21.6658 Miles Per Gallon
	Plymouth Fury I	21.0438 Miles Per Gallon
	Ford Custom	20.3407 Miles Per Gallon
Low Price Eight Cylinder	Chevrolet Impala	20.2960 Miles Per Gallon
	Chevrolet Caprice	19.0022 Miles Per Gallon
	Chevrolet Bel-Air	18.9484 Miles Per Gallon
Medium Price	Buick LeSabre	19.6082 Miles Per Gallon
	Pontiac Star Chief Executive	19.3155 Miles Per Gallon
	Pontiac Catalina	18.9594 Miles Per Gallon
Luxury Cars	Oldsmobile Toronado	17.5891 Miles Per Gallon
	Oldsmobile Ninety-Eight	17.0726 Miles Per Gallon
	Chrysler New Yorker	16.9514 Miles Per Gallon

Mike Jones in his Buick Riviera arrives at the Yankee Silversmith Inn in Wallingford, Connecticut, to a welcome by local area residents. After a short break, the drivers continued heading to Boston, marking the Economy Run's first visit to New England.

On March 19 the drivers would have a fast trip to Philadelphia that would be covered almost entirely on the Ohio and Pennsylvania Turnpikes. A time limit of 8 hours and 4 minutes was set for the 441.7-mile drive. After yesterday's adverse weather conditions, today would feature dry conditions and virtually no wind for the drivers to contend with. The first driver to leave at 6:30 a.m. was Bobby Spears as he guided his damaged Ford Fairlane 500 8 through Cleveland's downtown area toward the Ohio Turnpike. After covering 68 miles, the drivers crossed into Pennsylvania and picked up the Pennsylvania Turnpike heading for Murrysville, where all entries were topped off at Enick Kenesky Mobil Service. At this point two drivers encountered a problem. As one of the drivers, Frank Mowrer, later explained, "We missed the exit so just had to keep going. This was the first goof we'd made on the entire trip and weren't sure what we should do. There were areas in the median where we could have turned around, but they all had signs posted 'No U-Turn,' so we kept driving to the next Toll Station to turn around."[29]

Missing this exit caused them to drive an extra 22 miles. Another driver who missed this exit was John Rich in a Buick Special Deluxe 8. He was running a close second in the intermediate-size eight-cylinder class, but this error pretty much eliminated any chance for a first-place finish. After Murrysville and back on the Pennsylvania Turnpike, another time control section was on tap for the drivers as they "*MUST* average a minimum of 60 miles per hour between [this] point and the Toll Booth at exit #11. The total miles from the Gate #6 Toll Booth to the exit #11 Toll Booth is 89.7. The maximum running time is 1 hour, 29 minutes, 42 seconds." As was the custom, USAC timing stations were set up at both points, and penalties would apply for exceeding the maximum running time.[30] Four drivers, George Alsbury (Oldsmobile F-85 6), Mary Hauser (Chevrolet Chevelle Malibu 8), Harold Limbird (Mercury Comet Cyclone 8), and Gordon Madison (Chevrolet Chevy II 100 8), ran over on this timed section, from 3 seconds to 18 seconds, and were penalized. After they had completed this timed control section, it was a quick trip on U.S. Route 220 to Bedford for brunch at the Bedford Elks Country Club, with the local high school dance band on hand to provide entertainment. During this stop, two participants continued with traditions they began on the first day. Ambassador DPL co-driver and Kenosha Mayor Eugene Hammond presented city officials at each stop along the route with a "Key to the City," and Bedford's Mayor E.G. Studebaker was on hand to reciprocate. West Covina Chamber of Commerce's Stan Seeds, driver of the Cadillac Sedan deVille, presented certificates to local chambers along the way, and at this stop also honored Bedford's Junior Chamber of Commerce.[31] Back on the Pennsylvania Turnpike, the drivers continued heading eastward towards Gulph Mills for a refueling stop at Steve McLoughlin Mobil Service, followed by a short trip on Interstate 76 into Philadelphia and the impound at the Marriott Motor Hotel. At 3:20 p.m., Bobby Spears was the first to arrive in his Ford Fairlane 500 8, with members of Philadelphia's famed Ferko String Band providing an official welcome. Five drivers, Greg Bucilla (Oldsmobile F-85 Cutlass 8), Fran Foster (Ford Mustang 2+2), Bill Levy (Chevrolet Corvair Monza), Frank Mowrer (Mercury Comet Cyclone GT 8), and Les Viland (Rambler American 440), arrived late, ranging from 2 minutes and 10 seconds to 10 minutes and 55 seconds, and were penalized. Of the seven drivers who missed a turn and drove off course during the day, three were unable to make up the time and were among those arriving late.

Today's trip, with an average speed of 54.9024 miles per hour, was the fastest on this year's Run and naturally resulted in a decline in everyone's mileage figures. As expected, Les Viland continued to lead all entries with the best miles-per-gallon average with a figure of 24.0561. Although this was a drop of .3900 mile per gallon from yesterday's result, Les continued to maintain his lead over Byron Froelich's Ford Falcon. All of the class leaders continued to maintain their positions, although competition remained close in five. Two of these involved the intermediates. After John Rich's error earlier in the day, George Miller's Dodge Coronet Deluxe was all but guaranteed a first-place finish among the eight-cylinder models, as his lead stretched to .3236 mile per gallon. With the six-cylinders, a battle for second and third place continued as Paula Murphy eased her Buick Special past Don Francisco's Pontiac Tempest Custom by a slim .0551 mile per gallon. In the highly contested low-price class, Bob Cahill managed to move his Plymouth Fury I into third place, dropping Carolyn Thomas's Chevrolet Bel-Air to fourth. Bob now found himself a mere .0201 mile per gallon behind Paul Prior's Chevrolet Caprice in second. While Don Royer's Chevrolet Impala was still well out front by a margin of 1.2345 miles per gallon, it remained extremely close between second and fifth, with only .1599 mile per gallon separating these four competitors. The field saw its average drop to 19.6860 miles per gallon, with the teenagers dropping to 18.9257.

The top three performers in each class were:

Compact Six Cylinder	Rambler American 440	24.0561 Miles Per Gallon
	Ford Falcon	23.7002 Miles Per Gallon
	Chevrolet Corvair Monza	23.6270 Miles Per Gallon
Compact Eight Cylinder	Dodge Dart	21.6886 Miles Per Gallon
	Ford Mustang 2+2	20.7454 Miles Per Gallon
	Ford Falcon	19.7521 Miles Per Gallon
Intermediate Size Six Cylinder	Plymouth Belvedere I	23.2294 Miles Per Gallon
	Buick Special	22.9024 Miles Per Gallon
	Pontiac Tempest Custom	22.8473 Miles Per Gallon
Intermediate Size Eight Cylinder	Dodge Coronet Deluxe	21.1750 Miles Per Gallon
	Buick Special Deluxe	20.8514 Miles Per Gallon
	Ford Fairlane 500	19.5999 Miles Per Gallon
Standard Size Six Cylinder	Chevrolet Biscayne	21.2216 Miles Per Gallon
	Plymouth Fury I	20.7143 Miles Per Gallon
	Ford Custom	20.0054 Miles Per Gallon
Low Price Eight Cylinder	Chevrolet Impala	20.0084 Miles Per Gallon
	Chevrolet Caprice	18.7739 Miles Per Gallon
	Plymouth Fury I	18.7538 Miles Per Gallon
Medium Price	Buick LeSabre	19.3779 Miles Per Gallon
	Pontiac Star Chief Executive	19.0231 Miles Per Gallon
	Pontiac Catalina	18.7336 Miles Per Gallon
Luxury Cars	Oldsmobile Toronado	17.3737 Miles Per Gallon
	Oldsmobile Ninety-Eight	16.8005 Miles Per Gallon
	Chrysler New Yorker	16.7670 Miles Per Gallon

After covering a tough 3024.9 miles since leaving Los Angeles, the drivers expected to find an easy trip to Boston on March 20. This was not to be the case, as A.C. Pillsbury still had a few tricks up his sleeve. Although today's route did not include any of his famous two-lane country side roads, the opening paragraph of the Seventh Day Itinerary said it all: "Today is the final day of the 1966 Mobil Economy Run and you must be especially alert. You will be driving at maximum legal speed limits *prior to Sunrise* and the entire day. The course had many intricate turns and directional changes. Follow the itinerary VERY CAREFULLY and you will have no difficulty."[32] Fortunately this was Sunday and little traffic was expected through New Jersey, New York and into New England. Kay Kimes in a Ford Custom 500 8 was the first driver on the way at 5:00 a.m. for the 276.5-mile, 5-hour and 27-minute trip. After a short 12 miles on Interstates 76 and 676, they crossed into New Jersey and picked up the New Jersey Turnpike. Here foggy conditions were encountered which resulted in speed limits being reduced to 50 miles per hour for the next 57 miles. After crossing the Goethals Bridge into Staten Island, the drivers carefully made their way towards Brooklyn and Queens on Interstate 278. After crossing the Verrazano Narrows Bridge the drivers headed for the Long Island Expressway. After crossing the Throgs Neck Bridge they then followed the signs for "Interstate 95 North — New England" traveling on the Connecticut Turnpike towards Milford and the Wilbur Cross Parkway. Frank Mowrer and Bob Dunker missed another exit, which caused them to become lost once again, driving off-course for 5 miles. Frank later explained, "We were supposed to turn off at [Exit] #38. The first sign we saw was #39 so assumed that the next one would be #38. They went up! We turned off at #40 and jogged back, losing a few minutes."[33] Don Royer in his Chevrolet Impala, who was following Frank, also missed the exit but caught his error early, only driving off course for 2 miles. After a challenging morning, the drivers arrived in Wallingford, where they were ready for a short 30-minute brunch stop at the Yankee Silversmith Inn. Most of the California drivers were not accustomed to toll booths, but they had seen their share after stopping 10 times, dropping a total of $3.65 into the coin baskets. During this stop, USAC officials announced they were extending the time limit by 10 minutes due to the fog conditions in New Jersey. Leaving Wallingford, the drivers headed north on Interstates 91 and 84 towards the Massachusetts Turnpike's Charlton Service Area, where all entries were topped off for the final time. Just after 11:00 a.m., Kay Kimes's Ford Custom 500 8 was the first to arrive. Four drivers, Don Francisco (Pontiac Tempest Custom 6), Gail Hire (Pontiac Tempest 8), Frank Mowrer (Mercury Comet Cyclone GT) and Les Viland (Rambler American 440), arrived late, ranging in time from as little as 36 seconds up to 12 minutes, and were penalized. Two of the late arrivals had driven off course, with Gail Hire driving 17 miles extra before getting back on track. After their cars were topped off, they were parked nearby in a marshaling area, ready to make the final 55-mile drive to Boston.

At Boston's Prudential Center, Les Viland's Rambler American Rogue leads the convoy on drivers across the finish line, greeted by an enthusiastic J.C. Agajanian and a large crowd of spectators. Although their arrival in Boston was delayed for nearly an hour and a half, this did not dampen the enthusiasm of those on hand to welcome the participants. This marked the seventh year a Rambler American had achieved the best mileage for all entries, all but one of those years with Les behind the wheel (courtesy of Chrysler Group LLC).

Running an hour and a half behind schedule, the convoy of 50 competing drivers, with a police escort, continued eastbound on the Massachusetts Turnpike bound for the finish line at Boston's new Prudential Center, where a large crowd, estimated at 2500, waited for their arrival.[34] Around 2:30 p.m., Class A winner Les Viland and his Rambler American 440 was the first to receive a checkered-flag greeting from USAC's J.C. Agajanian. The seven remaining class winners followed in order, leading the remaining competitors in passing under the finish line banner. After Mary Faulkner in a Mercury Comet 202 6, the last car in the convoy, received his usual enthusiastic greeting, "Aggie" would roll up his green and checkered flags and retire after 16 years serving as the Run's official starter. With the class-winning cars on display at the Prudential Center, the drivers were presented with their trophies by Mobil Oil's President Fred H. Moore. Later in the day, a dinner and awards ceremony was held at the Sheraton-Boston Hotel. A most challenging, and for some competitors expensive, Economy Run was now history.

Despite a few hiccups along the way, Les Viland continued his strong performance, recording the best mileage for all entries with 23.8063 miles per gallon. Leading from the start, he continued to hold off Byron Froelich's Ford Falcon, finishing by a margin of .2333 mile per gallon. Crossing the finish line with the lowest miles per gallon, with a figure of 13.6802, was Frank Mowrer in a Mercury Comet Cyclone GT. With this performance, Frank had the distinction of recording the lowest mileage figure in the history of the Economy Runs. General Motors placed first in four classes, Chrysler Corporation placed first in three, and American Motors in one. It was an interesting "split" with American Motors and Chrysler taking the compact and intermediate size classes and General Motors taking those for full size models. Five classes were decided by less than a half-mile per gallon.

Class winners were:

CLASS	MAKE and MODEL	MILES PER GALLON
A — Compact Six Cylinder	Rambler American 440	23.8063
B — Compact Eight Cylinder	Dodge Dart	21.5761
C — Intermediate Size Six Cylinder	Plymouth Belvedere I	23.1019
D — Intermediate Size Eight Cylinder	Dodge Coronet Deluxe	21.0520
E — Standard Size Six Cylinder	Chevrolet Biscayne	21.0454
F — Low Price Eight Cylinder	Chevrolet Impala	19.8748
G — Medium Price	Buick LeSabre	19.2439
H — Luxury Cars	Oldsmobile Toronado Deluxe	17.3332

Detailed results for all competing cars can be found in the Appendix.

When the results were released by USAC officials, it was noted that this year's pace was the fastest in the Run's history with an average of 51.8367 miles per hour. This exceeded the prior record set in 1965 by a slim .1537 mile per hour. With a majority of models competing being equipped with eight-cylinder engines, a combination of factors, including speed, weather and terrain, all resulted in an overall miles per gallon average of 19.5614. This was a decline of 1.3267 miles per gallon from 1965's average, and was the first time below the 20-mile-per-gallon mark since 1959. Once again these figures reflected the efforts of both USAC and Mobil Oil to make these mileage results more realistic. The four teenage drivers averaged 18.7931 miles per gallon, which was a direct result of the larger and more powerful models they were piloting. Three of the four were in eight-cylinder models. This year the route was the longest in the Economy Run's history at 3301.4 miles and would not be surpassed in future years. The total driving time allowed was 64 hours and 2 minutes, which included the 10-minute extension on the last leg due to weather conditions. On a negative side, this year saw five competitors involved in collisions, one of which resulted not only in the injury to a USAC riding observer, but in the elimination of the Ford Fairlane 500 6 entry. Two of the damaged models were parked in the Boston winner's circle. No mechanical problems were noted, but one entry required a tire replacement.

Two drivers stood in the winner's circle for the first time, Bob Binder (Plymouth Belvedere I 6) and Don Royer (Chevrolet Impala 8), joining six veterans. For Les Viland, this was his fifth consecutive class win and ninth first-place finish. After disappointing results last year, Mel Alsbury, Jr. was back in good form, scoring his seventh first-place finish. George Miller and Pat Sawyer both notched second consecutive victories, and for Pat this was her fifth first-place award. Bob Checkley and Marta Retzlaff both captured first-place trophies for the second time.

From the start, it was certain that the class for six-cylinder compacts would be a battle between Les Viland's Rambler American and Byron Froelich's Ford Falcon. Renewing their rivalry from last year, Les was out in front from the start, although it was tight throughout the entire Run. At one point, the margin between the two was a mere .1252 mile per gallon. Bill Levy's Chevrolet Corvair Monza stayed close on their heels, coming in third with less than a half-mile per gallon separating the three entries. This was not an easy victory for Les Viland as he endured two late arrivals, one as a result of getting lost, which resulted in mileage penalties. Les estimated that "he would have exceeded 24 m.p.g. except for a penalty assessed Saturday when he got lost briefly, [and] called the run the most realistic in history." He also added that "the inclusion for the first time of California-enforced, exhaust anti-smog control systems cost his car about 2 m.p.g."[35]

One of the reasons contributing to the Rambler American's continued success, compared to its rivals, appears to deal with tire size. Prior to the start, Hugh Randolph of the *Oakland* (California) *Tribune* made the following observation:

> In the impound area, this writer discovered that the Rambler is equipped with 6.25 × 14 tires, while the Falcon will be running on 6.50 × 13-inch rubber. This means that, with the smaller tire, the Falcon's wheels must turn more times to travel a mile. When we queried race steward Art Rene about this theory, he said, "That's right. Unless the Falcon turns in an outstanding performance, the tire could mean the ballgame." The circumferential differences in the two wheels means approximately one gallon of gas difference per thousand miles. With the run stretching across 3300 miles, this means about 3.3 gallons of gas.[36]

Although not sure if his damaged Dodge Dart would make it to Boston, Bob Checkley maintained the comfortable lead he held from the start to take top honors in the compact car eight-cylinder class. With a baling-wire approach to secure the trunk lid, he finished .9988 mile per gallon ahead of Fran Foster's Ford Mustang 2+2 in second. Finishing in third, well off the pace, was Ed Dempsey's Ford Falcon, followed by both Chevrolet Chevy II entries.

Bob Binder and his Plymouth Belvedere I maintained his lead among the six-cylinder intermediate-sized models despite an issue with one of his tires which resulted in a penalty for a late arrival in Wichita. This was the only Plymouth model in the winner's circle this year, which marked its tenth consecutive first-place finish. Although he was unable to surpass Les Viland for top mileage honors, it was close, as Bob

Despite being involved in a minor accident in Ohio, Patricia Sawyer brought her Chevrolet Biscayne to a first-place finish in the standard-size six-cylinder class. Following Pat is Don Royer, driving a Chevrolet Impala, who placed first in the low-price eight-cylinder class, making this a double win for Chevrolet. These class wins also ended Plymouth's dominance among the "low-price three" that it had enjoyed since 1957.

came to within .0888 mile per gallon at one point. The real battle in this class was for second place between Paula Murphy in a Buick Special and Don Francisco in a Pontiac Tempest Custom. Although they swapped positions midway through the Run, Paula managed to finish in second place a slim .0733 mile per gallon ahead of Don. She was especially pleased at her performance this year after having to relinquish driving duties midway through last year's run due to a recurring back problem. After a strong third-place finish in 1965, Mary Faulkner and her Mercury Comet 202 struggled this year, finishing well off the pace in sixth.

Another class where competition was close from the start was among the eight-cylinder intermediate-size models. For the second consecutive year, a Dodge Coronet led from the start, facing a challenge from John Rich, who was looking for his fourth class win driving a Buick Special Deluxe. At one point, John came to within .1476 mile per gallon of the leader, only to miss an exit on the Pennsylvania Turnpike, resulting in his driving an additional 22 miles. This pretty much sealed his second-place finish. Despite a heavily damaged left front fender and hood, Bobby Spears managed to bring his Ford Fairlane 500 8 home in third place. This class had the largest field of competitors, two of which were equipped with high-performance engines. These models did not fare well at all, with Dick Harrell's Chevrolet Chevelle SS-396 in tenth and Frank Mowrer's Mercury Comet Cyclone GT in twelfth.

Competition among the popular full-size low-price models was especially keen and saw Chevrolet scoring a "one-two" punch. In an "all-girl" class, Pat Sawyer scored her second consecutive class win with her six-cylinder Chevrolet Biscayne, despite having been involved in a minor collision on the St. Louis-to-Cleveland leg. Drag racing champion Shirley Shahan had a good second-place finish, .4880 mile per gallon behind Pat, in a Plymouth Fury I. Ginny Sims placed third in her Ford Custom, over a mile per gallon off the pace. Among the eight-cylinder models, it was a strong finish for Don Royer's Chevrolet Impala with a 1.1852 margin over Paul Prior's Chevrolet Caprice in second. Don opened up with a strong lead from the start and never looked back, recording the Impala's first class victory since 1962. It was a different story for the four drivers battling for second and third place, as competition was extremely tight and positions changed frequently. After the final top-off, Paul Prior managed to keep his Chevrolet Caprice in second by a mere .0049 mile per gallon over Bob Cahill's Plymouth Fury I. Also at the end, Carl Diehl in a Dodge Polara 318 managed to pull ahead of Carolyn Thomas's Chevrolet Bel-Air, dropping her into fourth place by a mere .0051 mile per gallon. For Carolyn it was a disappointing ending after having been in both second and third place midway through the Run. In the end only .0831 mile per gallon separated second through fifth positions. Kay Kimes and his Ford Custom 500 finished in sixth place, almost a half-mile per gallon behind the others, and was never a factor.

For the first time since 1957, no full-sized six- or eight-cylinder Plymouth was in the winner's circle, bringing to an end its dominance over its traditional rivals in the low-price field. It was becoming clear that the newer, heavier Fury models were not as economical as the earlier, lighter-weight versions.

Among the medium-priced cars, Marta Retzlaff scored her second first-place finish in three years with her Buick LeSabre in another close contest. Due to a combination of factors, including Dodge entering its Polara 318 in the low-price class and the withdrawal of the Mercury Park Lane at the start, competition in the medium-price class was all General Motors. Art Arfons in a Pontiac Star Chief Executive finished a strong second, for the second year in a row, only .3231 mile per gallon behind the Buick LeSabre. Prior class winners Jean Calvin in a Pontiac Catalina finished third, with Stan Raymond in an Oldsmobile Jetstar 88 in fourth.

Veteran driver Mel Alsbury, Jr. and Oldsmobile's all-new Toronado not only took first place in the luxury car class but also ended Chrysler New Yorker's four-year string of victories. The Toronado's win marked the second time a front-wheel drive model both participated and placed first in its class. In the 1937 Gilmore Yosemite Economy Run, a Cord Beverly took home the top prize. Leading the class from the start, the Toronado maintained a fairly comfortable lead. The real competition was between second and fourth place, featuring two longtime veteran drivers. After sitting in third place for most of the Run, Hart Fullerton managed to pull in front of Pete Novotny's Oldsmobile Ninety-Eight by a slim margin of .0634 mile per gallon. Newcomer Mike Jones and his Buick Riviera was close behind in fourth, as only .1143 mile per gallon separated second from fourth places.

The four teenagers gave it their all but unfortunately did not finish as well as hoped. The bright star of this group was Carolyn Thomas in a Chevrolet Bel-Air, who managed to hold on to second or third place in a tight contest for most of the Run, only to end up finishing in fifth. The three who were competing in the same class for eight-cylinder intermediate-size cars were never much of a factor. Harold Limbird, driving a Mercury Comet Cyclone, had the best finish, placing fourth, followed by Jim Finnessy, driving a Chevrolet Chevelle 300, in sixth, and Frank Mowrer, driving a Mercury Comet Cyclone GT, in twelfth. Only Carolyn would return to compete in 1967.

When the Imperial Crown crossed the finish line, it would mark the end of its longtime involvement dating back to the 1940 Gilmore Yosemite Run, when it was considered a luxury model in the regular Chrysler lineup. The Imperial's most notable accomplishment was its three back-to-back Sweepstakes Awards (1956 to 1958) along with placing first in its class seven times. Imperial also placed second in its class seven times and third once. The last three years saw it place well down in the standings among the luxury models competing.

General Motors dominated the four classes for full-sized models in the 1966 Economy Run. In addition to the double win for Chevrolet and a win for the Oldsmobile Toronado, Marta Retzlaff brought her Buick LeSabre across the finish line, placing first in the medium-price class. This was Marta's and Buick's second win in this class in the last three years.

The end of this year's Run was a time to bid farewell to a few drivers and dealers who had remained active for a number of years. Among the drivers, Patricia Jones, who first drove in 1957 when women were first permitted to compete, is noted for being one of eight women to score a victory in their class. This came in 1959 as she drove a Dodge Coronet, competing in the low-medium-price class. Other drivers not returning included Art Arfons, Ed Dempsey, Mary Faulkner and George Miller. Two of the dealer entrants, Buick's Jim Retzlaff and Chevrolet's Melvin Hilliard, would also end their involvement.

Advertising and promotion showed another noticeable decline this year. The majority of advertising appeared in newspapers immediately following the Run, and with the exception of

Mobil Oil's efforts into mid–April, there was little else. Mobil Oil continued their tradition of announcing the results with two separate newspaper advertisements. The initial copy featured all text with the headline "1966 Mobil Economy Run proves gasoline mileage potential in cars like yours," and a brief thumbnail sketch of this year's event. This would also be one of the last times the "We proved it in public, too" tagline would be featured.[37] A second advertisement featured a photograph of a competing Chevrolet Impala with the headline: "How far can one gasoline go to show how far you can go?" It also listed the results while stressing the 3,310 miles the cars traveled this year to Boston. This was done once again to show a direct contrast to the different criteria used by Pure Oil in the economy portion of their Performance Trials. This advertisement also introduced a new tagline "Try Mobil / The Economy Run Gasoline."[38] A third newspaper advertisement highlighted the overall theme and conduct of the Economy Run. With a photograph of the Rambler Ambassador 880 crossing the Boston finish line, Mobil continued to stress that it proved it mileage claims in public and that there was "mileage potential built into Mobil gasolines."[39] An 8-page folder detailing the results was again available at Mobil service stations.

As expected, American Motors continued to publicize the Rambler American's economy performance. With a photograph of the winning American crossing the Boston finish line, the headline announced, "Rambler American Wins Mobil Economy Run — Class A ... and does it with new acceleration champ engine!" Readers were told that the American was equipped "with the newest, biggest, most powerful six in its class." Stressing performance along with economy, the ad also mentioned that this is "the same engine that powered us to an acceleration win over Falcon, Dart and Corvair in the Performance Trials at Daytona."[40] Since all advertising concerning the Economy Run required prior clearance from USAC, it is not surprising that the Pure Oil name was left out. This advertisement was reprinted in a large 33" × 50" poster for use by its dealers to display in their showroom windows. This same advertising theme was repeated in a 60-second television commercial featuring footage of the American 440 on the trip from Los Angeles to Boston.[41] American Motors also mentioned the American's continuing performance in its 1967 showroom and "X-Ray" comparison catalogs.

Chevrolet's double win was the theme of its advertising campaign. With the headline "Chevrolet's No. 1 in Mobil Economy Run!," a newspaper advertisement featured photographs of the winning Biscayne and Impala models while telling readers of story of this year's achievement.[42] These wins occurred in the midst of Chevrolet's "Double Dividend Days" sales campaign, and dealers were provided with several newspaper and radio advertising layouts to use in their local market.[43] A small 6-page folder was developed for Chevrolet salesmen in highlighting these twin wins.[44] Chevrolet's advertising also mentioned its earlier class wins with the Corvair, Chevelle SS396 and Caprice models in the Pure Oil Performance Trials.

Featuring a photograph of Marta Retzlaff and her Buick LeSabre, Buick announced, in reference to her victory in 1964, "Lady Driver Repeats! Buick LeSabre Wins First Place in Mobil Economy Run — Class G!" With both Specials placing second in their respective classes, Buick needed to note their efforts in a competing venue, noting that "a Buick Special V-6 — a car noted for economy — took the victor's laurels in its class of the Pure Oil Performance Trials."[45] Buick dealers were provided with a small 4-page folder titled "Now It's Hot News From Boston!," which highlighted its results along with a brief review of the Run's history.[46] Buick's showroom brochure for its 1967 models featured an interview with Marta on her Economy Run and driving experiences.[47]

Oldsmobile noted the achievement of the new Toronado, headlining, "First time entered — Toronado wins its class in Mobil Economy Run!," and giving its final mileage figure with only a brief mention of details about the Run.[48]

Choosing to stress its performance image over economy, both Dodge and Plymouth ignored the three individual class wins with no advertisements being produced.

Performance was the name of the game in 1966 with gasoline economy becoming passé. Newspapers continued to be the main source for coverage of the Run, with a continuing emphasis on the first day's collision between the class-winning Dodge Dart 8 and Ford Fairlane 6. A noticeable change seen in Southern California was the coverage by the *Los Angeles Times*. After years of providing daily coverage, only a single detailed story covering the final results was published after the Run's conclusion. After extensive coverage by enthusiast magazines in 1965, articles this year were virtually nonexistent. Only *Auto Topics* and *Car Craft* provided coverage.[49] In the July 1966 issue *Car Craft*, an extensive article appeared which focused on the experiences and trials of the Mercury Comet Cyclone GT team of Frank Mowrer and Bob Dunker. Chrysler-Plymouth and Dodge dealers read of their division's success in their publications, *Chrysler-Plymouth Times* and *Dodge Reporter*. Mobil Oil continued its practice of devoting an issue of *Mobil Dealer News* to the Economy Run, which this year focused on safe driving, with comments and suggestions on this topic from selected drivers.

While there continued to be a promotional benefit to staging the Economy Run each year, it was becoming an increasingly expensive proposition. As an example, Mobil Oil's expenditures for trophies alone were in excess of $5000.[50] Mobil Oil's Frank Meunier, nearing retirement age, continued to be the primary advocate for continuing the Economy Run, but found himself competing against a new generation of advertising executives who were beginning to have increased influence regarding corporate objectives. Studies were already underway to determine Mobil's future involvement in its sponsorship of the Indianapolis 500, which would end after this year's race.

After the conclusion of the Run, consumer research firms were consulted regarding the future value of the Economy Run. As noted by John Heitmann in an article appearing in the fall 2008 issue of *Automotive History Review*, the views of two researchers were at opposite ends of the spectrum:

> Consumer researchers, including Alfred Politz, were consulted on the value of the Run, and his views markedly contrasted with R.S. Brophy's in-house study. Politz argued for the Run's continuance, stating that "The motorists' fascination with gas mileage does not wear out — just as sex appeal has held its ground over the ages." Brophy countered this view by stating that "It is interesting to note that one area of greatest effort, the Economy Run, cannot be related to any meaningful difference...." Brophy also harnessed the results of a consumer survey citing that some 42 per cent thought that professional drivers were used, 35 percent concluded that cars were especially adjusted, and 16 percent doubted the miles per gallon results.[51]

Consumer perceptions on the Economy Run always showed a sizeable percentage who doubted the validity of the Run's results, and this latest survey confirmed that perception. What is interesting with these results deals with the perception that professional drivers were used and this was not the case. While some teams indeed used factory engineering or test personnel to pilot their entries, the majority of drivers came from varied backgrounds. In a review of the biographies of the 52 drivers competing in the 1966 Run, 8 were employed by manufacturers (American Motors, Chrysler Corporation and Ford Motor Company) and 7 came from varied racing-related mechanical or engineering positions. An additional 7 drivers came from automobile dealership backgrounds ranging from ownership to sales. The majority of this year's drivers did have prior Economy Run experience, with 36 veterans competing.[52]

The researchers also noted that "while performance events factored little into consumer gasoline purchases, getting more miles per gallon did. Ironically, and perhaps with some embarrassment, most consumers did not recognize Mobil as the best gasoline in terms of mileage; rather it was Shell."[53] As noted earlier, Shell Oil's advertising campaign highlighting Super Shell with Platformate, designed to improve mileage, was having an effect. Based on these results, a group of Mobil executives concluded that "at present the Run does not offer a favorable opportunity for our use in a national advertising campaign." Indeed, Mobil's ad agency, Doyle, Dane and Bernbach, was not interesting in using any performance events in their advertising strategy. Recommendations were made to reduce costs, shorten the route and cut the number of ads.[54] Approval was given to stage the 1967 Economy Run, although it would be on a slightly reduced scale.

This turned out to be an expensive year for both Ford Fairlane models competing in the Economy Run. Bobby Spears, driving a Ford Fairlane 500 8, was involved in a collision with a non-competing car in Ohio and sustained a good deal of damage to the left front. Despite the damage, Bobby was able to continue and placed third in the intermediate-size eight-cylinder class.

While discussions at a corporate level continued, another event would possibly cloud the future of the Economy Run. Arthur C. Pillsbury, who had guided the Economy Run from its beginning in 1936 as its Chief Steward, passed away on September 13, 1966. Art had been in declining health for most of the year and, while still calling the shots for USAC, was unable to make this year's trip to Boston. Reflecting on his impact on not only the Economy Run, but racing in general, a tribute which focused mainly on his involvement with the Bonneville Salt Flat Trials appeared in the December 1966 issue of *Road & Track*. It read in part:

> Arthur Pillsbury, whose recent death went almost completely unnoticed in the press, was one of the greatest, most important and most unappreciated figures in the long history of automotive competition. He began working with the American Automobile Association in 1915 and was a key man in founding the AAA Contest Board. A civil engineer by profession, he designed and built the country's best board speedways. During the depression he dipped deeply into his own reserves to keep first-class racing alive. He originated the Economy Run, which he masterminded for Mobil until his death.[55]

No one would doubt that he supervised the Economy Runs with a firm but fair hand, and his performance as chief steward was never questioned. As USAC looked for a new chief steward, one individual stood out among his contemporaries: Art Rene.[56] Appointed as a steward for the 1966 Run, Art brought to this assignment a long history of Economy Run involvement, beginning as a class-winning driver for Ford in 1950. During the years he had an association with Dodge and Plymouth before returning to the Ford team in 1963, later serving as its team captain. The transition would be seamless as other USAC officials, all with many years of experience, would continue working in their respective roles under Art's new leadership.

11

Detroit and New York

1967

Los Angeles to Detroit Via Lake Tahoe — Salt Lake City — Denver — Omaha — Chicago

April 4–9, 1967—2837.8 Miles

With Art Rene now at the helm as chief steward, preparations for a scaled-back Economy Run were finalized. On March 9, USAC officials announced that Detroit had been selected as this year's destination, returning after an absence of three years. They noted that this year's route was designed to embrace "road and weather conditions encountered by the average interstate motorist. The 31st annual Economy Run will stop overnight in Stateline (Lake Tahoe), Nev.; Salt Lake City, Utah; Denver, Colo.; Omaha, Neb.; and Chicago, Ill. The total distance will be nearly 3,000 miles, [climbing] from a few feet above sea level at the start to heights of more than 11,000 feet in the Colorado Rockies, testing drivers' skills as well as performance versatility of the competing American-made automobiles."[1] USAC officials also noted this six-day event would be over "the toughest Run course yet devised."[2] For those who had participated in 1963, they discovered that they would be traveling through some familiar territory.

With the rules firmly established and refined over time, the only changes necessary involved required equipment on the competing cars, along with a reduction of the competitive classifications to seven. Due to a decline in popularity, the class for standard-size six-cylinder models was eliminated, since the majority of these models were now being sold to fleet buyers. Air conditioning was a required option for cars competing in the luxury car class, and all cars were required to conform to California smog control regulations.[3]

When the "Competitive Divisions for the 1967 Mobil Economy Run" was released, it was apparent that the intentions of USAC officials "that cars competing in the Mobil Economy Run represent models being purchased and used by the public in substantial quantities" was being reinforced. In line with this, the provision in the eight-cylinder classes, where sports models could compete alongside their regular counterparts, was eliminated, resulting in only one entry being accepted for each eligible make and model.[4] This change was designed in an effort to make it easier for the public to better understand the results.

When the entries closed on February 17, 42 cars were set to compete, representing every U.S. make except Imperial and Lincoln. Again General Motors had the largest representation with 20 models entered, followed by Ford Motor Company with 11 and Chrysler Corporation with 10. Rounding out the field was a single American Motors Rambler American. Compact and intermediate-sized models continued to account for the bulk of entries with 27 entered. After sitting out the Run last year, the Plymouth Valiant and Barracuda returned to compete in both compact car classes. Making their debut this year were the Chevrolet Camaro and Mercury Cougar, competing against the established "pony cars" from Ford and Plymouth. Two-door models accounted for 33 of the entries, with the Ford Thunderbird returning for the first time since 1960. The Cadillac Calais, sponsored for the third year in a row by the West Covina Chamber of Commerce, was the only private entry.

This year saw a new location for the Pasadena impound, which was housed in a more modern and spacious building. Replacing the all-metal Quonset hut that had been in use since 1952, this new facility offered more space and better lighting, making it easier for USAC and each individual team to inspect and prepare each entry. Today this is the home of Phoenix Decorating, a well-known firm for designing floats for Pasadena's Tournament of Roses Parade. At its more visible location on Raymond Avenue, special signage now identified this as the Mobil Economy Run impound, accompanied by a large map highlighting the route to Detroit.

Most manufacturers continued their preparations and testing using the same team leaders and personnel from last year, although on a reduced scale as the manufacturers were splitting resources between the Mobil Economy Run and the Union/Pure Oil Performance Trials. There was a belief by some, and somewhat substantiated by press interest, that more publicity was now being realized from the Union/Pure Oil Trials.

At Los Angeles' Century Plaza Hotel, USAC Steward Fred Agabashian sends Al Johnson and his Ford Fairlane 500 8 on his way to Lake Tahoe. A smaller crowd at the start, composed mainly of those who were associated with the Economy Run, was again in evidence this year. Competing in the intermediate-size eight-cylinder class, Al's Ford Fairlane would place fifth.

This reduction in personnel was particularly evident for those associated with Chrysler Corporation and General Motors. Although Chrysler continued to have the largest team of factory engineers on-site, far fewer were seen this year than previously. General Motors scaled back their factory involvement, with Buick the only division providing the most direct support. Joining Chevrolet and Pontiac, Oldsmobile relied on dealers to provide preparation and support for their entries. Stepping in to handle the Oldsmobile's effort was West Covina dealer Mandy Williams, who was assisted by the experienced Alsbury brothers, Mel, Jr. and George. Only fielding a single entry, American Motors continued to rely on its factory team of Les Viland and Les Scott for preparation and competition responsibilities.

Dealer associations returning this year as sponsors represented American Motors, Dodge, and Mercury, along with single Buick and Plymouth entries. Dodge selected associations again based on the route and included those in Chicago and Detroit. With Mercury increasing its factory support, all of its entries were sponsored by the Mercury Dealers of U.S.A. Late in the impound period, a longtime participant, Inglewood's Bob Estes Lincoln-Mercury, was named as sponsor for its Cougar entry. This was designed to reflect Bob's longstanding association, dating back to 1950, with the Economy Run. Two dealers participating for the first time included Allan Brothers Motor Sales (Ford) in Port Clinton, Ohio, and Paul Harvey Ford in Indianapolis, Indiana.

With interest by manufacturers showing a decline, this year found more Economy Run regulars behind the wheel as drivers than in previous years. Six were participating for the first time and the most notable of this group was Sweden's Per "Pelle" Nystrom, named to drive a Ford Mustang 6. Well known in Europe for his successful participation in racing, rallies and economy runs, and in fact a European Economy Run champion for eight years, Per would be the only European-based driver in history to participate.[5] Per's daughter, Elizabeth, was named as co-driver, forming a father-daughter team. After three years, Mercury discon-tinued having a team of teenagers competing in one of its entries, relying on more experienced drivers. With 19-year-old Carolyn Thomas set to compete in a Chevrolet Camaro, this resulted in the widest age spread among drivers in Economy Run history. At the other end of the age scale was 65-year-old Pete Novotny driving an Oldsmobile Ninety-Eight. All nine women drivers had previously participated, including four class winners, Jean Calvin, Mary Hauser, Marta Retzlaff and Patricia Sawyer.

Two dealers with a history of participation returned in dual roles as entrants and drivers. Hart Fullerton (Chrysler 300) was joined by Mandy Williams (Oldsmobile F-85 Cutlass 8), returning after a two-year absence. Participating for the first time was Carl Bucky, owner of Hollywood Chrysler-Plymouth with a Plymouth Belvedere II 8. The four returning husband-and-wife teams, C.K. and Betty Enoch (Chevrolet Chevy II 100 6), Hart and Diana Fullerton (Chrysler 300), Gordon and Scherlee Madison (Chevrolet Corvair Monza), and Stan and Lois Raymond (Oldsmobile Delmont 88), were joined by George and Barbara Alsbury (Oldsmobile F-85 6). While George had been a regular participant since 1957, this was Barbara's first experience on the Run. Seven of last year's class-winning drivers returned, with George Miller the only absentee.

Random car selection and inspections began in early February. With the requirement that all entries conform to California's smog control regulations, USAC officials would no longer have the added expense of traveling out-of-state to locate an entry. Over the next five weeks each car was on the road for the standard 1500-mile break-in and test runs along with regular servicing and preparations. During this time, each entrant was again given the option of using either Mobil Regular or Mobil Premium gasoline. All elected to use Mobil Premium.

During this time, some teams would encounter problems, either real or perceived, with their cars. One area of concern involved both Buick Specials. In a letter to Art Rene, Buick's sales-engineering coordinator requested installation of overload springs or load-leveler shock absorbers in these models. Because of the extra weight of the special gas tanks, Buick's personnel were concerned about the lowered trim height "to a point where visibility through the rear view mirror is difficult for the driver." There was additional concern raised by Paula Murphy, driver of the Special 6 entry, who felt it was not only exasperating but unsafe as well. This request also stressed that these corrections would not affect the economy of characteristics of the car in any way.[6] Although no documentation has surfaced whether or not this request was approved, it is doubtful this was approved based on USAC's track record requiring absolute stock status for all competing cars. Additionally, photographic evidence does not show any abnormal look about either Special model when compared to other cars competing in their respective classes.

The first of many driving challenges to face the drivers occurred during the first leg. Heading into the Sierra Nevada Mountains, the drivers encountered snow and slush, which created dangerous driving conditions, although chains were not required. Bill Levy, driving a Mercury Capri, carefully navigates a slow 21.4-mile stretch in below-freezing temperatures as he makes his way to Lake Tahoe. Competing in the intermediate-size six-cylinder class, Bill would finish in first place, giving Mercury its first class win since 1952.

Coordination of publicity and press arrangements continued under the leadership of Mobil Oil's Frank Meunier and his staff. With years of experience under their belts, Roger Mahey and John Parkinson only needed to make slight refinements to their procedures, which mainly involved dealing with the establishment of daily press and hospitality rooms. By early February, arrangements

After arriving in South Lake Tahoe, the drivers' first stop was for refueling. Here Les Viland's Rambler American 220 6 and Carolyn Thomas's Chevrolet Camaro 8 wait for their turn. This year would see disappointing performances for both the Rambler American and Chevrolet Camaro entries. This would be the only year a Chevrolet Camaro would compete in the Economy Run.

had been completed to accommodate the press representatives traveling on the chartered United Airlines DC6-B, along with final arrangements for cars from various manufacturers to be used by additional press representatives who would be covering the Run from the perspective of the drivers and co-drivers.

The first of several traditional pre-start activities, the drawing of starting positions, was held at the Pasadena Impound on March 20. Two-time class winner Marta Retzlaff selected the No. 1 spot for her Buick LeSabre 400, and Bob Binder would be bringing up the rear when he selected No. 42 for his Plymouth Belvedere I 6. Two announcements were made by USAC officials at this time. This year's start would be at the recently opened Century Plaza Hotel, located in Los Angeles' famed Century City complex; and Edgar Marshall, who was injured in last year's opening-day accident, would be returning as a USAC riding observer.

A few days prior to the start, two changes occurred regarding driver and co-driver responsibilities. Carl Bucky was replaced as driver of the Plymouth Belvedere II 8 by Allan Brooks, who would participate for the first time. Stepping in to assist Allan as co-driver was Bert Sisson, bringing with him several years of experience. Les Scott, who had been slated to join Les Viland as co-driver in the Rambler American 220, stepped aside, and Tommy Thomas was named as his replacement.

On Monday, April 3, the competing cars were driven from the Pasadena impound to the Century Plaza Hotel. After their arrival, they were topped off and parked in their respective spots for tomorrow morning's start. This time provided an opportunity to talk with the drivers to gain insight on their chances against the competition. One driver who was expressing considerable apprehension was perennial winner Les Viland. In an interview with Hugh Randolph of the *Oakland Tribune*, Les remarked:

> "The Falcon, Dodge Dart, Valiant, and Chevy II are much improved over last year, and very well could give us a very hard time during the next six days," he said. "Axle ratios have been changed. Some of the cars have changed from 13" to 14" wheels, and the exhaust emission pump has been removed from the Falcon. Any or all of these changes can make a great

> deal of difference when mileage is measured to the ten-thousandths of miles per gallon." [Les] Viland also predicted that all compact cars in the run will be on radial tires within the next five years. "They ride harder and nosier, but they produce better than one mile per gallon on a 3,000 mile test."[7]

Les's concerns were well founded, as Chrysler Corporation was one of those fully capable of winning not only in its class but in achieving the best mileage for all entries. For the redesigned 1967 Plymouth Valiant and Dodge Dart models, Chrysler had switched to a part-throttle kickdown in the automatic transmission, which then resulted in the ability to use a lower 2.76 axle ratio with the 170-cubic-inch engines.[8] Comparing it to the Rambler American's 2.73 axle ratio with its 199-cubic-engine, Chrysler felt that the playing field was a little more level. Another area of interest within the compact six-cylinder class was the experience level of the drivers, with five (Bob Cahill, Bob Checkley, Byron Froelich, Patricia Sawyer, and Les Viland) having won their classes multiple times.

With the cars in place, the first day's itinerary and route description was distributed, giving the drivers a good idea that this year's Run was going to be tough. But before they could look at the routing, there was an opening tribute to A. C. Pillsbury which read:

> The 1967 Mobil Economy Run is dedicated to the memory of Art Pillsbury. The integrity, knowledge and expert advice he left us with has made it possible.[9]

Now it was down to business. Covering a distance of 488.3 miles, the first day's trip to Lake Tahoe was set with a time limit of 9 hours and 53 minutes, not including time for refueling or brunch stops. The drivers would learn quickly of a tight time schedule when USAC officials cautioned, "You must be on schedule upon your arrival in Sacramento as noted in the itinerary. There is no cushion and it is impossible to make up time *safely* between Sacramento and Lake Tahoe."[10] Weather conditions for tomorrow did not look favorable as rain and the possibility of snow flurries in the Sierra Nevada Mountains were being predicted.

On April 4, everyone on hand was ready for what would be a challenging opening leg. At 6:00 a.m., with a small group of spectators on hand under overcast skies, Marta Retzlaff in her Buick LeSabre 400 was the first to be sent on her way by USAC Steward Fred Agabashian. With J.C. Agajanian's retirement, the usual enthusiastic send-off was missing this year as each entry drove under the start banner and headed out of Century City bound for the San Diego Freeway. Heading north out of Los Angeles, the drivers followed U.S. Route 99 over Tejon Pass Summit, where they briefly encountered snow flurries, towards Bakersfield and Fresno. Driving through the San Joaquin Valley, the drivers then encountered rain and buffeting winds before arriving in Merced for the first stop of the day at Chet Adams Mobil Service, where all entries were topped off.[11] Continuing north on U.S. Route 99, an hour later they arrived in Modesto for a brunch stop at the Edgewater Inn, and then it was on to Sacramento. As it passed through Stockton, the route dipped to an elevation of 14 feet, marking the lowest point on this year's Run. After arriving in Sacramento, the drivers headed east on U.S. Route 50 into the Sierra Nevada mountains bound for Lake Tahoe, where more snow flurries were encountered. After they passed the town of Twin Bridges, driving snow and slush created dangerous driving conditions, although chains were not required. For the next 21.4 miles, in below-freezing temperatures, they slowly made their way over Echo Summit.[12] Shortly after the last competing car crossed the summit, the highway was closed to cars without chains.[13] The drivers continued to make their way to South Lake Tahoe for a refueling stop at Letterman Mobil Service, and after a short 5-mile drive they arrived in Stateline and the impound at the Sahara-Tahoe Hotel and Casino. Just before 5:00 p.m., Marta Retzlaff in her Buick LeSabre 400 was the first to arrive, with all of the drivers relieved to see this first leg completed. Later that evening one of the drivers from the east remarked, "It would be ironic if we encountered our worst weather of the trip in California."[14] With all of the competitors having arrived safely, USAC officials announced that today's time limit was extended 15 minutes due to adverse weather and road conditions. With this extension, everyone arrived within the adjusted 10-hour and 8-minute time limit.[15] Tomorrow's weather forecast for the trip across Nevada and Utah to Salt Lake City did not look encouraging.

When the results were released, the biggest surprise of the day was that Les Viland in his Rambler American was not found in his usual leading position. Bob Checkley, driving a Plymouth Valiant 100, was out in front of all entries with 26.2071 miles per gallon, as well as leading the compact six-cylinder class. A close second by only .6219 mile per gallon was Bob Cahill driving a Dodge Dart. Les's Rambler American 220 was found sitting in fifth place, behind the Ford Mustang and Falcon, 2.4570 miles per gallon off the pace. Competition was especially keen in two classes. The closest was in the low-price eight-cylinder class, with its history of close competition, as Kay Kimes's Ford Custom 500 held a mere .0061 mile-per-gallon lead over Don Royer's Chevrolet Impala SS. Another close contest was in the medium-price class where Marta Retzlaff's Buick LeSabre 400 had a slim .0520 mile-per-gallon margin over Stan Raymond's Oldsmobile Delmont 88. The remaining class leaders held comfortable leads, with one of the biggest surprises being the performance of the Mercury Capri. The Capri was leading the intermediate-size six-cylinder class, the first time Mercury led its class since 1952. Plymouth had a strong showing, leading in three classes, with Buick, Chrysler, Ford, and Mercury leading in one each. This was quite a reversal of fortune for General Motors, who had led in the three full-sized classes this time last year. The only first-time driver leading was Allan Brooks in a Plymouth Belvedere II 8. Among the models competing for the first time, the Mercury Cougar was in third place in its class, and disappointing results for the Chevrolet Camaro entries had both in last place. The field recorded an average of 20.5587 miles per gallon. With only one teenage driver competing, there was no reason to separate her average mileage figures. In completing the 488.3-mile trip within the adjusted time limit, the field averaged a speed of 49.2740 miles per hour.

The top three performers in each class were:

Compact Six Cylinder	Plymouth Valiant 100	26.2071 Miles Per Gallon
	Dodge Dart	25.5852 Miles Per Gallon
	Ford Mustang	24.6377 Miles Per Gallon
Compact Eight Cylinder	Plymouth Barracuda	23.2037 Miles Per Gallon
	Dodge Dart	22.0606 Miles Per Gallon
	Mercury Cougar	21.1019 Miles Per Gallon
Intermediate Size Six Cylinder	Mercury Capri	24.3991 Miles Per Gallon
	Plymouth Belvedere I	23.3259 Miles Per Gallon
	Buick Special	22.2878 Miles Per Gallon
Intermediate Size Eight Cylinder	Plymouth Belvedere II	20.8160 Miles Per Gallon
	Mercury Caliente	20.3846 Miles Per Gallon
	Dodge Coronet Deluxe	20.3447 Miles Per Gallon
Low Price Eight Cylinder	Ford Custom 500	19.8534 Miles Per Gallon
	Chevrolet Impala SS	19.8473 Miles Per Gallon
	Plymouth Fury II	19.1961 Miles Per Gallon
Medium Price	Buick LeSabre 400	19.1410 Miles Per Gallon
	Oldsmobile Delmont 88	19.0890 Miles Per Gallon
	Pontiac Catalina	18.9325 Miles Per Gallon
Luxury Cars	Chrysler 300	18.2523 Miles Per Gallon
	Buick Electra 225	17.5476 Miles Per Gallon
	Oldsmobile Ninety-Eight	16.8087 Miles Per Gallon

On April 5, the drivers prepared to head out on a 569.5-mile trek Salt Lake City within a time limit of 10 hours and 47 minutes. This would be the longest day in this year's Run in terms of both mileage and driving time. Shirley Shahan, driving a Dodge Coronet Deluxe 8, was the first driver on his way at 5:30 a.m., following U.S. Route 50 towards Carson City and Fernley. For the first 18 miles the drivers had slow going, having to contend with snow and ice on the road.[16] Leaving Fernley they headed east on Interstate 80 / U.S. Route 40, heading for their first stop of the day, and now encountering snow flurries as they made their way across Nevada. After 6 hours on the road they arrived in Elko for brunch at Stockmen's Motor Hotel and were pleased to see that there was no snow on the ground. Immediately after brunch they pulled into Parkside Mobil Service, where all entries were topped off. Back on U.S. Route 40 they continued east, crossing into Utah, entering the Mountain Time Zone, and passed the entrance to the Bonneville Salt Flats as they headed for Salt Lake City. More weather problems faced the drivers as they had to battle gusty winds, up to 45 miles per hour, across Utah, and then it was heavy rain for the last hour of driving. After their arrival in Salt Lake City, the first stop was for refueling at Redwood Mobil Service, followed by a short drive to the impound located at the Ramada Inn. This marked the fourth time Salt Lake City had hosted the Economy Run for an overnight stop. Just after 6:00 p.m., the first to arrive was Art Bertelsen driving a Dodge Polara 318. Despite today's weather and driving conditions, all drivers arrived within the time limit set by USAC.

Today's overall average of 20.4467 miles per gallon was only a fraction less than yesterday's results. Bob Checkley's Plymouth Valiant 100 continuing to record the best mileage with a figure of 25.5773. This was a drop of .6298 mile per gallon from yesterday's results, which reflected the difficult terrain and weather conditions. Close on Bob's heels, by .3559 mile per gallon, was Bob Cahill's Dodge Dart. Perennial leader Les Viland continued to fall off the pace in fifth place, 1.7775 miles per gallon behind the leader. A Mobil Oil executive remarked, "It would take a minor miracle for him to recover from his present position in Class A."[17] All of the class leaders stabilized their positions, although there was now close competition in three classes. In the intermediate-size eight-cylinder class, Allan Brooks saw his Plymouth Belvedere II's lead over Darrell Droke's Mercury Caliente shrink to a slim .1348 mile per gallon. In the low-price eight-cylinder class, Kay Kimes's Ford Custom 500 opened up a .1185 mile-per-gallon lead over Don Royer's Chevrolet Impala SS. Jim Latham's Plymouth Fury II was a close third, with a slim .2789 mile per gallon separating all three. In the medium-price class, Marta Retzlaff's Buick LeSabre 400 also opened up a .1764 mile-per-gallon lead over Stan Raymond's Oldsmobile Delmont 88. A battle was shaping up in this class for third place between Art Bertelsen's Dodge Polara 318 and Don Francisco's Pontiac Catalina: a mere .0043 mile per gallon separated the two. Scott Harvey's Plymouth Barracuda 8, Bill Levy's Mercury Capri 6 and Hart Fullerton's Chrysler 300 all maintained fairly comfortable leads. Four drivers — Byron Froelich (Ford Falcon 6), John Rich (Buick Special 8), Art Bertelsen (Dodge Polara 318), and Mel Alsbury, Jr. (Oldsmobile Toronado) — moved into third place. With a fairly direct highway route from Lake Tahoe, today saw one of the fastest average speeds yet with a figure of 53.0605 miles per hour.

The top three performers in each class were:

Compact Six Cylinder	Plymouth Valiant 100	25.5773 Miles Per Gallon
	Dodge Dart	25.2214 Miles Per Gallon
	Ford Falcon	24.6348 Miles Per Gallon
Compact Eight Cylinder	Plymouth Barracuda	22.8672 Miles Per Gallon
	Dodge Dart	21.7970 Miles Per Gallon
	Mercury Cougar	21.3791 Miles Per Gallon
Intermediate Size Six Cylinder	Mercury Capri	24.0090 Miles Per Gallon
	Plymouth Belvedere I	23.1130 Miles Per Gallon
	Buick Special	22.2583 Miles Per Gallon
Intermediate Size Eight Cylinder	Plymouth Belvedere II	20.5947 Miles Per Gallon
	Mercury Caliente	20.4599 Miles Per Gallon
	Buick Special	20.1575 Miles Per Gallon
Low Price Eight Cylinder	Ford Custom 500	19.8919 Miles Per Gallon
	Chevrolet Impala SS	19.7734 Miles Per Gallon
	Plymouth Fury II	19.6130 Miles Per Gallon
Medium Price	Buick LeSabre 400	19.0185 Miles Per Gallon
	Oldsmobile Delmont 88	18.8421 Miles Per Gallon
	Dodge Polara 318	18.4843 Miles Per Gallon
Luxury Cars	Chrysler 300	17.8152 Miles Per Gallon
	Buick Electra 225	17.1309 Miles Per Gallon
	Oldsmobile Toronado	16.5838 Miles Per Gallon

On April 6, another day of mountain driving faced the drivers as they began a 508.3-mile, 10-hour and 43-minute trip over the Rocky Mountains to Denver. With an early 4:00 a.m. start, Jack Kirkpatrick in a Chevrolet Chevy II 100 8 was the first to leave, heading out on U.S. Route 40 in light fog. After leaving Salt Lake City, one of the entries fell to a mechanical problem when a camshaft and valve train malfunction sidelined the Ford Futura 8 driven by Don Bailey.[18] This turned out to be Don's second string of bad luck: prior to last year's start, he was forced to withdraw after coming down with the flu. In two years of competition, Don had yet to reach the finish line.

Weather conditions would improve early in the day as clear skies with temperatures in the 40s and 50s would prevail. But there would be another obstacle facing the drivers when they encountered a construction zone near Duchesne, causing a slight delay along with rough gravel surfaces. After covering 202 miles, they crossed into Colorado bound for Craig and a stop at Hank Vallejo's Mobil Service, where all entries were topped off. As was a custom for years, the local high school band was on hand to entertain everyone during this stop. Further along U.S. Route 40, their next stop was in Steamboat Springs for brunch served by members of Holy Name Catholic Church, where students were dismissed from classes to welcome the competitors. As the drivers made their way to Denver, they crossed Berthoud Pass Summit, at 11307 feet the highest elevation on this year's Run, then began their descent into the Denver area. Although snow was still much in evidence in the mountain areas, the roads continued to remain dry. Their next stop was in Lakewood for refueling at Walt Brester Mobil Service, followed by a short 11-mile hop to the Denver impound at the Writer's Manor. Jim Latham in a Plymouth Fury II was the first to arrive around 5:30 p.m. C.K. Enoch, in a Chevrolet Chevy II 100 6, drove off-course for 4.7 miles and was unable to make up the lost time; arriving 4 minutes and 40 seconds late, he received a penalty. With the mountain driving conditions behind them and their trip now half over, the drivers were now looking forward

Ginny Sims brings her Mercury Cougar into the Sahara Tahoe Hotel and Casino impound after a 488.3-mile trip from Los Angeles. The drivers were relieved that for the time being the snow had stopped falling. This year would be the only appearance in the Economy Run for the Mercury Cougar, which would place third in the compact eight-cylinder class behind a Plymouth Barracuda and a Dodge Dart.

to some improving conditions in what had become by most accounts a very difficult three days.

With open highway driving and improving weather conditions, the drivers managed to show a slight improvement in their performance. The field averaged 20.6887 miles per gallon, a slight increase of .2420 mile per gallon from the previous day. Bob Checkley and his Plymouth Valiant 100 continued to lead all entries with the best miles per gallon figure of 25.7791. Continuing close on his heels was Bob Cahill in a Dodge Dart, closing the gap to .2420 mile per gallon. Six of the seven class leaders maintained their positions, although for some their margins over second place tightened. In the medium-price class, Stan Raymond's Oldsmobile Delmont 88 managed to squeak by Marta Rezlaff's Buick LeSabre 400, leading by an ever-so-slim .0013 mile per gallon. In two other classes the competition continued to remain close. In the intermediate-size eight-cylinder class, Allan Brooks in a Plymouth Belvedere II held onto his lead over Darrell Droke in a Mercury Caliente by a slim .1213 mile per gallon, while in the low-price eight-cylinder class, Kay Kimes in a Ford Custom 500 opened up a slightly larger .2078 mile-per-gallon lead over Don Royer in a Chevrolet Impala SS. In a tight battle for third place in the compact six-cylinder class, Per Nystrom in a Ford Mustang moved back into third place by a margin of .0496 mile per gallon over Byron Froelich in a Ford Falcon, now in fourth. At this stage, Chrysler Corporation officials were reported to be "jumping for joy as their cars went into the lead in four classes out of seven."[19] In addition to Bob Checkley's Plymouth Valiant 100 and Allan Brooks's Belvedere II, Scott Harvey's Plymouth Barracuda and Hart Fullerton's Chrysler 300 continued their leads by comfortable margins. With today's mountainous driving conditions, the average speed declined to 47.8365 miles per hour.

After two days of rough weather conditions, the drivers now had to contend with road construction. In eastern Utah, two eventual class winners, Hart Fullerton (Chrysler 300, front) and Bill Levy (Mercury Capri), encounter a traffic control delay before navigating rough gravel road surfaces along U.S. Route 40.

The top three performers in each class were:

Compact Six Cylinder	Plymouth Valiant 100	25.7791 Miles Per Gallon
	Dodge Dart	25.5371 Miles Per Gallon
	Ford Mustang	24.9781 Miles Per Gallon
Compact Eight Cylinder	Plymouth Barracuda	23.1491 Miles Per Gallon
	Dodge Dart	21.9859 Miles Per Gallon
	Mercury Cougar	21.6220 Miles Per Gallon
Intermediate Size Six Cylinder	Mercury Capri	24.2543 Miles Per Gallon
	Plymouth Belvedere I	23.2407 Miles Per Gallon
	Buick Special	22.6303 Miles Per Gallon
Intermediate Size Eight Cylinder	Plymouth Belvedere II	20.7941 Miles Per Gallon
	Mercury Caliente	20.6728 Miles Per Gallon
	Buick Special	20.2796 Miles Per Gallon
Low Price Eight Cylinder	Ford Custom 500	20.0793 Miles Per Gallon
	Chevrolet Impala SS	19.8715 Miles Per Gallon
	Plymouth Fury II	19.7144 Miles Per Gallon
Medium Price	Oldsmobile Delmont 88	18.9893 Miles Per Gallon
	Buick LeSabre 400	18.9880 Miles Per Gallon
	Dodge Polara 318	18.7004 Miles Per Gallon
Luxury Cars	Chrysler 300	18.0521 Miles Per Gallon
	Buick Electra 225	17.3846 Miles Per Gallon
	Oldsmobile Toronado	16.9354 Miles Per Gallon

On April 7, the drivers were ready for improved and more level driving conditions as they headed east towards Omaha. Everyone was pleased with today's forecast of dry conditions with temperatures expected in the 40s and low 50s. With a time limit of 10 hours and 3 minutes, most of the 553.6 miles would be fairly fast-paced on Interstate highways in eastern

Colorado and across Nebraska. Bob Cahill, driving a Dodge Dart 6, was the first to leave Denver at 5:30 a.m. following Interstate 80 / U.S. Route 118 towards the Nebraska state line. After crossing into Nebraska, the drivers picked up U.S. Route 30 as they headed towards North Platte, entering the Central Time Zone just east of Paxton. Arriving in North Platte, they made their first stop at Norm Mentzer Mobil Service, where all entries were topped off, quickly followed by brunch at the Chef Lalanne Kitchen. After a 40-minute break, the drivers were back on the road following Interstate 80, where they encountered light to gusty winds for several hours as they made their way towards the Pleasant Dale exit for refueling at Andy's Interstate Mobil Service. The drivers still had a 63-mile trip to Omaha and the impound at Schimmel's Indian Hills Inn, where they were greeted with light rain. This would be the second time Schimmel's hosted the Economy Run drivers, as this location was also an overnight stop in 1963. At 5:20 p.m., the first to arrive was Bob Cahill in a Dodge Dart 6. Three drivers—Tom Gillum, Jr. (Cadillac Calais), Pete Novotny (Oldsmobile Ninety-Eight) and Mandy Williams (Oldsmobile F-85 Cutlass 8)—arrived quite late, ranging from just over 10 minutes to 14½ minutes, and were penalized. In Mandy Williams's case, he drove off-course and was unable to make up the lost time. Although arriving within the time limit, Bill Cox in a Mercury Monterey also drove off-course and received a one-gallon penalty.

In Omaha, it was decided that a co-driver change was in order for Don Royer's Chevrolet Impala SS. While continuing to remain close to the class-leading Ford Custom 500, Don was not closing the gap as expected, and it was determined that a co-driver with more experience would make a difference. C.K. Enoch, driver of the Chevrolet Chevy II 100 6, who was sitting well down in the standings in the compact six-cylinder class, moved over to handle co-driver duties for Don. Betty Enoch took over as driver in the Chevy II, with Maxine Royer moving over from the Impala SS as co-driver.[20]

For the second day, everyone's mileage showed improvement, with the field now averaging 20.9137 miles per gallon. Bob Checkley's Plymouth Valiant 100 continued to lead all drivers with the best miles per gallon figure of 25.8885. Staying close on his heels was Bob Cahill's Dodge Dart with a figure of 25.7060 miles per gallon; a slim .1825 mile per gallon separated the two. In addition to Les Viland sitting in fifth place in the compact six-cylinder class, another veteran driver was also struggling with five-time class winner Patricia Sawyer in eighth (and last) place with her Chevrolet Camaro. Pat was averaging only 23.4762 miles per gallon, nearly 2½ miles per gallon off the pace being set by the Plymouth Valiant. In the medium-price class, Marta Retzlaff moved her Buick LeSabre 400 back into first place over Stan Raymond's Oldsmobile Delmont 88, although the margin between the two remained slim at .0338 mile per gallon. The remaining class leading drivers continued to stabilize their positions, although it remained very tight in the intermediate eight-cylinder class, with Allan Brooks's Plymouth Belvedere II holding onto a slim .1248 mile-per-gallon lead over Darrell Droke's Mercury Caliente. If Darrell could squeeze some extra mileage over the next two days, this would enable Mercury to have a first-place victory in both intermediate-size classes. Continuing the battle in the low-price eight-cylinder class, Kay Kimes's Ford Custom 500 maintained a slim .2277 mile-per-gallon lead over Don Royer's Chevrolet Impala SS. Those following the Run noted that after four days of driving, leaders in six out of the seven classes were running well ahead of the mileage figures posted by last year's class leaders. In some cases the improvement was "by as much as a gallon and a half."[21] Today's pace was fast with the drivers recording an average speed of 55.2245 miles per hour.

The top three performers in each class were:

Compact Six Cylinder	Plymouth Valiant 100	25.8885 Miles Per Gallon
	Dodge Dart	25.7060 Miles Per Gallon
	Ford Mustang	25.4154 Miles Per Gallon
Compact Eight Cylinder	Plymouth Barracuda	23.3262 Miles Per Gallon
	Dodge Dart	22.1109 Miles Per Gallon
	Mercury Cougar	21.9861 Miles Per Gallon
Intermediate Size Six Cylinder	Mercury Capri	24.6052 Miles Per Gallon
	Plymouth Belvedere I	23.4093 Miles Per Gallon
	Buick Special	22.9363 Miles Per Gallon
Intermediate Size Eight Cylinder	Plymouth Belvedere II	21.0058 Miles Per Gallon
	Mercury Caliente	20.8810 Miles Per Gallon
	Buick Special	20.5183 Miles Per Gallon
Low Price Eight Cylinder	Ford Custom 500	20.2529 Miles Per Gallon
	Chevrolet Impala SS	20.0252 Miles Per Gallon
	Plymouth Fury II	19.8778 Miles Per Gallon
Medium Price	Buick LeSabre 400	19.2814 Miles Per Gallon
	Oldsmobile Delmont 88	19.2476 Miles Per Gallon
	Pontiac Catalina	18.9231 Miles Per Gallon
Luxury Cars	Chrysler 300	18.2168 Miles Per Gallon
	Buick Electra 225	17.5657 Miles Per Gallon
	Oldsmobile Toronado	17.0512 Miles Per Gallon

After four days on the road, it was becoming evident to the experienced drivers that the tone of the Economy Run was evolving. As additional miles of the Interstate Highway System were being completed, more and more of their days would be spent traveling at a faster pace on these highways. The narrow two-lane country roads, known affectionately as "Pillsbury's Roads," were becoming largely a thing of the past, along with special timed control sections which the drivers had faced in the past two years. These changes were now reflecting America's evolving driving patterns as higher speeds and smooth road conditions were becoming more and more routine characteristics of cross-country travels.

In Steamboat Springs, a brunch stop was made at Holy Name Catholic Church, where students were on hand to welcome the drivers, including Fran Foster in a Ford Mustang 2 + 2 8. Although snow was very much in evidence, today's weather was clear and the roads remained dry as the drivers made their way to Denver.

On April 8, the drivers awoke to find heavy rain and gusty wind conditions as they prepared for the 513.5-mile trip to Chicago. With a time limit of 10 hours and 3 minutes, there was no doubt today would be a challenge. At 6:00 a.m., Jim Latham in a Plymouth Fury II 8 was the first to leave, following U.S. Route 275 and quickly crossing the Missouri River into Iowa. For the next hour the drivers followed Iowa 92 towards Interstate 80, bound for Tama-Montezuma for their first stop of the day at 63–80 Mobil Service, where all entries were topped off. After a short 45-mile drive, their next stop was in Amana for brunch at the Ox Yoke Inn. Instead of the usual 40-minute stop for brunch, USAC officials announced that this stop would be for 1 hour due to the uniqueness of this attraction. After battling the heavy rain and gusty winds up to 60 miles per hour for five hours, all of the participants welcomed this added time. During a downpour in Iowa, Marta Retzlaff had to contend with an electrical short in her Buick LeSabre that disabled the windshield wipers and caused her "to travel 65 miles an hour for more than 60 minutes peeping through a two-inch clear spot on [the] windshield."[22] After a short leg on Interstate 80, they picked up U.S. Route 30, crossing into Illinois as they made their way to Aurora for a refueling stop at Glen Smith Mobil Service. There still remained a 45-mile trek into Chicago, where the drivers would battle the evening commute traffic on the Eisenhower Expressway. After enduring wind and rain all day, the drivers faced another obstacle in the form of a large fire which leveled a six-story warehouse, spewing flaming debris on the roadway. This resulted in traffic coming to a standstill for nearly two hours, and while "some competitors slipped out of the on-course trouble area, others did not."[23] Slowly the drivers made their way to Chicago's Lake Shore Drive and the impound at the 50th on the Lake Motel, the second time this had been an overnight impound. Tom Gillum, Jr., in his Cadillac Calais was the first to arrive, with the majority of drivers arriving late after another long and difficult day on the road.

After arriving in Chicago, Celia Ross relinquished her co-driver duties for Stan Seeds and his Ford Thunderbird for reasons not disclosed. Replacing Celia as co-driver was Bill Watkins.[24] Later that evening, USAC's Chief Steward Art Rene made the following announcement:

> No penalties will be assessed today due to the unusual traffic condition on the inbound Eisenhower Expressway. Fuel will be computed to the Aurora (Refueling) Station. The 45.7 miles to the 50th on the Lake Motel overnight impound will be deleted. The cars have been topped off under the supervision of the United States Auto Club at the overnight impound. The mileage for the remaining leg will be computed from the start of the 6th day's log.[25]

As a result, the total distance was reduced to 467.8 miles, and the time limit was reduced to 9 hours and 11 minutes. This was the second time in the Run's history that a segment was eliminated from mileage calculations. The other occurrence was in 1963, when a flagman misdirected 12 cars in Lincoln, Nebraska, resulting in a 281.5-mile segment's being eliminated.

To no one's surprise, today's mileage figures showed a decline from the previous day, by well over a half-mile per gallon, with the field now averaging 20.4603 miles per gallon. Bob Checkley and his Plymouth Valiant 100 continued to lead all entries with 25.1517 miles per gallon. Bob also increased his lead over Bob Cahill's Dodge Dart, with the difference now .2412 mile per gallon. With one day remaining, six of the seven class-leading drivers continued to stabilize their positions, and barring any unforeseen problems they would be able to take home first-place trophies. Although Kay Kimes and his Ford Custom 500 continued to lead the low-price eight-cylinder class, the switch in co-drivers for Don Royer and his Chevrolet Impala SS appeared to be paying off.[26] Still sitting in second place, Don had closed the margin between the two to a mere .0624 mile per gallon. With Ford hoping for its first win since 1961 among the low-price models, it now appeared this would not be settled until the final top-off. Despite her problems on the road, Marta Retzlaff's Buick LeSabre 400 continued to lead in the medium-price class, although Stan Raymond's Oldsmobile Delmont 88 managed to close the gap to a slim .1964 mile per gallon. One driver ready to see the checkered flag tomorrow was Hart Fullerton, a regular competitor since 1950, who was on the verge of his first class win with his Chrysler 300. Today's pace slowed somewhat, with the field averaging 50.5013 miles per hour on the trip from Omaha to Chicago.

The top three performers in each class were:

Compact Six Cylinder	Plymouth Valiant 100	25.1517 Miles Per Gallon
	Dodge Dart	24.9105 Miles Per Gallon
	Ford Mustang	24.6614 Miles Per Gallon
Compact Eight Cylinder	Plymouth Barracuda	22.8500 Miles Per Gallon
	Dodge Dart	21.5869 Miles Per Gallon
	Mercury Cougar	21.4846 Miles Per Gallon
Intermediate Size Six Cylinder	Mercury Capri	23.9691 Miles Per Gallon
	Plymouth Belvedere I	22.8802 Miles Per Gallon
	Buick Special	22.4783 Miles Per Gallon
Intermediate Size Eight Cylinder	Plymouth Belvedere II	20.5298 Miles Per Gallon
	Mercury Caliente	20.3629 Miles Per Gallon
	Buick Special	20.2061 Miles Per Gallon
Low Price Eight Cylinder	Ford Custom 500	19.7802 Miles Per Gallon
	Chevrolet Impala SS	19.7178 Miles Per Gallon
	Plymouth Fury II	19.4759 Miles Per Gallon
Medium Price	Buick LeSabre 400	19.0985 Miles Per Gallon
	Oldsmobile Delmont 88	18.9021 Miles Per Gallon
	Pontiac Catalina	18.5994 Miles Per Gallon
Luxury Cars	Chrysler 300	17.8690 Miles Per Gallon
	Buick Electra 225	17.3358 Miles Per Gallon
	Oldsmobile Toronado	16.8177 Miles Per Gallon

Without a doubt, everyone was ready in the early morning hours of April 9 for the final 250.3-mile trip to finish line at Detroit's Cobo Hall. Today's time limit would be 4 hours and 44 minutes with no stops scheduled along the way. Being a Sunday, traffic was expected to be light for the entire trip. Bob Checkley in his Plymouth Valiant 100 6 was the first to leave at 5:30 a.m., bound for Lake Shore Drive and the Chicago Skyway (Interstates 90 and 94), heading towards the Indiana Toll Road (Interstates 80 and 90). After arriving in South Bend, the drivers headed north on U.S. Routes 31 and 33; crossing into Michigan and entering the Eastern Time Zone, they continued traveling on U.S. Route 131 towards Kalamazoo. Arriving in Kalamazoo, they headed east on Interstate 94 bound for Ann Arbor for the final top-off of the Run at Peterson-Holleran Mobil Service. On the trip from Chicago, the drivers continued to encounter light rain. Bob Checkley in his Plymouth Valiant 100 6 was the first to arrive, with no drivers exceeding today's time limit. Pat Sawyer, driving a Chevrolet Camaro 6, was off-course on this leg and ended up driving six fewer miles than shown on the itinerary. After each car was topped off, the drivers were directed to a marshaling area located at Ann Arbor's High School before beginning the final leg into Detroit.

While USAC officials computed the final results, at the marshaling area each class-winning car was placed at the head of the line. Here is where the roof fell in for the low-price eight-cylinder class leader Kay Kimes when his Ford Custom 500 was pulled out of line. After rechecking their calculations, USAC confirmed that Don Royer had managed to move his Chevrolet Impala SS into first place by a mere .0325 mile per gallon.[27]

Then, with a police escort, the convoy of 41 competing cars continued eastbound on Interstate 94 bound for Detroit's Cobo Hall, where several hundred spectators awaited their arrival. Just after 1:00 p.m., Class "A" Winner Bob Checkley in his Plymouth Valiant 100 was the first to receive the checkered flag from USAC Steward Fred Agabashian, followed by the six remaining class winners. While the remaining cars crossed the finish line, the winners were placed on display adjacent to Cobo Hall for pictures and other publicity events. That evening a dinner and formal awards ceremony was held at Detroit's Hotel Ponchartrain. Without a doubt, everyone concerned was pleased to see the most difficult and challenging Run in recent memory come to a safe conclusion. After his first time leading USAC's supervision of the Economy Run, no one could have been more relieved than Chief Steward Art Rene.

Putting the brakes on Rambler American's five-year domination of the Economy Run, Bob Checkley driving a Plymouth Valiant 100 posted the best mileage for all entries with 24.5722 miles per gallon. Leading from the first top-off, Bob managed to hold off a Bob Cahill in a Dodge Dart by a final margin of .2761 mile per gallon. Recording the lowest mileage figure was Stan Seeds in a Ford Thunderbird with 15.2610 miles per gallon. Chrysler Corporation dominated this year, placing first in four classes, which included Plymouth models in three. General Motors followed, placing first in two classes, with a single win for Ford Motor Company. Mercury captured its first win since Bill Stroppe took the Sweepstakes Award in 1952 with a Mercury Monterey. Both the Chevrolet Impala SS and Buick LeSabre 400 were the only ones to score back-to-back wins by repeating their performance from last year.

After the worst weather conditions in the history of the Economy Runs, rain continued to plague the drivers on the last leg from Chicago to Detroit. Arriving in Ann Arbor for the final top-off, Mel Alsbury, Jr.'s Oldsmobile Toronado waits its turn at the pumps. The adverse weather did not deter USAC officials as they donned rain gear, and in some instances covered the open trunks with plastic tarps, in order to refuel the entries.

Class winners were:

CLASS	MAKE and MODEL	MILES PER GALLON
A — Compact Six Cylinder	Plymouth Valiant 100	24.5722
B — Compact Eight Cylinder	Plymouth Barracuda	22.3198
C — Intermediate Size Six Cylinder	Mercury Capri	23.2905
D — Intermediate Size Eight Cylinder	Plymouth Belvedere II	20.0111
E — Low Price Eight Cylinder	Chevrolet Impala SS	19.2913
F — Medium Price	Buick LeSabre 400	18.7414
G — Luxury Cars	Chrysler 300	17.5932

Detailed results for all competing cars can be found in the Appendix.

When the results were released, USAC officials noted that this year's pace was the third fastest in history, with the field averaging 51.3612 miles per hour. This pace was nearly a half-mile per hour slower than 1966's record speed of 51.8367 miles per hour, and this decline was basically a result of adverse weather conditions experienced during the Run. Overall the field averaged 19.9884 miles per gallon. This year's route covered 2837.8 miles, shorter than the original route due to the elimination of a 45.7-mile segment between Aurora and Chicago, Illinois. The total adjusted driving time was 55 hours and 36 minutes, which took into account separate weather and traffic conditions. Despite some of the worst weather ever encountered, this was a safe event. There were no accidents or mishaps, although the Oldsmobile Ninety-Eight required a tire replacement. One entry, the Ford Falcon Futura 8, had a major mechanical problem, which forced it to withdraw shortly after leaving Salt Lake City.[28]

Two drivers stood in the winner's circle for the first time with their Economy Run experience at the opposite ends of the spectrum. Veteran Hart Fullerton (Chrysler 300) finally notched his first win after participating every year since 1950, while Allan Brooks (Plymouth Belvedere II 8) was victorious in his first attempt. This year saw second consecutive class wins for three drivers — Bob Checkley, Marta Retzlaff and Don Royer — while Scott Harvey notched his fourth class win and Bill Levy his third.

Those following the Run this year noted that the economy of the 1967 models was showing an overall improvement compared to their 1966 counterparts, which supported "a Detroit claim that the auto industry was paying extra attention to lower operational costs."[29] In spite of some of the worst weather conditions ever experienced, and a fast 51.3837-miles-per-hour pace, the entire field showed a .4370-mile-per-gallon increase over 1966. This reversed a six-year downward trend that began after the 1961 Run. A comparison of the class winners for these two years revealed that in four classes the mileage had shown an increase while in three there was a decrease. In recording the best mileage for all entries with 24.5722 miles per gallon, the Plymouth Valiant 100 6 exceeded the Rambler American's top performance in 1966 by a margin of .7659 mile per gallon.

Interest this year continued to focus on competition in the compact six-cylinder class with the strong one-two showing from the start by the Bob Checkley's Plymouth Valiant and Bob Cahill's Dodge Dart. Adding to this interest was the surprisingly poor performance by Les Viland's Rambler American. Les could only manage a fifth-place finish, thus ending his five-year streak in achieving the best mileage for all entries. A battle for third place between Byron Froelich's Ford Falcon and Per Nystrom's Ford Mustang remained close from the start and was not decided until the final top-off. Although they swapped positions a couple of times, in the end Byron managed to move into third by a slim .0245 mile per gallon. All three Chevrolet entries finished well off the pace. Disappointing too was the performance of the new Camaro, driven by five-time class winner Patricia Sawyer, who finished in eighth (and last) with 22.3668 miles per gallon.

From the first day there was never a doubt that Scott Harvey's Plymouth Barracuda would take home the trophy in the compact eight-cylinder class as he maintained a strong lead over Jean Calvin's Dodge Dart. The real contest here was for second place as Ginny Sims and her Mercury Cougar worked hard to close the gap. In the end, Jean's Dodge Dart prevailed, a slim .1344 mile per gallon separating the two. The new Camaro posted poor performance, coming in sixth (and last) with 19.2757 miles per gallon, 4.0441 miles per gallon behind the class-winning Barracuda.

In both intermediate-size classes it was a battle between Mercury and Plymouth. Among the six-cylinder entries, Bill Levy kept his Mercury Capri out front by a fairly comfortable margin from the start over last year's winner Bob Binder in a Plymouth Belvedere I. As noted earlier, this was Mercury's first class win in 15 years. Despite the concerns with her Buick Special during the impound period, Paula Murphy finished a strong third over two other General Motors entries, the Chevrolet Chevelle 300 and Oldsmobile F-85. It was a different story among the

At Detroit's Cobo Hall, Bob Checkley's Plymouth Valiant 100 takes the checkered flag from USAC Steward Fred Agabashian for a first-place finish in the compact six-cylinder class. The Valiant also achieved the highest mileage for all entries with 24.5722 miles per gallon. This win was one of three class wins for Plymouth this year, marking only the fifth time in Economy Run history this feat had been accomplished. This also was Plymouth's eleventh consecutive year to capture a first-place trophy in a streak that dated back to 1957.

After being active in the Economy Runs since 1950, West Los Angeles Chrysler-Plymouth dealer Hart Fullerton finally scored his first class win. Driving a Chrysler 300 in the luxury car class, Hart maintained a comfortable lead from the start. Two other veteran drivers, Mel Alsbury, Jr., and Pete Novotny, were also competing for top honors.

eight-cylinder models. Novice Allan Brooks managed to hold on to first place in his Plymouth Belvedere II over Darrell Droke's Mercury Caliente by a slim margin of .1970 mile per gallon. Competition between these two was close from the beginning, but Allan held on, ending any hopes for a sweep of both intermediate size classes by Mercury. John Rich recorded another strong performance, finishing third with his Buick Special. The Oldsmobile F-85 Cutlass, equipped with the largest engine in this class, placed seventh (and last), nearly 3 miles per gallon off the pace.

Continuing its recent trend of close competition, the battle in the low-price eight-cylinder class did not disappoint as it was a Ford and Chevrolet duel from the start. Barely ahead on the first day, Kay Kimes's Ford Custom 500 was leading Don Royer's Chevrolet Impala SS by a mere .0061 mile per gallon. Over the next four days, Kay was able to stretch his lead to .2277 mile per gallon. With two days remaining and after switching co-drivers, Don started to close the gap, but time was running out. After the final top-off the biggest surprise of the Run occurred with the announcement that Don Royer's Chevrolet Impala SS had indeed moved into first by a slim .0325 mile per gallon and spoiled the chance for Ford's first class win since 1961. In addition to being the second consecutive first-place finish for the Impala, this was its third in six years. Another novice driver, Jim Latham, placed a strong third with his Plymouth Fury II, with only .1405 mile per gallon separating the first three positions.

In the medium-price class, it was another close contest from the first day. In the end, Marta Retzlaff and her Buick LeSabre 400 held on for the win over Stan Raymond in an Oldsmobile Delmont 88 by a margin of .2238 mile per gallon. This was the second consecutive win for the LeSabre and its third in four years. Marta, who had led for the first two days, lost her advantage in the Rocky Mountains when Stan moved into first place by the slimmest of margins, .0013 mile per gallon. She then was able to regain the lead crossing the prairie (despite a malfunction with her windshield wipers during an Iowa rainstorm) and never looked back.[30] Marta later noted, "We've had bad weather on Economy Runs before but this is by far the worst we've ever seen."[31] Don Francisco kept his Pontiac Catalina in third place despite a steady challenge from Art Bertelsen's Dodge Polara 318, with only .2531 mile per gallon separating the two at the end. Never a factor in this class, the Mercury Monterey finished over 2 miles per gallon behind the class-leading Buick.

Competition in the luxury car class featured three long-time veterans competing for first-place honors. In the end, Hart Fullerton prevailed and placed first with his Chrysler 300, which marked the first time he took home a first-place trophy since his first appearance in 1950. From the start, Hart maintained a fairly comfortable lead over John Meyers in a Buick Electra 225 in second. The other two veterans, Mel Alsbury, Jr. and Pete Novotny, battled for third-place honors. Participating each year since 1955, Mel Alsbury, Jr. kept his Oldsmobile Toronado firmly in third place, holding off a challenge by Pete Novotny driving an Oldsmobile Ninety-Eight. Pete, who had been a regular participant since 1950, had to settle for fourth place, with a .5471 mile per gallon margin separating these two. Both the Cadillac Calais and Ford Thunderbird finished well off the pace.

Plymouth's three class wins constituted a feat had only been achieved four times previously. Studebaker recorded three wins in 1940 and 1954, along with Chrysler in 1952 and Buick in 1964.[32] With this year's strong performance, Plymouth captured a first-place trophy for eleven consecutive years in a streak that dated back to 1957. After the results were announced, Chrysler-Plymouth Division's General Manager Robert Anderson remarked to the *Detroit Free Press*, "I don't remember a time when one car performed as well as the Plymouth did on the stock racing circuit last year and then came back and performed so well in the Economy Run."[33]

After a disappointing performance this year in a Chevrolet Camaro 6, five-time class winner Patricia Sawyer decided it was time to retire. Patricia had driven a Chevrolet entry every year since 1960, with the exception of 1964, scoring three consecutive class wins in a Corvair Monza along with back-to-back wins in a Biscayne. Three Buick dealers with a long record of participation would also see their involvement come to an end as dealer associations would take over sponsorship next year. Tracing their participation back to 1956, these were Burbank's Barnett Buick, Culver City's Bill Murphy Buick, and Glendale's Colonial Buick. Colonial Buick saw its Special entries place first in its class on three occasions.

Advertising and promotion declined again, mainly resulting from Plymouth's strong performance coupled with the poor showing of the Rambler American. Appearing mainly in newspapers, virtually all of the advertising was in the days immediately following the completion of the Run. Mobil Oil continued its longstanding tradition of announcing the results, this year under the headline, "How does the mileage in the 1967 Mobil Economy Run compare with yours? Read it and weep." The text went on to detail the preparation of the cars, participants who drove each entry, and driving hints to improve gasoline mileage. This layout was in an effort to fully answer those consumer perceptions on the validity of these results. Intent on putting the responsibility for good gas mileage squarely on the driver's shoulder, Mobil advised, "Even though no one's watching *you*, try driving as if you were in the Mobil Economy Run. You'll find that the more you use your head while driving, the less you use your wallet."[34] Mobil dealers distributed a 6-page folder which noted that "America's Mileage Classic" was in its 31st year. Featuring a photograph of the class-winning Plymouth Valiant, in addition to listing the results, it contained a brief thumbnail sketch of this year's routing.[35]

Chrysler Corporation wasted no time in announcing its strong performance with a full-page ad announcing, "Plymouth Stuns Automotive World in Mobil Economy Run." This went on to highlight its winning three out of seven classes along with the Valiant's delivering the best overall mileage, and also noted "A win by Chrysler 300 in the luxury class give Chrysler-Plymouth Division a fourth title."[36] Chrysler-Plymouth dealers were encouraged to promote this success in a special merchandising bulletin with the theme "Valiant: New Economy Champ." This packet included special newspaper layouts and radio advertising scripts for use in their local market. Dealers were also provided with small window displays which highlighted each class winner's mileage results.[37] Throughout the remainder of the year, only a brief mention of the Valiant's top mileage performance was made in magazine advertisements. The economy performance of both the Valiant and Barracuda models was highlighted in their 1968 showroom brochures. Plymouth also produced a special performance folder which highlighted 1967's success in stock car racing, road racing, rallying and economy.

Appearing in newspapers, Buick ads announced, "Marta Retzlaff drove 2837.8 miles in the Mobil Run to demonstrate Buick LeSabre's winning economy." Featuring a photograph of Marta Retzlaff with her winning LeSabre, the text also highlighted the second-place performance of the Electra 225 and third place for both the six-cylinder and eight-cylinder Specials.[38] Buick dealers were provided with a small 4-page folder "Buick Wins Again!," which highlighted the LeSabre's win against other models in the medium price field.[39]

As expected, American Motors completely ignored the 1967 Economy Run in print advertising, although one television commercial was produced that claimed, "Not only is the American priced to compete with the economy imports, but it continues to deliver good gas mileage." With clips of the competing American 220 on the open road to Detroit, its actual miles per gallon result was highlighted without mentioning its fifth-place showing.[40]

No advertising or promotional materials were created which highlighted the class-winning performance for either Chevrolet or Mercury.

Most enthusiast magazines continued to devote their editorial efforts to the Union/Pure Oil Performance Trials over the Economy Run, so the main source for coverage was still newspapers from coast-to-coast. The emphasis on the reporting this year dealt with the challenging weather and road conditions, along with the struggling performance by perennial champion Les Viland and his Rambler American 220. Adding to each day's interest was the close battle between Ford and Chevrolet for top honors in the low-price eight-cylinder class. The only two articles found in enthusiast magazines were from totally different perspectives.[41] Writing in her column "The Rally Scene" in *Sports Car Graphic*, Jean Calvin briefly told of her experiences driving a Dodge Dart 8 in which she placed second behind Scott Harvey's Plymouth Barracuda 8. Taking a completely opposite approach, Brock Yates wrote a somewhat satirical article in *Car and Driver* that was based on an interview with a fictional driver named L.T. "Balloonfoot" Bodine. In a question-and-answer format, this dealt more with the preparations for and history of the Run than with covering this year's challenges. Appearing prior to the Run, *Ford Times* told its readers about the background and preparation for competing in this year's Run. In separate articles, aimed at the female readers, economy driving tips were discussed by veteran driver Ginny Sims, while another looked at the special wardrobe Ford designed for all of its drivers and co-drivers.[42] Chrysler-Plymouth dealers read of their division's success in the publication *Chrysler-Plymouth Times*.

After the safe and successful conclusion to one of the most challenging Economy Runs in history, discussions continued at the corporate level regarding its future. Even though corporate interest continued to show a decline for promoting performance events, Mobil Oil executives consented to continue sponsorship the Economy Run in 1968. No doubt, Art Rene's performance in his first year as chief steward, in which he faced several challenges along the way, was a factor in this decision.

1968

Los Angeles to New York Terminated at Indianapolis Via Grand Canyon — Colorado Springs — Kansas City — Indianapolis — Washington, D.C.

APRIL 2–5, 1968—2272.25 MILES

Looking for ways to renew the public's interest in the Economy Run, at a time when performance over economy was paramount, USAC and Mobil Oil officials unveiled on March 5 a "Visit to Vacationland USA" travel theme. Returning to New York City after a two-year absence, the "competing cars will start from Anaheim, home of one of America's most famous tourist destinations, Disneyland. Overnight stops will be Grand Canyon, Ariz.; Colorado Springs, Colo.; Kansas City, Mo.; Indianapolis, Ind.; and Washington, D.C. en route to the Times Square destination. Drivers will demonstrate how motoring families can travel safely, comfortably and economically on a transcontinental trip." In line with this, USAC officials pointed out that another innovation this year would be daily results based on varying terrain, road conditions and temperature. Covering over 3,000 miles over a five-and-a-half day period, drivers would be required to maintain an average speed of more than 54 miles per hour, with a range in altitude from sea level to more than 10,000 feet in the Rockies. This pace would be the fastest in the Economy Run's history, as it would cover nearly the same distance as previous trips to New York within a shorter timeframe.[43] It was becoming apparent that Chief Steward Art Rene was beginning to make his own mark on the conduct of the Economy Run.

This year saw the most changes to the rules in several years, which basically dealt with required equipment along with the requirement that competing cars conform to U.S. Exhaust Emission Control regulations. Power steering and brakes were required in all classes except for the two designed for compact cars. Power steering was now required on all eight-cylinder compact cars, and reflecting increased popularity, air conditioning was required in both the medium-price and luxury car classes. Radial tires were now becoming more readily available, although still representing a low percentage of tire production. With this in mind, USAC officials specified that any car selected that was equipped with these tires would have them exchanged for standard tires after impounding and inspection. A more significant change involved the conduct of break-in runs. Concerned over the treatment of the cars during this time, USAC mandated that "all break-in miles will be accomplished on a course specified by USAC."[44]

The "Competitive Divisions for the 1968 Mobil Economy Run" were basically the same as those used in 1967, although one change was made regarding the intermediate-size models from General Motors. These models, which were completely redesigned for 1968, included separate wheelbases for two-door and four-door models. In an attempt to keep these models comparable in size with those from Chrysler Corporation and Ford Motor Company, eligibility was initially restricted to four-door models with a wheelbase of 116 inches.[45] This was later changed to allow the two-models, with their 112-inch wheelbase, to compete.

When the entries closed on February 16, a field of 41 cars was set, representing a cross section of the industry's most popular models. General Motors continued to have the largest representation with 17 models, followed by Chrysler Corporation with 12, Ford Motor Company with 10, and American Motors 2.[46] Joining the Imperial and Lincoln in choosing not to compete were three members of the now popular "pony car" group, the Chevrolet Camaro, Mercury Cougar and Pontiac Firebird. The two private entries competed in the luxury car class. Returning for the fourth year was the West Covina Chamber of Commerce entering an Oldsmobile Toronado, joined by Los Angeles GMC truck dealer Pete Novotny with a Cadillac Coupe de Ville.

There were no significant changes this year in the effort by manufacturers in the preparation and testing of their entries. On a reduced scale, engineers from Chrysler Corporation were on hand focusing their efforts not only on repeating last year's performance by the Valiant in achieving the best mileage, but on the performance of all the Plymouth models. Factory representatives from American Motors, Ford Motor Company and General Motors' Buick Division duplicated their efforts from prior years. Although relying on local dealers for preparation and testing of their entries, General Motors' Oldsmobile Division stepped in by providing several cars for use by USAC officials.

This year saw a slight increase in dealer associations as sponsors representing American Motors, Buick, and Dodge, along with three Plymouth entries. The three Buick entries were represented by its New York dealer group, while Plymouth selected those associations based on the route, including those in Kansas City, Washington, D.C., and New York. Dealers participating for the first time from outside of Southern California included Colorado Springs' Perkins Chrysler-Plymouth and Philadelphia's Roger Penske Chevrolet.

Although the majority of drivers had previous Economy Run experience, they were joined by seven who would be participating for the first time. Notable among the newcomers was USAC Championship Car Series driver Ronnie Duman, whose racing record of over 60 starts included the last four Indianapolis 500 Mile Races. On the ladies' side, Mary Ann Foss, well known in drag racing circles for driving the Dodge "Go Hummer," would be competing against Shirley Shahan, another well-known drag racer in her own right.[47] Although not competing in the same class, both would be driving for Dodge, and they made a friendly bet over who would end up in, or closest to, first place in their respective classes.[48] For the first time since 1963, there would be no engineers from Chrysler Corporation involved as either drivers or co-drivers for the three Dodge entries. Of the seven women who were drivers this year, five had participated previously, including two class winners, Mary Hauser and Marta Retzlaff.

Two Economy Run veterans, with their involvement dating back to 1950, were set to compete against each other in the luxury car class. West Los Angeles Chrysler-Plymouth dealer Hart Fullerton would again drive his own entry, a Chrysler New Yorker, looking to repeat last year's first-place finish. During these years Hart had placed second in his class seven times. With one exception, he always drove his own entries, beginning with Hudson in 1950, then switching to DeSoto in 1957. After DeSoto ended production in late 1960, Hart drove a Dodge Lancer in 1961 for Santa Monica's Claude Short Dodge; and after being awarded a Chrysler-Plymouth franchise, he returned driving a Chrysler in 1962. Pete Novotny had been participating either as a driver or co-driver each year except for 1959, when he was sidelined by a leg injury. Teaming up in 1950 with Leonard DeBell's Southern California Plating Company, one of the few private entrants to participate, Pete was behind the wheel of a Chevrolet through 1956 before switching to Studebaker for 1957 and then to a Cadillac in 1960. After a three-year stint as co-driver, he returned to driving duties with Oldsmobile in 1964. This year Pete would drive a Cadillac Coupe deVille sponsored by his Los Angeles GMC Truck dealership. At age 66, Pete was the oldest driver competing this year. In his years competing, Pete had finished in second place on two occasions.

Joining Hart Fullerton and Pete Novotny in dual roles as entrants and drivers was Mandy Williams, this year driving an Oldsmobile Ninety-Eight. The three husband-and-wife teams returning this year, Hart and Diana Fullerton (Chrysler New Yorker), Gordon and Scherlee Madison (Chevrolet Corvair 500), and Stan and Lois Raymond (Oldsmobile Delta 88), were joined by newcomers Jim and Jessie Latham (Chrysler Newport). All of last year's class-winning drivers returned, looking to repeat their performance.

The Disneyland Hotel was the scene for the start of the 1968 Economy Run. Assisting USAC's Fred Agabashian in sending the drivers on their way was Miss Anaheim, waving the green flag for Al Johnson and his Mercury Montego. An early 5:30 a.m. starting time, with light rain conditions, kept down the number of spectators who were on hand to see the drivers on their way.

After encountering rain and windy conditions across the California desert, the drivers' first stop was at the Tropicana Hotel complex in Las Vegas. After refueling, Mary Ann Foss and her Dodge Dart 6 continued to the Aladdin Hotel for a brunch stop. Later in the day, the drivers would have to contend with icy road conditions before they arrived at the Grand Canyon impound.

With random car selection and inspections beginning in mid–February, the scene at the Pasadena impound was basically a repeat from prior years. The only change that faced the competitors was a more defined procedure for break-in runs, for which a specific route was to be followed. The route selected by USAC took the competing cars north on Interstate 5 / U.S. Route 99 to Bakersfield. Here the drivers were instructed to "turn off at any one of the Bakersfield exits and immediately return to the freeway. You must return to the impound over the same route." Drivers were also instructed that "under no circumstance can you pull off an off-ramp and immediately back onto the freeway on an on-ramp. If you take an off-ramp and turn around, you must return to the impound." Additionally, downshifting using the gear shift lever was not permitted under any circumstances.[49]

For years there had been suspicions that some teams had been able to make modifications to their cars without being discovered by USAC officials. In the vast majority of cases any unauthorized modifications were noticed by USAC and the cars brought back to standard specifications. One of these suspected modifications dealt with the aerodynamics of the Plymouth entries. Writing years later in *High Performance MoPar*, Chrysler Corporation's Team Captain Bob Cahill revealed their attempts for improved aerodynamics. As he explained, their expert at front alignment "had learned how to set [the] cars to the proper height, and he was able, by means known only to him, to set the front ends lower than normal, even with the USAC [official] watching him. By 1968, however, a number of stock car racing people had become involved with the competitive teams. One of their sharper-eyed mechanics was able to spot the difference between our lower-nosed cars and theirs." The weekend before the start, a protest was made to USAC. After an inspection that showed that the cars were "a bit low in the front," USAC insisted that the front ends be reset to the correct height.[50]

Again this year each entrant was given the option of using either Mobil Regular or Mobil Premium gasoline, with all entries using Premium.

At the same time the entries were undergoing preparations for the Run, Mobil Oil's publicity and press efforts

The first driver to arrive at the Grand Canyon was Tommy Thomas in an American Motors Javelin. Using a different driving approach this year, Tommy would stay close to the speed limit during the run, averaging three miles per hour more than the rest of the field. This would have a definite effect on his mileage as he would finish fifth (and last) in the compact eight-cylinder class with 16.3720 miles per gallon.

were moving into high gear. Under the continued leadership of Frank Meunier, arrangements were completed for over 50 press representatives who would be covering the Run, representing newspapers located in cities from Southern California east to New England. These were joined by reporters from two all-news radio stations, Los Angeles' KNX (part of the CBS Radio Network) and New York's WINS (part of the Westinghouse Broadcasting), who would provide live reports to their listeners during the six days of the Run. While most of the press contingent would be following the Run in a chartered United Airlines DC6-B, opportunities would again be available for those who wished to drive along the route.

After sitting on the sidelines for two years, Dodge was back with another unique promotion tied to the Run. Electing to capitalize on its domination of the New York taxi market, Dodge selected one of the more interesting local drivers, Mrs. Gertrude DeWitt, to set out on a cross-country goodwill junket with Odra "Chuck" Chandler, the former mayor of Anaheim, California, as her passenger. Mrs. DeWitt, who at age 65 had been driving a New York taxi for over 20 years after a varied career which included a stint as a showgirl in the mid–1920s, would follow the competing cars in her Dodge Coronet taxi with the meter running just for fun. While riding along, Chuck Chandler would be presenting "Keys to the City of Anaheim" to local officials in cities and towns along the route. At the conclusion of the Run, a special presentation was planned at the Times Square finish line, where New York City's Mayor John V. Lindsay would also welcome Mrs. DeWitt back home from her cross-country travels.[51]

The traditional drawing of starting positions was again held at the Pasadena impound, and was one of the few times all of the drivers would be together before the start. Veteran driver and two-time class winner Mary Hauser selected the 1st spot for her Chevrolet Chevy II Nova 8, with another veteran, Fran Foster, bringing up the rear with her Ford Mustang 8 starting in the 41st spot. There was really no suspense this year regarding the location of the start, as the Disneyland Hotel had been announced as this site weeks earlier. Another effort by Mobil Oil's publicity team was to have an increased involvement by local officials from Anaheim and representatives from Disneyland in the pre-start ceremonies. In this they succeeded.

On Monday, April 1, the competing cars were driven from the Pasadena impound to the Anaheim Convention Center, where several events were planned. With members of the Anaheim High School band providing entertainment, a crowd of several hundred was on hand as the participants were welcomed by various city officials and representa-

Weather conditions at Colorado's Wolf Creek Pass Summit had deteriorated to such an extent that chain control was established for the 20.4 miles. In stark contrast to the scene from 1954's Economy Run, assistance was now provided from various sources to handle chain installation. The Plymouth Barracuda 6 was one entry that was hoisted by a tow truck for a quick chain installation by the truck's driver.

tives from Disneyland. As the drivers were introduced, they were presented on stage with the first day's itinerary and route summary for the trip to the Grand Canyon. After the conclusion of the ceremonies it was a short drive to the Disneyland Hotel, where all entries were topped off and parked for tomorrow morning's start.

Covering a distance of 561.2 miles, the first leg to the Grand Canyon was set with a time limit of 10 hours and 25 minutes, and would be the first visit to this area since 1952. Once the drivers arrived in San Bernardino, they would cover virtually the same route used in the 1941 Gilmore Grand Canyon Run. That year drivers had a 14½-hour time limit to cover the 599.3-mile trip from Los Angeles, in which they averaged 43.01 miles per hour. This year's shorter time limit and estimated speed of nearly 54 miles per hour was a reflection on improvements made to the highways over the years. Missing this year was the route description, used since 1964, which provided the participants with a word picture of the terrain and scenery they would be encountering along the way. This change highlighted that today most cross-country travelers would basically use only maps to guide their way, and this was an attempt to make the Run even more realistic.

Just before sunrise on April 2, a small group of spectators were on hand when Mary Hauser in a Chevrolet Chevy II Nova was the first driver on her way at 5:30 a.m. With light rain falling, the first few drivers were given the green flag by USAC Steward Fred Agabashian, who then turned over this honor to several local officials. Here the elevation was 194 feet, which was the lowest point on this year's Run. As the drivers headed out they passed under a large "good luck horseshoe" erected by the City of Anaheim.

A change was immediately visible dealing with USAC's riding observers. Traditionally sitting up front next to the driver, they would now be making their observations from the back seat with the co-driver now up front. With the ceremonies now behind them, it was down to business as they followed the Riverside Freeway towards San Bernardino, where they picked up Interstate 15 and continued to encounter rain along the way. Driving through California's Mojave Desert, they crossed into Nevada and headed for Las Vegas for the first stop of the day at Tropicana Mobil Service, where all entries were topped off. This was immediately followed by a 40-minute brunch stop at the Aladdin Hotel before they continued south towards Hoover Dam, crossing into Arizona and entering the Mountain Time Zone. During the day the drivers would battle quartering winds up to 35 miles per hour. Arriving in Kingman, they headed east on U.S. Route 66 to Williams, then north on U.S. Route 180 towards the Grand Canyon. The drivers would now have to contend with snow and icy roads before arriving in Tusayan for refueling at Grand Canyon Mobil Service, then continuing on to Grand Canyon Village and the impound at the El Tovar Hotel. With patches of snow along the sides of the road, Tommy Thomas in an American Motors Javelin was the first to arrive at 5:45 p.m., which reflected a different driving approach. Leaving the work for another economy record to teammate Les Viland and his Rambler American Rogue, Tommy drove "with an eye on a transcontinental speed record rather than economy, [and] deliberately crowded the speed limit at all times, averaging three miles per hour more than the rest of the pack."[52] One driver, Diana Lowe in an Oldsmobile Delmont 88, arrived nearly 3 minutes late and was penalized.

As the drivers settled in at the El Tovar, they soon discovered that tomorrow's leg to Colorado Springs would be the most difficult to date as they would face over 600 miles of driving with a forecast of challenging weather conditions.

When the results were released, Les Viland and his Rambler American Rogue 6 was back in his usual spot with the highest miles per gallon average for all entries. With 26.2198 miles per gallon, Les also had a nearly one-mile-per-gallon lead over second-place Jerry Gross' Plymouth Barracuda in the compact six-cylinder class. Competition was close in five classes, with the biggest margin between first and second in the luxury car class, where Hart Fullerton's Chrysler New Yorker had a .3707 mile-per-gallon margin. The closest was in the compact eight-cylinder class, where Jack Kirkpatrick's Plymouth Barracuda held a mere .0033 mile-per-gallon lead over Mary Hauser's Chevrolet Chevy II Nova. In the low-price eight-cylinder class, Don Royer's Chevrolet Impala had a slim .0615 mile-per-gallon lead over Bill Keller's Plymouth Fury III. Slim leads were also found in both intermediate-size classes, with Al Johnson's Mercury Montego 6 and Shirley Shahan's Dodge Coronet Deluxe 8 sitting in first place. Chrysler Corporation was shown leading in four classes, with American Motors, Ford Motor Company and General Motors each leading in one. The entire field recorded an average mileage figure of 20.6403 miles per gallon. Today's pace was fast, with an average of 54.7055 miles per hour.

Crossing into Kansas, the drivers saw an end to adverse weather concerns as they made their way across the state to the Prom Sheraton Motor Inn in Kansas City. With Chief Steward Art Rene pointing the way to the parking entrance, Jim Whipple arrives in his dirty snow-streaked Dodge Polara. As the drivers were arriving, word was being flashed that Dr. Martin Luther King had been assassinated in Memphis.

After the results were released, USAC officials began to question the mileage figures reported in the medium-price class, especially with regards to Jim Latham's Chrysler Newport. The initial figures showed the Chrysler Newport with a commanding lead, recording 23.6456 miles per gallon.[53] This was a 3.6185 miles-per-gallon margin over Ted Block's Pontiac Catalina sitting in second. After USAC officials reviewed their calculations, Chief Steward Art Rene announced that mileage figures in the medium-price class "would be based only on gasoline consumed between Las Vegas, Nev., and New York City, rather than from the start at Disneyland, because of a human error in the operation of one of the recording devices at the Las Vegas refuel."[54] This also resulted in an adjustment the overall mileage average for today to 20.9069 miles per gallon. With constant efforts to ensure total accuracy, along with their track record over the years, this was an embarrassment for USAC, as this situation had never occurred before. More challenges and problems would again face USAC officials in the coming days.

The top three performers in each class were:

Compact Six Cylinder	Rambler American Rogue	26.2918 Miles Per Gallon
	Plymouth Barracuda	25.3375 Miles Per Gallon
	Ford Mustang	25.1557 Miles Per Gallon
Compact Eight Cylinder	Plymouth Barracuda	21.0639 Miles Per Gallon
	Chevrolet Chevy II Nova	21.0606 Miles Per Gallon
	Ford Mustang	20.8429 Miles Per Gallon
Intermediate Size Six Cylinder	Mercury Montego	24.8550 Miles Per Gallon
	Ford Fairlane	24.5256 Miles Per Gallon
	Plymouth Belvedere	23.7872 Miles Per Gallon
Intermediate Size Eight Cylinder	Dodge Coronet Deluxe	21.7986 Miles Per Gallon
	Plymouth Belvedere	21.6268 Miles Per Gallon
	Mercury Montego	20.6660 Miles Per Gallon
Low Price Eight Cylinder	Chevrolet Impala	20.2402 Miles Per Gallon
	Plymouth Fury III	19.6249 Miles Per Gallon
	Ford Custom	19.0476 Miles Per Gallon
Medium Price Results invalidated due to human error in calculation at Las Vegas Top-off.		
Luxury Cars	Chrysler New Yorker	18.0058 Miles Per Gallon
	Buick Electra 225 Custom	17.6351 Miles Per Gallon
	Ford Thunderbird	17.3392 Miles Per Gallon

On April 3, the drive to Colorado Springs would be a record-setting 632.8-mile, 12-hour and 8-minute trek through three states, crossing the San Juan Mountain Range. This would also be the Run's first return visit to this city since 1956. At 4:30 a.m., Jack Kirkpatrick in a Plymouth Barracuda 8 was the first driver on his way, following Arizona 64 and U.S. Route 164 through the Painted Desert towards New Mexico. Traveling on U.S. Route 550, the drivers arrived in Farmington, where all entries were topped off at Blake Hick's Mobil Service, followed by a stop in Aztec for brunch at C.V. Koogler Junior High School. Continuing on U.S. Route 550, they crossed into Colorado, where they picked up U.S. Route 160 and headed for Wolf Creek Pass Summit, at 10,850 feet the highest elevation on this year's Run. Weather conditions were now deteriorating, being described as a "Colorado Rockies spring blizzard," with snow and ice on the road. Plows worked all morning to keep the road open over the summit.[55] With no improvement in conditions, chain control was established for the 20.4-mile trip over the summit for only the second time in the Run's history. In a contrast to the scene in 1954, when only the drivers and co-drivers could handle chain installation, this year saw assistance from various sources to handle this task. After a slow trip over Wolf Creek Pass Summit, the weather slowly began to clear as the drivers continued through Alamosa and crossed North La Veta Pass Summit with an elevation of 9459 feet. Arriving in Walsenberg, the drivers headed north on Interstate 25 bound for Pueblo and a refueling stop at Sunrise Mobil Service. Diane Lowe, driving an Oldsmobile Delmont 88, had the misfortune of running out of fuel 7 miles before arriving in Pueblo and was disqualified.[56] In her first experience driving in the Run, Diane struggled over the first two days, sitting seventh (and last) in her class. A combination of yesterday's late-arrival penalty coupled with today's mountainous road conditions was too much to overcome. From Pueblo it was a 42-mile drive to Colorado Springs and the impound at Antler's Plaza Hotel. Just before 6:00 p.m., Scott Harvey in a Plymouth Belvedere 8 was the first to arrive, close to 30 minutes later than expected, after a long, exhausting day on the road. After all the competitors had arrived, USAC officials announced that today's time limit was being extended by 14 minutes due to the chain control at Wolf Creek Pass. Even with this extension, five drivers — Tom Gillum, Jr. (Oldsmobile Toronado), Jack Kirkpatrick (Plymouth Barracuda 8), Jim Latham (Chrysler Newport), Marta Retzlaff (Buick LeSabre Custom), and John Rich (Buick Special Deluxe 8) — arrived late, by as much as 13 minutes and 12 seconds, and were penalized.

All the drivers saw their mileage figures decline, which was a reflection on today's road and weather conditions. Les Viland and his Rambler American Rogue 6 continued their strong performance, recording the highest miles-per-gallon average for all entries with 25.8231. In the compact six-cylinder class, Les maintained a strong .9946 mile-per-gallon lead over Jerry Gross' Plymouth Barracuda, which was the largest lead for all classes. There was one change among the class leaders: in the intermediate-size six-cylinder class, Ginny Sims moved her Ford Fairlane past Al Johnson's Mercury Montego by a margin of .2696 mile per gallon. In the first official results in the medium-price class, Marta Retzlaff in her Buick LeSabre Custom was shown in first place by a margin of .2450 mile per gallon over Jim Wright's Dodge Polara. With two exceptions, the class-leading drivers slightly extended their lead over second-place challengers. The closest contest again was in the compact eight-cylinder class, where Jack Kirkpatrick's Plymouth Barracuda held a slim .1250 mile-per-gallon lead over Bob Checkley's Plymouth Valiant Signet, with Mary Hauser's Chevrolet Chevy II Nova dropping to fourth place. The other close contest was in the intermediate-size eight-cylinder class, where Shirley Shahan maintained a .1718 mile-per-gallon lead in her Dodge Coronet Deluxe over Scott Harvey's Plymouth Belvedere. Both Don Royer's Chevrolet Impala and Hart Fullerton's Chrysler New Yorker maintained comfortable leads in their classes. In the luxury car class, Mandy Williams moved his Oldsmobile Ninety-Eight into second place by a slim .0114 over Paula Murphy's Buick Electra 225 in third. The drivers averaged 51.7057 miles per hour over today's route, which was a direct reflection on the weather and road conditions they faced.

When the results were released, another problem was reported regarding the refueling stop at Farmington, New Mexico. USAC officials said that "a member of the service station crew failed to reset one of the precision metering devices" for one of the cars competing in the compact six-cylinder class. As a result, Chief Steward Art Rene announced, "The resultant error made it necessary for officials to eliminate from their calculations all fuel added for Class A for compact 6 cars at the Farmington, N.M. fuel stop. No other classes were affected. The 288 miles from Grand Canyon to Farmington will not be included as part of the course for Class A cars." USAC officials also announced they "are adding a 'back-up' gasoline recording system to check their present measuring equipment at refueling stops."[57] As USAC officials worked to put these problems behind them, the press turned their attention to tomorrow's trip to Kansas City, focusing on close contests developing in several classes.

The top three performers in each class were:

Compact Six Cylinder	Rambler American Rogue	25.8231 Miles Per Gallon
	Plymouth Barracuda	24.8285 Miles Per Gallon
	Ford Mustang	24.5665 Miles Per Gallon
Compact Eight Cylinder	Plymouth Barracuda	20.6455 Miles Per Gallon
	Plymouth Valiant Signet	20.5205 Miles Per Gallon
	Ford Mustang	20.4410 Miles Per Gallon
Intermediate Size Six Cylinder	Ford Fairlane	24.1730 Miles Per Gallon
	Mercury Montego	23.9034 Miles Per Gallon
	Plymouth Belvedere	23.3248 Miles Per Gallon
Intermediate Size Eight Cylinder	Dodge Coronet Deluxe	21.6579 Miles Per Gallon
	Plymouth Belvedere	21.5722 Miles Per Gallon
	Mercury Montego	20.0666 Miles Per Gallon
Low Price Eight Cylinder	Chevrolet Impala	19.5542 Miles Per Gallon
	Plymouth Fury III	19.1553 Miles Per Gallon
	Ford Custom	18.6291 Miles Per Gallon
Medium Price	Buick LeSabre Custom	19.0765 Miles Per Gallon
	Dodge Polara	18.8315 Miles Per Gallon
	Pontiac Catalina	18.5095 Miles Per Gallon
Luxury Cars	Chrysler New Yorker	17.5324 Miles Per Gallon
	Oldsmobile Ninety-Eight	17.1480 Miles Per Gallon
	Buick Electra 225 Custom	17.1366 Miles Per Gallon

On April 4, the drivers faced another long day on the road with a 591-mile trek to Kansas City. Today would feature open highway driving through Colorado and Kansas; the drivers would have few interruptions along the way, with "five stop lights out of Colorado Springs and two boulevard stops enroute to Kansas City."[58] Gordon Madison in a Chevrolet Corvair 500 was the first to leave at 6:00 a.m. in less than ideal conditions for the 10-hour and 6-minute trip. With temperatures hovering near zero, the drivers encountered ice conditions, with accumulations up to six inches deep, on the roads for the first 75 miles through Eastern Colorado as they followed U.S. Route 24 and Interstate 70.[59] Crossing into Kansas, entering the Central Time Zone, brought an end to weather concerns as they continued to drive at a fairly rapid pace towards Wakeeney, where all entries were topped off at Roy Muckholz Mobil Service. Then it was a short trip to Hays for a brunch stop at the Ramada Inn before continuing on Interstate 70. Passing through Topeka, they entered the Kansas Turnpike bound for the Lawrence Service Area for refueling. Then it was a short 35-mile drive across the Kansas River into Missouri and the impound at Kansas City's Prom-Sheraton Motor Inn. Just before 6:00 p.m., Byron Froelich in a Mercury Monterey was the first of the dirty, snow-streaked cars to arrive. All drivers arrived within the allowable time limit.

As the competitors were arriving at the Prom Sheraton impound, news bulletins began to flash the word that Dr. Martin Luther King, Jr. had been shot in Memphis. Minutes after the last car arrived came the word that he had died, and riots began to erupt in dozens of cities across the country. Kansas City and Indianapolis remained calm. The situation was becoming quite serious in Washington, D.C., where the Run was scheduled to arrive in two days. As the evening progressed, concerned USAC and Mobil officials watched as events unfolded before making the decision that it was safe to continue tomorrow to Indianapolis.

With a combination of mountain terrain and rough weather, everyone's mileage figures took a tumble, with the field now averaging 19.2947 miles per gallon. Still recording the best mileage for all entries was Les Viland's Rambler American Rogue 6 with 24.5174 miles per gallon. While Les continued his comfortable lead in the compact six-cylinder class over Jerry Gross' Plymouth Barracuda, a real battle here was taking place between third and fifth. Bob Cahill moved his Plymouth Valiant 100 up to third, dropping Kay Kimes' Ford Mustang 2+2 to fourth and Gordon Madison's Chevrolet Corvair 500 to fifth. A slim .0596 mile per gallon separated these three drivers. All of the class-leading drivers, with one exception, maintained their leads. In the medium-price class, Jim Wright in a Dodge Polara moved into first place, dropping Marta Retzlaff's Buick LeSabre Custom to second by a slim .0275 mile per gallon. Marta was one of the drivers penalized for arriving late into Colorado Springs, which in part accounted for her losing the lead.[60] The only other close contest was in the intermediate-size eight-cylinder class, where Shirley Shahan's Dodge Coronet Deluxe held

For the second time in the Economy Run's history, arrangements were made for the drivers to make one lap around the Indianapolis Motor Speedway before heading to the overnight impound. Ted Block, driving a Pontiac Catalina, is one of the first to pull through the Speedway's gates. This would not be the only time this year they would complete a lap around the famed speedway.

on to a .0673 lead over Scott Harvey's Plymouth Belvedere. In the luxury car class, Nelson Stacy moved his Ford Thunderbird back into third place by a margin of .1363 mile per gallon, dropping Paula Murphy's Buick Electra 225 Custom to fourth. Today's trip was the fastest in Economy Run history as the drivers averaged 58.8777 miles per hour. With a speed of 61.8272 miles per hour, Tommy Thomas, in his American Motors Javelin, became the only driver to date to record an average speed over 60 miles per hour. This pace was having a definite effect on Tommy's mileage, with a figure of 16.4257 miles per gallon. Putting him in fifth (and last) place in the compact eight-cylinder class, his average was 3.8619 miles per gallon behind the class-leading Plymouth Barracuda.

The top three performers in each class were:

Compact Six Cylinder	Rambler American Rogue	24.5174 Miles Per Gallon
	Plymouth Barracuda	23.6703 Miles Per Gallon
	Plymouth Valiant 100	22.9016 Miles Per Gallon
Compact Eight Cylinder	Plymouth Barracuda	20.2876 Miles Per Gallon
	Plymouth Valiant Signet	19.9680 Miles Per Gallon
	Ford Mustang	19.8911 Miles Per Gallon
Intermediate Size Six Cylinder	Ford Fairlane	23.1306 Miles Per Gallon
	Mercury Montego	22.6929 Miles Per Gallon
	Plymouth Belvedere	22.4365 Miles Per Gallon
Intermediate Size Eight Cylinder	Dodge Coronet Deluxe	20.8667 Miles Per Gallon
	Plymouth Belvedere	20.7994 Miles Per Gallon
	Mercury Montego	19.6754 Miles Per Gallon
Low Price Eight Cylinder	Chevrolet Impala	18.9235 Miles Per Gallon
	Plymouth Fury III	18.5482 Miles Per Gallon
	Ford Custom	18.0469 Miles Per Gallon
Medium Price	Dodge Polara	18.3935 Miles Per Gallon
	Buick LeSabre Custom	18.3660 Miles Per Gallon
	Pontiac Catalina	17.9261 Miles Per Gallon
Luxury Cars	Chrysler New Yorker	17.1953 Miles Per Gallon
	Oldsmobile Ninety-Eight	16.6563 Miles Per Gallon
	Ford Thunderbird	16.5239 Miles Per Gallon

On April 5, with the country still in a state of shock over yesterday's events, it was time to press on to Indianapolis. At 6:00 a.m., Ted Block in a Pontiac Catalina was the first driver to leave for a 492.3-mile, 8-hour and 48-minute trip. Heading out of Kansas City on Interstate 70, the drivers made the first stop of the day in St. Charles at Elmo's Mobil Service, where all entries were topped off. Continuing "along the southern border of Mark Twain Country and through towns which played a key role in the westward movement of the 19th Century," they approached the St. Louis area, bypassing the city using Interstate 270.[61] Then, crossing the Mississippi River into Illinois, they headed for Edwardsville for a brunch stop at the Holiday Inn. From Edwardsville it was back on Interstate 70 and U.S. Route 40 towards the Indiana state line, entering the Eastern Time Zone. As the drivers neared Indianapolis, their first stop was in Speedway for refueling at Speedway Mobil Service, which would be quickly followed by another stop at the Indianapolis Motor Speedway. For only the second time in the Run's history, the drivers made one lap around the famous Brickyard before arriving at the Indianapolis Motor Speedway Motel impound. "For the first time since the Run began, weather did not play a prominent part" in today's travels.[62] The first to arrive was Scott Harvey, in a Plymouth Belvedere 8, pulling in just after 4:30 p.m. Dr. Tom Evans, driving a Chevrolet Chevelle Malibu 8, arrived nearly 4½ minutes late and was penalized. After arriving, everyone continued to watch developments surrounding Dr. King's assassination, including the earlier announcement of President Johnson's proclamation designating Sunday, April 7, as a national day of mourning. They were beginning to suspect that this year's Economy Run would end prematurely.

Les Viland and his Rambler American Rogue 6 continued his strong performance with the best mileage figure for all entries with 24.1912 miles per gallon. At the same time he maintained a comfortable lead in the compact six-cylinder class over Jerry Gross' Plymouth Barracuda. It continued to be interesting in the battle between third and fifth place. Gordon Madison jumped into third place with his Chevrolet Corvair 500 dropping Bob Cahill's Plymouth Valiant 100 to fourth and Kay Kimes's Ford Mustang 2 + 2 to fifth. A very slim .1387 mile per gallon separated the three drivers. Another battle between second and fourth place was developing in the compact eight-cylinder class. Fran Foster moved her Ford Mustang into second place, with Mary Hauser's Chevrolet Chevy II in third and Bob Checkley's Plymouth Valiant Signet dropping to fourth. A mere .0200 mile per gallon separated the three drivers. In the medium-price class, a very tight contest remained as Marta Retzlaff moved her Buick LeSabre Custom back into first place over Jim Wright's Dodge Polara. The difference between the two was now a mere .0182 mile per gallon. In the other six classes, the leading drivers continued to maintain their leads, while in both classes for intermediate-size models, close contests were developing. Looking for her first win, Ginny Sims in a Ford Fairlane 6 had a slim .1844 mile-per-gallon lead over Al Johnson's Mercury Montego 6, while Shirley Shahan's Dodge Coronet Deluxe 8 maintained another slim .1642 mile-per-gallon lead over Scott Harvey's Plymouth Belvedere 8. The field averaged 19.2061 miles per gallon as they covered today's distance at a fast 56.3758 miles per hour.

At 7:00 p.m. came the announcement from Chief Steward Art Rene: "In deference to President Johnson's proclamation designating Sunday, the day the Mobil Economy Run

After the early termination of the 1968 Economy Run, arrangements were made for the drivers to complete a ceremonial lap around the Indianapolis Motor Speedway before crossing the brick start-finish line. Leading the convoy is Les Viland and his Rambler Rogue as he receives the checkered flag from USAC's Fred Agabashian. Les would once again achieve the highest mileage for all entries with 24.0917 miles per gallon.

was scheduled to end in New York, as a day of mourning for Dr. Martin Luther King, the run has been terminated here."[63] The announcement went on to explain that "final results will be calculated tonight and announced tomorrow. [The] schedule of Saturday's events through Saturday evening will be posted as soon as available."[64] Although details had been kept confidential, plans were well underway 24 hours earlier, during the stop in Kansas City, for an early termination at Indianapolis. Among the concerns were the riots that were occurring in both Washington, D.C., and Baltimore as the route "through Washington included five blocks involved in the rioting."[65]

The top three performers in each class were:

Compact Six Cylinder	Rambler American Rogue	24.1912 Miles Per Gallon
	Plymouth Barracuda	23.3732 Miles Per Gallon
	Chevrolet Corvair 500	22.8586 Miles Per Gallon
Compact Eight Cylinder	Plymouth Barracuda	20.0902 Miles Per Gallon
	Ford Mustang	19.6587 Miles Per Gallon
	Chevrolet Chevy II Nova	19.6514 Miles Per Gallon
Intermediate Size Six Cylinder	Ford Fairlane	22.8397 Miles Per Gallon
	Mercury Montego	22.6553 Miles Per Gallon
	Plymouth Belvedere	22.1210 Miles Per Gallon
Intermediate Size Eight Cylinder	Dodge Coronet Deluxe	20.6758 Miles Per Gallon
	Plymouth Belvedere	20.5116 Miles Per Gallon
	Mercury Montego	19.4990 Miles Per Gallon
Low Price Eight Cylinder	Chevrolet Impala	18.8655 Miles Per Gallon
	Plymouth Fury III	18.4095 Miles Per Gallon
	Ford Custom	17.9780 Miles Per Gallon
Medium Price	Buick LeSabre Custom	18.4679 Miles Per Gallon
	Dodge Polara	18.4497 Miles Per Gallon
	Pontiac Catalina	17.9691 Miles Per Gallon
Luxury Cars	Chrysler New Yorker	17.1474 Miles Per Gallon
	Oldsmobile Ninety-Eight	16.6990 Miles Per Gallon
	Ford Thunderbird	16.5696 Miles Per Gallon

Mike Cook of Mobil Oil's Special Promotions Department described the scene thirty-seven years later in an article in the December 2005 issue of *Hemmings Sports & Exotic Cars*:

> With Frank [Meunier] we brainstormed what hard decisions had to be made before we hit Indianapolis the next day. President Johnson declared Sunday a national day of mourning, eliminating our Times Square finish. Washington, with riots and fires, was too dangerous. Frank contacted Indy officials and hastily planned a finish at the Brickyard on Saturday morning. The mileage results through Friday were declared official. The [Mobil] Jetstar was dispatched to New York to pick up the finish line banners, trophies, etc. and he had the awards banquet [scheduled for] the Speedway Motel.[66]

Working late into the night, in addition to the job of computing the final results, officials finalized details for the finish line ceremonies and awards ceremonies to be held on April 6. As the banners and trophies were already on their way from New York, another task at hand was the relocation of nearly 300 participants from the Indianapolis Motor Speedway Motel to the Holiday Inn Northwest. Overlooked in the press reports was the added task of canceling earlier reservations for an April 6 brunch stop in St. Clairsville, Ohio, and at hotels in Washington, D.C., and New York. As noted in *Competition Press and Autoweek*, "The Mobil people turned in a fantastic job when faced with cancelling out at Indianapolis."[67]

On the morning of April 6, the final results were announced by USAC, noting that the mileage calculations were made through the final refueling stop in Indianapolis. There were no changes in the seven class leaders, with Marta Retzlaff holding off a strong challenge from Jim Wright's Dodge Polara in the medium-price class by a mere .0244 miles per gallon. Preparations were underway for a ceremonial trip around the Indianapolis Motor Speedway scheduled for 2:00 p.m. Under gray skies with chilly temperatures, only a handful of factory participants, officials and press were on hand for this event. Led by Class A Winner Les Viland in his Rambler American Rogue 6, the competitors drove under the finish line banner, each receiving a checkered-flag greeting from USAC Steward Fred Agabashian. After the remaining cars crossed the finish line and headed for the Holiday Inn, the class winners remained behind for individual pictures. The most eventful Economy Run was now history, finishing not in front of a huge crowd as planned in New York's Times Square, but in front of an empty spectator stand at Indianapolis Motor Speedway. Later an awards presentation and dinner was held at the Holiday Inn, although it was reported some teams passed on this event and headed directly home.[68]

Lost in all of this was Dodge's promotion with New York taxi driver Mrs. Gertrude DeWitt and her passenger, former Anaheim Mayor Chuck Chandler. After completing a lap around the Indianapolis Motor Speedway, following the competing cars, Mrs. DeWitt turned off her meter, which showed a fare of $535.15 for the 2272.25-mile trip from Disneyland.

For the seventh time in nine years, Les Viland and a Rambler American posted the best mileage for all entries with 24.0917 miles per gallon. With a strong lead from the start, Les was never challenged as he captured his tenth first-place trophy ... the most of any driver in Economy Run history. Pete Novotny in a Cadillac Coupe deVille had the distinction of recording the lowest mileage figure with 15.4621 miles per gallon. Chrysler Corporation again had a strong showing, placing first in three classes, followed by General Motors with two. American Motors and Ford Motor Company rounded out the field with one each. Women placed first in three classes, the most since 1957 when they were first permitted to compete. Four models — Plymouth Barracuda 8, Chevrolet

Impala, Buick LeSabre Custom and Chrysler New Yorker—scored back-to-back wins in repeating their 1967 performance.

Class winners were:

CLASS	MAKE and MODEL	MILES PER GALLON
A—Compact Six Cylinder	Rambler American Rogue	24.0917
B—Compact Eight Cylinder	Plymouth Barracuda	20.0078
C—Intermediate Size Six Cylinder	Ford Fairlane	22.7680
D—Intermediate Size Eight Cylinder	Dodge Coronet Deluxe	20.5744
E—Low Price Eight Cylinder	Chevrolet Impala	18.8162
F—Medium Price	Buick LeSabre Custom	18.4883
G—Luxury Cars	Chrysler New Yorker	17.1340

Detailed results for all competing cars can be found in the Appendix.

When USAC announced the results, it reported that the field averaged 19.1671 miles per gallon over the 2272.25-mile route to the final top-off near Indianapolis. This was the lowest since 1958, when the field averaged 18.7529 miles per gallon, and was a direct reflection on the speed, weather and road conditions encountered over the four days. The field averaged 55.0317 miles per hour, which is the fastest in the Economy Run's history, and was just over three miles per hour faster than the previous record of 51.8367 set in 1966. The total adjusted driving time to cover the route was 41 hours and 41 minutes. Again this year, it was another safe event, as no accidents or mishaps were reported. One entry failed to finish when the Oldsmobile Delmont 88 ran out of fuel on the second day.

Three drivers, each with varied Economy Run experience, celebrated their first wins. Veteran participant Ginny Sims (Ford Fairlane 6) finally took home a first class trophy as a driver. Participating since 1957, the year women were first allowed to compete, Ginny was co-driver that year for class-winning driver Mary Davis in a Plymouth Belvedere 8. Shirley Shahan (Dodge Coronet Deluxe 8) was victorious in her third year as a driver, with Jack Kirkpatrick (Plymouth Barracuda 8) victorious in his second effort. This year also saw a second consecutive class win for Hart Fullerton and the third consecutive class wins for Marta Retzlaff and Don Royer.

After the first day, there was never any doubt that Les Viland and his Rambler American Rogue would reclaim the honor of achieving the best mileage for all entries. In the compact six-cylinder class, he maintained a comfortable lead over Jerry Gross' Plymouth Barracuda as the real battle was between third and fifth place. By a margin of .1477 mile per gallon, Gordon Madison's Chevrolet Corvair 500 placed third, holding off a challenge by Bob Cahill's Plymouth Valiant 100 and Kay Kimes's Ford Mustang. Finishing well off the pace by over two miles per gallon were both the Dodge Dart and the Ford Falcon. The Falcon's performance was especially disappointing with its worst finish since 1960.

A big question was what happened to the Plymouth Valiant after its success a year earlier. This would not be answered for another twenty-eight years. When the cars were ordered reset to their correct height, the team's expert was unavailable and a mechanic from a local dealership was hired for the

Jack Kirkpatrick crosses the finish line in his Plymouth Barracuda, taking the first-place trophy in the compact eight-cylinder class. In each of the three years a Plymouth Barracuda had competed in the Economy Run, it had placed first in its class. This win also marked the twelfth consecutive year a Plymouth had a first-place finish. Plymouth also holds the record for the most class victories for all makes with 19.

job. Without revealing that he didn't know how to operate the "antique wheel-alignment stand that USAC had at the impound," he reset the heights and wheel alignment and the changes were accepted. From the start, the Valiant was not doing as well as it should, based on the preliminary figures obtained during break-in runs. Finally, on the Kansas City to Indianapolis leg, it was noted by another team member that the wheel alignment did not look right, which would result in the poorer mileage figures. After arriving at the Indianapolis impound, an inspection of the tires revealed that "the front tires were half worn out, after 2,000 miles of low speed driving!"[69]

This is not the scene Mobil Oil and USAC officials had envisioned for the finish of the 1968 Economy Run. The assassination two days earlier of Dr. Martin Luther King and its aftermath caused the Economy Run's termination at Indianapolis. Instead of seeing a large crowd gathered in Times Square for the finish, the drivers now drove past empty spectator stands at the Indianapolis Motor Speedway.

Barely in first place on the first day, Jack Kirkpatrick slowly expanded his lead in his Plymouth Barracuda competing in the compact eight-cylinder class. After dropping to fourth place on the second day, Mary Hauser battled her way back to a second-place finish with her Chevrolet Chevy II Nova. Here again the real battle was between third and fourth place, where the separation between Fran Foster's Ford Mustang and Bob Checkley's Plymouth Valiant Signet was a mere .0021 mile per gallon. Tommy Thomas, in his American Motors Javelin, recorded the highest average speed for all drivers with 57.8153 miles per hour. Running 2 hours and 15 minutes faster than the time limit set by USAC, he ended up with a final mileage figure of 16.3270 miles per gallon. Sitting in fifth (and last) place, he was 3.6358 miles per gallon behind the class-leading Barracuda.

In taking the top honors in the intermediate-size six-cylinder class, Ginny Sims and her Ford Fairlane edged out Al Johnson's Mercury Montego by a slim .1079 mile per gallon. They were leading on the first day, but Mercury's hopes for a back-to-back win in this class were dashed when Ginny moved into the lead and recorded Ford's second win in this class since 1965. Carl Diehl's Plymouth Belvedere was never a factor in placing third. Although equipped with the largest engine (230 cubic inch) in this class, the Chevrolet Chevelle 300 finished well off the pace, 3.5318 miles per gallon behind the leading Ford Fairlane.

In the intermediate-size eight-cylinder class, Shirley Shahan and her Dodge Coronet Deluxe held off a constant challenge from Scott Harvey and his Plymouth Belvedere. In one of the closest contests in this year's Run, a slim .0590 mile per gallon separated these two Chrysler Corporation entries. Veteran racer Ronnie Duman, driving a Mercury Montego, finished in third, ahead of Darrell Droke's Ford Torino GT by a fairly comfortable margin. After the finish of the Run, Ronnie Duman remarked "that the strain of keeping his foot OFF the [gas] pedal proved to be worse than any race he's ever been in." He added that with this stress "he was smoking three packs of cigarettes a day."[70] After several years of fairly strong performance, the four General Motors models finished well off the pace by as much as 3.6820 miles per gallon behind the leading Dodge. Disappointed with their performance were two class-winning veteran drivers. John Rich brought his Buick Special Deluxe home to a fifth-place finish, while Mel Alsbury, Jr., could do no better than seventh in his Oldsmobile F-85. Both of these models, along with the Pontiac Tempest Custom, were equipped with the largest engines (350 cubic inch) in this class.

In the low-price eight-cylinder class, competition was not as close as in previous years, with .8600 mile per gallon separating the first three places. For the third year in a row,

Don Royer placed first with a Chevrolet Impala. Bill Keller in his Plymouth Fury III finished a strong second over John Allen in a Ford Custom by a margin of .3770 mile per gallon. C.K. Enoch and his Chevrolet Caprice were never a factor, finishing fourth, 1.4463 miles per gallon off the pace.

The closest contest of this year's Run occurred in the medium-price class. After a problem during the refueling process in Las Vegas, which invalidated the first day's results, Marta Retzlaff in her Buick LeSabre Custom opened up a small .2450 mile-per-gallon lead over Jim Wright in a Dodge Polara. One of the drivers penalized after a late arrival in Colorado Springs, Marta dropped into second place by a mere .0275 mile per gallon when the field reached Kansas City. On the final leg to Indianapolis, she battled back, placing first by a margin of .0244 mile per gallon. Ted Block, driving in his first Economy Run, finished a strong third in his Pontiac Catalina. The rest of the field, Jim Latham's Chrysler Newport, Byron Froelich's Mercury Monterey and Stan Raymond's Oldsmobile Delta 88, were never a factor and finished well off the pace.

In the luxury car class, Hart Fullerton and his Chrysler New Yorker opened up a comfortable lead from the start and never looked back. Here again the real competition was between second and fourth place. Mandy Williams's Oldsmobile Ninety-Eight held off a challenge from Nelson Stacy's Ford Thunderbird to finish in second, with Paula Murphy's Buick Electra 225 Custom finishing in fourth. A slim .1733 mile per gallon separated these three drivers. Tom Gillum's Oldsmobile Toronado and Pete Novotny's Cadillac Coupe deVille placed well down in the standings and were never a factor.

Significant milestones for class winners included:

- Buick recorded its fifth class win in six years when the LeSabre Custom placed first in the medium-price class. This was the third year in a row the LeSabre topped all models in its class.
- Chevrolet scored its fourth consecutive win when the Impala placed first place in the class for low-price eight-cylinder models. This also marked the third year in a row the Impala topped its class.
- Chrysler marked its tenth class win in twelve years when the New Yorker placed first in the luxury car class. This also was the second consecutive win for both Chrysler and its driver, West Los Angeles Chrysler-Plymouth dealer Hart Fullerton.
- Dodge returned to the winner's circle for the fifth time in six years when a Coronet took top honors in the intermediate-size eight-cylinder class. In the four years

Throughout most of the Economy Run's history, the best competition was usually found between Chevrolet, Ford and Plymouth models. Known in the industry as "The Low-Price Three," each make shared the spotlight over the years in topping this class, first by Ford and then Plymouth. For the third consecutive year, Don Royer and his Chevrolet Impala took top honors in the low-price eight-cylinder class after it had been dominated for a number of years by Plymouth.

With their first-place trophy between them, co-driver Les Scott (left) and driver Les Viland stand under the finish line banner with their Rambler Rogue. If one driver stands out above the rest in Economy Run history, it is Les Viland. With a record of ten class wins, the most of any driver, Les also has the distinction of achieving the highest miles per gallon figure for all entries in seven of the last nine Economy Runs. Les also has two Sweepstakes Awards to his credit and he set the mark for the highest ton miles per gallon figure in 1951 driving a Lincoln.

the Coronet had competed in this class, it had placed first on three occasions.

- Plymouth recorded its twelfth consecutive class win, dating back to 1957, when its Barracuda placed first in the compact eight-cylinder class. Not only was this a back-to-back win for the Barracuda, in the three years this model had competed, it had placed first each time.
- Rambler American's class win competing against other compact models was its eighth in ten years. In each of these years the American had the distinction of recording the highest miles-per-gallon figure for all entries. For Les Viland, this win was his tenth, including two Sweepstakes Awards, the most for any driver to date in the Economy Runs.

Advertising and promotion were extremely limited this year, no doubt due in part to the events surrounding the early termination of the Run. Mobil Oil continued its longstanding tradition of announcing the results with the headline, "Results of the 1968 Mobil Economy Run. Your car can do it. Can *you?*" Continuing to follow the theme used in 1967's advertising campaign, the text emphasized safe driving by noting: "It's no good making excuses. You too can do it if you drive carefully. And any driver in the Run will tell you, economy and safety go hand in hand. So if you don't, it may cost you more than your money."[71] This was followed by a second advertisement which showed a man walking down the road carrying a gas can under the headline, "What you learn from the Mobil Economy Run could save you a long walk." The text provided hints for economical driving while at the same time promoting Mobil's new Detergent Gasoline.[72] Noting that the Economy Run had been "America's only definitive mileage test for 32 years," Mobil dealers distributed a 4-page folder detailing the results and providing a thumbnail sketch of obstacles the competitors faced. This included photographs showing drivers installing chains at Wolf Creek Pass and the parade across the Indianapolis Motor Speedway finish line.[73]

Appearing in newspapers, Buick ads announced the "Fourth victory for Buick LeSabre in the Mobil Economy Run," which highlighted not only the LeSabre's performance but also the driving skills exhibited by Marta Retzlaff.[74] Following the trend of the past two years, Buick provided its dealers with a small 6-page folder titled "Wouldn't You Really Rather Drive A Winner?," which again highlighted both the LeSabre and its driver Marta Retzlaff. In addition there was mention of the mileage performance by the Special Deluxe and Electra 225.[75]

In one of the only advertisements appearing in general interest publications, Ford announced that "Fairlane — Success Car '68 — Tops All Intermediate Sizes in Mobil Economy Run!" Featuring an illustration of a Fairlane hardtop, the text highlighted the basic details of the Run.[76]

In a surprise move, American Motors did not promote the Rambler American Rogue's performance during the remaining months of the 1968 model year. When the 1969 models were introduced, value was the overriding theme in promoting the American as a "compact car that could compete in price with the imports." In the text it was pointed out, "This is the only car that has won its class in the Mobil Economy run for six of the past seven years."[77]

Focusing on the performance aspect of its models, Chrysler Corporation did not produce any advertising on its performance in the Economy Run. Later in the year, the 1969 Barracuda showroom brochure told readers that "a '68 Barracuda equipped with a 318 V-8 won its class in last year's Mobil Economy Run with an average of 20.01 miles per gallon."[78] In export markets, where the price of gasoline generally is more expensive, the Plymouth Barracuda and Dodge Coronet's achievements were mentioned in both Plymouth and Dodge export-only showroom brochures. Dodge pointed out, "A Coronet V-8 won first place over all competition in the Intermediate-Size Eight class of the U.S.A.'s 1968 Mobil Economy Run. More proof that careful Chrysler Corporation engineering produces cars that are winners —*for all the world's drivers!*"[79]

No advertising or promotional materials were created which highlighted the class-winning performance of the Chevrolet Impala.

Enthusiast magazines completely ignored this year's Economy Run while continuing to devote their editorial efforts to the Union/Pure Oil Performance Trials. This again made newspapers from across the country the main source of information. This year's emphasis was on the challenging weather and road conditions in Colorado, later overshadowed by Dr. Martin Luther King's assassination and the early termination of the Run. The only magazine article dealing with the Economy Run, "Cars on Trial" by Don MacDonald, appeared in the Automobile Club of Southern California's magazine *Westways*.[80] Using the same theme found in earlier years, MacDonald gave readers a detailed look at the background and preparation of the Economy Run. The article also included some driving techniques used by the drivers. Both the *Dodge Reporter* and *Chrysler-Plymouth Times* told their dealers of each division's success in their May 1968 issues.

The jury was still out as to whether or not Mobil Oil would continue the Economy Run in 1969, as it continued to be shown that performance was becoming more important to motorists than gasoline economy. Even with the uncertain future, USAC officials reviewed the problems that surfaced this year and made plans to improve in those areas. During the summer of 1968, preparations began for the 33rd edition of the Economy Run.

12
Cancellation

Although rumored for years, the end came swiftly and without warning. On December 20, 1968, Mobil Oil Corporation announced it had made a decision to discontinue the Mobil Economy Run in the United States. In its statement, "Richard F. Tucker, vice president of marketing for the company's North American division, said changes in emphasis on automotive performance factors, in the attitudes of motorists, and in Mobil's advertising strategy all contributed to the decision." Mr. Tucker added, "Mobil is deeply indebted to USAC for their continually skillful handling of the Run, to the automobile manufacturers for their interest and cooperation, and to the many other dedicated people who helped conduct the event according to the highest professional standards."[1]

Times had changed, and the public was "more concerned about acceleration and horsepower, two enemies of fuel economy."[2] This was clearly evident in the high-performance models, such as the Dodge Super Bee, Plymouth Road Runner, and Pontiac GTO, manufactured by Detroit's Big Three.

Years later, Mike Cook (who at the time worked in Mobil Oil's Special Promotions Department) noted that earlier "the car manufacturers had notified Mobil that they would not enter or promote another Run."[3] Whether or not this information was passed on to USAC officials is not known. At the time of the December 20 announcement, entries had been open for nearly a month and preparations were well into the final stages.

Everyone outside of Mobil Oil was apparently caught off-guard. Four days earlier, Chief Steward Art Rene had made the initial announcement of the 1969 Run, noting several changes that were made which would have a direct impact on the drivers and co-drivers. Among the changes was the elimination of special signage or signalmen on the route: "Many people who are accustomed to these will have to pay attention to the log book. If they get lost, they are on their own." The route logs were greatly simplified so as to be "used primarily to get off and back on the route in gas, food and overnight stops."[4]

When asked if he was surprised by Mobil's decision, Art Rene replied that he was "Stunned. We've never been so well prepared this early before. Everything was done — the timing runs, the logs, 1000 license frames, jackets and the rent was paid."[5]

Not everyone applauded Mobil's decision. Hugh Randolph of the *Oakland Tribune*, a strong supporter of the Economy Run, wrote:

> Aside from the warm, personal attachment that we had for the event, this writer is of the opinion that Mobil's board of directors, in cancelling the run, made an error in judgment that could come back to haunt them. The Mobil Economy Run probably generated more national publicizing for their firm than any other promotion by any other petroleum company in the country. With the combination of their years of identification with the Indianapolis "500" and the Economy Run, Mobil "had it made," publicity wise, for years.
>
> Cancellation of the run also automatically makes the annual Union — Pure Oil Performance Trials the No. 1 event of its kind in the country. We can only hope that discontinuance of this event is a temporary condition, and that the economy run will be back one of these years — soon.[6]

The Economy Run would not return, although another gauge of fuel mileage would make an appearance within a few years when the U.S. government's Environmental Protection Agency (EPA) began a program of reporting fuel economy test results. These results were obtained during each vehicle's Clean Air Act certification tests. This program was expanded to require manufacturers to publicize these test results and further prohibited them from providing results from alternate sources. This sealed the fate of any possible revival of the Economy Run. The EPA tests created some of the same confusion with the public as did the mileage figures obtained in the Economy Runs, and the question continued to be asked: "Why can't I obtain the same gas mileage as those reported?" With this, a new term entered America's vocabulary: "Your mileage may vary."

What would the 1969 Economy Run have looked like? From April 8 to 13, the drivers would face a six-day, 3172-mile journey from Los Angeles to New York's Times Square. *Los Angeles Times* auto writer Bob Thomas provided readers with each day's planned routing:

Tuesday, April 8 — Cars were to gather at dawn at the new International Hotel near Universal Studios for a 6 a.m. departure. The first leg was a 547.4-mile run to Tucson, via San Diego, a 9-hour, 43-minute drive at an average speed of 58.25 m.p.h. Despite the miles, it was a relatively easy opener with only 10 stop lights and no boulevard stops along the way.
April 9 — this was to be the second longest one-day leg in the events history, 633.7 miles from Tucson to Lubbock, with a brunch stop near Las Cruces, N.M., at Mesilla.
April 10 — And this was the longest ... 645.7 miles from Lubbock to Springfield, Mo., nearly 12 hours of driving time at an average speed of 57.7 m.p.h.
April 11 — Today the troupe was to head for Cincinnati, a 590-mile trek that includes a St. Louis stop for brunch on a Mississippi riverboat.
April 12 — This is the toughest stretch of all ... 535 miles, sans Interstate Highways. Average speed slumps to 45 m.p.h. as the caravan pokes through 200 little towns in Ohio, West Virginia, Maryland, West Virginia again and Virginia en route to Washington, D.C. A dozen hours of driving was to begin at 5:30 a.m. with 93 interruptions for stop lights, plus two boulevard stops. "It's hard driving all day," [Art] Rene said about the leg, "over nothing but hills. We would have had many changes in the standings."
April 13 — This was to be a breeze, a 219.5-mile Sunday drive into Times Square, most of it on expressway.[7]

One particular item stands out in this planned itinerary: the routing through San Diego. Despite its proximity to Los Angeles, this would have been the Run's first visit to that city.

A Look Back at a Few Highlights and Records

Since the inaugural 1936 Gilmore Yosemite Economy Run, the competitors traveled a total of 1,596,354.8 miles through 29 states and 1 Canadian province. An exceptionally safe event, only four collisions occurred over the years which resulted in four minor injuries, only one of which required hospitalization. With total of 875 cars entered, only 20 failed to finish due to a variety of reasons, the most common being mechanical in nature. The largest field was in 1961 with 67 cars competing, while the smallest was in 1954 with only 20 cars.

Virtually every make that was produced between 1936 and 1968 participated in the Economy Runs. Plymouth holds the distinction being the only one to participate in every Economy Run.

The longest distance covered was 3301.4 miles from Los Angeles to Boston in 1966. The longest one day's leg was 632.8 from Grand Canyon to Colorado Springs in 1968, while the shortest was 149.3 in 1965 from Albany to New York City. Eleven different locations — Boston, Chicago, Colorado Springs, Detroit, Galveston, the Grand Canyon, Kansas City, Minneapolis, New York City, Sun Valley and Yosemite National Park — were selected over the years as destinations. The most popular destination was at Yosemite National Park, selected five times from 1936 to 1940. The Grand Canyon, Detroit and Sun Valley each hosted the terminus activities on three separate occasions.

Mileage Records, 1936–1968

Highest Average Miles Per Gallon — All Cars ... 23.923 set in 1951
Lowest Average Miles Per Gallon — All Cars ... 18.7529 set in 1958

Best Miles Per Gallon — Standard Transmission Cars:

4-Cylinder ... Willys 77	33.2100 in 1936
6-Cylinder ... Ford Falcon	32.6800 in 1961
8-Cylinder ... Ford 60	28.8500 in 1938
12-Cylinder ... Lincoln Zephyr	23.4700 in 1938

Best Miles Per Gallon — Automatic Transmission Cars:

4-Cylinder ... Pontiac Tempest	27.4406 in 1961
6-Cylinder ... Chevrolet Corvair Monza	29.3539 in 1961
8-Cylinder ... Studebaker Commander	27.4366 in 1955

Best Ton Miles Per Gallon — Standard Transmission Cars:

4-Cylinder ... Henry J Corsair	51.3749 in 1952
6-Cylinder ... Nash Ambassador Custom	58.6461 in 1950
8-Cylinder ... Lincoln	66.4840 in 1951
12-Cylinder ... Lincoln Custom	57.8270 in 1941

Best Ton Miles Per Gallon — Automatic Transmission Cars:

6-Cylinder ... Nash Ambassador Super	58.5560 in 1955
8-Cylinder ... Imperial Crown	64.5153 in 1957

Class Winners, 1936–1968

Plymouth	19	Nash	9	Mercury	5
Chrysler	18	Cadillac	8	Imperial	4
Studebaker	18	Oldsmobile	7	Henry J	3
Ford	14	Willys	7	Kaiser	2
Chevrolet	11	Packard	6	Terraplane	2
Rambler	11	Pontiac	6	Cord	1
Dodge	10	Graham	5	DeSoto	1
Buick	9	Hudson	5	Frazer	1
Lincoln	9				

Plymouth holds the record for the total number of class victories, as well as a consecutive twelve-year streak of victories from 1957 through 1968. The Rambler American holds the record for achieving the highest actual miles per gallon among all entries with eight.

Sweepstakes Winners, 1936–1958

Studebaker	4	Imperial	3	Mercury	2
Graham	3	Lincoln	2	Ford	1

Both Graham and Imperial recorded consecutive three-year Sweepstakes Awards ... Graham in 1936 to 1938 and Imperial in 1956 to 1958. The highest Ton Miles Per Gallon average was set by a Lincoln in 1951 with a figure of 66.484.

Class-Winning Drivers, 1936–1968

Over 330 drivers competed over the duration of the Gilmore and Mobilgas Economy Runs, although an exact number remains unknown at this time due to incomplete records for the 1936 and 1941 Gilmore Economy Runs. Of the 333 confirmed drivers, 172 participated for one year only and 107 drivers placed first in their respective classes. Les Viland holds the record for the most class wins with 10. Drivers with two or more class wins include:

Les Viland	10	Bill Levy	3	Mary Hauser	2
Mel Alsbury, Jr.	7	Clay Moore	3	Dick Johnson	2
William Martin	6	Art Rene	3	Horace Jones	2
Patricia Sawyer	5	John Rich	3	William Knopp	2
Pierce Venable	5	Don Royer	3	Howard Kyser	2
George Alsbury	4	Bill Stroppe	3	George Magee	2
Woody Bell	4	Jack Van Sant	3	Marshall Martin	2
Bob Cahill	4	Bob Burns	2	Willard McArthy	2
Scott Harvey	4	Mary Davis	2	George Miller	2
Andy Henderson	4	Danny Eames	2	Bub Ridings	2
Marta Retzlaff	4	Art Elliott	2	Clay Smith	2
Les Butts	3	Allan Fleming	2	Homer Thompson	2
Bob Checkley	3	Hart Fullerton	2	Osborne Wedgeworth	2
Byron Froelich	3	Lee Hamer	2		

Sweepstakes-Winning Drivers, 1936–1958

Mel Alsbury, Jr.	3 — 1956–1958
Clay Moore	3 — 1936–1938
William Martin	2 — 1939–1940
Bill Stroppe	2 — 1950 & 1952
Les Viland	2 — 1951 & 1953
Dick Griffith	1 — 1954
Dick Johnson	1 — 1955
Marshall Martin	1 — 1941

Mel Alsbury, Jr. and Clay Moore both recorded consecutive three-year Sweepstakes Awards with Clay driving a Graham from 1936 to 1938 and Mel, Jr. driving an Imperial from 1956 to 1958. Les Viland recorded the highest Ton Miles Per Gallon in 1951 driving a Lincoln.

After hanging up their driving gloves, several drivers began an association as officials with the AAA Contest Board and USAC. These included John Bodine, Lee Hamer, Andy Henderson and Art Rene, all of whom had at least one class win to their credit. Both John Bodine and Andy Henderson traced their involvement back to 1937, driving entries for Oldsmobile and Nash in the Gilmore Yosemite Runs.

13

Mobil Mileage Rally, 1958–1961

During the mid–fifties a small but growing market for more economical cars that were imported from Europe (France, Germany and Great Britain), and to a much lesser extent from Japan, was developing. During this period most of this activity was in the Southern California and New York metropolitan areas where their larger population base and access to ports made establishment of distributorships and dealers economically viable.[1] As the Mobilgas Economy Run continued to expand in scope, the question was asked as early as 1956: "Why are small imported cars not allowed to run?"[2] Then, in 1958, the United States was in its worst economic setback since World War II, which saw new car sales drop 31.4 percent from the previous year. At this same time, imports made up 8.1 percent of the new car market.[3] Against this backdrop, General Petroleum and USAC decided to stage a one-day Economy Run for imported models.

Information regarding the Mobil Mileage Rallies is not extensive and some areas remain open to question. They were held in late October or early November, around the time of the yearly new model introductions, and the little documentation available indicates that calendar year model imports competed, i.e., those competing in the 1958 Mobil Mileage Rally were 1958 models, etc.

1958

Southern California

Downtown Los Angeles to Los Angeles' Ambassador Hotel

October 19, 1958—258 Miles

With the procedures used in the Mobilgas Economy Run as a basis, rules for the inaugural Mobil Mileage Rally were developed under the direction of USAC's Chief Steward Joe Petrali. Joe, who had recently been named a steward for the 1959 Mobilgas Economy Run, brought with him years of experience, and also had assisting him individuals knowledgeable in the supervision of events of this type. After each entry was accepted, it was impounded the day prior to the Rally. USAC officials would conduct an inspection to certify stock status, and each entry was decorated on each side with banners identifying the Mobil Mileage Rally, entrant's name and model identification. This was followed by fueling and sealing of the gas tank and hood to prevent tampering. At this time, USAC officials held a drivers' meeting where they were given a map of the route, followed by a drawing for starting positions. In addition to the driver, each entry would also carry an observer, named by one of the competing entrants, to ensure all rules were followed. The observers would also assist each driver as needed by acting as a navigator.[4] The same procedures would continue in future events. The Rally would introduce the use of a miles-per-gallon only formula in determining class winners, which would be adopted by the Economy Run in 1959.

All entries had to be regular production models of imported cars regularly sold in the United States. Entries were divided into four competition classes based on engine size. Class "A" comprised those equipped with an engine under 750cc. Class "B" entries were equipped with engines between 750 and 1099cc, with Class "C" having engines between 1100 and 1499cc. Class "D" entries were equipped with engines over 1500cc.[5] When the entries closed, the smallest car was a Goggomobil T-400, equipped with a two-cylinder 392cc engine; a Rolls-Royce Silver Cloud, equipped with an eight-cylinder 4887cc engine, was the largest. Dealers entered the majority of the 37 cars competing. These included three dealers with prior Economy Run participation who also held import franchises — Inglewood's Bob Estes Lincoln-Mercury, Glendale's Art Frost DeSoto-Plymouth, and San Fernando's Lee Hamer Volvo. Previously Lee had been a Hudson dealer. Two private entries were also on hand to participate. *Hot Rod Magazine* entered a Peugeot 403 driven by Bob Pendergast, and former stock car driver, now a wine salesman, Merv Straley entered and drove a Goggomobil T-400.[6]

The majority of drivers were experienced in various types of sports car rallies or sports car races, although most were new at economy-style events. Several drivers had previous Economy Run experience, including Al Cottle, Mary Davis,

Danny Eames, Joan Fischal, Lee Hamer and Boyd Trombella, and were joined by several others who had experience as relief drivers. One of these was Ginny Sims, who was Mary Davis's relief driver when they became the first women's team in 1957 to take home a first-place trophy. Other drivers would go on to become regular participants in future Economy Runs.

At 5:00 a.m., the first driver took the green flag to begin a 258-mile route that would take them on a circular trip around the Los Angeles region with mountain driving making up 60 percent of the total distance. USAC officials established a time limit of 7 hours and 15 minutes. One driver, W. Halstead, did not arrive in time for the start, and his Opel Rekord entry was withdrawn. Leaving Los Angeles, the drivers headed for San Bernardino, then into the San Bernardino Mountains towards Crestline and Wrightwood. During the early stages of the Rally, J. Green was involved in an accident with her Lloyd 600 Alexander and was forced to withdraw from competition. At Cajon Junction, a refueling station was available, although most drivers chose not to stop. Leaving Wrightwood, the drivers crossed Dawson Saddle, at 7901 feet the highest elevation on this year's Rally, then continued on the Angelus Crest Highway towards La Canada. A forest fire had erupted which blocked the Angelus Crest Highway for two hours, and California Highway Patrol officers "forced some cars to retract several miles of the course."[7] Not only would this delay the competitors, but for three this would prove to be problematic. After arriving in La Canada, the drivers picked up the Ventura Freeway, heading west towards Malibu Canyon, and then onto the Pacific Coast Highway to Santa Monica. Delayed by the forest fire, the drivers had to contend with late Sunday afternoon traffic on Wilshire Boulevard towards the finish line located at the Ambassador Hotel. Each driver on arrival received a checkered-flag greeting from movie star Frank Lovejoy and was directed to special refueling platforms, where entries were topped off by USAC officials.[8] Mary Faulkner, driving a Ford Anglia, ran out of gas 50 feet from the finish line and was pushed across the line. Mary was not penalized by USAC officials, as she had been given a credit for the extra fuel used in the 2½-mile detour caused by the forest fire.[9] Two other drivers, Bill Levy in a Fiat 500 and Ginny Sims in a Hillman Minx, ran out of gas miles before the finish and were disqualified. Also disqualified, for driving off course, was J. Gobel, driving a Datsun 1000. At 5:00 p.m., the results were announced, followed by an awards dinner hosted by General Petroleum.

John Rich, driving a Fiat 750, achieved the highest mileage figure for all entries with an impressive 50.5404 miles per gallon. The 32 cars which completed the Rally recorded an average of 33.4840 miles per gallon. Not surprisingly, due to its weight and engine size, the Rolls-Royce Silver Cloud recorded the lowest mileage figure with 14.283 miles per gallon. This would be the Rolls-Royce's only appearance in the Rally.

Class winners were:

CLASS	MAKE and MODEL	MILES PER GALLON
A—Under 750cc	Fiat 750	50.5404
B—750 to 1099cc	Renault Dauphine	43.3548
C—1100 to 1499cc	Alfa Romeo Sprint	33.3673
D—Over 1500cc	Volvo PV-444	36.0672

Detailed results for all competing cars can be found in the Appen-

Both General Petroleum and USAC officials were generally pleased with the first Mileage Rally, despite some problems along the way, mainly caused by the forest fire in the Angelus National Forest.

Out of the four classes, only one featured close competition, with Jim Parkinson's Alfa Romeo Sprint coming out on top of Al Harmon's Volkswagen by a slim .0962 miles per gallon.

Newspaper articles summarizing the class winners and highlights of the Rally appeared mainly in Southern California area newspapers. Several enthusiast magazines reported in some detail, focusing on the high mileage figures obtained by the smaller imports.[10] In one of the more detailed articles to appear, Bob Pendergast, writing in *Hot Rod Magazine*, covered the Rally's events from his experience driving a Peugeot 403.[11]

"The first nationally recognized fuel economy competition for imported cars held in the United States" was now history.[12] Credit was already being given to General Petroleum and USAC for staging this event, and within months preliminary plans were underway for the 1959 Mobil Mileage Rally.

1959

Southern California

Downtown Los Angeles to Santa Monica Civic Auditorium

November 8, 1959—345.6 Miles

With positive reactions to the inaugural Mobil Mileage Rally, officials from General Petroleum and USAC completed plans for a second year, which saw an expansion of both competition classes and number of entries competing. All of the basic rules and procedures in force for the 1958 Rally remained in place.

All entries were regular production models of imported cars regularly sold in the United States and were now divided into six competition classes based on engine size and passenger capacity. Four classes were designed for passenger sedans carrying four or more passengers, with two new classes for sports cars. Class "A" entries were sedans equipped with an engine under 750cc. Class "B" entries were also sedans and equipped with engines between 750 and 1099cc. Class "C" entries were sedans having engines between 1100 and 1599cc, and Class "D" entries were sedans with engines over 1600cc. The new classes added for sports cars were Class "E" with engines under 1600cc and Class "F" with engines over 1600cc. When the entries closed, a total of 47 cars had been entered, which by coincidence was the same number that competed in the "regular" Economy Run held the previous April. The smallest engine size competing was in a Berkeley 3–60 sports car equipped with a two-cylinder 328cc engine, while the largest was in an Austin-Healey equipped with a six-cylinder 2912cc engine. Participating for the first time were Austin Healey, Berkeley, Lancia, Lotus, MG, NSU and Subaru.

The majority of drivers were again experienced in various types of sports car rallies or sports car races, with an increased number with having Economy Run experience. Participants in their first Rally from this group included Woody Bell, Ruth Doushkess, Art Rene, Pierce Venable and Lillian Warner. All of last year's class winners were back, looking to repeat their performances.

All entries were impounded by USAC officials the day prior, following the inspection and fueling procedures used in 1958. Each entry was again decorated on each side with banners identifying the Mobil Mileage Rally, entrant's name and model identification. Maps of the route were distributed to the drivers, with some taking advantage of the time remaining before the start for a practice run.

USAC officials laid out a different and challenging route which would cover 345.6 miles, traveling through 36 towns, all the while contending with 203 traffic signals.[13] The drivers faced a time limit of 9 hours to complete the route. Before dawn, the first driver took the green flag and headed south to Long Beach, then turned north towards La Verne and onto the Ridge Route to Maricopa. Leaving Maricopa, the drivers headed south towards Ojai, reaching an elevation of 5200 feet along the way before arriving in Ventura. From here the drivers followed the Pacific Coast Highway south through Malibu and on to the finish at the Santa Monica Civic Auditorium. The first driver crossed the finish line around 2:00 p.m. and was directed to special refueling platforms, where entries were topped off by USAC officials. Two entries, a Berkeley 3–60 and a Subaru 360, encountered mechanical difficulties along the route, forcing them to withdraw.

Eclipsing the 50.5404 miles per gallon record set by John Rich's Fiat 750 in 1958, Jim Parkinson in an Austin Healey Sprite achieved the highest mileage figure for all entries with an impressive 56.4975. Three other entries surpassed the 50-mile-per-gallon mark, including Vern Loveland's Berkeley 60 with 55.8453 miles per gallon, followed by Bill Levy's Fiat 500 with 55.2989 and Art Rene's Fiat Abarth Zagato with 52.9660. The 45 cars which completed the Rally recorded an average of 38.4256 miles per gallon. This was an increase of 4.9416 miles per gallon over 1958's average of 33.4840. The sports cars averaged 42.53 miles per gallon, while the sedans averaged 37.10. The lowest mileage recorded for all entries was 22.4046 miles per gallon, set by Jim Bell in a Simca Vedette.

Class winners were:

CLASS	MAKE and MODEL	MILES PER GALLON
A — Under 750cc	Fiat 500	55.2989
B — 750 to 1099cc	Renault Dauphine	46.2458
C — 1100 to 1599cc	Volkswagen	43.6698
D — Over 1600cc	Citroen ID-19	35.3425
E — Sports Cars Under 1600cc	Austin-Healey Sprite	56.4975
F — Sports Cars Over 1600cc	Triumph TR-3	33.9321

Detailed results for all competing cars can be found in the Appendix.

In reporting the results, USAC officials noted that the field averaged a speed of 38.73 miles per hour over the 345.6-mile route. In addition to the two entries who withdrew due to mechanical problems, USAC reported that John Rich's Fiat Abarth 750 exhausted its water supply at Maricopa, and after a delay was able to continue.[14]

Jim Parkinson topped off a successful year, competing in both the Mobil Mileage Rally and Mobilgas Economy Run. Not only did he achieve the highest mileage for all entries, Jim scored back-to-back class wins in the Mileage Rally, which came on the heels of his earlier performance placing first in the Economy Run's high-price class with a Cadillac Sixty-Two. This year also saw back-to-back class wins for the Renault Dauphine. In the four classes for sedan models, competition was not close, with the separation between first and second as high as 6.14 miles per gallon. The closest competition in any of the classes was between Jack Draper's Triumph TR-3 and Bill Brandes' Mercedes-Benz 190SL in the Sports Cars Over 1600cc, where the two entries were separated by .32 mile per gallon.

British Motor Corporation was quick to advertise the achievements of its Austin-Healey Sprite. Appearing in newspapers after the Rally, the headline announced "56½ miles per gallon! Austin-Healey Sprite Gets Most Miles Per Gallon In Mobil Mileage Rally." The ad provided readers with a brief overview of the event. The mileage results for other BMC models, including the Austin A-40, Austin A-55 Cambridge, Austin-Healey 100, MG Magnette, MG 1600 and Morris Minor 1000, were also highlighted.[15] The John Green Corporation, distributor for Renault models in California, Arizona, Nevada and Utah, advertised the success of the Renault

Dauphine, noting it won its class "with an amazing mileage of 46.25 mpg."[16]

Newspaper articles again this year only summarized the class winners and highlights of the Rally, mainly appearing in Southern California area newspapers. Several enthusiast magazines reported on the Rally, while not in the same detail as last year; these articles focused on the achievement of Jim Parkinson's Austin-Healey Sprite.[17] In its article, *Small Car Parade* made the following observation:

> The results of the run are interesting when compared with those of the 1959 Mobilgas Domestic car run held last April. By coincidence both runs had 47 starters and the overall average of domestic cars was 19.44 mpg, compared with 38.42 mpg overall average for the imports. In the April domestic car run, the overall mileage was increased by allowing the then-available compacts cars such as Rambler and Studebaker to complete. Without the compact cars running, the overall average would have been close to the previous year's 18.75-mpg average, which is about half the average of the imports.[18]

With a second successful Mileage Rally now history, preliminary plans were underway by officials for a major change in 1960. With the merger of General Petroleum Corporation into Socony-Mobil Corporation, the newly formed Mobil Oil Company was looking to expand the scope of the Rally. Later in the year came the announcement that the 1960 Mobil Mileage Rally would be held in New York, and would alternate between the East and West Coasts in future years.

1960

New York

New York Coliseum at Columbus Circle to New Rochelle's Glen Island Casino

October 30, 1960—201 Miles

Looking to capitalize on the East Coast import market, officials with Mobil Oil decided to hold the 1960 Mobil Mileage Rally in the New York metropolitan area. With guidance from current USAC officials, additional personnel were recruited locally to handle an expanded inspection process. The rules from the previous two Rallies remained intact, with the only change now requiring all entries to be impounded four days prior to the start for inspections.

All entries continued to be regular production models of imported cars regularly sold in the United States and divided into six competition classes based on engine size and passenger capacity. Engine sizes for individual classes remained identical to those from 1959. When the entries closed, a total of 35 cars had been entered and represented most of imported models currently being sold. The smallest engine size was in a Vespa 400 sports car equipped with a two-cylinder 394cc engine, while the largest again was in the Austin-Healey 3000 equipped with a six-cylinder 2912cc engine. Two makes, DAF and Vespa, participated for the first time, although several who were active in the previous two Rallies were absent. These included Alfa Romeo, Datsun, Lancia, Mercedes-Benz, and MG.

All but two of drivers were new to economy competition, although most had previous experience in various types of sports car rallies or sports car races. Drivers came from the New York, New Jersey, Massachusetts, and Washington, D.C., areas. One of Europe's top drivers, Serge Meunier, traveled from Paris to drive the Renault Caravelle entry. The two drivers with previous Rally experience, Lee Hamer and John Lance, traveled from California to participate. John Lance, again driving a Renault Dauphine, was looking for a repeat of last year's first-place performance.

After the entries were impounded, USAC officials began a thorough inspection, ensuring that all entries were typical of those sold to the public and components were completely stock. This year each entry was devoid of all identification for the Mobil Mileage Rally, apparently to conform to advertising regulations on certain parkways and roads in the region. Maps of the route were distributed to the drivers a few hours prior to the start, which eliminated any opportunity to make a practice run.

USAC officials designed a challenging route this year which would cover 201 miles, traveling through New York, New Jersey and Connecticut. Within a time limit of 6½ hours, the drivers would pass through 46 cities and towns on a blustery fall day.[19] Starting at dawn in midtown Manhattan, the drivers made a loop through Brooklyn and Long Island before crossing the George Washington Bridge into New Jersey. Traveling north along the Hudson River, the drivers crossed back into New York over the Bear Mountain Bridge. Their next challenge was a climb over Bear Mountain before making their way south to the Connecticut Turnpike. Arriving in New Rochelle, the drivers made their way to the finish line at Glen Island Casino. The drivers started arriving around 1:00 p.m. and were directed to special refueling platforms, where entries were topped off by USAC officials. An Opel Caravan encountered difficulties along the route and did not finish.

With an impressive figure of 62.29 miles per gallon, an NSU Sport Prinz driven by Henry Fanelli recorded the highest mileage for all entries. This figure was also the highest attained in the three-year history of the Mobil Mileage Rally. Three entries surpassed the 50-mile-per-gallon mark, including Dwight Holbrook's Morris Mini-Minor with 55.44 miles per gallon, followed by Lawrence Friedman's Vespa 400 with 51.93 and Serge Meunier's Renault Caravelle with 50.13. The 34 cars which completed the Rally recorded an average of 40.2351 miles per gallon, an increase of 1.8095 miles per gallon over

1959's average. One explanation for this increase was that this year's route lacked tough mountain stretches encountered in the two previous Rallies held in Southern California. The lowest mileage recorded for all entries was 24.57 miles per gallon set by John Norwood in a Fiat 2100. Only four entries recorded mileage figures under 30 miles per gallon, and all were equipped with engines over 1600cc.

Class winners were:

CLASS	MAKE and MODEL	MILES PER GALLON
A — Under 750cc	Renault 4CV	49.18
B — 750 to 1099cc	Morris Mini-Minor	55.44
C — 1100 to 1599cc	Fiat 1200	40.55
D — Over 1600cc	Citroen ID-19	35.98
E — Sports Cars Under 1600cc	NSU Sport Prinz	62.29
F — Sports Cars Over 1600cc	Triumph TR-3	32.23

Detailed results for all competing cars can be found in the Appendix.

The Citroen ID-19 and Triumph TR-3 scored back-to-back wins in their respective classes, with Fiat and Renault scoring their third consecutive class wins. Competition was not close in any of the classes. The closest was in Class "C," where Ray Cuomo's Fiat 1200 finished with a 2.52 miles-per-gallon margin over Ken Norton's Peugeot 403.

For the first time in the history of the Rally, Mobil advertised the results with the headline "34 Foreign Cars Average 40.24 Miles Per Gallon In Mobil Mileage Rally." The ad also provided readers with the background and route of the event.[20] Renault was the only importer to heavily advertise the results and featured photographs of its three competing models.[21] Newspaper articles were few and far between this year and continued to summarize the class winners and highlights of the Rally; most appeared in the New York metropolitan area. Very few articles on this year's Rally were seen in magazines; these focused mainly on reporting the results, along with noting the top mileage achievement of NSU Sports Prinz.[22] It was becoming apparent that press interest in the Rally was declining.

1961

Southern California

Downtown Los Angeles Returning to Downtown Los Angeles

November 12, 1961—300 Miles

Returning to Southern California, officials from Mobil Oil were looking for a renewed interest in the Mobil Mileage Rally. With the basic rules and procedures remaining in place from the previous three Rallies, everything was set to challenge the high mileage figures obtained in last year's Rally.

All entries were now divided into five competition classes based on engine size and passenger capacity. Three classes were designed for four-passenger models and two classes for two-passenger models. Engine size requirements for individual classes were modified, with Class "A" for four-passenger entries equipped with an engine under 1000cc, Class "B" having engines between 1000 and 1600cc, and Class "C" with engines over 1600cc. In the two classes for two-passenger models, Class "D" entries were equipped with an engine under 1400cc, while Class "E" entries had engines over 1400cc. When the entries closed, a total of 34 cars had been entered and represented most of imported models currently being sold. Making its debut this year was the Israeli-made Sabra. The entry list also saw a return of several makes who sat out the 1960 Rally, including Datsun, Jaguar, MG and Sunbeam. The smallest engine size this year was in an NSU Sport Prinz sports car equipped with a four-cylinder 583cc engine, and the largest, for the third year, was in an Austin-Healey 3000 equipped with a six-cylinder 2912cc engine.

Regular participants from the 1958 and 1959 Rallies returned this year, with several also active in the Economy Runs. Two drivers, Lee Hamer and John Lance, had competed in each Rally since its inception. Three class-winning drivers from 1959 — Jack Draper, John Lance and Bill Levy — returned, looking to repeat their earlier performance.

All entries were impounded by USAC officials four days beforehand to allow for a thorough inspection and refueling following procedures established in prior years. In contrast to the 1960 Rally, each entry was again decorated on each side with banners identifying the Mobil Mileage Rally, the entrant's name and the car model.

Another challenging route, which resembled a figure eight design, faced the drivers. Covering 300 miles, it passed through 44 towns in five Southern California counties, all within an eight-hour time limit. Starting before dawn in downtown Los Angeles, the drivers headed east towards San Bernardino, then south through Riverside and down to Perris. Leaving Perris, the drivers climbed to an elevation of 2600 feet as they made their way to San Juan Capistrano, then took the Santa Ana Freeway back into downtown Los Angeles. Here drivers could refuel at their own discretion. After driving through Santa Monica, Thousand Oaks and Moorpark, they continued over the Santa Susana Pass into the San Fernando Valley. Back in downtown Los Angeles, the drivers made their way to the finish line located at the intersection of Eighth and Francisco Streets.[23] With a small group of spectators on hand, they began arriving around 2:00 p.m. and were directed to special refueling platforms, where entries were topped off by

USAC officials. Four entries — Datsun Bluebird, Facel Vega Facellia, Saab 96 and Sabra — were withdrawn for reasons not known.

For the second straight year, a NSU Sport Prinz recorded the highest mileage for all entries with 59.76 miles per gallon. It was driven by Bill Levy, whose figure was 2.53 miles per gallon below 1960s result of 62.29 miles per gallon. This lower figure was a result of this year's longer and more challenging route, which traveled through a larger and more mountainous urban area. In addition to the NSU Sport Prinz, three other entries surpassed the 50-mile-per-gallon mark, including Marty Evelhock's Fiat Abarth 747 with 53.65, followed by Jim Parkinson's Morris Mini-Minor with 51.399 miles per gallon and Fran Foster's Lotus Elite with 51.04. The 30 cars which completed the Rally recorded an average of 40.1854, which was a slim decline of .0497 mile per gallon over 1960s average. The lowest mileage recorded for all entries was 23.90 miles per gallon set by Ray Wallace in an Austin Healey 3000. Only three entries recorded mileage figures under 30 miles per gallon, and all were equipped with engines over 1600cc.

Class winners were:

CLASS	MAKE and MODEL	MILES PER GALLON
A — Four Passenger Under 1000cc	Morris Mini-Minor	51.399
B — Four Passenger 1000 to 1600cc	Volkswagen	40.384
C — Four Passenger Over 1600cc	Citroen ID-19	35.68
D — Two Passenger Under 1600cc	NSU Sport Prinz	59.76
E — Two Passenger Over 1600cc	Triumph TR-3	38.99

Detailed results for all competing cars can be found in the Appendix.

Several familiar faces were seen in this year's winner's circle, with Jim Parkinson scoring his third class win. Jack Draper and Bill Levy each scored their second win. Mary Hauser, with two Economy Run victories to her credit, scored her first class win in the Rally. Paula Murphy, driving in her first Rally, also placed first in her class. Both the Citroen ID-19 and Triumph TR-3 scored their third consecutive class wins, with back-to-back class wins for the Morris Mini-Minor and NSU Sport Prinz. Only one class featured competition that was fairly close, and that was Class "E," where Jack Draper's Triumph TR-3 had a margin of .84 mile per gallon over Mary Davis's Porsche 356B. In all other classes the margin between first and second ranged from 1.06 to 6.11 miles per gallon.

Both advertising and articles on the 1961 Rally were virtually nonexistent. The only coverage in enthusiast magazines was in the January 1962 issue of *Automobile Topics*, which only listed the final results along with a photograph of the class-winning entries and their drivers.[24]

After four years, it was clear the Mobil Mileage Rally was not drawing either the press or public's interest as Mobil Oil had envisioned. This situation was especially evident when compared to the more extensive coverage nationwide by newspapers and enthusiast-oriented publications of the regular and more established Economy Runs. Without any formal announcement, 1961 would be the last Mobil Mileage Rally held, as it quietly faded into history.

Appendix: Results

1936 Gilmore Yosemite Economy Run

January 9, 1936 — 30 Cars — 352 Miles

SWEEPSTAKES WINNER — GRAHAM SUPERCHARGER 6

Place	Ton M.P.G.	M.P.G.	No.	Car	Driver	Entrant
CLASS "1-A" — UNDER $700						
1	49.48	33.21	31	Willys 77 4	Howard R. Gwynn	Gwynn Motor Co. — Los Angeles, CA
2	45.52	32.00	32	Willys 77 4	Otis M. King	Crown City Motor Co.
3	33.77	25.88	33	Willys 77 4	L.T. Sollender	Fred Fudge — Los Angeles, CA
CLASS "2-A" — $701 to $900						
1	47.61	25.33	13	Graham Crusader 6	O.J. Wedgeworth	Ralph Hamlin Motors, Inc. — Los Angeles
2	43.10	22.42		Plymouth Deluxe 6	Don Langmo	Sweitzer & Shortell — Los Angeles, CA
3	41.43	25.70		Chevrolet Master Deluxe 6	W.A. Smith	Chevrolet Dealers — Los Angeles, CA
CLASS "3-A" — $901 to $1000						
1	47.89	23.95	26	Pontiac 6	Fred H. Miller	Savage-Halderman Co. — Los Angeles, CA
2	45.37	23.95	15	Terraplane Deluxe 6	Babe Stapp	Earle C. Anthony, Inc. — Los Angeles, CA
CLASS "4-A" — $1001 to $1100						
1	50.98	24.28	29	Studebaker Dictator 6	William C. Martin	Paul G. Hoffman Co., Inc. — Los Angeles
2	46.11	22.28		Nash 400 6	Homer Thompson	Thompson Nash Motor Co. — Los Angeles
3	45.10	22.00		Pontiac Deluxe 8	Pat Wheeland	Savage-Halderman Co. — Los Angeles, CA
4	44.47	23.78		Oldsmobile Six		
CLASS "5-A" — $1101 to $1300						
1	55.47	26.67	12	Graham Supercharger 6	H. Clay Moore	Ralph Hamlin Motors, Inc. — Los Angeles
2	48.11	23.16		DeSoto Airstream 6	W.D. Walker	Sweitzer & Shortell — Los Angeles, CA
3	46.62	20.95		Chrysler Airstream 6	R.W. Arnold	Angelus Motors, Inc. — Los Angeles, CA
4	46.40	22.85	14	Hudson Deluxe 8	Wilbur Shaw	Earle C. Anthony, Inc. — Los Angeles, CA
5	46.07	23.62		Auburn 6		
6	45.96	22.42		Hupmobile 6		
7	45.61	19.55		Nash Ambassador 6		Thompson Nash Motor Co. — Los Angeles
8	38.68	19.34		Oldsmobile Eight		
9	31.99	15.30		Reo Flying Cloud 6		
CLASS "6-A" — $1301 to $1600						
1	48.63	20.35	30	Studebaker President 8	James E. Van Sant	Paul G. Hoffman Co., Inc. — Los Angeles
2	48.26	21.60		DeSoto Airflow 6	R.W. Washburn	Sweitzer & Shortell — Los Angeles, CA
3	45.84	19.34		Hupmobile 8	G.W. Pritchard	R. Whitsell — Los Angeles, CA
4	44.49	19.78		Chrysler Airstream 8		Angelus Motors, Inc. — Los Angeles, CA
5	44.46	18.53		Nash Ambassador 8		Thompson Nash Motor Co. — Los Angeles
6	44.40	20.00		Packard One-Twenty 8		Earle C. Anthony, Inc. — Los Angeles, CA
7	42.40	20.00		Auburn 8		
CLASS "7-A" — $1601 to $1800						
1	53.35	18.33	5	Chrysler Airflow 8	Allen A. MacLean	Angelus Motors, Inc. — Los Angeles, CA
2	38.91	17.69		Lincoln Zephyr 12	Ed Maddux	Maddux, Inc. — Hollywood, CA

As of publication date, some car numbers, drivers' and entrants' names remain unknown.

1937 Gilmore Yosemite Economy Run

January 7, 1937 — 26 Cars — 352 Miles

Sweepstakes Winner — Graham Supercharger 6

Place	Ton M.P.G.	M.P.G.	No.	Car	Driver	Entrant
CLASS "A"						
1	49.2	28.16	2	Willys 4	Howard R. Gwynn	Willys Distributors, Inc.— Los Angeles, CA
2	46.4	27.29	1	Willys 4	Otis M. King	Willys Distributors, Inc.— Los Angeles, CA
Disqualified Off Course			3	Willys 4	Jack Stotts	Willys Distributors, Inc.— Los Angeles, CA
CLASS "B"						
1	41.8	19.45	5	Chevrolet Master Deluxe 6	Roy H. Woods	Los Angeles Chevrolet Dealers Association
2	41.5	20.83	4	Plymouth Deluxe 6	Jimmy Walker	Harry A. Shortell — Los Angeles, CA
CLASS "C"						
1	48.4	22.00	6	Terraplane Deluxe 6	Austin Elmore	Dick Klein, Inc.— Huntington Park, CA
2	47.5	21.60	7	Pontiac Deluxe 6	Fred H. Miller	Los Angeles Pontiac Dealers
3	45.5	20.23	10	DeSoto 6	Kenneth Dobbs	Harry A. Shortell — Los Angeles, CA
4	44.8	20.47	8	Nash Lafayette 6	Andy Henderson	Thompson Nash Motor Co.— Los Angeles
Did Not Start			9	Dodge 6	Joseph Bozzani	Pelton Motor Co.— Los Angeles, CA
CLASS "D"						
1	52.246	24.62	14	Graham Cavalier 6	O.J. Wedgeworth	Ralph Hamlin Motors, Inc.— Los Angeles
2	41.3	18.33	28	Pontiac Deluxe 8	Roland L. Amberg	Los Angeles Pontiac Dealers
3	40.2	18.53	12	Oldsmobile Six	S.A. Gillis	Kaiser Brothers — Los Angeles, CA
CLASS "E"						
1	53.4	23.95	19	Graham Supercharger 6	H. Clay Moore	Ralph Hamlin Motors, Inc.— Los Angeles
2	52.232	22.71	18	Hudson Deluxe 8	Dick Klein	Dick Klein, Inc.— Huntington Park, CA
3	49.59	22.14	20	Chrysler Royal 6	Bill Comstock	Angelus Motors, Inc.— Los Angeles, CA
4	47.3	21.60	17	Nash Ambassador 6	Johnny Clark	Thompson Nash Motor Co.— Los Angeles
5	44.5	19.89	15	Packard Six	Frank E. Podus	Earle C. Anthony, Inc.— Los Angeles, CA
6	41.0	17.17	16	Oldsmobile Eight	John C. Bodine	Kaiser Brothers — Los Angeles, CA
CLASS "F"						
1	49.6	20.35	21	Nash Ambassador 8	Homer Thompson	Thompson Nash Motor Co.— Los Angeles
2	47.6	19.45	24	Chrysler Imperial 8	C.J. Housley	Angelus Motors, Inc.— Los Angeles, CA
3	41.7	18.05	22	Lincoln Zephyr 12	George W. Magee	Deaton & Roberts Co., Inc.— Beverly Hills
DNF — Mechanical			23	Graham Supercharger 6	E.G. Baker	Ralph Hamlin Motors, Inc.— Los Angeles
CLASS "G"						
1	43.6	18.05	25	Packard Eight	Allan W. Fleming	Earle C. Anthony, Inc.— Los Angeles, CA
CLASS "H"						
1	45.6	18.24	26	Cord Beverly 8	N. C. Williams	Auburn Motor Sales Corporation
2	43.6	15.64	27	Chrysler Airflow 8	Allen A. MacLean	Angelus Motors, Inc.— Los Angeles, CA

As of publication date, price class breakdowns remain unknown.

1938 Gilmore Yosemite Economy Run

January 13, 1938 — 26 Cars — 314.5 Miles

Sweepstakes Winner — Graham Supercharger 6

Place	Ton M.P.G.	M.P.G.	No.	Car	Driver	Entrant
CLASS "A"						
1	48.900	30.534	2	Willys Deluxe 4	Wayne E. Burdick	Willys Distributors, Inc.— Los Angeles, CA
2	47.175	29.392	1	Willys Deluxe 4	R.R. Rice	Willys Distributors, Inc.— Los Angeles, CA
3	44.929	28.080	3	Willys Deluxe 4	Curly Crawford	Willys Distributors, Inc.— Los Angeles, CA
CLASS "B"						
1	50.767	28.850	4	Ford "60" 8	Eddie Pullen	Al Stuebing — Hollywood, CA

2	43.491	21.993	5	Chevrolet Master Deluxe 6	Roy H. Woods	Los Angeles Chevrolet Dealers Association
3	40.460	21.250	6	Plymouth Deluxe 6	Ben Cole	Harry A. Shortell — Los Angeles, CA
CLASS "C"						
1	50.461	25.779	8	Hudson Terraplane 6	Otto Butts	Boswell & Gratiot Co. — Glendale, CA
2	48.837	22.305	10	Nash Lafayette Deluxe 6	Andy Henderson	Homer C. Thompson, Inc. — Los Angeles
3	47.087	21.840	9	Pontiac Deluxe 6	Fred H. Miller	Pontiac Dealers of Southern CA
4	42.109	21.107	7	Dodge 6	Joseph Bozzani	Pelton Motor Co. — Los Angeles, CA
CLASS "D"						
1	46.756	20.967	11	Pontiac Deluxe 8	Roland L. Amberg	Pontiac Dealers of Southern CA
2	42.871	20.691	12	DeSoto 6	Don Langmo	Harry A. Shortell — Los Angeles, CA
CLASS "E"						
1	51.807	24.380	14	Studebaker Six	William C. Martin	Paul G. Hoffman Co., Inc. — Los Angeles
2	41.595	19.177	15	Oldsmobile Six	J. Duquette	Kurtz Motor Co. — Los Angeles, CA
CLASS "F"						
1	53.559	23.647	18	Nash Ambassador 6	Homer Thompson	Homer C. Thompson, Inc. — Los Angeles
2	52.982	24.764	20	Graham 6	O.J. Wedgeworth	Ralph Hamlin Motors, Inc. — Los Angeles
3	48.070	21.840	16	Hupmobile 6	Tom Moore	Greer-Robbins Co. — Los Angeles, CA
4	45.227	20.290	19	Packard Six	Frank E. Podus	Earle C. Anthony, Inc. — Los Angeles, CA
5	43.315	19.061	17	Oldsmobile Eight	Ted Chestek	Kurtz Motor Co. — Los Angeles, CA
CLASS "G"						
1	49.409	22.790	21	Hudson Deluxe 8	Walter Jennings	Boswell & Gratiot Co. — Glendale, CA
2	47.568	19.650	23	Nash Ambassador 8	B. Green	Homer C. Thompson, Inc. — Los Angeles
3	46.968	20.691	22	Studebaker President 8	James E. Van Sant	Paul G. Hoffman Co., Inc. — Los Angeles
CLASS "H"						
1	55.927	25.779	24	Graham Supercharger 6	H. Clay Moore	Ralph Hamlin Motors, Inc. — Los Angeles
2	49.272	20.556	25	Hupmobile 8	Ray Logan	Greer-Robbins Co. — Los Angeles, CA
CLASS "I"						
1	54.744	23.470	26	Lincoln Zephyr 12	George W. Magee	Deaton & Roberts Co., Inc. — Beverly Hills
CLASS "J"						
1	46.543	16.211	27	Packard Eight	Les P. Butts	Earle C. Anthony, Inc. — Los Angeles, CA

As of publication date, price class breakdowns remain unknown.

1939 Gilmore Yosemite Economy Run

January 5, 1939 — 29 Cars — 314.5 Miles

SWEEPSTAKES WINNER — STUDEBAKER COMMANDER 6

Place	Ton M.P.G.	M.P.G.	No.	Car	Driver	Entrant
CLASS "A" — UNDER $800						
1	42.720	26.21	1	Willys Overland Deluxe 4	Bennett Hill	Willys-Overland Motors, Inc.
2	40.936	25.36	3	Willys Overland Deluxe 4	Babe Stapp	Willys-Overland Motors, Inc.
DIS	43.398	26.88	2	Willys Overland Deluxe 4	Walter Hersley	Willys-Overland Motors, Inc.
CLASS "B" — $801 to $1000						
DIS	48.649	24.57	6	Ford "85" Deluxe 8	Charles Rocheville	Al Stuebing — Hollywood, CA
DIS	46.980	24.38	5	Hudson 112 Deluxe 6	Walter Jennings	Ralph Hamlin Motors, Inc. — Los Angeles
DIS	46.628	23.30	7	Chevrolet Master Deluxe 6	Roy H. Woods	Los Angeles Chevrolet Dealers Association
DIS	43.700	22.79	4	Plymouth Deluxe 6	Ben Cole	Harry A. Shortell — Los Angeles, CA
CLASS "C" — $1001 to $1060						
1	43.763	21.40	12	Oldsmobile "60" 6	John C. Bodine	Oldsmobile Division — General Motors
2	42.038	20.97	9	Dodge Special 6	Herb Ford	Pelton Motor Co. — Los Angeles, CA
DIS	44.936	21.84	8	Pontiac Quality Deluxe 6	Austin Elmore	Pontiac Motor Division — General Motors
CLASS "D" — $1061 to $1130						
1	47.388	21.25	11	Nash Lafayette Deluxe 6	Andy Henderson	Nash California Co. — Los Angeles, CA
DIS	48.114	23.47	10	Dodge Deluxe 6	Joseph Bozzani	Pelton Motor Co. — Los Angeles, CA
CLASS "E" — $1131 to $1200						
1	42.808	20.03	17	Chrysler Royal 6	Ray E. Shafer	Ray E. Shafer — Santa Monica, CA

DIS	48.914	23.83	16	Mercury 8	William R. Knopp	J. E. Coberly, L.A.—Boyd Gibbons, L.A.–Ham Nerney, L.A.—Al Stuebing, Hollywood
DIS	46.644	21.25	14	Pontiac Deluxe 8	Fred H. Miller	Pontiac Motor Division—General Motors
DIS	45.502	22.31	15	Hudson 6	O.J. Wedgeworth	Ralph Hamlin Motors, Inc.—Los Angeles
CLASS "F"—$1201 to $1300						
1	55.875	25.78	21	Studebaker Commander 6	William C. Martin	Paul G. Hoffman Co., Inc.—Los Angeles
2	50.426	23.47	18	Chrysler Royal (with OD) 6	Eddie Seward	Lysle Greenman—Los Angeles, CA
3	46.192	20.69	22	Hupmobile Senior 6	Tom Moore	Greer-Robbins Co.—Los Angeles, CA
4	45.483	19.91	19	Nash Ambassador 6	H. Floyd Brown	Nash California Co.—Los Angeles, CA
DIS	49.730	23.13	20	DeSoto Deluxe 6	Don Langmo	Harry A. Shortell—Los Angeles, CA
CLASS "G"—$1301 to $1350						
DIS	46.275	20.69		Oldsmobile "80" 8	J. W. Schiller	Oldsmobile Division—General Motors
CLASS "H"—$1351 to $1500						
1	52.914	22.96	25	Studebaker President 8	James E. Van Sant	Paul G. Hoffman Co., Inc.—Los Angeles
2	50.419	21.99	28	Packard Six	Frank E. Podus	Earle C. Anthony, Inc.—Los Angeles, CA
3	46.222	19.06		Nash Ambassador 8	Jules Ellingbee	Nash California Co.—Los Angeles, CA
4	44.847	20.42		Hudson Country Club 8	Norman Williams	Ralph Hamlin Motors, Inc.—Los Angeles
DIS	46.312	19.18	29	Hupmobile Senior 8	Ray Logan	Greer-Robbins Co.—Los Angeles, CA
CLASS "I"—$1501 to $1750						
1	52.963	21.84	30	Packard One-Twenty 8	Les P. Butts	Earle C. Anthony, Inc.—Los Angeles, CA
CLASS "J"—$1751 and UP						
1	47.490	19.66	31	Lincoln Zephyr 12	George W. Magee	Deaton & Roberts Co., Inc.—Beverly Hills

DIS—Disqualified under entry rules because it did not complete the Run in the required 11 hours.
As of publication date, some car number information remains unknown.

1940 Gilmore Yosemite Economy Run

January 4, 1940—28 Cars—306.5 Miles

SWEEPSTAKES WINNER—STUDEBAKER COMMANDER 6

Place	Ton M.P.G.	M.P.G.	No.	Car	Driver	Entrant
CLASS "A"—UNDER $700						
1	48.124	30.05	20	Willys Speedway 4	Curly Crawford	Willys Distributors, Inc.—Los Angeles, CA
2	47.656	29.76	1	Willys Deluxe 4	Walter Hersley	Willys Distributors, Inc.—Los Angeles, CA
3	47.655	29.47	18	Willys Deluxe 4	Wayne E. Burdick	Willys Distributors, Inc.—Los Angeles, CA
CLASS "B"—$701 to $850						
1	53.448	29.19	3	Studebaker Champion 6	James E. Van Sant	Paul G. Hoffman Co., Inc.—Los Angeles, CA
2	51.420	24.92	25	Ford "85" Deluxe 8	Charles Rocheville	Al Stuebing—Hollywood, CA
3	48.022	22.87	22	Chevrolet Master 6	Roy H. Woods	Los Angeles Chevrolet Dealers Association
4	45.164	22.54	27	Plymouth Roadking 6	Ben Cole	Harry A. Shortell—Los Angeles, CA
CLASS "C"—$851 to $950						
1	52.699	23.76	6	Nash Lafayette Deluxe 6	Andy Henderson	Nash California Co.—Los Angeles, CA
2	50.703	23.76	12	Dodge Deluxe 6	Joseph Bozzani	Bozzani Motors Ltd.—Los Angeles, CA
3	49.795	23.22	19	Oldsmobile "60" 6	John C. Bodine	Oldsmobile Division—General Motors
4	47.055	22.05	2	Pontiac Special 6	Fred H. Miller	Pontiac Motor Division—General Motors
CLASS "D"—$951 to $1000						
1	50.822	23.76	11	Mercury 8	William R. Knopp	Ham Nerney, Los Angeles, CA and Al Stuebing, Hollywood, CA
2	49.941	23.10	17	DeSoto Custom 6	William Claus	Harry A. Shortell—Los Angeles, CA
CLASS "E"—$1001 to $1050						
1	54.750	24.72	9	Studebaker Commander 6	William C. Martin	Paul G. Hoffman Co., Inc.—Los Angeles
2	53.159	23.95	7	Packard One-Ten 6	Frank E. Podus	Earle C. Anthony, Inc.—Los Angeles, CA
3	52.405	23.05	26	Nash Ambassador 6	Arthur Florence	Nash California Co.—Los Angeles, CA
4	50.992	22.70	14	Oldsmobile "70" 6	James Hemstreet	Oldsmobile Division—General Motors
CLASS "F"—$1051 to $1200						
1	54.573	23.40	13	Studebaker President 8	Howard N. Kyser	Paul G. Hoffman Co., Inc.—Los Angeles
2	49.327	21.28	15	Pontiac Torpedo 8	Roland L. Amberg	Pontiac Motor Division—General Motors

Place	Ton M.P.G.	M.P.G.	No.	Car	Driver	Entrant
3	46.407	21.58	4	Chrysler Windsor 6	Jack K. Shafer	Ray E. Shafer — Santa Monica, CA
4	45.599	18.80	24	Oldsmobile "90" 8	J.W. Schiller	Oldsmobile Division — General Motors
CLASS "G"—$1201 to $1400						
1	51.655	21.43	10	Nash Ambassador 8	J.D. Bowersock	Nash California Co. — Los Angeles, CA
2	50.896	21.58	8	Packard One-Twenty 8	Les P. Butts	Earle C. Anthony, Inc. — Los Angeles,CA
3	45.149	18.35	28	Chrysler New Yorker 8	Eddie Seward	Lysle Greenman — Los Angeles, CA
CLASS "H"—$1401 to $1600						
1	49.329	19.90	23	Lincoln Zephyr 12	Horace M. Jones	Deaton Motor Co. — Beverly Hills, CA
2	41.532	17.12	5	Chrysler Saratoga 8	Van Spaulding	Ray E. Shafer — Santa Monica, CA
CLASS "I"—$1601 and UP						
1	47.660	19.04	21	Packard One-Sixty 8	Allan W. Fleming	Fleming Motor Co. — Los Angeles, CA
2	46.574	17.12	16	Chrysler Crown Imperial 8	Ray E. Shafer	Ray E. Shafer — Santa Monica, CA

1941 Gilmore Grand Canyon Economy Run

January 9, 1941 — 24 Cars — 599.3 Miles

SWEEPSTAKES WINNER — LINCOLN CUSTOM 12

Place	Ton M.P.G.	M.P.G.	No.	Car	Driver	Entrant
CLASS "A"— FOUR-CYLINDER						
1	48.624	29.06	17	Willys Plainsman 4	J.C. Marbe	Willys-Overland Motors, Inc.
2	47.319	28.40	10	Willys Plainsman 4	Henry Amon	Willys-Overland Motors, Inc.
3	41.334	25.18	15	Willys Americar 4		Willys-Overland Motors, Inc.
CLASS "B"—$750 to $809						
1	49.557	23.05	24	Ford Super Deluxe 8	William R. Knopp	Al Stuebing — Hollywood, CA
2	43.417	21.71	8	Plymouth Special Deluxe 6		
CLASS "C"—$810 to $914						
1	48.651	25.81	19	Nash Ambassador 600 6	Andy Henderson	Nash California Co. — Los Angeles, CA
2	45.003	24.61	18	Studebaker Champion 6	James E. Van Sant	Studebaker Pacific Corporation
CLASS "D"—$915 to $984						
1	50.969	24.96	20	Hudson Deluxe 6	Glenn White	Hudson Associated Dealers
2	50.895	23.35	11	Mercury 8		W. I. Tupman, Inc. — Los Angeles, CA
3	48.425	22.47	16	Dodge Deluxe 6		
CLASS "E"—$985 to $1069						
1	52.975	24.36	12	Studebaker Commander 6	William C. Martin	Studebaker Pacific Corporation
2	51.512	23.16	9	Nash Ambassador 6		Nash California Co. — Los Angeles, CA
3	46.632	20.04	6	DeSoto Deluxe 6		
CLASS "F"—$1070 to $1189						
1	51.988	22.53	5	Studebaker President 8	Howard N. Kyser	Studebaker Pacific Corporation
2	48.579	21.18	4	Nash Ambassador 8	L.R. Denny	Nash California Co. — Los Angeles, CA
3	44.060	19.71	2	Chrysler Royal 6		
CLASS "G"—$1190 to $1299						
1	44.988	20.14	21	Chrysler Windsor 6	Lee Winchester	Lysle Greenman — Los Angeles, CA
CLASS "H"—$1300 to $1389						
1	46.846	19.78	1	Packard One-Twenty 8	Les P. Butts	Earle C. Anthony, Inc. — Los Angeles, CA
2	45.906	20.18	7	Hudson Commodore 8		Hudson Associated Dealers
DNF — Accident			23	Chrysler Saratoga 8		
CLASS "I"—$1390 to $1699						
1	57.749	22.96	13	Lincoln Zephyr 12	Horace M. Jones	Deaton Motor Co. — Beverly Hills, CA
2	42.956	17.47	14	Chrysler New Yorker 8		
CLASS "J"—$1700 to $4000						
1	57.827	21.03	22	Lincoln Custom 12	Marshall Martin	Deaton Motor Co. — Beverly Hills, CA
2	46.265	16.46	3	Chrysler Crown Imperial 8		

As of publication date, some drivers' and entrants' names remain unknown.

1950 Mobilgas Grand Canyon Economy Run
Los Angeles to Grand Canyon
February 15 and 16, 1950—31 Cars—751.3 Miles
Sweepstakes Winner—Mercury 8

Place	Ton M.P.G.	M.P.G.	No.	Car	Driver	Entrant
CLASS "A"—FOUR-CYLINDER						
1	44.3039	26.099	25	Willys Jeepster 4	Don Kenower	Willys-Overland Motors, Inc.
2	43.1300	25.371	12	Willys Jeepster 4	August Schutze	Willys-Overland Motors, Inc.
3	43.0882	25.458	1	Willys Jeepster 4		Willys-Overland Motors, Inc.
CLASS "B"—$1450 to $1750						
1	48.5759	23.326	23	Ford Deluxe 6	Art Rene	Al Stuebing—Hollywood, CA
2	44.0392	21.071	20	Chevrolet Fleetline Deluxe 6	Pete Novotny	So. California Plating Co.—Los Angeles
3	43.8372	21.254	8	Plymouth Special Deluxe 6		Andrew J. Crevolin Co.—Alhambra, CA
CLASS "C"—$1751 to $2100						
1	53.0400	23.946	19	Kaiser Special 6	George McLeod	Kaiser-Frazer Corporation
2	51.8399	26.551	16	Studebaker Champion 6	Vic Del Coma	L.A. Studebaker Metropolitan Dealers Association
3	51.3636	25.522	5	Nash Statesman Custom 6	Andy Henderson	Nash Motors
4	47.3679	21.385	28	Dodge Coronet 6	Joseph Bozzani	Bozzani Motors Ltd.—Los Angeles, CA
CLASS "D"—$2101 to $2275						
1	61.2711	26.524	7	Mercury 8	Bill Stroppe	Art Hall, Inc.—Long Beach, CA
2	58.4641	26.424	10	Nash Ambassador Custom 6	James Fosdick	Nash Motors
3	52.6445	23.794	3	Studebaker Commander 6	Ed DeBerry	L.A. Studebaker Metropolitan Dealers Association
4	52.2521	22.595	17	Hudson Pacemaker 6	Ivan Hunter	Hudson Dealers of Southern CA
5	47.6037	20.192	11	Oldsmobile "88" 8		Oldsmobile Division—General Motors
6	43.4248	18.778	31	DeSoto Custom 6		McNeil-Stanley, Inc.—East Los Angeles
CLASS "E"—$2276 to $2475						
1	55.6855	24.887	2	Studebaker Land Cruiser 6	Bill Heimberger	L.A. Studebaker Metropolitan Dealers Association
2	48.1304	19.447	15	Oldsmobile "98" 8		Oldsmobile Division—General Motors
3	47.3004	18.920	29	Packard 8	Les P. Butts	Earle C. Anthony, Inc.—Los Angeles, CA
4	47.2987	19.853	24	Chrysler Windsor 6		Paul Bobst Co., Inc.—Hollywood, CA
CLASS "F"—$2476 to $2950						
1	54.3303	23.907	6	Frazer Manhattan 6	W. Doonan	Kaiser-Frazer Corporation
2	52.2895	21.386	30	Hudson Commodore 8	Hart Fullerton	Hart Fullerton—Santa Monica, CA
3	48.1428	18.150	9	Lincoln 8		Bob Estes—Inglewood, CA
4	45.2508	17.108	27	Chrysler New Yorker 8	Melvin Lewis	Lloyd Gregg, Inc.—North Hollywood, CA
5	41.7709	16.004	4	Packard Super Eight	Erwin Baker	Earle C. Anthony, Inc.—Los Angeles, CA
CLASS "G"—$2951 to $3150						
1	58.5212	22.972	22	Cadillac Series 61 8	A.E. "Bud" Ridings	Ridings Motors—Long Beach, CA
2	55.7783	23.965	26	Kaiser Virginian 6		Kaiser-Frazer Corporation
CLASS "H"—$3151 to $3750						
1	58.5651	22.525	18	Cadillac Series 62 8	Dan Hoferer	Ridings Motors—Long Beach, CA
2	49.3410	17.559	32	Lincoln Cosmopolitan 8	Marshall Martin	Deaton Motor Co.—Beverly Hills, CA
CLASS "I"—$3751 to $4500						
1	59.1179	22.080	14	Cadillac Series 60 Special 8	Art Elliott	Ronald E. Moran, Inc.—Hermosa Beach, CA
CLASS "J"—$4501 to $6000						
1	51.0874	17.245	21	Cadillac Series 75 8	Willard McArthy	Ridings Motors—Long Beach, CA

As of publication date, some drivers' names remain unknown.

1951 Mobilgas Economy Run
Los Angeles to Grand Canyon
March 6 and 7, 1951—32 Cars—840.05 Miles
Sweepstakes Winner—Lincoln 8

Place	Ton M.P.G.	M.P.G.	No.	Car	Driver	Entrant
CLASS "A"—$1400 to $1750						
1	54.587	25.994	11	Ford Deluxe 8	Al Cottle	Ford Dealers of Southern CA
2	54.321	28.621	20	Studebaker Champion 6	Vic Del Coma	L.A. Studebaker Metropolitan Dealers Association
3	53.838	25.915	1	Ford Deluxe 6	Art Rene	Al Stuebing—Hollywood, CA
4	47.934	22.990	7	Plymouth Cranbrook 6	Eddie Bishop	Art Frost, Inc.—Glendale, CA
5	45.956	22.041	32	Chevrolet Styleline Deluxe 6	Pete Novotny	So. California Plating Co.—Los Angeles
CLASS "B"—$1751 to $1950						
1	58.173	28.001	18	Studebaker Commander 8	Stanley Van Horn	L.A. Studebaker Metropolitan Dealers Association
2	52.637	26.122	25	Nash Statesman Custom 6	Andy Henderson	Nash Motors
CLASS "C"—$1951 to $2175						
1	59.868	25.945	14	Mercury 8	Bill Stroppe	Art Hall, Inc.—Long Beach, CA
2	58.744	27.644	5	Studebaker Land Cruiser 8	Bill Heimberger	L.A. Studebaker Metropolitan Dealers Association
3	58.268	24.926	26	Nash Ambassador Custom 6	James Fosdick	Nash Motors
4	52.828	24.713	17	Kaiser Deluxe 6	Art Elliott	De Paolo Car Co.—Glendale, CA
5	51.135	21.622	2	DeSoto Deluxe 6	Elmer Johnson	DeSoto Dealers of Southern CA
CLASS "D"—$2176 to $2450						
1	53.020	22.023	31	Packard 200 8	Harold Michelson	Earle C. Anthony, Inc.—Los Angeles, CA
2	52.268	20.886	4	Chrysler Windsor 6	Walter Nelson	Paul Bobst Co., Inc.—Hollywood, CA
3	47.760	19.921	10	DeSoto Custom 6	John Rasmussen	DeSoto Dealers of Southern CA
4	46.723	19.590	30	Hudson Commodore 6	Joseph Ricketts	Storey-Ricketts—Long Beach, CA
CLASS "E"—$2451 to $2700						
1	66.484	25.448	28	Lincoln 8	Les Viland	Bob Estes—Inglewood, CA
2	53.785	22.623	29	Hudson Hornet 6	Hart Fullerton	Hart Fullerton—Santa Monica, CA
CLASS "F"—$2701 to $3000						
1	55.492	21.719	9	Cadillac Series 61 8	J.E. Woodruff	Ridings Motors—Long Beach, CA
2	52.196	20.941	24	Packard 300 8	William Parkinson	Earle C. Anthony, Inc.—Los Angeles, CA
CLASS "G"—$3001 to $3300						
1	59.457	21.178	22	Chrysler Imperial 8	Melvin Lewis	Lloyd Gregg, Inc.—North Hollywood, CA
2	56.412	21.531	6	Cadillac Series 62 8	Rod Wilcoxin	Ridings Motors—Long Beach, CA
3	47.601	17.123	8	Lincoln Cosmopolitan 8	Charles Nelson	Deaton Motor Co.—Beverly Hills, CA
CLASS "H"—$3301 to $4000						
1	58.795	21.979	16	Cadillac Series 60 Special 8	A.E. "Bud" Ridings	Ridings Motors—Long Beach, CA
CLASS "I"—$4001 to $6000						
1	63.289	19.208	3	Chrysler Crown Imperial 8	Bob Burns	Andrew J. Crevolin Co.—Alhambra, CA
2	58.513	19.869	12	Cadillac Series 75 8	Willard McArthy	Clarence Dixon Cadillac—Hollywood, CA
SPECIAL LIGHTWEIGHT 4-CYLINDER						
1	49.153	30.109	23	Henry J 4	Arne Andersen	De Paolo Car Co.—Glendale, CA
2	46.110	26.769	27	Willys Jeepster 4	Don Kenower	Willys-Overland Motors, Inc.
SPECIAL LIGHTWEIGHT 6-CYLINDER						
1	53.489	31.053	15	Nash Rambler 6	Pierce Venable	Nash Motors
2	48.954	24.145	13	Plymouth Concord 6	Lee Winchester	Greene-Halderman—Los Angeles, CA
3	48.340	28.860	19	Henry J 6	Woodward Bell	De Paolo Car Co.—Glendale, CA
4	43.266	24.973	21	Willys Jeepster 6	Wayne Burdick	Willys-Overland Motors, Inc.

1952 Mobilgas Economy Run
Los Angeles to Sun Valley
April 14–16, 1952 — 29 Cars —1415.4 Miles
SWEEPSTAKES WINNER — MERCURY MONTEREY 8

Place	Ton M.P.G.	M.P.G.	No.	Car	Driver	Entrant
CLASS "A"—$1600 to $1950						
1	53.8551	25.4634	15	Ford Mainline 6	Bill Hooks	Al Stuebing — Hollywood, CA
2	52.4444	27.8220	5	Studebaker Champion 6	Vic Del Coma	L.A. Studebaker Metropolitan Dealers Association
3	49.0433	23.5220	21	Plymouth Cranbrook 6	Eddie Bishop	Art Frost, Inc. — Glendale, CA
4	48.1744	22.1492	19	Ford Mainline 8	Al Cottle	Ford Dealers Association of Southern CA
5	42.8915	20.5714	12	Chevrolet Styleline Deluxe 6	Pete Novotny	So. California Plating Co. — Los Angeles
CLASS "B"—$1951 to $2225						
1	53.6094	24.6480	25	Kaiser Deluxe 6	Art Elliott	So. CA Kaiser-Frazer Dealers Association
2	52.9854	25.5968	23	Studebaker Commander 8	J.R. Van Tright	L.A. Studebaker Metropolitan Dealers Association
Did Not Start			11	Nash Statesman Super 6	Andy Henderson	Nash California Co. — Los Angeles, CA
CLASS "C"—$2226 to $2350						
1	59.7118	24.4093	14	Mercury Monterey 8	Bill Stroppe	Bob Estes — Inglewood, CA
2	53.5586	25.3832	28	Studebaker Land Cruiser 8	Bill Heimberger	L.A. Studebaker Metropolitan Dealers Association
Did Not Start			22	Nash Ambassador Super 6	James Fosdick	Nash California Co. — Los Angeles, CA
CLASS "D"—$2351 to $2500						
1	47.7830	20.4638	8	Hudson Wasp 6	Jim Abbott	Southern CA Hudson Dealers Association
2	45.9798	19.3599	2	Chrysler Windsor 6	Rudy Resnik	Max Barish, Inc. — Los Angeles, CA
CLASS "E"—$2501 to $2900						
1	54.7368	21.2777	30	DeSoto Firedome 8	Verne Holmquist	Southern CA DeSoto Dealers Association
2	49.8816	20.8274	16	Hudson Hornet 6	Hart Fullerton	Hart Fullerton — Santa Monica, CA
3	48.8516	20.3973	4	Hudson Commodore 8	Joseph Ricketts	Storey-Ricketts — Long Beach, CA
4	44.9930	19.2278	3	Packard 200 8	Harold Michelson	Earle C. Anthony, Inc. — Los Angeles, CA
CLASS "F"—$2901 to $3450						
1	49.1762	19.0237	29	Chrysler Saratoga 8	John Gerfen	Mel Alsbury, Inc. — Hollywood, CA
2	40.8065	16.4211	1	Packard 300 8	Les P. Butts	Earle C. Anthony, Inc. — Los Angeles, CA
CLASS "G"—$3451 to $3750						
1	58.9085	22.3562	18	Lincoln Capri 8	Art Rene	Deaton Motor Co. — Beverly Hills, CA
2	52.1844	19.0802	6	Chrysler Imperial 8	Danny Eames	Esquire Motors — Los Angeles, CA
CLASS "H"—$3751 to $4500						
1	50.5729	17.6520	9	Chrysler Saratoga 8-Pass. 8	Lewis Jabro	Lew Jabro, Inc. — Los Angeles, CA
2	43.1400	16.9509	17	Packard Patrician 400 8	William Parkinson	Earle C. Anthony, Inc. — Los Angeles, CA
Disqualified			24	DeSoto Firedome 8-Pass. 8	Randy Maahs	Murphy Motors Co. — Culver City, CA
CLASS "I"—$4501 to $7000						
1	53.1736	16.2362	10	Chrysler Crown Imperial 8	Bob Burns	Greene-Halderman — Los Angeles, CA
SPECIAL LIGHTWEIGHT 4-CYLINDER						
1	51.3749	30.8558	20	Henry J Corsair 4	Woodward Bell	So. CA Kaiser-Frazer Dealers Association
SPECIAL LIGHTWEIGHT 6-CYLINDER						
1	47.0250	23.0797	26	Plymouth Concord 6	Bill Cameron	Andrew J. Crevolin Co. — Alhambra, CA
2	44.8908	26.3676	7	Henry J Corsair Deluxe 6	Willard McArthy	So. CA Kaiser-Frazer Dealers Association
Did Not Start			27	Nash Rambler Custom 6	Ron Mezzantto	Shavey & Langford — Compton, CA

DeSoto Firedome 8-Passenger 8 was automatically disqualified on the second day after running out of fuel before reaching the Beaver, Utah, refueling station.

1953 Mobilgas Economy Run
Los Angeles to Sun Valley
April 20–22, 1953 — 26 Cars —1206.1 Miles
Sweepstakes Winner — Ford Mainline 6

Place	Ton M.P.G.	M.P.G.	No.	Car	Driver	Entrant
CLASS "A"— LOW PRICE ($1500 to $2050)— Standard and Overdrive						
1	56.7028	27.0335	12	Ford Mainline 6	Les Viland	Al Stuebing — Hollywood, CA
2	50.8770	26.8622	5	Studebaker Champion 6	Vic Del Coma	L.A. Studebaker Metropolitan Dealers Association
3	48.5494	22.5183	7	Ford Mainline 8	Bill Hooks	Southern CA Ford Dealers Association
4	47.9825	25.4279	21	Hudson Super Jet 6	Rolfe S. Lowdon	Rolfe S. Lowdon — Bakersfield, CA
5	46.9501	22.8301	9	Plymouth Cranbrook 6	Eddie Bishop	Art Frost, Inc. — Glendale, CA
Disqualified			6	Chevrolet Two-Ten 6	Pete Novotny	So. California Plating Co. — Los Angeles
CLASS "B"— LOW PRICE ($1500 to $2050)— Automatic Drive						
1	42.4606	22.0575	2	Hudson Jet 6	Joseph Ricketts	Storey-Ricketts — Long Beach, CA
CLASS "C"— LOW MEDIUM PRICE ($2051 to $2500)— Standard and Overdrive						
1	52.8565	23.4189	20	Dodge Coronet 8	Danny Eames	Los Angeles Area Dodge Dealers Association
2	52.5629	23.1554	24	Mercury Monterey 8	Bill Stroppe	Bob Estes — Inglewood, CA
3	50.9767	24.5080	8	Studebaker Commander 8	Dick Griffith	L.A. Studebaker Metropolitan Dealers Association
4	49.7755	23.3962	3	Studebaker Land Cruiser 8	C. Fred James	L.A. Studebaker Metropolitan Dealers Association
CLASS "D"— LOW MEDIUM PRICE ($2051 to $2500)— Automatic Drive						
1	49.3476	22.8885	17	Studebaker Land Cruiser 8	Clarke Sanders	L.A. Studebaker Metropolitan Dealers Association
2	45.1229	19.0956	4	Hudson Super Wasp	Lee Hamer	Hamer Motors, Inc. — San Fernando, CA
3	41.5321	19.4439	15	Nash Statesman Custom 6	Robert J. Steiner	Degnan Motors, Inc. — Los Angeles, CA
CLASS "E"— UPPER MEDIUM PRICE ($2501 to $3000)— Standard and Overdrive						
1	51.3140	22.5457	1	Nash Ambassador Custom 6	Andy Henderson	Nash California Co. — Los Angeles, CA
2	50.9843	20.9209	26	DeSoto Firedome 8	Verne Holmquist	Southern CA DeSoto Dealers Association
3	49.0847	22.2707	16	Kaiser Manhattan 6	Harry Greenwell	So. CA Kaiser-Frazer Dealers Association
4	45.2838	18.6737	18	Packard Clipper Deluxe 8	Les P. Butts	Earle C. Anthony, Inc. — Los Angeles, CA
CLASS "F"— UPPER MEDIUM PRICE ($2501 to $3000)— Automatic Drive						
1	48.9037	21.1156	22	Nash Ambassador Custom 6	Archie Bowen	Hills Motor Co. — Pasadena, CA
2	47.9835	21.4499	23	Kaiser Dragon 6	Thomas Nicholson	Southern California Kaiser-Frazer Dealers Association
3	46.2409	18.9473	11	Hudson Hornet 6	Hart Fullerton	Hart Fullerton — Santa Monica, CA
CLASS "G"— HIGH PRICE ($3001 to $4500)— Automatic Drive						
1	52.3467	19.9416	25	Lincoln Capri 8	Clay Smith	Deaton Motor Co. — Beverly Hills, CA
2	48.1553	17.2909	14	Chrysler Imperial 8	John Gerfen	Mel Alsbury, Inc. — Hollywood, CA
3	45.7009	17.7479	19	Chrysler New Yorker 8	Jim Ickes	Lew Jabro, Inc. — Los Angeles, CA
SPECIAL LIGHTWEIGHT CLASS						
1	48.5768	28.2587	10	Henry J Corsair 4	Woodward Bell	Southern California Kaiser-Frazer Dealers Association
2	45.2432	25.3748	13	Nash Rambler Custom 6	Pierce Venable	Severin Motors, Inc. — Long Beach, CA

Chevrolet Two-Ten 6 was automatically disqualified on the second day after getting off-course outside Winnemucca, Nevada, and then running out of fuel before reaching the Boise, Idaho, impound.

1954 Mobilgas Economy Run
Los Angeles to Sun Valley
April 5–7, 1954 — 20 Cars — 1335 Miles
SWEEPSTAKES WINNER — STUDEBAKER LAND CRUISER 8

Place	Ton M.P.G.	M.P.G.	No.	Car	Driver	Entrant
CLASS "A" — LOW PRICE ($1500 to $2050) — Standard and Overdrive						
1	56.5138	29.5806	7	Studebaker Champion 6	Dick Johnson	L.A. Studebaker Metropolitan Dealers Association
2	55.5773	26.1479	2	Ford Mainline 6	Les Viland	Al Stuebing — Hollywood, CA
3	50.7528	23.3884	6	Ford Mainline 8	Bill Hooks	Southern California Ford Dealers Association
4	47.7067	22.4397	16	Plymouth Belvedere 6	LeRoy Schlick	Plymouth Dealers of Southern California
5	46.9655	22.3167	19	Chevrolet Two-Ten 6	Pete Novotny	So. California Plating Co. — Los Angeles
CLASS "B" — LOW PRICE ($1500 to $2050) — Automatic Drive						
1	43.1249	21.6382	5	Hudson Jet-Liner 6	Lee Hamer	Hamer Motors, Inc. — San Fernando, CA
CLASS "C" — LOW MEDIUM PRICE ($2051 to $2400) — Standard and Overdrive						
1	58.0456	25.3973	20	Dodge Royal 8	Danny Eames	Los Angeles Area Dodge Dealers Association
2	49.9795	21.4781	4	Mercury Monterey 8	Bill Stroppe	L.A. District Lincoln-Mercury Dealers Association
CLASS "D" — LOW MEDIUM PRICE ($2051 to $2400) — Automatic Drive						
1	49.7122	19.7506	13	Oldsmobile Super 88 8	Willard McArthy	DeBell Auto Enterprises — Los Angeles, CA
2	48.5951	21.3511	18	Dodge Royal 8	Art Rene	Frank E. Beckett Co. — Fresno, CA
CLASS "E" — UPPER MEDIUM PRICE ($2401 to $3000) — Standard and Overdrive						
1	60.8465	28.1046	9	Studebaker Land Cruiser 8	Dick Griffith	Paramount Motors, Inc. — San Francisco, CA
2	49.5558	20.3514	8	DeSoto Firedome 8	Verne Holmquist	McNeil-Stanley, Inc. — East Los Angeles
CLASS "F" — UPPER MEDIUM PRICE ($2401 to $3000) — Automatic Drive						
1	53.9942	24.5763	12	Studebaker Land Cruiser 8	Robert Donkin	Don Rasmussen Co. — Portland, OR
2	47.7721	20.6582	1	Nash Ambassador Super 6	Archie Bowen	Mansell Sutton — Pasadena, CA
3	47.7346	19.8481	17	Hudson Hornet 6	Hart Fullerton	Hart Fullerton — Santa Monica, CA
4	44.1334	17.9770	11	DeSoto Firedome 8	Boyd Trombella	Art Frost, Inc. — Glendale, CA
5	43.6744	17.8957	14	Packard Clipper Deluxe 8	Harold Michelson	Earle C. Anthony, Inc. — Los Angeles, CA
CLASS "G" — HIGH PRICE ($3001 to $5000) — Automatic Drive						
1	52.2894	19.7542	15	Lincoln Capri 8	Clay Smith	L.A. District Lincoln-Mercury Dealers Association
2	45.5792	17.8322	10	Chrysler New Yorker 8	Lewis Jabro	Lew Jabro, Inc. — Los Angeles, CA
3	45.1370	16.4463	3	Chrysler Imperial 8	John Gerfen	Mel Alsbury, Inc. — Hollywood, CA

1955 Mobilgas Economy Run
Los Angeles to Colorado Springs
March 14–16, 1955 — 22 Cars — 1323 Miles
SWEEPSTAKES WINNER — STUDEBAKER COMMANDER CUSTOM 8

Place	Ton M.P.G.	M.P.G.	No.	Car	Driver	Entrant
CLASS "A" — LOW PRICE ($1500 to $2100)						
1	51.2789	27.4733	8	Rambler Super 6	Les Viland	Hudson Dealers of Southern California and Nash Dealers Association of Southern California
2	48.7201	22.8197	17	Ford Fairlane 6	Marshall Martin	Ford Dealers Advertising Association of Southern California
3	47.3203	21.5631	1	Chevrolet Bel-Air 6	Pete Novotny	Southern California Plating Co. — Los Angeles
4	45.7892	20.5979	20	Ford Fairlane 8	Bill Hooks	Ford Dealers Advertising Association of Southern California
5	43.7723	19.3683	4	Plymouth Belvedere 8	Ak Miller	Plymouth Dealers
6	42.2862	19.6406	22	Chevrolet Bel-Air 8	James L. Rush	Rush Chevrolet — Gardena, CA

CLASS "B"—LOW MEDIUM PRICE ($2101 to $2500)						
1	58.6320	27.4366	23	Studebaker Commander 8	Dick Johnson	J. R. Townsend Co., Inc.—San Diego, CA
2	51.9297	21.6149	10	Pontiac Chieftain 870 8	Archie Bowen	Savage-Halderman Co.—Los Angeles, CA
3	51.9157	21.5866	14	Mercury Monterey 8	Art Elliott	Bob Estes—Inglewood, CA Sponsored by: Mercury Dealers Advertising Association
4	49.9957	22.1220	15	Hudson Wasp 6	Lee Hamer	Hamer Motors, Inc.—San Fernando, CA
Disqualified			18	Dodge Coronet 8	William Losher	Joe Fisher—Portland, OR
CLASS "C"—UPPER MEDIUM PRICE ($2501 to $3300)						
1	58.5560	25.1854	12	Nash Ambassador Super 6	Pierce Venable	Nash Dealers Association of Southern California
2	56.8707	22.5543	3	Oldsmobile "88" 8	Woody Bell	Danny DePaolo—Glendale, CA
3	55.1957	25.2901	16	Studebaker President 8	Robert Donkin	Charles J. Ketcham—Las Vegas, NV
4	52.2267	22.3287	5	Dodge Custom Royal 8	Danny Eames	Mikes Motor Co.—Clinton, IN
5	48.7857	20.6544	7	Hudson Hornet Custom 6	Hart Fullerton	Fullerton-Loadville, Inc.—Santa Monica
6	47.0683	18.2294	21	Chrysler Windsor Deluxe 8	Mel Alsbury, Jr.	Mel Alsbury, Inc.—Hollywood, CA
7	43.8635	17.8962	9	DeSoto Firedome 8	Verne Holmquist	McNeil-Stanley, Inc.—East Los Angeles
CLASS "D"—HIGH PRICE (OVER $3301)						
1	53.7596	19.7827	11	Buick Roadmaster 8	Al Cottle	Abbott-Kiple Co.—West Los Angeles, CA
2	51.6322	18.7515	19	Packard Patrician 8	Harold Mickelson	Earle C. Anthony, Inc.—Los Angeles, CA
3	50.3666	18.8780	6	Lincoln Capri 8	Art Rene	Deaton Motor Co.—Beverly Hills, CA Sponsored by: Lincoln Dealers Advertising Association
4	50.1099	19.0170	2	Chrysler New Yorker 8	Lewis Jabro	Lew Jabro, Inc.—Los Angeles, CA

Dodge Coronet 8 was automatically disqualified on the first day after running out of fuel before reaching the Tucson, Arizona, impound.

1956 Mobilgas Economy Run Los Angeles to Colorado Springs

March 19–22, 1956—21 Cars—1468.8 Miles

SWEEPSTAKES WINNER—IMPERIAL 8

Place	Ton M.P.G.	M.P.G.	No.	Car	Driver	Entrant
SPECIAL LIMITED DISPLACEMENT						
1	48.2218	24.3545	1	Rambler Deluxe 6	Tommy Thomas	Nash and Hudson Dealers of Southern California
2	39.7426	20.0467	6	Studebaker Champion 6	Bob Russo	Universal Specialties—Denver, CO
CLASS "A"—LOW PRICE ($1500 to $2000)						
1	47.7643	20.5217	22	Ford Customline 8	Marshall Martin	Ford Dealers Advertising Association of Southern California
2	47.7381	20.6793	15	Plymouth Belvedere 8	Art Rene	W. R. Shadoff, Inc.—Pomona, CA
3	47.2124	21.1715	8	Chevrolet Bel-Air 6	Pete Novotny	Southern California Plating Co.—Los Angeles
4	46.3752	20.7032	20	Chevrolet Bel-Air 8	James L. Rush	Rush Chevrolet—Gardena, CA
5	44.2490	18.7496	16	Ford Fairlane 8	Don Rice	Ford Dealers Advertising Association of Southern California
Disqualified			9	Ford Fairlane 6	Bill Hooks	Ford Dealers Advertising Association of Southern California
CLASS "B"—LOW MEDIUM PRICE ($2101 to $2400)						
1	51.7297	21.1098	12	Pontiac Chieftain 870 8	Pierce Venable	Greer-Halderman—Los Angeles, CA
2	50.1419	20.9012	18	Dodge Coronet 500 8	William Losher	Los Angeles Area Dodge Dealers Association
3	47.6965	19.1552	19	Mercury Montclair 8	Art Elliott	Bob Estes—Inglewood, CA and Mercury Dealers Advertising Association
CLASS "C"–MEDIUM PRICE ($2401 to $2700)						
1	50.8077	19.7044	10	Oldsmobile "88" 8	Woody Bell	Bob Yeakel "Wilshire" Oldsmobile—L. A. & Paola Oldsmobile—La Crescenta, CA
2	47.9907	20.7124	11	Nash Ambassador Special 8	Carl Chakmakian	Carl Chakmakian—Dearborn, MI
3	47.4944	20.4894	2	Hudson Hornet Special Super 8	Lee Hamer	Hamer Motors, Inc.—San Fernando, CA
DNF—Accident			21	Nash Ambassador Super 6	Les Viland	Nash Dealers of Southern CA

CLASS "D"—UPPER MEDIUM PRICE ($2701 to $3300)						
1	50.9488	18.6694	17	Oldsmobile "98" 8	Ray Brock	Paola Oldsmobile—La Crescenta, CA & Bob Yeakel "Wilshire" Oldsmobile—L.A.
2	48.9566	17.7798	5	Buick Super 8	Don Bridges	Bill Murphy Buick, Inc.—Culver City, CA
3	47.3496	17.8341	3	DeSoto Fireflite 8	Danny Eames	Hart Fullerton—Santa Monica, CA
CLASS "E"—HIGH PRICE ($3301 to $10,000)						
1	61.3796	21.0420	4	Imperial 8	Mel Alsbury, Jr.	Mel Alsbury, Inc.—Hollywood, CA
2	50.3207	18.0750	14	Packard Four-Hundred 8	Dick Johnson	Earle C. Anthony, Inc.—Los Angeles, CA
3	48.2510	17.4191	7	Buick Roadmaster 8	Al Cottle	Bill Murphy Buick, Inc.—Culver City, CA

Ford Fairlane 6 was automatically disqualified on the second day after reaching the Salt Lake City, Utah, impound 2 minutes and 20 seconds late.

1957 Mobilgas Economy Run
Los Angeles to Sun Valley

April 14–18, 1957—24 Cars—1568.3 Miles

SWEEPSTAKES WINNER—IMPERIAL CROWN 8

Place	Ton M.P.G.	M.P.G.	No.	Car	Driver	Entrant
CLASS "A"—LOW PRICE (UNDER $2400)						
1	52.6211	21.3907	15	Plymouth Belvedere 8	Mary Davis	W.R. Shadoff, Inc.—Pomona, CA
2	52.5181	22.2534	17	Ford Fairlane 500 6	Marshall Martin	Ford Dealers Advertising Association of Southern California
3	49.4378	21.2636	16	Chevrolet Bel-Air 8	Vince Piggins	Rush Chevrolet—Gardena, CA
4	49.0080	21.4948	11	Chevrolet Bel-Air 6	James L. Rush	Rush Chevrolet—Gardena, CA
5	47.8914	21.6214	9	Rambler Rebel 8	Les Viland	Nash and Hudson Dealers of Southern California
6	45.4013	19.1567	18	Ford Fairlane 500 8	Al Cottle	Ford Dealers Advertising Association of Southern California
—	52.3140	22.1201	14	Ford Fairlane 500 6	Ina Mae Overman	Ford Dealers Advertising Association of Southern California
—	51.4061	20.8968	6	Plymouth Belvedere 8	Dick Griffith	W.R. Shadoff, Inc.—Pomona, CA
—	47.5606	20.7236	21	Chevrolet Bel-Air 8	Betty Skelton	Rush Chevrolet—Gardena, CA
—	44.9959	18.9456	4	Ford Fairlane 500 8	Mopsy Pagan	Ford Dealers Advertising Association of Southern California
CLASS "B"—LOW MEDIUM PRICE ($2401 to $2800)						
1	55.8920	22.0047	2	Dodge Coronet 500 8	William Losher	Los Angeles Area Dodge Dealers Association of Southern California
2	52.7877	19.5149	5	Oldsmobile "88" 8	Lorraine Bell	Oldsmobile Dealers Association of Southern California
3	50.2384	20.4221	8	Pontiac Chieftain 870 8	Pierce Venable	Greer-Halderman—Los Angeles, CA
4	44.8769	19.9453	7	Studebaker President 8	Pete Novotny	Southern California Plating Co.—Los Angeles
—	54.4509	21.7803	22	Dodge Coronet 500 8	Patricia Jones	Los Angeles Area Dodge Dealers Association
—DNF—Mechanical			19	Oldsmobile "88" 8	Link Paola	Oldsmobile Dealers Association of Southern California
CLASS "C"–UPPER MEDIUM PRICE ($2801 to $4000)						
1	56.7267	20.7032	23	Chrysler Saratoga 8	George Alsbury	Mel Alsbury, Inc.—Hollywood, CA
2	56.4464	20.9838	10	DeSoto Firedome 8	Hart Fullerton	Hart Fullerton—Santa Monica, CA
3	53.5177	19.2164	24	Oldsmobile Starfire 98 8	Marilyn Miller	Oldsmobile Dealers Association of Southern California
—	52.1482	18.7247	20	Oldsmobile Starfire 98 8	Ray Brock	Oldsmobile Dealers Association of Southern California
—	48.0985	18.4994	13	DeSoto Firedome 8	Myra Buchanan	McNeil-Stanley, Inc.—East Los Angeles
CLASS "D"—HIGH PRICE (OVER $4000)						
1	64.5153	20.9465	1	Imperial Crown 8	Mel Alsbury, Jr.	Mel Alsbury, Inc.—Hollywood, CA
2	52.3466	18.6287	12	Buick Roadmaster 8	Don Bridges	Bill Murphy Buick, Inc.—Culver City, CA
—	61.7535	19.9572	3	Imperial Crown 8	Mildred Alsbury	Mel Alsbury, Inc.—Hollywood, CA

Nine entrants exercised an option to enter a second car of the same make and model provided that the second car was operated by a qualified woman driver and woman relief driver. The official score was determined by whichever of the two cars achieved the higher Ton Miles Per Gallon over the entire course.

1958 Mobilgas Economy Run
Los Angeles to Galveston
April 13–17, 1958 — 29 Cars —1883.2 Miles
Sweepstakes Winner — Imperial Crown 8

Place	Ton M.P.G.	M.P.G.	No.	Car	Driver	Entrant
CLASS "A"— LOW PRICE (UNDER $2575)						
1	48.3264	20.0088	5	Plymouth Belvedere 8	Pierce Venable	W.R. Shadoff, Inc.— Pomona, CA
2	48.1801	19.9897	28	Plymouth Belvedere 8	Mary Davis	W.R. Shadoff, Inc.— Pomona, CA
3	47.0738	20.6441	24	Chevrolet Delray 6	Gus Retzlaff	Holiday Chevrolet — Chandler, AZ
4	46.8827	19.7775	12	Ford Fairlane 8	Marshall Martin	Ford Dealers Advertising Association of Southern California
5	46.2543	20.8870	10	Ford Custom 300 6	Ivan Lewis	Ford Dealers Advertising Association of Southern California
6	44.3255	18.7284	17	Chevrolet Bel-Air 8	John Hartman	Rush Chevrolet — Gardena, CA
7	41.2717	17.1787	14	Chevrolet Impala 8	Dorothy Retzlaff	Holiday Chevrolet — Chandler, AZ
CLASS "B"— LOW MEDIUM PRICE ($2576 to $3300)						
1	50.5951	19.2377	27	Oldsmobile Dynamic 88 8	Link Paola	Yeakel Oldsmobile — Los Angeles, CA
2	49.2339	18.9471	21	DeSoto Firedome 8	Myra Buchanan	Tom Dalby DeSoto — Huntington Park, CA
3	48.9044	19.2518	29	Edsel Pacer 8	Don Rice	Edsel Dealers Advertising Association
4	48.3955	18.8017	13	Pontiac Super Chief 8	Lynn Paulette	Jim Retzlaff Co.— Barstow, CA
5	48.2209	19.0052	2	Edsel Pacer 8	Loretta Colange	Edsel Dealers Advertising Association
6	48.1613	18.1621	25	Mercury Montclair 8	Fran Hernandez	Lincoln-Mercury Dealers Advertising Association
7	45.7969	16.9445	15	Buick Special 8	Joe Hill	Al Parker Buick Co.— Houston, TX
8	44.2404	17.8137	26	Dodge Custom Royal 8	Robert Donkin	Southern CA Dodge Dealers Association
9	43.4382	17.5490	8	Dodge Custom Royal 8	Joan Fischal	Southern CA Dodge Dealers Association
10	43.3224	19.2608	9	Studebaker President 8	Pete Novotny	So. California Plating Co.— Los Angeles
CLASS "C"–UPPER MEDIUM PRICE ($3301 to $4500)						
1	58.4928	21.0217	1	Chrysler New Yorker 8	George Alsbury	Mel Alsbury, Inc.— Hollywood, CA
2	51.9198	19.6573	16	DeSoto Fireflite 8	Hart Fullerton	Hart Fullerton — Santa Monica, CA
3	51.0536	18.8390	4	Mercury Turnpike Cruiser 8	Bill Stroppe	Lincoln-Mercury Dealers Advertising Association
4	48.4367	17.4186	19	Mercury Park Lane 8	Vern Houle	Lincoln-Mercury Dealers Advertising Association
5	48.2714	17.4849	22	Oldsmobile Ninety-Eight 8	Ray Brock	Yeakel Oldsmobile — Los Angeles, CA
6	47.9027	17.6698	20	Buick Century 8	Robert Flanagan	Al Parker Buick Co.— Houston, TX
7	47.1114	17.2474	3	Edsel Citation 8	Art Rene	Edsel Dealers Advertising Association
8	47.0751	18.2869	23	DeSoto Fireflite 8	Ruth Doushkess	McNeil-Stanley, Inc.— East Los Angeles
CLASS "D"— HIGH PRICE (OVER $4500)						
1	62.7188	20.5821	6	Imperial Crown 8	Mel Alsbury, Jr.	Mel Alsbury, Inc.— Hollywood, CA
2	55.8884	18.0737	11	Lincoln Continental Mk III 8	Danny Eames	Lincoln-Mercury Dealers Advertising Association
3	55.3927	17.9638	18	Lincoln Continental Mk III 8	Nickey Griffen	Lincoln-Mercury Dealers Advertising Association
4	50.3234	17.4024	7	Buick Roadmaster 8	Don Bridges	Bill Murphy Buick, Inc.— Culver City, CA

1959 Mobilgas Economy Run
Los Angeles to Kansas City
April 5–9, 1959 — 47 Cars —1894.4 Miles

Place	M.P.G.	No.	Car	Driver	Entrant
CLASS "A"— COMPACT CAR					
1	25.2878	26	Rambler American Deluxe 6	Woody Bell	Rambler Dealers — Western Division
2	22.9572	40	Rambler Deluxe 6	Les Viland	Rambler Dealers — Western Division
3	22.4422	42	Studebaker Lark Deluxe 6	Bill Corey	Lloyd Pearson — Pasadena, CA
4	22.2837	2	Studebaker Lark Regal 8	Dick Griffith	Danyluk Motors, Inc.— Glendale, CA

CLASS "B"—LOW PRICE—SIX CYLINDER (UNDER $2500)

1	22.3788	8	Chevrolet Biscayne 6	Mary Hauser	C. S. Mead Chevrolet Co.—Pasadena, CA
2	21.9439	32	Chevrolet Biscayne 6	Gus Retzlaff	Holiday Chevrolet—Chandler, AZ
3	20.3288	46	Ford Custom 300 6	Danny Eames	Ford Dealers Advertising Association of Southern California
4	19.9244	33	Ford Custom 300 6	Ruth Doushkess	Ford Dealers Advertising Association of Southern California
5	19.6585	9	Edsel Ranger 6	Art Rene	L.A. Mercury-Edsel-Lincoln Dealers Advertising Association
6	19.5051	16	Edsel Ranger 6	Nickey Griffen Ward	L.A. Mercury-Edsel-Lincoln Dealers Advertising Association

CLASS "C"—LOW PRICE—EIGHT CYLINDER (UNDER $2500)

1	21.1548	38	Plymouth Belvedere 8	Pierce Venable	W.R. Shadoff, Inc.—Pomona, CA
2	20.9880	47	Plymouth Belvedere 8	Mary Davis	W.R. Shadoff, Inc.—Pomona, CA
3	19.6698	15	Ford Custom 300 8	Ak Miller	Ford Dealers Advertising Association of Southern California
4	19.4401	21	Ford Custom 300 8	Marian Lowe	Ford Dealers Advertising Association of Southern California
5	19.2760	12	Ford Fairlane 500 8	Jim Fosdick	Ford Dealers Advertising Association of Southern California
6	19.2577	4	Chevrolet Impala 8	Vince Piggins	Howell Chevrolet—Glendale, CA
7	19.1643	6	Chevrolet Bel-Air 8	James L. Rush	Rush Chevrolet—Gardena, CA
8	19.0279	39	Chevrolet Bel-Air 8	Betty Skelton	Holiday Chevrolet—Chandler, AZ

CLASS "D"—LOW MEDIUM PRICE ($2501 to $3225)

1	21.7454	37	Dodge Coronet 8	Patricia Jones	Southern CA Dodge Dealers Association
2	21.0164	36	Dodge Coronet 8	Link Paola	Southern CA Dodge Dealers Association
3	19.6454	5	Chrysler Windsor 8	George Alsbury	Mel Alsbury, Inc.—Hollywood, CA
4	19.2266	10	Rambler Ambassador Super 8	Tommy Thomas	Rambler Dealers—Western Division
5	18.9015	22	Mercury Monterey 8	Byron Froelich	L.A. Mercury-Edsel-Lincoln Dealers Advertising Association
6	18.8608	3	DeSoto Firesweep 8	Myra Buchanan	Tom Dalby DeSoto—Huntington Park, CA
7	18.8290	13	Buick LeSabre 8	Joe Hill	Al Parker Buick Co.—Houston, TX
8	18.6791	28	Buick LeSabre 8	Barbara Nieland	Colonial Buick, Inc.—Glendale, CA
9	18.6613	25	Pontiac Catalina 8	Marta Retzlaff	Jim Retzlaff Co.—Barstow, CA
10	18.5558	29	Pontiac Star Chief 8	Andy Klein	Andy Klein Pontiac—Overland Park, KS
11	18.3060	20	Pontiac Catalina 8	Jim Retzlaff	Jim Retzlaff Co.—Barstow, CA
12	18.2165	11	Oldsmobile Dynamic 88 8	Lillian Lee Warner	Paola Oldsmobile—La Crescenta, CA
13	18.1062	31	Oldsmobile Dynamic 88 8	Tom Neal	Tom Neal Oldsmobile—Burbank, CA
14	17.8975	44	Edsel Corsair 8	Bill Stroppe	L.A. Mercury-Edsel-Lincoln Dealers Advertising Association
15	16.7708	24	Oldsmobile Super 88 8	Dwight Hinton	Hinton Oldsmobile Co.—Mt. Vernon, WA

CLASS "E"—UPPER MEDIUM PRICE ($3226 to $5000)

1	19.1301	7	Ford Thunderbird 8	Art Chrisman	Ford Dealers Advertising Association of Southern California
2	19.0701	17	Chrysler Saratoga 8	Mel Alsbury, Sr.	Mel Alsbury, Inc.—Hollywood, CA
3	19.0632	27	DeSoto Firedome 8	Hart Fullerton	Hart Fullerton—Santa Monica, CA
4	18.9949	41	Ford Thunderbird 8	Loretta Colange	Ford Dealers Advertising Association of Southern California
5	18.7781	14	Mercury Montclair 8	Don Rice	L.A. Mercury-Edsel-Lincoln Dealers Advertising Association
6	18.5755	43	Buick Electra 8	Don Bridges	Bill Murphy Buick, Inc.—Culver City, CA
7	18.2003	23	Buick Invicta 8	Judy Allred	Claude L. Allred Buick—Visalia, CA
8	17.8735	45	Buick Invicta 8	Robert Flanagan	Al Parker Buick Co.—Houston, TX
9	17.5405	19	Oldsmobile Ninety-Eight 8	Stan Raymond	Hawk Oldsmobile Co.—Bakersfield, CA
10	16.9411	1	Pontiac Bonneville 8	Bob Russo	Jim Retzlaff Co.—Barstow, CA
11	16.8728	34	Oldsmobile Ninety-Eight 8	June Galbraith	Hawk Oldsmobile Co.—Bakersfield, CA

CLASS "F"—HIGH PRICE (OVER $5001)

1	19.0352	18	Cadillac Sixty-Two 8	Jim Parkinson	Italiano Motors—Burbank, CA
2	18.2027	30	Imperial Custom 8	Mel Alsbury, Jr.	Mel Alsbury, Inc.—Hollywood, CA
3	17.2739	35	Lincoln 8	Vern Houle	L.A. Mercury-Edsel-Lincoln Dealers Advertising Association

1960 Mobilgas Economy Run
Los Angeles to Minneapolis
April 2–6, 1960—56 Cars—2061.4 Miles

Place	M.P.G.	No.	Car	Driver	Entrant
CLASS "A"—COMPACT CAR					
1	28.3533	14	Rambler American Custom 6	Les Viland	Twin City Rambler Dealer Group
2	27.2992	10	Valiant V-100 6	Ginny Sims	Los Angeles Valiant Advertising Association
3	27.0315	5	Chevrolet Corvair 500 6	Vince Piggins	Chevrolet Dealers Association of Southern California
4	27.4700	17	Valiant V-100 6	Art Rene	Los Angeles Valiant Advertising Association
5	26.2162	43	Chevrolet Corvair 500 6	Patricia Sawyer	Chevrolet Dealers Association of Southern California
6	25.6421	28	Ford Falcon 6	Ruth Doushkess	Ford Dealers Advertising Association of Southern California
7	25.6343	30	Ford Falcon 6	Bill Stroppe	Ford Dealers Advertising Association of Southern California
8	23.2838	29	Studebaker Lark Deluxe 8	Dick Griffith	Danyluk Motors, Inc.—Glendale, CA and Southern CA Studebaker Dealers Association
9	23.2076	12	Rambler Deluxe 6	Tommy Thomas	Rambler Dealers—Western Division
10	21.8585	8	Studebaker Lark Deluxe 6	Bill Corey	Lloyd Pearson—Pasadena, CA and Southern CA Studebaker Dealers Association
CLASS "B"—LOW PRICE—SIX CYLINDER (UNDER $2500)					
1	24.8265	48	Plymouth Savoy 6	Mary Hauser	Plymouth Dealers of Greater Detroit
2	24.7429	9	Dodge Dart Seneca 6	Woody Bell	Frank G. Elliott, Inc.—Bedford, OH
3	24.7360	55	Plymouth Savoy 6	Pierce Venable	W.R. Shadoff, Inc.—Pomona, CA
4	23.9037	15	Dodge Dart Seneca 6	Vicki Wood	Hetfield-Queenan—St. Paul, MN
5	22.4843	6	Ford Fairlane 6	Dan Jones	Ford Dealers Advertising Association of Southern California
6	21.4874	18	Chevrolet Biscayne 6	Betty Skelton	Chevrolet Dealers Association of Southern California
7	21.2904	35	Chevrolet Biscayne 6	John Fitch	Chevrolet Dealers Association of Southern Californiaa
CLASS "C"—LOW PRICE—EIGHT CYLINDER (UNDER $2500)					
1	22.8899	4	Plymouth Belvedere 8	Mary Davis	W.R. Shadoff, Inc.—Pomona, CA
2	22.5256	47	Plymouth Belvedere 8	Jim Fosdick	Plymouth Dealers Association of Southern California
3	22.2885	23	Dodge Dart Seneca 8	Lute Eldridge	New York-New Jersey Dodge Dealers Association
4	21.7326	51	Dodge Dart Seneca 8	Patricia Jones	Taylor Motor Co., Inc.—King William, VA
5	21.5802	1	Ford Fairlane 8	Ak Miller	Ford Dealers Advertising Association of Southern California
6	21.0054	22	Ford Starliner 8	Loretta Colange	Ford Dealers Advertising Association of Southern California
7	20.9914	49	Ford Starliner 8	Art Chrisman	Ford Dealers Advertising Association of Southern California
8	20.4236	53	Chevrolet Bel-Air 8	Chuck Daigh	Chevrolet Dealers Association of Southern California
9	20.2345	45	Chevrolet Bel-Air 8	Ruth Levy	Chevrolet Dealers Association of Southern California
10	18.6132	41	Chevrolet Impala 8	Sunnie Baker	Nickey Chevrolet—Chicago, IL
11	18.4537	40	Chevrolet Impala 8	Melvin Hilliard	Melvin Hilliard Chevrolet—Kansas City, MO
CLASS "D"—LOW MEDIUM PRICE ($2501 to $3200)					
1	22.9899	2	Studebaker Hawk 8	James Peterson	Frank M. Darling—Laguna Beach, CA and Southern CA Studebaker Dealers Association
2	21.1214	36	Dodge Matador 8	Link Paola	Southern CA Dodge Dealers Association
3	21.0696	37	Mercury Monterey 8	Byron Froelich	L.A. Lincoln-Mercury Dealers Advertising Association
4	21.0139	16	DeSoto Fireflite 8	Al Cottle	Harlow DeSoto—Corona, CA
5	20.7221	27	Mercury Monterey 8	Nickey Griffen Ward	L.A. Lincoln-Mercury Dealers Advertising Association
6	20.2698	42	Rambler Ambassador Custom 8	Jim Moore	Rambler Dealers—Western Division
7	19.7329	24	Pontiac Catalina 8	Marta Retzlaff	Jim Retzlaff Co.—Barstow, CA
8	19.3778	3	Pontiac Catalina 8	Mickey Thompson	Andy Klein Pontiac—Overland Park, KS

9	19.0788	26	Buick LeSabre 8	Walter Vandenbos	Jim Retzlaff Co.— Barstow, CA
10	19.0670	32	Buick LeSabre 8	Betty Shutes	Butler Buick — Manhattan Beach, CA
11	19.0424	39	Pontiac Star Chief 8	Smokey Yunick	Hine Pontiac — Dallas, TX
12	18.2401	54	Oldsmobile Dynamic 88 8	Jim Parkinson	Tom Neal Oldsmobile — Burbank, CA
13	18.2095	25	Oldsmobile Dynamic 88 8	Lillian Lee Warner	Paola Oldsmobile — La Crescenta, CA
14	18.0729	52	Pontiac Star Chief 8	Frank Foster	Gillman Pontiac — Houston, TX
15	18.0209	20	Chrysler Windsor 8	Grant Slife	Danny McGroo, Inc.— Culver City, CA
CLASS "E"— UPPER MEDIUM PRICE ($3201 to $5000)					
1	20.8715	50	Chrysler New Yorker 8	Mel Alsbury, Jr.	Mel Alsbury, Inc.— Hollywood, CA
2	20.3627	34	DeSoto Adventurer 8	Hart Fullerton	Hart Fullerton — Santa Monica, CA
3	19.8912	33	Ford Thunderbird 8	Vern Houle	Ford Dealers Advertising Association of Southern California
4	19.8165	46	Chrysler Saratoga 8	Eleanor Edwards	Twin Cities Chrysler Dealers
5	18.4910	11	Buick Invicta 8	John Rich	Colonial Buick, Inc.— Glendale, CA
6	17.9581	31	Buick Electra 8	Barbara Nieland	Barnett Buick — Burbank, CA
7	17.9076	21	Buick Electra 8	Don Bridges	Bill Murphy Buick, Inc.— Culver City, CA
8	17.6966	56	Buick Invicta Custom 8	Judy Allred	Claude L. Allred Buick — Visalia, CA
9	17.1785	7	Oldsmobile Ninety-Eight 8	L.E. Jann	Malkerson Oldsmobile — Minneapolis, MN
10	17.1469	13	Pontiac Bonneville 8	Judy Thompson	Ulrich Pontiac — Culver City, CA
11	16.7715	44	Pontiac Bonneville 8	Don Francisco	A. E. England Pontiac — Hollywood, CA
CLASS "F"— HIGH PRICE (OVER $5001)					
1	20.5036	38	Imperial Custom 8	George Alsbury	Mel Alsbury, Inc.— Hollywood, CA
2	18.8170	19	Cadillac Sixty-Two 8	Pete Novotny	DeBell Auto Sales — Glendale, CA

1961 Mobilgas Economy Run
Los Angeles to Chicago
March 11–16, 1961— 67 Cars — 2560.7 Miles

Place	M.P.G.	No.	Car	Driver	Entrant
CLASS "A"— COMPACT CAR — MANUAL TRANSMISSION					
1	32.6800	47	Ford Falcon 6	Dan Jones	Chicago Metropolitan Ford Dealers Association
2	31.6199	32	Ford Falcon 6	Dick Lyman	Southern CA Ford Dealers Association
3	30.7907	53	Chevrolet Corvair 500 6	Ronnie Kaplan	Nickey Chevrolet — Chicago, IL
4	28.6435	61	Mercury Comet 6	Bill Levy	Lincoln-Mercury Dealer Association
5	27.8342	49	Mercury Comet 6	Patricia Jones	Lincoln-Mercury Dealer Association
6	26.6707	66	Studebaker Lark Deluxe 6	Bill Corey	Southern CA Studebaker Dealers Association
7	25.4695	1	Oldsmobile F-85 Deluxe 8	Stan Raymond	Paola Oldsmobile — La Crescenta, CA
8	25.3260	14	Oldsmobile F-85 8	Edward J. Wilder	Ray Nolting Oldsmobile — Kirkwood, MO
9	25.2353	45	Pontiac Tempest Custom 4	Darrell Droke	Los Angeles Pontiac Dealer Club
10	24.7312	33	Pontiac Tempest 4	John Price	Los Angeles Pontiac Dealer Club
11	24.7075	42	Buick Special 8	Bob Russo	Bill Murphy Buick, Inc.— Culver City, CA
DNF — Mechanical		3	Chevrolet Corvair 700 6	Vince Piggins	Doane Chevrolet — Dundee, IL
CLASS "B"— COMPACT CAR — FOUR and SIX CYLINDER					
1	29.3539	50	Chevrolet Corvair Monza 6	Patricia Sawyer	Nickey Chevrolet — Chicago, IL
2	29.3267	10	Chevrolet Corvair Monza 6	Bill Horton	Doane Chevrolet — Dundee, IL
3	27.4406	18	Pontiac Tempest 4	Jim Nelson	Los Angeles Pontiac Dealer Club
4	26.3467	7	Ford Falcon 6	Loretta Colange	Tuttle Broadway Ford — Tucson, AZ
5	26.1946	24	Ford Falcon 6	Leon Funderburk	Desmond Ford. Co.— Roswell, NM
6	26.1339	52	Plymouth Valiant V-100 6	Ginny Sims	Plymouth-Valiant Dealers of Southern California
7	26.0641	20	Mercury Comet 6	Ruth Doushkess	Lincoln-Mercury Dealer Association
8	25.8244	56	Mercury Comet 6	Joe Tuchel	Lincoln-Mercury Dealer Association
9	25.8114	37	Pontiac Tempest Custom 4	Judy Thompson	Los Angeles Pontiac Dealer Club
10	25.0882	23	Plymouth Valiant V-100 6	Carmin Adams	Plymouth-Valiant Dealers of Southern California
11	24.8055	26	Dodge Lancer 770 6	Hart Fullerton	Claude Short Dodge — Santa Monica, CA
12	24.7963	26	Dodge Lancer 770 6	Bill Yeager	Brookhurst Dodge — Garden Grove, CA
13	23.6804	60	Studebaker Lark Deluxe 6	Bill Krause	North Park Sales & Service — Chicago, IL
Did Not Start		11	Rambler American Custom 8	Gene Marx	Rambler Dealers of America
Did Not Start		17	Rambler American Custom 8	Les Viland	Rambler Dealers of America
CLASS "C"— COMPACT CAR — EIGHT CYLINDER					
1	25.0929	44	Buick Special 8	John Rich	Colonial Buick, Inc.— Glendale, CA
2	24.6756	2	Buick Special Deluxe 8	Barney Navarro	Barnett Buick — Burbank, CA
3	24.0747	25	Pontiac Tempest Custom 8	Joanne Cole	Jim Retzlaff Co.— Barstow, CA
4	23.9405	36	Pontiac Tempest Custom 8	Dee Hildebrand	Hine Pontiac — Dallas, TX

5	23.2086	63	Oldsmobile F-85 8	Link Paola	Paola Oldsmobile — La Crescenta, CA
6	23.0533	4	Studebaker Lark Deluxe 8	James Peterson	Ben Lindenbusch, Inc. — St. Louis, MO
7	22.6383	67	Oldsmobile F-85 Deluxe 8	Carol Paola	Paola Oldsmobile — La Crescenta, CA
CLASS "D" — LOW PRICE — SIX CYLINDER (UNDER $2800)					
1	23.1540	59	Plymouth Savoy 6	Art Rene	W.R. Shadoff, Inc. — Pomona, CA
2	23.0156	35	Dodge Dart Seneca 6	Woody Bell	L.A.-Orange County Dodge Dealers Association
3	22.5656	15	Mercury Meteor 600 6	Frank Hainley	Lincoln-Mercury Dealer Association
4	22.3364	58	Chevrolet Biscayne 6	Sunnie Baker	Melvin Hilliard Chevrolet — Kansas City, MO
5	22.2631	5	Mercury Meteor 600 6	Nickey Griffen Ward	Lincoln-Mercury Dealer Association
6	22.2342	65	Dodge Dart Seneca 6	Bob Trowbridge	L.A.-Orange County Dodge Dealers Association
7	22.1710	55	Plymouth Savoy 6	Bob Cahill	Plymouth-Valiant Dealers of Southern California
8	22.0959	28	Ford Fairlane 6	Ruth Levy	Rebsamen Motor Co. — Little Rock, AR
9	21.8455	27	Chevrolet Biscayne 6	Mel Schroeder	Doane Chevrolet — Dundee, IL
10	21.6154	64	Ford Fairlane 6	Bob Cannon	Moore Ford Co. — Little Rock, AR
CLASS "E" — LOW PRICE — EIGHT CYLINDER (UNDER $2800)					
1	21.3314	40	Ford Fairlane 8	Byron Froelich	St. Louis Metropolitan Ford Dealers Association
2	21.2129	39	Plymouth Savoy 8	Mary Davis	W.R. Shadoff, Inc. — Pomona, CA
3	21.1649	6	Ford Fairlane 8	Robert Rice	Dallas Metropolitan Ford Dealers Association
4	21.0703	12	Chevrolet Bel-Air 8	Melvin Hilliard	Melvin Hilliard Chevrolet — Kansas City, MO
5	20.9745	54	Mercury Meteor 600 8	Myra Buchanan	Lincoln-Mercury Dealer Association
6	20.8529	62	Mercury Meteor 600 8	Bob Snider	Lincoln-Mercury Dealer Association
7	20.8379	41	Pontiac Catalina 8	Smokey Yunick	Andy Klein Pontiac — Overland Park, KS
8	20.8377	34	Chevrolet Bel-Air 8	Doug Roe	Doane Chevrolet — Dundee, IL
9	20.3612	57	Pontiac Catalina 8	Paul Russo	Gillman Pontiac — Houston, TX
10	20.3122	51	Dodge Dart Seneca 8	Eugene Beck	Cassens and Sons, Inc. — Edwardsville, IL
11	19.8593	13	Plymouth Savoy 8	Bob McDaniel	Plymouth-Valiant Dealers of Southern California
CLASS "F" — MEDIUM PRICE ($2801 to $4500)					
1	19.9954	48	Chrysler Newport 8	Mel Alsbury, Jr.	Mel Alsbury, Inc. — Hollywood, CA
2	19.6093	8	Pontiac Star Chief 8	Mel Chastian	Los Angeles Pontiac Dealer Club
3	19.5897	43	Pontiac Star Chief 8	Mickey Thompson	Los Angeles Pontiac Dealer Club
4	19.3988	16	Chrysler Windsor 8	Marty Gordon	Danny McGroo, Inc. — Culver City, CA
5	18.8093	46	Buick LeSabre 8	Marta Retzlaff	Jim Retzlaff Co. — Barstow, CA
6	18.6184	21	Oldsmobile Dynamic 88 8	C.G. Cockman	Newbill Motor Co., Inc. — Benton, AR
7	18.4574	19	Oldsmobile Dynamic 88 8	Lute Eldridge	Paola Oldsmobile — La Crescenta, CA
8	18.4311	22	Mercury Monterey 8	Ak Miller	Lincoln-Mercury Dealer Association
CLASS "G" — HIGH PRICE (OVER $4501)					
1	18.9253	9	Cadillac Sedan deVille 8	Lee Hamer	Hamer Motors — Mission Hills, CA
2	18.7360	29	Imperial Custom 8	George Alsbury	Mel Alsbury, Inc. — Hollywood, CA
3	17.2329	38	Lincoln Continental 8	Jim Fosdick	Lincoln-Mercury Dealer Association
4	16.6684	31	Lincoln Continental 8	Jack Hills	Lincoln-Mercury Dealer Association

1962 Mobil Economy Run
Los Angeles to Detroit
March 31–April 5, 1962 — 46 Cars — 2497.6 Miles

Place	M.P.G.	No.	Car	Driver	Entrant
CLASS "A" — COMPACT CARS — MANUAL TRANSMISSION					
1	31.1105	13	Rambler American 400 6	Les Viland	Rambler Dealers of America
2	30.4941	7	Ford Falcon (144) 6	Dan Jones	Detroit Metropolitan Ford Dealers Association
3	30.0354	29	Chevrolet Chevy II 300 4	Bill Horton	Don Allen Midtown Chevrolet — New York
4	28.8987	43	Chevrolet Corvair 700 6	Ronnie Kaplan	Nickey Chevrolet — Chicago, IL
5	27.4286	18	Ford Falcon Deluxe (170) 6	Fran Foster	Fred Jones Co. — Tulsa, OK
6	27.3411	28	Pontiac Tempest Custom 4	Jim Wasson	Packer Pontiac — Detroit, MI
7	25.7912	34	Ford Fairlane 6	Ginny Sims	Southern CA Ford Dealers Association
8	24.6467	3	Chevrolet Chevy II 300 6	Dee Hildebrand	Doane Chevrolet — Dundee, IL
9	24.1889	36	Studebaker Lark Deluxe 6	Bill Corey	Danyluk Motors, Inc. — Glendale, CA and Bill Corey — Pasadena, CA
CLASS "B" — SMALL ENGINE COMPACT CARS (UNDER 160 CUBIC INCH ENGINE)					
1	27.0178	17	Chevrolet Corvair Monza 6	Patricia Sawyer	Don Steves Chevrolet — La Habra, CA
2	25.6400	22	Ford Falcon Deluxe (144) 6	Leon Funderburk	Mason Motor Co. — Tucumcari, NM

3	25.6329	26	Chevrolet Chevy II 300 4	Mary Hauser	Kelley Chevrolet—Detroit, MI
CLASS "C"—LARGE ENGINE COMPACT CARS (161 through 200 CUBIC INCH ENGINE)					
1	27.2950	23	Pontiac Tempest Custom 4	Mickey Thompson	Salta Pontiac—Long Beach, CA
2	27.0927	20	Rambler American 400 6	Tommy Thomas	Rambler Dealers of America
3	24.8718	25	Plymouth Valiant V-100 6	William McNulty	Skyland Chrysler-Plymouth—Dallas, TX
4	23.9714	2	Mercury Comet Custom 6	Mary Faulkner	George Moyer, Inc.—Compton, CA
5	23.9020	27	Chevrolet Chevy II Nova 6	Don Edmunds	Gates Chevrolet Co.—South Bend, IN
6	23.8410	16	Buick Special 6	Bob Russo	Bill Murphy Buick, Inc.—Culver City, CA
7	22.8304	6	Ford Falcon Deluxe (170) 6	Loretta Colange	St. Louis Metropolitan Ford Dealers Association
CLASS "D"—LUXURY COMPACT CARS (201 through 260 CUBIC INCH ENGINE)					
1	24.8525	41	Plymouth Valiant V-100 6	Joe Strum	Dick Green Chrysler-Plymouth, Inc.—Detroit
2	24.4003	8	Dodge Lancer 770 6	Danny Eames	L.A.–Orange County Dodge Dealers Association
3	23.6752	30	Buick Special Deluxe 8	John Rich	Colonial Buick, Inc.—Glendale, CA
DNF—Mechanical		40	Oldsmobile F-85 Deluxe 8	Edward J. Wilder	Ray Nolting Oldsmobile—Kirkwood, MO
DNF—Mechanical		31	Ford Fairlane 8	Sunnie Baker	Bill Doenges—Tulsa, OK
CLASS "E"—STANDARD SIZE—SIX CYLINDER					
1	24.6149	45	Plymouth Savoy 6	Bob Cahill	W.R. Shadoff, Inc.—Pomona, CA
2	22.8350	12	Dodge Dart 6	Woody Bell	Dodge Dealers of Detroit
3	21.6210	14	Chevrolet Biscayne 6	Paula Murphy	Johnson Chevrolet Co.—Dallas, TX
4	20.6407	21	Ford Galaxie 6	Myra Buchanan	Southern California Ford Dealers Association
CLASS "F"—LOW PRICE—EIGHT CYLINDER					
1	20.7788	4	Chevrolet Impala 8	Byron Froelich	Bob Rosenthal Chevrolet—Arlington, VA
2	20.7478	19	Ford Galaxie 8	Bill Levy	Chicago Metropolitan Ford Dealers Association
3	20.7141	5	Plymouth Savoy 8	Bob McDaniel	Monroe-Zeder Chrysler-Plymouth—North Miami
4	19.2010	46	Pontiac Catalina 8	Roy Swanson	John Hine Pontiac—San Diego, CA
5	18.9969	44	Dodge Dart 330 8	Art Rene	San Francisco Bay Area Dodge Dealers Association
6	18.5502	35	Chevrolet Bel-Air 8	Melvin Hilliard	Melvin Hilliard Chevrolet—Kansas City, MO
7	18.4887	42	Rambler Ambassador 400 8	Les Scott	Rambler Dealers of America
CLASS "G"—MEDIUM PRICE					
1	19.4797	38	Pontiac Star Chief 8	Judy Thompson	Andy Klein Pontiac—Overland Park, KS
2	19.0518	11	Oldsmobile Dynamic 88 8	Stan Raymond	Paola Oldsmobile—La Crescenta, CA
3	18.9585	33	Chrysler Newport 8	Mel Alsbury, Jr.	Mel Alsbury, Inc.—Hollywood, CA
4	17.8496	1	Chrysler 300 8	Hart Fullerton	Hart Fullerton Chrysler-Plymouth—West Los Angeles
5	17.6456	15	Mercury Monterey Custom 8	Al Cottle	Bob Estes Lincoln-Mercury—Inglewood
6	17.1496	9	Buick LeSabre 8	Marta Retzlaff	Jim Retzlaff Co.—Barstow, CA
CLASS "H"—LUXURY CARS					
1	18.1139	39	Chrysler New Yorker 8	George Alsbury	Mel Alsbury, Inc.—Hollywood, CA
2	17.5183	37	Oldsmobile Ninety-Eight 8	Link Paola	Mandy Williams Oldsmobile—West Covina
3	17.2467	10	Buick Electra 225 8	Barney Navarro	Barnett Buick—Burbank, CA
4	17.1872	32	Imperial Crown 8	Byron Clouse	Briley Motor Co.—Omaha, Nebraska
5	16.3640	24	Lincoln Continental 8	Lee Hamer	Frank Adam Lincoln—Groose Pointe Farms

1963 Mobil Economy Run
Los Angeles to Detroit

March 30–April 4, 1963—36 Cars—2373.2 Miles

Place	M.P.G.	No.	Car	Driver	Entrant
CLASS "A"—SMALL ENGINE COMPACT CARS (144 through 153 CUBIC INCH ENGINE)					
1	27.90	26	Chevrolet Corvair Monza 6	Patricia Sawyer	Don Steves Chevrolet—La Habra, CA
2	27.52	8	Chevrolet Chevy II 300 4	Mary Hauser	Kelley Chevrolet—Detroit, MI
3	25.70	22	Ford Falcon Deluxe 6	Irma Payne	Paul Freed, Inc.—Waynesboro, VA
CLASS "B"—MEDIUM ENGINE COMPACT CARS (169 through 170 CUBIC INCH ENGINE)					
1	26.29	9	Ford Falcon Deluxe 6	Bill Levy	Wm. L. Hughson Ford, Inc.—San Francisco
2	26.26	36	Plymouth Valiant V-100 6	Bob Checkley	Dick Green Chrysler-Plymouth, Inc.—Detroit
3	25.75	18	Mercury Comet Custom 6	Mary Faulkner	Bob Estes Lincoln-Mercury—Inglewood
4	25.59	4	Dodge Dart 170 6	George Miller	L.A.–Orange County Dodge Dealers Association

CLASS "C"—LARGE ENGINE COMPACT CARS (194 through 225 CUBIC INCH ENGINE)

1	28.61	13	Rambler American 440 6	Les Viland	Rambler Dealers of America
2	27.31	2	Pontiac Tempest Custom 4	Judy Thompson	Pat Clark Pontiac — Las Vegas, NV
3	24.03	20	Chevrolet Chevy II 300 6	Ronnie Kaplan	Nickey Chevrolet — Chicago, IL

CLASS "D"—INTERMEDIATE SIZE—EIGHT CYLINDER

1	23.81	25	Buick Special Deluxe 8	John Rich	Colonial Buick, Inc.— Glendale, CA
2	21.47	5	Ford Fairlane 8	Jim Fosdick	Stark-Hickey, Inc.— Detroit, MI
3	21.13	27	Pontiac Tempest 8	Jim Nelson	John Hine Pontiac — San Diego, CA
4	21.10	35	Oldsmobile F-85 8	Edward J. Wilder	Ray Nolting Oldsmobile — Kirkwood, MO
5	20.56	32	Studebaker Lark Regal 8	Bill Corey	Bill Corey — Pasadena, CA

CLASS "E"—STANDARD SIZE—SIX CYLINDER

1	24.33	24	Plymouth Savoy 6	Scott Harvey	W.R. Shadoff, Inc.— Pomona, CA
2	23.69	29	Dodge 330 6	Jim Whipple	Detroit Dodge Dealers Association
3	22.61	17	Chevrolet Biscayne 6	Roy Swanson	Doane Chevrolet — Dundee, IL
4	21.50	15	Ford 300 6	Paula Murphy	Litsinger Ford — Chicago, IL

CLASS "F"—LOW PRICE—EIGHT CYLINDER

1	21.20	16	Dodge 330 8	Bob Cahill	Cleveland Dodge Dealers Association
2	21.04	7	Plymouth Savoy 8	Bob McDaniel	Tom O'Brien Co.— Indianapolis, IN
3	20.97	6	Chevrolet Bel-Air 8	Don Edmunds	Don Allen Midtown Chevrolet — New York
4	19.72	1	Chevrolet Impala 8	Melvin Hilliard	Melvin Hilliard Chevrolet — Kansas City, MO
5	18.75	12	Ford 300 8	Art Rene	Tom Frost Ford — Warrenton, VA

CLASS "G"—MEDIUM PRICE

1	20.01	19	Oldsmobile Dynamic 88 8	Stan Raymond	Paola Oldsmobile — La Crescenta, CA and Irving Air Chute Co.— Glendale, CA
2	19.68	34	Chrysler Newport 8	Mel Alsbury, Jr.	Mel Alsbury, Inc.— Hollywood, CA
3	19.38	10	Pontiac Catalina 8	Jim Wasson	Salta Pontiac — Long Beach, CA
4	19.27	11	Pontiac Star Chief 8	Mickey Thompson	Packer Pontiac — Detroit, MI
5	18.35	30	Dodge Custom 880 8	Don Bridges	Don Bridges' Nickelodeon — Santa Monica
6	18.33	14	Buick LeSabre 8	Marta Retzlaff	Jim Retzlaff Co.— Barstow, CA
7	17.39	3	Mercury Monterey Custom 8	Al Cottle	Bob Estes Lincoln-Mercury — Inglewood
8	17.27	21	Chrysler 300 8	Hart Fullerton	Hart Fullerton Chrysler-Plymouth — West Los Angeles

CLASS "H"—LUXURY CARS

1	18.43	33	Chrysler New Yorker 8	Mildred Alsbury	Mel Alsbury, Inc.— Hollywood, CA
2	17.40	23	Imperial Custom 8	Marty Gordon	Danny McGroo, Inc.— Culver City, CA
3	17.18	28	Oldsmobile Ninety-Eight 8	Mandy Williams	Mandy Williams Oldsmobile — West Covina
4	16.60	31	Buick Electra 225 8	Barney Navarro	Barnett Buick — Burbank, CA

1964 Mobil Economy Run
Los Angeles to New York

April 3–9, 1964 — 45 Cars — 3243.8 Miles

Place	M.P.G.	No.	Car	Driver	Entrant
CLASS "A"—SMALL ENGINE COMPACTS (160 through 170 CUBIC INCH ENGINE)					
1	26.1191	34	Dodge Dart 170 6	Bob Checkley	Los Angeles Dodge Dealers Association
2	25.8921	20	Plymouth Valiant V-100 6	Carl Diehl	Dick Green Chrysler-Plymouth, Inc.— Detroit
3	23.8341	30	Ford Falcon Futura 6	Art Rene	Tom Frost Ford — Warrenton, VA
4	23.2714	9	Chevrolet Corvair 700 6	Earl Schilling	Chevy Teen Team
CLASS "B"—LARGE ENGINE COMPACTS (190 through 327 CUBIC INCH ENGINE)					
1	27.8336	42	Rambler American 440 6	Les Viland	Rambler Dealers of America
2	23.2182	7	Chevrolet Chevy II 100 6	Jean Fitzgerald	Chevy Teen Team
3	23.0851	33	Plymouth Valiant V-100 8	Bob Knoll	Airport Motors, Inc.— Jackson Heights, NY
4	22.8755	37	Dodge Dart 270 8	George Miller	New York Dodge Dealers Association
5	21.1906	15	Chevrolet Chevy II Nova 8	Ronnie Cherry	Chevy Teen Team
6	21.1420	25	Ford Falcon Deluxe 8	Don Royer	Hatfield Motors, Inc.— Indianapolis, IN
CLASS "C"—INTERMEDIATE SIZE—SIX CYLINDER					
1	25.2986	38	Buick Special Deluxe 6	Bill Levy	Bill Murphy Buick, Inc.— Culver City, CA
2	25.2494	31	Pontiac Tempest 6	Ginny Sims	Salta Pontiac — Long Beach, CA

3	24.5321	24	Mercury Comet 202 6	Julian "Bo" Jett	Bob Estes Lincoln-Mercury — Inglewood
4	24.3462	1	Chevrolet Chevelle 300 6	Richard Doyle	Chevy Teen Team
5	23.4678	43	Chevrolet Chevelle Malibu 6	James Sny	Chevy Teen Team
6	23.2917	44	Ford Fairlane 6	Fran Foster	Litsinger Ford — Chicago, IL
7	23.2406	28	Oldsmobile F-85 Deluxe 6	W.S. "Red" Eason	Mandy Williams Oldsmobile — West Covina and Irving Air Chute Co. — Glendale, CA

CLASS "D" — INTERMEDIATE SIZE — EIGHT CYLINDER

1	23.7441	5	Buick Special Deluxe 8	John Rich	Colonial Buick, Inc. — Glendale, CA
2	22.4083	39	Oldsmobile F-85 Deluxe 8	Stan Raymond	Paola Oldsmobile — La Crescenta, CA
3	22.3413	35	Ford Fairlane 500 8	Patricia Jones	Wm. L. Hughson Ford, Inc. — San Francisco
4	21.9861	4	Mercury Comet Caliente 8	Mary Faulkner	Bob Estes Lincoln-Mercury — Inglewood
5	19.9949	8	Pontiac Tempest 8	Mary Davis	Andy Klein Pontiac — Overland Park, KS
6	19.3480	6	Chevrolet Chevelle Malibu 8	Shirley Palmer	Chevy Teen Team

CLASS "E" — FULL SIZE — SIX CYLINDER

1	25.2955	16	Plymouth Savoy 6	Scott Harvey	Armory Garage, Inc. — Albany, NY
2	25.0732	11	Dodge 330 6	Jim Whipple	Connecticut Dodge Dealers Association
3	22.2577	40	Chevrolet Biscayne 6	Karole Roth	Chevy Teen Team
4	20.6849	29	Ford Custom 500 6	Gordon Madison	Flinn Motor Corp. — Larchmont, NY

CLASS "F" — LOW PRICE — EIGHT CYLINDER

1	22.1963	22	Plymouth Savoy 8	Bob Cahill	Melrose Motors — Oakland, CA
2	21.4649	32	Dodge 330 8	Herman "Hy" Plotkin	New Jersey Dodge Dealers Association
3	20.7486	21	Ford Custom 500 8	Jim Fosdick	King Motors — Huntington Park, CA
4	20.6873	36	Chevrolet Impala 8	Ronald Olbernolte	Chevy Teen Team
5	19.7589	41	Chevrolet Bel-Air 8	Larry Booth	Chevy Teen Team

CLASS "G" — MEDIUM PRICE

1	21.3670	27	Buick LeSabre 8	Marta Retzlaff	Jim Retzlaff Co. — Barstow, CA
2	20.1058	45	Oldsmobile Jetstar 88 8	Pete Novotny	Mandy Williams Oldsmobile — West Covina
3	19.8584	2	Pontiac Star Chief 8	Mary Hauser	Rogers Pontiac — Brooklyn, NY
4	19.5069	14	Pontiac Catalina 8	Woody Bell	Rex Pontiac — Jackson Heights, NY
5	19.4769	18	Chrysler Newport 8	George Alsbury	Mel Alsbury, Inc. — Hollywood, CA
6	17.9747	17	Mercury Monterey Marauder 8	Al Cottle	Bob Estes Lincoln-Mercury — Inglewood
7	17.8837	3	Oldsmobile Dynamic 88 8	Edward J. Wilder	Ray Nolting Oldsmobile — Kirkwood, MO
8	17.4872	12	Chrysler 300 8	Hart Fullerton	Hart Fullerton Chrysler-Plymouth — West Los Angeles

CLASS "H" — LUXURY CARS

1	18.7737	10	Chrysler New Yorker Salon 8	Mel Alsbury, Jr.	Mel Alsbury, Inc. — Hollywood, CA
2	18.6509	26	Cadillac Sixty-Two 8	Ed Dempsey	Concrete Coring Co. — Los Angeles, CA
3	18.1171	23	Oldsmobile Ninety-Eight 8	Mandy Williams	Mandy Williams Oldsmobile — West Covina
4	17.4537	19	Buick Electra 225 8	Paula Murphy	Barnett Buick — Burbank, CA
5	17.2633	13	Imperial Custom 8	Marty Gordon	Danny McGroo, Inc. — Culver City, CA

1965 Mobil Economy Run Los Angeles to New York

April 5–11, 1965 — 49 Cars — 3266.1 Miles

CLASS "A" — SMALL ENGINE COMPACTS (160 through 170 CUBIC INCH ENGINE)

Place	M.P.G.	No.	Car	Driver	Entrant
1	25.6164	22	Ford Falcon 6	Byron Froelich	Gotham Ford — New York City
2	24.0943	34	Plymouth Valiant 100 6	Carl Diehl	Dick Green Chrysler-Plymouth, Inc. — Detroit
3	23.5016	42	Dodge Dart 6	Marty Payne	California Dodge Dealers Association
4	20.2960	2	Chevrolet Corvair Monza 6	Melvin Hilliard	Melvin Hilliard Chevrolet — Kansas City, MO

CLASS "B" — LARGE ENGINE COMPACTS (190 through 290 CUBIC INCH ENGINE)

Place	M.P.G.	No.	Car	Driver	Entrant
1	25.6524	44	Rambler American 440 6	Les Viland	Rambler Dealers of America
2	24.3585	14	Ford Falcon Futura 6	Dave Apple	Niagara Frontier Ford Dealers Association
3	22.5122	30	Chevrolet Chevy II Nova 6	Gordon Madison	Oscar Gregory Chevrolet — Paramount, CA

CLASS "C"—INTERMEDIATE SIZE—SIX CYLINDER

1	24.7134	20	Ford Fairlane 6	Darrell Droke	Tom Frost Ford—Warrenton, VA
2	23.3984	29	Dodge Coronet 6	Cass Szpieg	New Jersey Dodge Dealers Association
3	23.2333	32	Mercury Comet 202 6	Mary Faulkner	Bob Estes Lincoln-Mercury—Inglewood
4	23.1156	13	Plymouth Belvedere I 6	Bob Binder	Hunter Brothers Motors—Pueblo, CO
5	23.5433	33	Pontiac Tempest 6	Don Francisco	Andy Klein Pontiac—Overland Park, KS
6	22.0422	8	Oldsmobile F-85 Deluxe 6	Ed Dempsey	Oldsmobile Dealers of U.S.A.
7	21.2546	49	Chevrolet Chevelle Malibu 6	Scherlee Madison	Oscar Gregory Chevrolet—Paramount, CA
8	20.8468	11	Buick Special Deluxe 6	Jan Detrick	Bill Murphy Buick, Inc.—Culver City, CA
9	20.2630	4	Rambler Ambassador 880 6	Tommy Thomas	Rambler Dealers of America

CLASS "D"—INTERMEDIATE SIZE—EIGHT CYLINDER

1	20.7682	43	Dodge Coronet 8	Bob Cahill	New York Dodge Dealers Association
2	20.6102	36	Plymouth Belvedere II 8	Art Hesskamp	E. Koeppel, Inc.—Jamaica, Long Island, NY
3	20.3071	17	Buick Special Deluxe 8	John Rich	Colonial Buick, Inc.—Glendale, CA
4	19.8068	37	Oldsmobile F-85 Deluxe 8	Stan Raymond	Oldsmobile Dealers of U.S.A.
5	19.5889	48	Mercury Comet 404 8	Ned Aberly	Bob Estes Lincoln-Mercury—Inglewood
6	19.2850	18	Ford Fairlane 8	Al Johnson	Kansas City Ford Dealers Association
7	19.1124	23	Pontiac Tempest Custom 8	Kay Kimes	Salta Pontiac—Long Beach, CA
8	18.4486	10	Rambler Ambassador 880 8	Don Bridges	Rambler Dealers of America
9	18.1707	41	Chevrolet Chevelle Malibu 8	Art Taraldsen	Emerald Chevrolet—Beverly Hills, CA

CLASS "E"—FULL SIZE—SIX CYLINDER

1	21.1748	19	Chevrolet Biscayne 6	Patricia Sawyer	Nickey Chevrolet—Chicago, IL
2	21.0757	28	Plymouth Fury II 6	Bob Checkley	Bill Snethkamp, Inc.—Highland Park, MI
3	20.7759	26	Ford Custom 6	Ginny Sims	Chicago Ford Dealers Association

CLASS "F"—LOW PRICE—EIGHT CYLINDER

1	19.8292	3	Plymouth Fury II 8	Scott Harvey	Armory Garage, Inc.—Albany, New York
2	19.2840	25	Ford Galaxie 500 8	Fran Foster	Jess Hunter Motors–Pueblo, CO
3	18.6145	39	Chevrolet Impala 8	Don Royer	Melvin Hilliard Chevrolet—Kansas City, MO

CLASS "G"—MEDIUM PRICE

1	19.8423	46	Pontiac Catalina 8	Jean Calvin	Rogers Pontiac—Brooklyn, NY
2	19.6474	5	Pontiac Star Chief 8	Art Arfons	Rex Pontiac—Jackson Heights, NY
3	19.6320	6	Buick LeSabre 8	Marta Retzlaff	Jim Retzlaff Co.—Barstow, CA
4	18.1148	31	Oldsmobile Jetstar 88 8	Pete Novotny	Oldsmobile Dealers of U.S.A.
5	17.8404	35	Chrysler Newport 8	Hart Fullerton	Hart Fullerton Chrysler-Plymouth—West Los Angeles
6	17.6870	12	Dodge Custom 880 8	Patricia Jones	Dodge Town, Inc.—San Diego, CA
7	17.1135	27	Oldsmobile Dynamic 88 8	George Alsbury	Oldsmobile Dealers of U.S.A.
8	16.1618	24	Mercury Monterey 8	Al Cottle	Bob Estes Lincoln-Mercury—Inglewood

CLASS "H"—LUXURY CARS

1	17.2609	9	Chrysler New Yorker 8	Art Bertelsen	Sites Brothers Motors—Kansas City, MO
2	16.4073	38	Cadillac Sedan deVille 8	Stan Seeds	Chamber of Commerce—West Covina, CA
3	16.3466	16	Buick Electra 225 8	John Meyers	Barnett Buick—Burbank, CA
4	15.7911	45	Oldsmobile Ninety-Eight 8	Mel Alsbury, Jr.	Oldsmobile Dealers of U.S.A.
Disqualified		1	Imperial Crown 8	Glen "Ole" Olsen	Beverly Hills Chrysler—Beverly Hills

CLASS "S"—SPECIAL SPORTS

1	21.4231	15	Plymouth Barracuda 8	George Miller	Wilson Motors—Flagstaff, AZ
2	20.7464	40	Ford Mustang 2+2 8	Bill Levy	Vel's Ford—Torrance, CA
3	20.5399	21	Dodge Dart GT 8	Dareth Rich	L.A.–Orange County Dodge Dealers Association
4	19.3766	7	Mercury Comet Caliente 8	Edward "Budd" Liburdi	Lynch Lincoln-Mercury—Santa Monica
5	18.4831	47	Chevrolet Chevelle Malibu SS 8	E.L. "Kirk" Kirkpatrick	Hastings Chevrolet—Santa Monica, CA

Imperial Crown 8 was automatically disqualified on the fifth day after getting off-course on the previous day outside Chicago and running out of fuel before reaching the Toledo, Ohio, refueling station.

1966 Mobil Economy Run
Los Angeles to Boston
March 14–20, 1966 — 52 Cars — 3301.4 Miles

Place	M.P.G.	No.	Car	Driver	Entrant
CLASS "A" — COMPACT — SIX CYLINDER					
1	23.8063	29	Rambler American 440 6	Les Viland	American Motors / Rambler Dealers
2	23.5730	38	Ford Falcon (170) 6	Byron Froelich	Homer Heller Ford — Escondido, CA
3	23.3534	8	Chevrolet Corvair Monza 6	Bill Levy	Sun-Air Chevrolet — Hemet, CA
4	23.0542	1	Chevrolet Chevy II 100 (194) 6	C.K. Enoch	Groth Chevrolet — Huntington Beach, CA
5	21.9072	15	Ford Falcon Futura (200) 6	Carl Westervelt	E.D. Babbitt Motor Co. — Flagstaff, AZ
6	21.4750	50	Chevrolet Chevy II Nova (230) 6	Carole Stroud	Washburn Chevrolet — Santa Barbara, CA
CLASS "B" — COMPACT — EIGHT CYLINDER					
1	21.5761	20	Dodge Dart 8	Bob Checkley	Boston Dodge Dealers Association
2	20.5773	46	Ford Mustang 2+2 8	Fran Foster	Vel's Ford — Torrance, CA
3	19.6550	36	Ford Falcon 8	Ed Dempsey	Frank Galpin Ford — San Fernando, CA
4	19.0832	9	Chevrolet Chevy II Nova SS 8	Jim Randall	Melvin Hilliard Chevrolet — Kansas City, MO
5	18.9294	44	Chevrolet Chevy II 100 8	Gordon Madison	Nickey Chevrolet — Chicago, IL
CLASS "C" — INTERMEDIATE SIZE — SIX CYLINDER					
1	23.1019	41	Plymouth Belvedere I 6	Bob Binder	Dick Green Chrysler-Plymouth, Inc. — Detroit
2	22.7379	34	Buick Special 6	Paula Murphy	Bill Murphy Buick, Inc. — Culver City, CA
3	22.6646	27	Pontiac Tempest Custom 6	Don Francisco	Pontiac Village, Inc. — Brookline, MA
4	22.3440	48	Oldsmobile F-85 6	George Alsbury	Oldsmobile Dealers of U.S.A.
5	22.2630	23	Chevrolet Chevelle 300 6	Ed Miller	Don Steves Chevrolet — La Habra, CA
6	20.9582	18	Mercury Comet 202 6	Mary Faulkner	Bob Estes Lincoln-Mercury — Inglewood
7	20.6030	24	Ambassador 880 6	Tommy Thomas	American Motors / Rambler Dealers
DNF — Accident		17	Ford Fairlane 500 6	Al Johnson	Tom Frost Ford — Warrenton, VA
CLASS "D" — INTERMEDIATE SIZE — EIGHT CYLINDER					
1	21.0520	25	Dodge Coronet Deluxe 8	George Miller	Philadelphia Dodge Dealers Association
2	20.6054	28	Buick Special Deluxe 8	John Rich	Colonial Buick, Inc. — Glendale, CA
3	19.4100	10	Ford Fairlane 500 8	Bobby Spears	Bob Tasca Ford — E. Providence, RI
4	19.2780	3	Mercury Comet Cyclone 8	Harold Limbird	Bob Estes Lincoln-Mercury — Inglewood
5	19.0083	40	Oldsmobile F-85 8	Patricia Jones	Oldsmobile Dealers of U.S.A.
6	18.4948	37	Chevrolet Chevelle 300 8	Jim Finnessy	Bernard Chevrolet, Inc. — Libertyville, IL
7	18.2960	32	Chevrolet Chevelle Malibu 8	Mary Hauser	Kirk Chevrolet — Santa Monica, CA
8	18.0981	30	Pontiac Tempest 8	Gail Hire	Majestic Pontiac, Inc. — Los Angeles, CA
9	17.8367	6	Oldsmobile F-85 Cutlass 8	Greg Bucilla	Prestige Oldsmobile, Inc. — Los Angeles
10	17.6704	42	Chevrolet Chevelle SS-396 8	Dick Harrell	Nickey Chevrolet — Chicago, IL
11	17.4205	11	Ambassador DPL 8	George Sorenson	City of Kenosha, WI
12	13.6802	33	Mercury Comet Cyclone GT 8	Frank Mowrer	Lynch Lincoln-Mercury — Santa Monica
CLASS "E" — STANDARD SIZE — SIX CYLINDER					
1	21.0454	26	Chevrolet Biscayne 6	Patricia Sawyer	Nickey Chevrolet — Chicago, IL
2	20.5574	51	Plymouth Fury I 6	Shirley Shahan	Harris Auto Sales, Providence, RI
3	19.9196	19	Ford Custom 6	Ginny Sims	Bob Ford — Dearborn, MI
CLASS "F" — LOW PRICE — EIGHT CYLINDER					
1	19.8748	31	Chevrolet Impala 8	Don Royer	Melvin Hilliard Chevrolet — Kansas City, MO
2	18.6896	13	Chevrolet Caprice 8	Paul Prior	Emerald Chevrolet — Beverly Hills, CA
3	18.6849	16	Plymouth Fury I 8	Bob Cahill	Domenich-Basso, Inc. — Los Angeles, CA
4	18.6113	22	Dodge Polara 318 8	Carl Diehl	Newark Dodge Dealers Association
5	18.6065	35	Chevrolet Bel-Air 8	Carolyn Thomas	Cone Chevrolet — Fullerton, CA
6	18.1885	7	Ford Custom 500 8	Kay Kimes	Robert H. Loud Ford — Pasadena, CA
CLASS "G" — MEDIUM PRICE					
1	19.2439	5	Buick LeSabre 8	Marta Retzlaff	Jim Retzlaff Co. — Barstow, CA
2	18.9208	49	Pontiac Star Chief Executive 8	Art Arfons	Salta Pontiac — Long Beach, CA
3	18.6181	43	Pontiac Catalina 8	Jean Calvin	Sea Crest Pontiac, Inc. — Lynn, MA
4	18.5169	45	Oldsmobile Jetstar 88 8	Stan Raymond	Oldsmobile Dealers of U.S.A.
Did Not Start		14	Mercury Park Lane 8	Don Bailey	Bob Estes Lincoln-Mercury — Inglewood

CLASS "H"—LUXURY CARS					
1	17.3332	39	Oldsmobile Toronado Deluxe 8	Mel Alsbury, Jr.	Oldsmobile Dealers of U.S.A.
2	16.7489	47	Chrysler New Yorker 8	Hart Fullerton	Hart Fullerton Chrysler-Plymouth—West Los Angeles
3	16.6835	2	Oldsmobile Ninety-Eight 8	Pete Novotny	Oldsmobile Dealers of U.S.A.
4	16.6346	4	Buick Riviera 8	Mike Jones	Los Angeles Metropolitan Buick Dealers Association
5	16.3212	52	Cadillac Sedan deVille 8	Stan Seeds	Chamber of Commerce—West Covina, CA
6	16.2682	21	Buick Electra 225 8	John Meyers	Barnett Buick—Burbank, CA
7	16.0873	12	Imperial Crown 8	Bert Parker	Rotary Club of Baldwin Park, CA

1967 Mobil Economy Run
Los Angeles to Detroit
April 4–9, 1967—42 Cars—2837.8 Miles

Place	M.P.G.	No.	Car	Driver	Entrant
CLASS "A"—COMPACT—SIX CYLINDER					
1	24.5722	35	Plymouth Valiant 100 6	Bob Checkley	Dick Green Chrysler-Plymouth, Inc.—Detroit
2	24.2961	30	Dodge Dart 6	Bob Cahill	L.A.–Orange County Dodge Dealers Association
3	24.0663	38	Ford Falcon 6	Byron Froelich	Tom Frost Ford—Warrenton, VA
4	24.0418	10	Ford Mustang 6	Per Nystrom	Paul Harvey Ford—Indianapolis, IN
5	23.3681	34	Rambler American 220 6	Les Viland	American Motors / Rambler Dealers
6	22.8318	11	Chevrolet Corvair Monza 6	Gordon Madison	Bill Bader Chevrolet, Inc.—Fontana, CA
7	22.5269	13	Chevrolet Chevy II 100 6	Betty Enoch	Don Steves Chevrolet—La Habra, CA
8	22.3668	8	Chevrolet Camaro 6	Patricia Sawyer	Nickey Chevrolet—Chicago, IL
CLASS "B"—COMPACT—EIGHT CYLINDER					
1	22.3198	5	Plymouth Barracuda 8	Scott Harvey	Los Angeles Plymouth Dealers Association
2	21.1269	16	Dodge Dart 8	Jean Calvin	L.A.–Orange County Dodge Dealers Association
3	20.9925	3	Mercury Cougar 8	Ginny Sims	Bob Estes Lincoln-Mercury—Inglewood
4	20.7281	21	Ford Mustang 2+2 8	Fran Foster	Galpin Ford—Sepulveda, CA
5	20.5785	2	Chevrolet Chevy II 100 8	Jack Kirkpatrick	Nickey-Bill Thomas—Anaheim, CA
6	19.2757	41	Chevrolet Camaro 8	Carolyn Thomas	Nickey Chevrolet—Chicago, IL
DNF—Mechanical		31	Ford Falcon Futura 8	Don Bailey	Allan Bros. Motor Sales—Port Clinton, OH
CLASS "C"—INTERMEDIATE SIZE—SIX CYLINDER					
1	23.2905	25	Mercury Capri 6	Bill Levy	Mercury Dealers—U.S.A.
2	22.3026	42	Plymouth Belvedere I 6	Bob Binder	Gage Chrysler-Plymouth—Chicago, IL
3	21.9204	33	Buick Special 6	Paula Murphy	Bill Murphy Buick, Inc.—Culver City, CA
4	21.4600	14	Chevrolet Chevelle 300 Deluxe 6	Mary Hauser	Nickey Chevrolet—Chicago, IL
5	20.0466	22	Oldsmobile F-85 6	George Alsbury	Century Oldsmobile—Van Nuys, CA
CLASS "D"—INTERMEDIATE SIZE—EIGHT CYLINDER					
1	20.0111	9	Plymouth Belvedere II 8	Allan Brooks	Hollywood Chrysler-Plymouth—Hollywood, CA
2	19.8141	24	Mercury Caliente 8	Darrell Droke	Mercury Dealers—U.S.A.
3	19.7196	19	Buick Special 8	John Rich	Colonial Buick, Inc.—Glendale, CA
4	18.9898	36	Dodge Coronet Deluxe 8	Shirley Shahan	Detroit Dodge Dealers Association
5	18.8938	4	Ford Fairlane 500 8	Al Johnson	Robert H. Loud Ford—Pasadena, CA
6	18.6012	20	Chevrolet Chevelle Malibu 8	George Decker	Bill McGuire Chevrolet—Buena Park, CA
7	17.0613	23	Oldsmobile F-85 Cutlass 8	Mandy Williams	Mandy Williams Oldsmobile—West Covina
CLASS "E"—LOW PRICE—EIGHT CYLINDER					
1	19.2913	32	Chevrolet Impala SS 8	Don Royer	Nickey-Bill Thomas—Anaheim, CA
2	19.2588	40	Ford Custom 500 8	Kay Kimes	Bob Tasca Ford—E. Providence, RI
3	19.1508	6	Plymouth Fury II 8	Jim Latham	Mackaig & Sons—Monrovia, CA
4	18.6473	7	Chevrolet Bel-Air 8	Tom Evans	Ed Taylor Chevrolet—Whittier, CA
CLASS "F"—MEDIUM PRICE					
1	18.7414	1	Buick LeSabre 400 8	Marta Retzlaff	Detroit Metropolitan Buick Dealers Association
2	18.5176	12	Oldsmobile Delmont 88 8	Stan Raymond	Mandy Williams Oldsmobile—West Covina
3	18.1949	18	Pontiac Catalina 8	Don Francisco	Bryant Pontiac—Covina, CA

4	17.9418	37	Dodge Polara 318 8	Art Bertelsen	Chicago Dodge Dealers Association
5	16.6837	15	Mercury Monterey 8	Billy Cox	Mercury Dealers — U.S.A.
CLASS "G" — LUXURY CARS					
1	17.5932	27	Chrysler 300 8	Hart Fullerton	Hart Fullerton Chrysler-Plymouth — West Los Angeles
2	16.9970	29	Buick Electra 225 8	John Meyers	Barnett Buick — Burbank, CA
3	16.4821	17	Oldsmobile Toronado 8	Mel Alsbury, Jr.	Bill Romel Oldsmobile — Redondo Beach
4	15.9350	28	Oldsmobile Ninety-Eight 8	Pete Novotny	Mandy Williams Oldsmobile — West Covina
5	15.6230	26	Cadillac Calais 8	Tom Gillum, Jr.	Chamber of Commerce — West Covina, CA
6	15.2610	39	Ford Thunderbird	Stan Seeds	Homer Heller Ford — Escondido, CA

1968 Mobil Economy Run
Los Angeles to New York
Run Terminated at Indianapolis

April 2–7, 1967 — 41 Cars — 2272.25 Miles

Place	M.P.G.	No.	Car	Driver	Entrant
CLASS "A" — COMPACT — SIX CYLINDER					
1	24.0917	9	Rambler Rogue 6	Les Viland	American Motors Dealers
2	23.3179	31	Plymouth Barracuda 6	Jerry Gross	Northside Chrysler-Plymouth — Indianapolis
3	22.8535	4	Chevrolet Corvair 500 6	Gordon Madison	Bill Bader Chevrolet, Inc. — Fontana, CA
4	22.7058	18	Plymouth Valiant 100 6	Bob Cahill	Dick Green Chrysler-Plymouth, Inc. — Detroit
5	22.6487	33	Ford Mustang 2+2 6	Kay Kimes	Robert H. Loud Ford — Pasadena, CA
6	21.9158	6	Dodge Dart 6	Mary Ann Foss	L.A.–Orange County Dodge Dealers Association
7	21.1303	19	Ford Falcon 6	Bill Levy	Homer Heller Ford — Escondido, CA
CLASS "B" — COMPACT — EIGHT CYLINDER					
1	20.0078	40	Plymouth Barracuda 8	Jack Kirkpatrick	Hollywood Chrysler-Plymouth — Hollywood, CA
2	19.6260	1	Chevrolet Chevy II Nova 8	Mary Hauser	Nickey-Bill Thomas — Anaheim, CA
3	19.5429	41	Ford Mustang 2+2 8	Fran Foster	Paul Harvey Ford — Indianapolis, IN
4	19.5408	21	Plymouth Valiant Signet 8	Bob Checkley	Perkins Chrysler-Plymouth — Colorado Springs, CO
5	16.3720	2	Javelin 8	Tommy Thomas	American Motors Dealers
CLASS "C" — INTERMEDIATE SIZE — SIX CYLINDER					
1	22.7680	29	Ford Fairlane 6	Ginny Sims	Tom Frost Ford — Warrenton, VA
2	22.6601	36	Mercury Montego 6	Al Johnson	Bob Estes Lincoln-Mercury — Inglewood
3	21.9822	14	Plymouth Belvedere 6	Carl Diehl	New York Plymouth Dealers Association
4	19.2543	12	Chevrolet Chevelle 300 6	Ed Miller	Nickey Chevrolet — Chicago, IL
CLASS "D" — INTERMEDIATE SIZE — EIGHT CYLINDER					
1	20.5744	16	Dodge Coronet Deluxe 8	Shirley Shahan	L.A.–Orange County Dodge Dealers Association
2	20.5154	39	Plymouth Belvedere 8	Scott Harvey	Greater Kansas City Plymouth Dealers Association
3	19.4283	32	Mercury Montego 8	Ronnie Duman	O'Connor Lincoln-Mercury — Los Angeles
4	19.0656	37	Ford Torino GT 8	Darrell Droke	Allan Bros. Motor Sales — Port Clinton, OH
5	18.5357	25	Buick Special Deluxe 8	John Rich	New York Buick Dealers Group
6	17.8811	23	Chevrolet Chevelle Malibu 8	Tom Evans	Bill McGuire Chevrolet — Buena Park, CA
7	17.0889	8	Oldsmobile F-85 8	Mel Alsbury, Jr.	Mandy Williams Oldsmobile — West Covina
8	16.8924	28	Pontiac Tempest Custom 8	Don Francisco	Bryant Pontiac — Covina, CA
CLASS "E" — LOW PRICE — EIGHT CYLINDER					
1	18.8162	10	Chevrolet Impala 8	Don Royer	Nickey-Bill Thomas — Anaheim, CA
2	18.3332	7	Plymouth Fury III 8	Bill Keller	Washington, D.C., Plymouth Dealers Association
3	17.9562	11	Ford Custom 8	John Allen	Bob Tasca Ford — E. Providence, RI
4	17.3699	20	Chevrolet Caprice 8	C.K. Enoch	Roger Penske Chevrolet — Philadelphia, PA
CLASS "F" — MEDIUM PRICE					
1	18.4883	22	Buick LeSabre Custom 8	Marta Retzlaff	New York Buick Dealers Group
2	18.4639	17	Dodge Polara 8	Jim Wright	L.A.–Orange County Dodge Dealers Association

3	17.9815	35	Pontiac Catalina 8	Ted Block	Bryant Pontiac — Covina, CA
4	17.3538	15	Chrysler Newport 8	Jim Latham	Mackaig & Sons — Monrovia, CA
5	16.9084	3	Mercury Monterey 8	Byron Froelich	Sachs & Son Lincoln-Mercury — Downey
6	16.3248	13	Oldsmobile Delta 88 8	Stan Raymond	Mandy Williams Oldsmobile — West Covina
Disqualified		26	Oldsmobile Delmont 88	Diane Lowe	Bill Romel Oldsmobile — Redondo Beach
CLASS "G" — LUXURY CARS					
1	17.1340	38	Chrysler New Yorker 8	Hart Fullerton	Hart Fullerton Chrysler-Plymouth — West Los Angeles
2	16.7003	30	Oldsmobile Ninety-Eight 8	Mandy Williams	Mandy Williams Oldsmobile — West Covina
3	16.5735	34	Ford Thunderbird 8	Nelson Stacy	Vel's Ford — Torrance, CA
4	16.5270	5	Buick Electra 225 Custom 8	Paula Murphy	New York Buick Dealers Group
5	15.9898	27	Oldsmobile Toronado 8	Tom Gillum, Jr.	Chamber of Commerce — West Covina, CA
6	15.4621	24	Cadillac Coupe DeVille	Pete Novotny	Novotny, Inc. — Los Angeles, CA

Oldsmobile Delmont 88 8 was automatically disqualified on the second day after running out of fuel before reaching the Pueblo, Colorado, refueling station

Mobil Mileage Rally Results

1958 Mobil Mileage Rally
Downtown Los Angeles to Los Angeles' Ambassador Hotel

October 19, 1958 — 37 Cars — 258 Miles

Place	M.P.G.	No.	Car	Driver	Entrant
CLASS "A" — UNDER 750cc					
1	50.5404	15	Fiat 750 4	John Rich	Italiano Motors — Burbank, CA
2	48.6911		BMW Isetta 660 2	Dale Alexander	Santa Monica Imports — Santa Monica
3	42.8557		Citroen 2CV 2	William Lynn	Citroen Cars Corp. — Beverly Hills, CA
4	41.0225		Renault 4CV 4	Barbara Nieland	Nick Pastor Imports — South Gate, CA
5	40.2017		Goggomobil T-400 2	Merv Straley	Merv Straley — El Monte, CA
6	39.8102		Fiat 600 4	Gene Johnson	Bill Hunt — Santa Monica, CA
DNF — Off Course			Datsun 1000 4	J. Gobel	Lloyd Pearson — Pasadena, CA
DNF — Out of Gas			Fiat 500 4	Bill Levy	Italiano Motors — Burbank, CA
DNF — Accident			Lloyd 600 Alexander 2	J. Green	Manhattan Motors, Inc. — Manhattan Beach
CLASS "B" — 750 to 1099cc					
1	43.3548		Renault Dauphine 4	Bill Likes	Yeakel Plymouth — Downey, CA
2	41.6434		Panhard Dyna 2	Bob Murphy	Citroen Cars Corp. — Beverly Hills, CA
3	37.7902		Triumph 4	Bill Rohrer	Cal Sales, Inc. — Gardena, CA
4	37.4098		Fiat 1100 4	Walter McKelvy	Italiano Motors — Burbank, CA
5	34.9593		DKW 3=6 3	Gunther Draheim	Peter Satori Co. Ltd. — Pasadena, CA
6	34.8749		Morris Minor 1000 4	Bob Nichols	Henry S. Perren — Los Angeles, CA
7	31.4605		Skoda S440 4	Bob Russo	Fisher Imports, Inc. — Los Angeles, CA
8	27.5413		Goliath 1100 4	Danny Eames	Goliath Company of Southern CA
CLASS "C" — 1100 to 1499cc					
1	33.3673		Alfa Romeo Sprint 4	Jim Parkinson	Italiano Motors — Burbank, CA
2	33.2711		Volkswagen 4	Al Harmon	Competition Motors — Hollywood, CA
3	32.1293		Simca Aronde Elysee 4	Zaz Haan	Bob Smith, Inc. — Glendale, CA
4	30.6233		Austin A-55 Cambridge 4	Steve Geraghty	Gough Industries Inc. — Compton, CA
5	29.6451		Sunbeam Rapier 4	Joan Fischal	Rootes Motors, Inc. — Los Angeles, CA
6	28.3150		Borgward Isabella 4	Rudy Cleye	Fisher Imports, Inc. — Los Angeles, CA
7	27.8472		Ford Anglia 4	Mary Faulkner	Bob Estes — Inglewood, CA
8	27.8171	29	Peugeot 403 4	Bob Pendergast	Hot Rod Magazine — Hollywood, CA
9	26.4792		Toyopet Crown 4	Boyd Trombella	Art Frost of Glendale — Glendale, CA
Did Not Start			Opel Rekord 4	W. Halstead	Colonial Imports — Glendale, CA
DNF — Out of Gas			Hillman Minx 4	Ginny Sims	Rootes Motors, Inc. — Los Angeles, CA
CLASS "D" — OVER 1500cc					
1	36.0672		Volvo PV-444 4	Lee Hamer	Hamer Motors, Inc. — San Fernando, CA
2	35.3445		Porsche 1600 4	Mary Davis	Precision Motor Cars — Beverly Hills, CA
3	30.8247		Ford Consul 4	Al Cottle	George Byrum Lincoln-Mercury — Burbank

4	30.5419		Vauxhall Victor 4	Bill Luther	Tom Ray Pontiac — Glendale, CA
5	29.3998		Citroen ID-19 4	Bill Jones	Citroen Cars Corp. — Beverly Hills, CA
6	27.9402		Ford Taunus 17M 4	Walt Glassett	Foothill Motors — Pasadena, CA
7	26.2377		Jaguar 3–4 6	Bill Corey	Bill Corey Sports Car Service — Pasadena
8	21.3500		Mercedes-Benz 219 6	Dick Moon	Mercedes-Benz Sales, Inc. — Hollywood
9	14.2083	13	Rolls-Royce Silver Cloud 6	Walter Fife	Len H. Roos — Burbank, CA

As of publication date, some car number information remains unknown.

1959 Mobil Mileage Rally Downtown Los Angeles to Santa Monica Civic Auditorium

November 8, 1959 — 47 Cars — 345.6 Miles

Place	M.P.G.	No.	Car	Driver	Entrant
CLASS "A" — UNDER 750cc					
1	55.2989	2	Fiat 500 2	Bill Levy	Italiano Motors — Burbank, CA
2	49.1636			Fiat Abarth 750 4	John Rich
3	48.6696		NSU Prinz 2	Bill Likes	
4	42.7201	21	Renault 4CV 4	Barbara Nieland	John Green Corp. — Los Angeles, CA
5	38.5074		Citroen 2CV 2	Egon Haag	Citroen Cars Corp. — Beverly Hills, CA
DNF — Mechanical			Subaru 360 4		
CLASS "B" — 750 to 1099cc					
1	46.2458	22	Renault Dauphine 4	John Lance	John Green Corp. — Los Angeles, CA
2	42.8009		Lancia Appia 4	Jim Downey	
3	42.2406		Austin A-40 4	Louis Calvo	
4	41.3395		Morris Minor 1000 4	Bob Nichols	
5	38.6084		Triumph 4	D.R. Feuerhelm	
6	38.4333		Fiat 1100 4	Walt McKelvy	
7	36.6863		DKW 3=6 3	Gunther Draheim	Peter Satori Co. Ltd. — Pasadena, CA
8	34.9105		Goliath 1100 4	Dick Yeo	
9	33.7296		Skoda 440 4	William Doushkess	
CLASS "C" — 1100 to 1599cc					
1	43.6698	29	Volkswagen 4	Lillian Warner	Competition Motors — Hollywood, CA
2	40.9398		Porsche 356A 4	Mary Davis	
3	39.2115		Austin A-55 Cambridge 4	Ocee Ritch	
4	34.3740		Datsun 4	Bob Russo	
5	37.2508		Simca Aronde 4	Betty Shutes	
6	36.9867		Fiat 1200 4	Marty Evelhoch	
7	36.4213		Opel Rekord 4	Milan Micka	
8	34.3796		Morris Oxford 4	Dave Long	
9	34.3274		Peugeot 403 4	John Barrett	
10	34.0687		Vauxhall Victor 4	Bill Luther	
11	32.6259		MG Magnette 4	Clyde Thigpen	
12	32.4539		Volvo PV-544 4	Lee Hamer	Hamer Motors, Inc. — San Fernando, CA
13	32.0248		Hillman Minx 4	Sunnie Baker	
14	25.4490		Toyopet Crown Custom	Don Flitton	
CLASS "D" — OVER 1600cc					
1	35.3425	38	Citroen ID-19 4	Lance Rocke	Citroen Cars Corp. — Beverly Hills, CA
2	31.8593		Ford Taunus 17M 4	Pierce Venable	
3	29.1473		Ford Consul 4	Al Cottle	
4	28.3429		Mercedes-Benz 180 4	Bill Corey	
5	27.6269		Lancia Flaminia 4	Roger Burnett	
6	22.4046		Simca Vedette 4	Jim Bell	
CLASS "E" — SPORTS CARS UNDER 1600cc					
1	56.4975	18	Austin-Healey Sprite 4	Jim Parkinson	Burbank Sports Car Center — Burbank
2	55.8453		Berkeley 60 3	Vern Loveland	
3	52.9660		Fiat Abarth Zagato 4	Art Rene	
4	49.7040	20	Lotus Elite 4	Jay Chamberlain	Jay Chamberlain, Inc. — North Hollywood
5	47.3777	40	Volkswagen Karmann Ghia 4	Ginny Sims	Competition Motors — Hollywood, CA
6	40.1640		Alfa Romeo Sprint Veloce 4	Woody Bell	

7	33.4633	MG-A 4	Ray Pickering	
8	32.3147	Simca Aronde 4	Ruth Doushkess	
DNF—Mechanical			Berkeley 3-60 2	
CLASS "F"—SPORTS CARS OVER 1600cc				
1	33.9321	Triumph TR-3 4	L. Jack Draper	The Auto Works—Granada Hills, CA
2	33.6125	Mercedes-Benz 190SL 6	William C. Brandis	
3	31.9795	Austin-Healey 100 Six	C. W. Vandagriff	

As of publication date, some car numbers, drivers' and entrants' names remain unknown.

1960 Mobil Mileage Rally
New York Coliseum at Columbus Circle to New Rochelle's Glen Island Casino
October 30, 1960—35 Cars—201 Miles

Place	M.P.G.	No.	Car	Driver	Entrant
CLASS "A"—UNDER 750cc					
1	49.18	29	Renault 4CV 4	Henry J. Dahl	Dolphin Motors—Shoreham, NY
2	44.72		BMW 700 2	Gerhard Pilz	BMW—Fadex Corporation—New York City
3	36.37		DAF Deluxe 2	Saul Kaminsky	International Car Exchange
CLASS "B"—750 to 1099cc					
1	55.44		Morris Mini-Minor 4	Dwight N. Holbrook	Road-Air Enterprises, Inc.—Bellerose, NY
2	48.54	22	Renault Dauphine 4	John Lance	Renault, Inc.—New York City
3	46.97		Morris Minor 850 4	Dick Marco	International Motor Sales
4	46.77		Austin Seven 4	Fred L. Herdeen	Hambro Automotive—New York City
5	46.41		Austin A-40 4	Frank W. Studnicha	Road-Air Enterprises, Inc.—Bellerose, NY
6	44.78		Panhard PL-17 2	Michel LeBidois	Bastille Motors Corporation
7	41.06		Ford Anglia 4	Louis J. Goss	Julius E. Mayer
8	39.89		Morris Minor 1000 4	L. Richardson	Flushing Imported Motors—Flushing, NY
9	39.87		Saab 96 3	Robert H. Wehman	Saab Motors, Inc.—New York City
10	37.22		Triumph Herald 4	Michael Jackling	Standard-Triumph Motor Co.—New York City
11	36.67		Borgward Hansa 4	Murray W. Sperling	Foreign Motor Distribution Company
CLASS "C"—1100 to 1599cc					
1	40.55		Fiat 1200 4	Ray Cuomo	Franklin D. Roosevelt, Jr.—New York City
2	38.03		Peugeot 403 4	Ken Norton	Don Kirkpatrick
3	34.86		Volvo PV 544 4	Lee Hamer	Auto Imports, Inc.
4	34.77		Volkswagen 4	Maurice McAuliffe	Maurice McAuliffe
Did Not Finish			Opel Caravan 4	Claude	Gaylin Buick, Inc.—Union Township, NJ
CLASS "D"—OVER 1600cc					
1	35.98		Citroen ID-19 4	J.O. Berry	Citroen Cars Corp.—New York City
2	28.75		Ford Consul 4	Nat Scolnick	Scolnick Brothers English Ford
3	25.26		Mercedes Benz 190 4	John Macron	Milano Motors, Inc.—Union City, NJ
4	24.57		Fiat 2100	John Norwood	Franklin D. Roosevelt, Jr.—New York City
CLASS "E"—SPORTS CARS UNDER 1600cc					
1	62.29	6	NSU Sport Prinz 4	Henry Fanelli	Transcontinental Motors, Inc.
2	51.93		Vespa 400 2	Lawrence Friedman	Vescony, Inc.
3	50.13	11	Renault Caravelle 4	Serge Meunier	Renault, Inc.—New York City
4	42.83		Volkswagen Karmann-Ghia 4	Konrad Ortner	Riverbrook Motor Corp.—Stamford, Conn.
5	42.05		Triumph Herald 4	Stephen Wilder	Standard-Triumph Motor Co.—New York City
6	40.58		Austin-Healey Sprite 4	Mervyn A. Hegarty	Hegarty Motors—Bronx, NY
7	39.80		Fiat Abarth 850 4	George Spagna	Franklin D. Roosevelt, Jr.—New York City
8	35.86		Borgward Isabella 4	Kenneth E. Tarsia	Kenneth E. Tarsia
9	34.66		Borgward Hansa Tiger 4	David T. Latsha	Foreign Motor Distribution Company
10	33.11		Porsche 356B 4	Thomas Stamp	Blanchard Automobile—Stamford, CT
CLASS "F"—SPORTS CARS OVER 1600cc					
1	32.23	26	Triumph TR-3 4	George Waltman	North Country Motors—Great Neck, NY
2	25.80		Austin-Healey 3000 6	Robert Herman	Road-Air Enterprises, Inc.—Bellerose, NY

As of publication date some car numbers and entrants' location information remains unknown.

1961 Mobil Mileage Rally Downtown Los Angeles Returning to Downtown Los Angeles

November 12, 1961—34 Cars—300 Miles

Place	M.P.G.	No.	Car	Driver	Entrant
CLASS "A"—FOUR PASSENGER UNDER 1000cc					
1	51.399	7	Morris Mini-Minor 4	Jim Parkinson	Burbank Sports Cars—Burbank, CA
2	50.34		Renault Gorgini	John Lance	Renault of Southern California
3	49.699		Fiat 600D 4	John Rich	Rich Motors, Inc.—Glendale, CA
4	45.72		Austin 850 4	Chick Vandagriff	Hollywood Sports Cars—Hollywood, CA
5	41.88		Austin A-40 4	Jim Alexander	Hollywood Sports Cars—Hollywood, CA
6	41.76	15	Triumph Herald 4	Jim Groves	The Auto Works—Granada Hills, CA
Withdrawn			Saab 96	John Halliwell	John Halliwell
CLASS "B"—FOUR PASSENGER 1000 to 1600cc					
1	40.384	25	Volkswagen 4	Mary Hauser	Precision Motor Cars—Beverly Hills, CA
2	35.42		Volvo PV 544	Lee Hamer	Hamer Imports—Mission Hills, CA
3	34.027		Simca 4	Danny Eames	20th Century Imports
4	32.068		Austin A-55 Cambridge 4	Bill Stucker	Hollywood Sports Cars—Hollywood, CA
5	31.656		MG Magnette 4	Doane Spencer	Hollywood Sports Cars—Hollywood, CA
Withdrawn			Datsun Bluebird 4	Bob Russo	Nissan Motors—Los Angeles, CA
CLASS "C"—FOUR PASSENGER OVER 1600cc					
1	35.68		Citroen ID-19 4	Paula Murphy	The Auto Works—Granada Hills, CA
2	33.03		Peugeot 404 4	Cliff Rucker	Burbank Imports—Burbank, CA
3	25.21		Fiat 2100 6	Felipe Turich	Rich Motors, Inc.—Glendale, CA
Withdrawn			Facel Vega Facellia 4	Lee Whitely	Peter Satori, Inc.—Pasadena, CA
CLASS "D"—TWO PASSENGER UNDER 1600cc					
1	59.76		NSU Sport Prinz 4	Bill Levy	Rich Motors, Inc.—Glendale, CA
2	53.65		Fiat Abarth 747 4	Mary Evelhock	Italiano Motors = Burbank, CA
3	51.04		Lotus Elite Series II 4	Fran Foster	Rich Motors, Inc.—Glendale, CA
4	48.36		MG Midget 4	Stanley Paul	Burbank Sports Cars—Burbank, CA
5	46.67		Austin-Healey Sprite 4	Ronnie Bucknum	Hollywood Sports Cars—Hollywood, CA
6	43.75	18	Triumph Herald 4	Duane Feuerhelm	The Auto Works—Granada Hills, CA
7	41.76		Volkswagen Karmann Ghia 4	Jack McAfee	Jack McAfee Motors—Sherman Oaks, CA
8	40.49		Renault Caravelle 4	Jane Nerpel	Jane Nerpel
CLASS "E"—TWO PASSENGER OVER 1600cc					
1	38.99	26	Triumph TR-3 4	Jack Draper	The Auto Works—Granada Hills, CA
2	38.15		Porsche 356B 4	Mary Davis	Precision Motor Cars—Beverly Hills, CA
3	37.89		Volvo P1800 4	Bill Carroll	Hamer Imports—Mission Hills, CA
4	36.98		MG-A 4	Chuck Wood	Burbank Sports Cars—Burbank, CA
5	35.77		Triumph TR-4 4	John Barrett	Burbank Imports—Burbank, CA
6	34.35		Sunbeam Alpine 4	Barbara Nieland	Nick Pastor Imports—South Gate, CA
7	25.74		Jaguar XK-E 6	Bill Corey	Bill Corey—Pasadena, CA
8	23.90		Austin-Healey 3000 6	Roy Wallace	Henry S. Perren—Los Angeles, CA
Withdrawn			Sabra 4	Jay Mintz	Sabra Sales Corporation

As of publication date, some car numbers and entrants' location information remains unknown.

Chapter Notes

Chapter 1

1. John A. Heitmann, October 21, 2009, "Los Angeles to Camp Curry Yosemite Economy Runs"; October 22, 2009, "Los Angeles to Camp Curry Yosemite Economy Runs"; and October 23, 2009, "1925 L.A. to Yosemite Economy Run—Star and Flint Vehicles Dominate Run and "More on the L.A. to Camp Curry Yosemite Economy Runs," *Automobile and American Life Blogspot*, http://automobileandamericanlife.blogspot.com.
2. "'Stingy Run' Record Set," *Los Angeles Times*, January 19, 1936.
3. Ibid.
4. Ibid.
5. Ibid.
6. Ibid.
7. Ibid.
8. Ibid.
9. "Thrift Run Lauded by Studebaker Official," *Los Angeles Times*, January 26, 1936.
10. Graham-Paige Motors Corporation, print advertisement, *Collier's*, May 2, 1936, 35.
11. The Studebaker Corporation, print advertisement, *The Saturday Evening Post*, April 11, 1936.
12. Pontiac Motor Company, Division of General Motors Corporation, print advertisement, *Motor*, March 1936, 82.
13. "1937 Automotive Economy News," brochure, Pontiac Motor Company, Division of General Motors Corporation, Vol. 1, No. 1, 2.
14. Gilmore Oil Company, print advertisement, *Los Angeles Times*, January 19, 1936.
15. "Amazing Results of the Gilmore Yosemite Run," *Gilmore Cub*, Gilmore Oil Company, February–March 1936, 2; "Fuel Consumption on 30 Popular Cars," *Motor*, April 1936; "Graham Wins the Yosemite Economy," *Motor West*, February 1936.
16. "Graham Wins the Yosemite Economy," *Motor West*, February 1936.
17. "Economy Run Rules Now Perfected," *Oakland Tribune*, December 5, 1936.
18. "Economy Run Next Thursday," *Oakland Tribune*, January 3, 1937.
19. "Graham Economy Outstrips All Other Contestants," *Graham Supercharger News*, Volume 1, Number 2, January 1937, Graham-Paige Motors Corporation, 2–3.
20. John H. Weiser, "Yosemite Economy Run Proves Sensational," *Los Angeles Times*, January 17, 1937.
21. Ibid.
22. Ibid.
23. Ibid.; "Graham Repeats in Yosemite Contest," *Motor West*, February 1937, 8.
24. "Graham Economy Outstrips All Other Contestants," *Graham Supercharger News*, Volume 1, Number 2, January 1937, Graham-Paige Motors Corporation, 2–3.
25. Ibid.
26. Ibid.
27. "Gilmore–Yosemite Test Proves Red Lion Economy," *Gilmore Cub*, Gilmore Oil Company, February 1937, 2.
28. "23 Cars Complete Grueling Contest," *San Francisco Call Bulletin*, January 17, 1937.
29. "Gilmore–Yosemite Test Proves Red Lion Economy," *Gilmore Cub*, Gilmore Oil Company, February 1937, 2.
30. John H. Weiser, "Yosemite Economy Run Proves Sensational," *Los Angeles Times*, January 17, 1937.
31. Graham-Paige Motors Corporation, print advertisement, *Los Angeles Times*, January 17, 1937.
32. Graham-Paige Motors Corporation, print advertisement, *Life*, February 15, 1937, 65.
33. Gilmore Oil Company, print advertisement, *Los Angeles* Times, January 17, 1937.
34. "Graham Again Wins Economy Test Run," *Automobile Topics*, January 18, 1937; "Supercharged Graham Wins Economy Run," *Motor*, February 1937; "Graham Repeats in Yosemite Contest," *Motor West*, February 1937, 8.
35. "Economy Run Record Set," *Los Angeles Times*, January 23, 1938.
36. "New Economy Mark in Yosemite Jaunt," *Motor West*, February 1938.
37. Lynn J. Rogers, "Automotive High Lights," *Los Angeles Times*, January 23, 1938.
38. "New Economy Mark in Yosemite Jaunt," *Motor West*, February 1938.
39. "Economy Run Record Set," *Los Angeles Times*, January 23, 1938.
40. Ford Motor Company, print advertisement, *Los Angeles* Times, January 23, 1938.
41. Lincoln Motor Company, print advertisement, *Los Angeles* Times, January 23, 1938.
42. "Yosemite Winners Gilmore Economy Run," *Ford News*, Ford Motor Company, February 1938.
43. Gilmore Oil Company, print advertisement, *Los Angeles Times*, January 23, 1938.
44. Champion Spark Plug Company, print advertisement, *Life*, June 27, 1938, 2.
45. "Yosemite Takes Gilmore Yosemite Economy Run," *Automobile Topics*, January 24, 1938; "Red Lion Averages 22.88," *Gilmore Cub*, March 1938, 2; "New Economy Mark in Yosemite Jaunt," *Motor West*, February 1938.
46. Lynn J. Rogers, "Automotive High Lights," *Los Angeles Times*, January 23, 1938.
47. Lynn J. Rogers, "Economy Run Records Set in Face of Rough Going," *Los Angeles Times*, January 15, 1939.
48. Lynn J. Rogers, "Automotive High Lights," *Los Angeles Times*, January 15, 1939.
49. "Studebaker Wins in Gilmore-Yosemite," *Motor West*, February 1939, 11.
50. Lynn J. Rogers, "Automotive High Lights," *Los Angeles Times*, January 15, 1939.
51. "Thirty Yosemite Economy Run Cars Await Go Signal," *Los Angeles Times*, January 1, 1939.
52. Lynn J. Rogers, "Economy Run Records Set in Face of Rough Going," *Los Angeles Times*, January 15, 1939.
53. "Studebaker Gets High Mileage," *Los Angeles Times*, January 22, 1939.
54. "Studebaker Wins in Gilmore-Yosemite," *Motor West*, February 1939, 11.
55. "Officials And Drivers All Good Sportsmen," *Los Angeles Times*, January 22, 1939.
56. Lynn J. Rogers, "Economy Run Records Set in Face of Rough Going," *Los Angeles Times*, January 15, 1939.
57. "Studebaker, Packard Get Yosemite Awards," *Los Angeles Examiner*, January 15, 1939.
58. "Studebaker Wins in Gilmore-Yosemite," *Motor West*, February 1939, 11.
59. "Nash Wins Class Honors in Annual Stingy Run," *Los Angeles Times*, January 22, 1939.
60. "Red Lion Averages 21.9 Miles," *Gilmore Cub*, Gilmore Oil Company, February 1939, 2.
61. "Studebaker Wins in Gilmore-Yosemite," *Motor West*, February 1939, 11.
62. "Officials and Drivers All Good Sportsmen," *Los Angeles Times*, January 22, 1939.
63. "Studebaker Wins in Gilmore-Yosemite," *Motor West*, February 1939, 11.
64. "Studebaker Gets High Mileage," *Los Angeles Times*, January 22, 1939.
65. "Studebaker Gets High Mileage," *Los Angeles Times*, January 22, 1939.
66. The Studebaker Corporation, print advertisement, *Los Angeles Times*, January 15, 1939.
67. The Studebaker Corporation, print advertisement, *Los Angeles Times*, January 22, 1939.
68. The Studebaker Corporation, print advertisement, *The Saturday Evening Post*, March 11, 1939.
69. The Studebaker Corporation, print advertisement, *Motor*, February 1939.
70. "Studebaker Wins Yosemite Sweepstakes!" *The Studebaker Wheel*, Studebaker Corporation, May 1939.
71. Nash Motors Division, Nash-Kelvinator Corporation, "Night Flight," print advertisement, circa 1939.

72. "Here Begins a New Adventure," 1940 model year brochure, Nash Motors Division, Nash-Kelvinator Corporation, F-3447C, 1940, 14.

73. Oldsmobile Division, General Motors Corporation, print advertisement, *Chicago Sunday Tribune*, June 4, 1939.

74. Gilmore Oil Company, print advertisement, *Los Angeles Times*, January 15, 1939.

75. Gilmore Oil Company, "Well! After This Use Red Lion!" print advertisement, circa 1939.

76. "Gilmore Run," *Automobile Trade Journal*, February 1939; "Red Lion Averages 21.9 Miles," *Gilmore Cub*, February 1939, 2; "Bad Weather Hit Economy Run Contestants," *Motor Age*, February 1939, 36; "Studebaker Wins in Gilmore-Yosemite," *Motor West*, February 1939, 11.

77. "Economy Run Won with Champions!" *The Spirit of a Champion*, Champion Spark Plug Company, February 1939, 8.

78. Don Butler, *The History of Hudson* (Sarasota, FL: Crestline Publishing, 1982), 226.

79. "Yosemite Stingy Run Won By Studebakers," *Motor West*, February 1940, 14.

80. "Economy Run Records Fall," *Los Angeles Times*, January 14, 1940.

81. Peter Winnewisser, "Watching the Fords Go By with Peter Winnewisser," *Old Cars Weekly*, June 2002.

82. "Yosemite Stingy Run Won By Studebakers," *Motor West*, February 1940, 14.

83. The Studebaker Corporation, print advertisement, *Los Angeles* Times, January 14, 1940.

84. Ford Motor Company, print advertisement, *Los Angeles* Times, January 14, 1940.

85. Lincoln Motor Company, print advertisement, *Los Angeles Times*, January 14, 1940.

86. Gilmore Oil Company, print advertisement, *Los Angeles Times*, January 14, 1940.

87. The Studebaker Corporation, "Enjoy this distinction and save 10% to 25% every mile," print advertisement, circa 1940.

88. "Studebaker Scores Clean Sweep in Gilmore Economy Run," and "How *you* can get better gas mileage," *The Studebaker Wheel*, Studebaker Corporation, February 1940, 2.

89. "See America First ... In America's Stand-Out Car," 1941 model year brochure, The Studebaker Corporation, circa 1940, 11.

90. Nash Motors Division, Nash-Kelvinator Corporation, "The One You've Always Wanted," print advertisement, circa 1940.

91. "What Happened to Me ... When I Changed to Nash," 1940 model year brochure, Nash Motors Division, Nash-Kelvinator Corporation, F-3494, 1940, 8.

92. Ford Motor Company, "Count the Cylinders, Check the Features, Compare the Price!" print advertisement, circa 1940.

93. "Ford and Lincoln-Built Cars Set New Records in Rain and Sleety in the Famous Gilmore-Yosemite Economy Run," *Ford News*, Ford Motor Company, February 1940.

94. Champion Spark Plug Company, "Put Your Car on the Rainbow Trail with New Champion Spark Plugs," print advertisement, circa 1940.

95. *Gilmore Graphic*, Gilmore Oil Company, February 1940, 4.

96. "Gilmore Yosemite Economy Run," *Automobile Trade Journal*, February 1940; "Extra! Gilmore-Yosemite Run Results," *Gilmore Cub*, Gilmore Oil Company, February 1940, 1; "Studebaker Takes Honors in Yosemite Economy Run," *Motor Age*, February 1940, 31; "Yosemite Stingy Run Won By Studebakers," *Motor West*, February 1940, 14.

97. "Red Lion Gasoline Breaks Mileage Records in 1940 Gilmore Yosemite Run," *Gilmore Graphic*, January 1940, 1.

98. "History Is Made!!" *The Studebaker News* 19, No. 2, The Studebaker Corporation, January 19, 1940, 1.

99. "Nash Wins '2 Firsts' in Gilmore Economy Run!" *Nash Merchandising News* 6, No. 1, Nash Motors Division, Nash-Kelvinator Corporation, February 1940, 1.

100. "Ford Again Demonstrates Better Economy and Stamina Than Chevrolet or Plymouth," *Do You Know? 1940 Confidential Bulletin*, No. 4 1940 Series, Sales Development Corporation, 1940, 1; "Lincoln-Zephyr Winner in Its Class for Third Consecutive Year," *The Class Teller Confidential Bulletin*, 1940 Series, Sales Development Corporation, 1940, 1; "Mercury Wins," *Confidentially Yours Bulletin*, 1940 Series No. 1, Sales Development Corporation, 1940, 1.

101. "Test Proves High Gasoline Economy," *Motor Age*, February 1941.

102. "First Entries Named for 1940 Gilmore Economy Run," *Oakland Tribune*, December 15, 1940.

103. "Gilmore Test Finished in Triumph by 23 Cars," *Los Angeles Times*, January 19, 1941.

104. "Test Proves High Gasoline Economy," *Motor Age*, February 1941; "Lincoln One-Two in Sweepstakes," *Motor West*, February 1941, 1.

105. "Gilmore Test Finished in Triumph by 23 Cars," *Los Angeles Times*, January 19, 1941.

106. Ibid.

107. "Economy Run Attracts 500," *Arizona Republic*, January 10, 1941.

108. Lynn J. Rogers, "Automotive Highlights and Facts," *Los Angeles Times*, January 19, 1941.

109. Ibid.

110. "Test Proves High Gasoline Economy," *Motor Age*, February 1941.

111. Ibid.

112. "Gilmore-Grand Canyon Run," Studebaker Confidential Sales Bulletin No. 621, January 21, 1941.

113. Ford Motor Company, print advertisement, *Los Angeles Times*, January 19, 1941.

114. The Studebaker Corporation, "Studebaker Triumphs Again in Gilmore-Grand Canyon Run," print advertisement, circa 1941.

115. The Studebaker Corporation, print advertisement, *Los Angeles Times*, January 19, 1941.

116. Ibid.

117. Nash Motors Division, Nash-Kelvinator Corporation, "Nash Proves Its Amazing Economy!" print advertisement, circa 1941.

118. Willys-Overland Motors, Inc., "Averaging 29.06 Miles Per Gallon!" print advertisement, circa 1941.

119. "The People's Car," Willys-Overland Motors, Inc., 1942 model year brochure circa 1941, 4.

120. Willys-Overland Motors, Inc., "Again! Willys Americar Beats All Other Cars," dealership showroom poster, circa 1941.

121. Gilmore Oil Company, print advertisement, *Los Angeles Times*, January 19, 1941.

122. Hudson Motor Car Company, print advertisement, *The Saturday Evening Post*, March 22, 1941, 53; Hudson Motor Car Company, print advertisement, *The Saturday Evening Post*, April 12, 1941, 71; Hudson Motor Car Company, "All Cars Alike? Not on Your Tintype!" print advertisement, circa 1941.

123. Hudson Motor Car Company, 1941 model year features brochure, March 1, 1941, 25.

124. Nash Motors Division, Nash-Kelvinator Corporation, "It's Talk Like This — That's Switching Thousands to Nash!" print advertisement, circa 1941.

125. Nash Motors Division, Nash-Kelvinator Corporation, "Nash Proves Its Amazing Economy!" print advertisement, circa 1941

126. Nash Motors Division, Nash-Kelvinator Corporation, 1941 model year brochure, F-3570, 1941, 7.

127. The Studebaker Corporation, print advertisement, *Life*, February 17, 1941, 1.

128. The Studebaker Corporation, "Studebaker President Eight — Commander Six Spring and Summer Series 1941," 1941 model year brochure, 8; "1942 Studebakers — The President, The Commander," 1942 model year brochure, circa 1941, 8.

129. The Studebaker Corporation, "1942 Studebaker Champion," 1942 model year brochure, circa 1941, 15.

130. Ford Motor Company advertisement, *Automobile Topics*, January 20, 1941.

131. "Ford Wins Annual Economy Run," *Ford News*, Ford Motor Company, March 1941, 65.

132. Champion Spark Plug Company, "Choose Your Spark Plugs for Greater Mileage and You'll Choose Champions!" print advertisement, circa 1941.

133. "Overdrives Okay," *Automobile Topics*, January 20, 1941, 6; "Lincoln Wins Gilmore Fuel Contest," *Motor*, February 1941; "Test Proves High Gasoline Economy," *Motor Age*, February 1941; "Lincoln One-Two in Sweepstakes," *Motor West*, February 1941, 1.

134. "Tricks You Can Learn from the World's Stingiest Drivers," *Popular Science*, May 1941, 133.

135. "Studebaker Wins Again!! Takes High Honors in 4th Consecutive Gilmore Run!" *The Studebaker News*, Studebaker Corporation, January 20, 1941, 1.

136. "Gilmore-Grand Canyon Run," Studebaker Confidential Sales Bulletin No. 621, January 21, 1941.

137. "Ford Wins in Its Price Class Gilmore-Grand Canyon Economy Run," *Do You Know 1941 Confidential Bulletin*, No. 5, 1941 Series, Sales Development Corporation, 1941, 1.

138. "1-2-3 For Champions in 1941 Gilmore Economy Sweepstakes," *The Spirit Of Champion*, Champion Spark Plug Company, February 1941, 2.

139. "Overdrives Okay," *Automobile Topics*, January 20, 1941, 6.

140. "Gilmore Run Set for January 8," *The Seattle Times*, December 7, 1941.

141. Gilmore Oil Company, "The 1942 Gilmore Grand Canyon Run Postponed," print advertisement, circa 1941.

142. "Merger of G.P. and Gilmore Announced Here," *Torrance Herald*, March 15, 1945, 2-B.

Chapter 2

1. "Mercury Wins Economy Event," *Motor Trend*, April 1950, 6.

2. Contest Board of the American Automobile Association, "Official Rules 1950 Mobilgas Grand Canyon Run," circa 1949.

3. Lynn Rogers, "Grand Canyon Run Proves Economy of '50 Automobiles," *Los Angeles Times*, March 12, 1950.

4. Contest Board of the American Automobile Association, "Official Rules 1950 Mobilgas Grand Canyon Run," circa 1949.

5. Floyd Clymer, "Covering the Run with Clymer," *The 1950 Mobilgas Grand Canyon Economy Run* (Floyd Clymer, 1950), 8.

6. Lynn Rogers, "Grand Canyon Run Proves Economy of '50 Automobiles," *Los Angeles Times*, March 12, 1950.

7. "Summary of Remarks by C.S. Beesemyer, executive vice-president, General Petroleum Corporation," 1950 Mobilgas Grand Canyon Run Press Release, General Petroleum Corporation, for release February 26, 1950.

8. "Summary of Remarks by A.C. Pillsbury, Regional Director of the Contest Board of the American Automobile Association," 1950 Mobilgas Grand Canyon Run Press Release, General Petroleum Corporation, for release February 26, 1950.

9. Floyd Clymer, "Highlights of the Grand Canyon Contest," *The 1950 Mobilgas Grand Canyon Economy Run* (Floyd Clymer, 1950), 34.

10. Socony-Vacuum Oil Company, Inc., "Mobilgas Grand Canyon Run Proves," print advertisement, circa 1950.

11. Mercury Division, Ford Motor Company, "26.52 miles per gallon Mercury Wins Sweepstakes in Mobilgas Grand Canyon Run," 1950 model year brochure.

12. "Winning Driver Tells How to Save Gasoline," *Los Angeles Examiner*, March 5, 1950.

13. "Autos: Test Run," *Time*, February 27, 1950, "Mercury Wins Economy Event," *Motor Trend*, April 1950, 6; "Mercury High in Economy," *Motor West*, March 1950, 22; "Economy Run Details," *Road and Track*, April 1950, 12; "Grand Canyon Economy Run," *Speed Age*, May 1950, 24.

14. "Grand Canyon Run —1950," *The Motor*, February 22, 1950, 102.

15. "Winning Ford Six," *Ford Times*, Ford Motor Company, June 1950, 17; "Sweepstakes Champion, *Lincoln-Mercury News,* Ford Motor Company, May–June 1950, 27.

16. "Summary of Remarks by A.C. Pillsbury, Regional Director of the Contest Board of the American Automobile Association," 1950 Mobilgas Grand Canyon Run Press Release, General Petroleum Corporation, for release February 26, 1950.

17. "AAA Okays 1951 Run," *Motor West*, March 1950, 22.

18. "Stingy Run Proves '51 Motor Economy," *Los Angeles Times*, March 18, 1951.

19. AAA Contest Board, American Automobile Association, "Official Rules 1951 Mobilgas Economy Run Final Edition," circa 1950, 11.

20. "First Car Picked for Economy Run," *Nash News*, Nash Motors Division, Nash-Kelvinator Corporation, March 1951, 2.

21. AAA Contest Board, American Automobile Association, "Official Rules 1951 Mobilgas Economy Run Final Edition," circa 1950, 17.

22. William V. Wright, "The Economy Run: A Report from One of Caltech's Student Observers," *Caltech Engineering & Science*, April 1951, 30; "Tech Students Monitor Mobilgas Economy Run," *California Tech*, California Institute of Technology, March 30, 1951, 1.

23. AAA Contest Board, American Automobile Association, "Official Rules 1951 Mobilgas Economy Run Final Edition," circa 1950, 17.

24. Merlin Blais, "Economy Run Offers Gas-Saving Ideas," *The Oregonian*, March 18, 1951.

25. "Get All the Mileage Your Car Can Deliver," *The Flying Red Horse*, Socony-Vacuum Oil Company, Inc., May–June 1951, 12.

26. "Record Field to Compete in Economy Run," *Automotive News*, February 19, 1951.

27. AAA Contest Board, American Automobile Association, "Official Rules 1951 Mobilgas Economy Run Final Edition," circa 1950, 13.

28. Slim Barnard, "Automotive — Slim's Pickin's," *Los Angeles Examiner*, March 18, 1951; "Packard '200' Wins Its Class in Gas Economy Run," *Packard News*, Packard Motor Car Company, April 1951.

29. "Stingy Run Proves '51 Motor Economy," *Los Angeles Times*, March 18, 1951.

30. Slim Barnard, "Automotive — Slim's Pickin's," *Los Angeles Examiner*, March 18, 1951.

31. "Stingy Run Proves '51 Motor Economy," *Los Angeles Times*, March 18, 1951.

32. Slim Barnard, "Automotive — Slim's Pickin's," *Los Angeles Examiner*, March 18, 1951; Merlin Blais, "Economy Run Offers Gas-Saving Ideas," *The Oregonian*, March 18, 1951; Bob Finlay, "High Factory Interest Marks Economy Run," *Automotive News*, March 19, 1951, 2; "Lincoln Performance 1949–1951," *Collectible Automobile*, October 1990.

33. Slim Barnard, "Automotive — Slim's Pickin's," *Los Angeles Examiner*, March 18, 1951.

34. AAA Contest Board, American Automobile Association, "Official Rules 1951 Mobilgas Economy Run Final Edition," circa 1950, 34.

35. Lincoln Division, Ford Motor Company, "Smashing Triumph!" print advertisement, circa 1951.

36. Mercury Division, Ford Motor Company, "Mercury Wins Again in Mobilgas Economy Run!" print advertisement, circa 1951, The Studebaker Corporation; "Three Studebakers finished 1st, 2nd and 3rd in actual miles per gallon," print advertisement, circa 1951.

37. Socony-Vacuum Oil Company, Inc., "All the Gasoline Mileage Your Car Can Deliver," print advertisement, circa 1951.

38. Ibid.

39. Mercury Division, Ford Motor Company, "Nothing Like It for Staying Power!— Nothing Like It for Plus Value!— Nothing Like It for Going Places!" print advertisements, circa 1951.

40. "Strength for Thousands of Tomorrows..."; "And for the Years to Come..."; "Over the Miles and Through the Years...," print advertisements, circa 1951.

41. Nash Motors Division, Nash-Kelvinator Corporation, "First in Miles to the Gallon!" print advertisement, circa 1951.

42. Kaiser-Frazer Sales Corporation, "All-Round Proof! Kaiser-Frazer's Henry J Is America's All-Round Car!" print advertisement, circa 1951.

43. The Studebaker Corporation, "Top Value of the Top 4 Lowest Price Cars..." and "New Studebaker Commander V-8," print advertisements, circa 1951.

44. Borg-Warner, "First 15 Cars in Mobilgas Economy Run Equipped with B-W Overdrive!" print advertisement, circa 1951.

45. "Autos: Road Test," *Time*, March 19, 1951; "Kilted Car Competition Lincoln — Nash Scotch Gas to Top Field of 32 Cars in Mobilgas Economy Run," *Motorsport*, June 1951, 16; "Lincoln Wins Economy Run," *Motor West*, April 1951, 16; Stanley Gardner, "Willow Run Monopolizes," *Speed Age*, July 1951, 22; "Get All the Mileage Your Car Can Deliver," *The Flying Red Horse*, May–June 1951, 12; Burgess H. Scott, "Ford Takes the Run Again," *Ford Times*, Ford Motor Company, September 1951, 48.

46. "Drivers in 1951 Mobilgas Economy Run Tell Ways to Get Best Economy," *Motor Trend*, June 1951, 16; Les Viland, "How I Save Gas ... By This Year's Stingiest Driver," *Popular Science*, June 1951, 168; Robert M. Hodesh, "Lincoln and Mercury Thrive on Economy," *Lincoln-Mercury Times*, Ford Motor Company, May–June 1951, 35.

47. "Drivers in 1951 Mobilgas Economy Run Tell Ways to Get Best Economy," *Motor Trend*, June 1951, 16.

48. Floyd Freel, "32 Cars Compete in Annual Trek to Grand Canyon," *Hollywood Citizen-News*, March 20, 1951.

Chapter 3

1. "1952 Mobilgas Economy Run to Sun Valley," *Automobile Topics*, March 1952, 34.

2. "Gallon Stretchers," *Motorsport*, July 1952, 10.

3. Contest Board of the American Automobile Association, "Official Rules 1952 Mobilgas Economy Run, circa 1951," 25.

4. "A Run for Your Money," *The Flying Red Horse*, May–June 1952, 2.

5. General Petroleum Corporation, print advertisement, *Automotive News*, April 14, 1952, 21.

6. "To Be a 1000 Mile Event," *Automobile Topics*, March 1952, 34.

7. Contest Board of the American Automobile Association, "Official Rules 1952 Mobilgas Economy Run, circa 1951, 8.

8. Jim Abbott and Bob Kiple, "Mobilgas Economy Run," *Speed Age*, July 1952, 72.

9. "Economy Run Slated Apr. 14–16 To Sun Valley," *Automotive News*, March 1952.

10. Contest Board of the American Automobile Association, "Official Rules 1952 Mobilgas Economy Run, circa 1951, 14.

11. Russ Catlin, "The Run As I Saw It," *1952 Mobilgas Economy Run* (Floyd Clymer, 1952), 36.

12. "Two Enter Mobilgas Test Run," *Torrance Herald*, March 20, 1952.

13. Jim Abbott and Bob Kiple, "Mobilgas Economy Run," *Speed Age*, July 1952, 72.

14. Contest Board of the American Automobile Association, "Official Rules 1952 Mobilgas Economy Run, circa 1951, 16.

15. "45 Caltech Students Officiate as Observers for AAA," *Los Angeles Examiner*, April 27, 1952.

16. Contest Board of the American Automobile Association, "Official Rules 1952 Mobilgas Economy Run," circa 1951, 22.

17. Russ Catlin, "The Run As I Saw It," *1952 Mobilgas Economy Run* (Floyd Clymer, 1952), 36.

18. Bill Stroppe, "Winning the Economy Run," *Motor Trend*, July 1952, 17.

19. "A Run for Your Money," *The Flying Red Horse*, May–June 1952, 2.
20. Russ Catlin, "The Run As I Saw It," *1952 Mobilgas Economy Run* (Floyd Clymer, 1952), 36.
21. "Economy run shorts," *Los Angeles Mirror*, April 25, 1952.
22. Bob Williams, "Economy Run Greeted in Capitol Rites," *Salt Lake Tribune*, April 15, 1952.
23. Keith Monroe, "The Long Long Run for the Big Big Money," *True's Automobile Yearbook* No. 2, 1953, 17.
24. Russ Catlin, "The Run As I Saw It," *1952 Mobilgas Economy Run* (Floyd Clymer, 1952), 36.
25. Contest Board of the American Automobile Association, "1952 Mobilgas Economy Run Route Description and Map," circa 1952, 12.
26. Jim Abbott and Bob Kiple, "Mobilgas Economy Run," *Speed Age*, July 1952, 72; Slim Barnard, "1952 Economy Run Hailed as One of Best," *Los Angeles Examiner*, April 27, 1952.
27. "Sick Driver Emerges as Sweepstakes Hero," *Los Angeles Mirror*, April 25, 1952.
28. Bruce Kerr, "26 New Cars Complete Grueling 3-day Run from Los Angeles to Sun Valley," *Los Angeles Daily News*, April 29, 1952.
29. "Out of Gas — Still Wins," *Los Angeles Mirror*, April 25, 1952.
30. Slim Barnard, "1952 Economy Run Hailed as One of Best," *Los Angeles Examiner*, April 27, 1952.
31. Mercury Division, Ford Motor Company, "Again Mercury Wins Sweepstakes," print advertisement, circa 1952, Lincoln Division, Ford Motor Company, "Again Lincoln Wins Class G," print advertisement, circa 1952.
32. The Studebaker Corporation, "Studebaker Finished 1st and 2nd in Actual Miles Per Gallon," print advertisement, circa 1952.
33. Ford Division, Ford Motor Company, "Ford Wins Again!" print advertisement, circa 1952.
34. Chrysler Sales Division, Chrysler Corporation, "180HP Chrysler V-8 First in 3 Classes in Economy Run," print advertisement, circa 1952.
35. Socony-Vacuum Oil Company, Inc., "3rd Straight Year of Proof," print advertisement, circa 1952.
36. Mercury Division, Ford Motor Company, "Can't Be Beaten for Economy — Won't Be Dated for Years!" print advertisement, circa 1952.
37. Mercury Division, Ford Motor Company, print advertisement, *Motor Trend*, July 1952, 3.
38. "The Mobilgas Economy Run from Los Angeles to Sun Valley, Idaho," *Automobile Topics*, May 1952, 19; "Gallon Stretchers," *Motorsport*, July 1952, 10; Bill Stroppe, "Winning the Economy Run," *Motor Trend*, July 1952, 17; "Mercury Wins Economy Run," *Motor West*, May 1952, 11; Andrew R. Boone, "Champion Gas Savers Boost Mileage 50%," *Popular Science*, June 1952, 148; Jim Abbott and Bob Kiple, "Mobilgas Economy Run," *Speed Age*, July 1952, 72.
39. "Winner and Still Champion —1952 Mercury," *Lincoln-Mercury Times*, Ford Motor Company, July–August 1952.
40. "Summary of Remarks Made by Mr. Minckler," *1952 Mobilgas Economy Run* (Floyd Clymer, 1952), 28.
41. Keith Monroe, "The Long Long Run for the Big Big Money," *True's Automobile Yearbook* No. 2, 1953, 17.
42. General Petroleum Corporation, 1953 Mobilgas Economy Run to Sun Valley Press Release, for release Sunday, March 22, or thereafter.
43. "Mobilgas Economy Run," *Auto Racing Yearbook 1953*, 44.
44. Contest Board of the American Automobile Association, "Official Rules 1953 Mobilgas Economy Run," circa 1952, 7.
45. General Petroleum Corporation, 1953 Mobilgas Economy Run to Sun Valley Press Release, for release Sunday, March 22, or thereafter.
46. Contest Board of the American Automobile Association, "Official Rules 1953 Mobilgas Economy Run," circa 1952, 16.
47. Ibid., 22.
48. Ibid., 7.
49. "Mobilgas Economy Run," *Auto Racing Yearbook 1953*, 44.
50. Ibid., 44.
51. John A. Heitmann, "The Race That Wasn't a Race: The Mobilgas Economy Run 1936–1968," *Automotive History Review* (Fall 2008): 20.
52. Contest Board of the American Automobile Association, "Official Rules 1953 Mobilgas Economy Run," circa 1952, 12.
53. Ibid., 23; Contest Board of the American Automobile Association, "1953 Mobilgas Economy Run Route Description and Map," circa 1953, 4.
54. Jack Weed, "Ford 6 Wins Economy Run," *Automotive News*, April 27, 1953, 1.
55. "Is The Economy Run Honest?" *Speed Age*, July 1953, 56.
56. "All Cars End Initial Day of Economy Run," *Twin Falls Times-News*, April 22, 1953.
57. "Is the Economy Run Honest?" *Speed Age*, July 1953, 56.
58. Jack Weed, "Ford 6 Wins Economy Run," *Automotive News*, April 27, 1953, 1.
59. R.A. Van Allen, "1953 Economy Run Observations" *1953 Mobilgas Economy Run* (Floyd Clymer, 1953), 38.
60. "Economy Run Autos Cleared Through T.F." *Twin Falls Times-News*, April 23, 1953.
61. R.A. Van Allen, "1953 Economy Run Observations" *1953 Mobilgas Economy Run* (Floyd Clymer, 1953), 38.
62. General Petroleum Corporation, 1953 Mobilgas Economy Run to Sun Valley Press Release, April 23, 1953.
63. Jack Weed, "Ford 6 Wins Economy Run," *Automotive News*, April 27, 1953, 1.
64. 1953 Mobilgas Economy Run to Sun Valley Press Release, General Petroleum Corporation, April 23, 1953.
65. Ibid.
66. "Nash Only Make to Score in 2 Economy Run Classes," *Los Angeles Mirror*, May 1, 1953.
67. Ford Division, Ford Motor Company, "Ford Economy Wins Again!" print advertisement, circa 1953.
68. Kaiser-Frazer Sales Corporation, "28.2 m.p.g. Henry J Wins in Mobilgas Economy Run!" print advertisement, circa 1953.
69. Dodge Division, Chrysler Corporation, "Dodge Sweeps the Field," print advertisement, circa 1953.
70. Lincoln Division, Ford Motor Company, "Winner! Lincoln Again Sweeps Its Class," print advertisement, circa 1953.
71. The Studebaker Corporation, "Studebaker Commander V-8 With Overdrive Finished 1st in Actual Miles Per Gallon," print advertisement, circa 1953.
72. Socony-Vacuum Oil Company, Inc., "High Quality Economy Gasoline" and "Unmatched On-the-Road Results," print advertisements, circa 1953.
73. Champion Spark Plug Company, "Champions That Deliver Mileage Like This," print advertisement, circa 1953.
74. Champion Spark Plug Company, "Champion Campeon de Economica," print advertisement, Circa 1953.
75. Dodge Division, Chrysler Corporation, print advertisement, *The Saturday Evening Post*, July 18, 1953, 72.
76. Dodge Division, Chrysler Corporation, print advertisement, *Dodge News* 18, No. 6, 8–9.
77. The Studebaker Corporation, 1954 model year brochure, D-197 11–53, 9 and 17.
78. "The Mobilgas-AAA Economy Run to Sun Valley," *Automobile Topics*, May 1953, 20; "Mobilgas Economy Run," *Auto Racing Yearbook 1953*, 44; "How Did You Do It?" *Motor Trend*, June 1953, 64; Walt Woron and Bob Scollay, "America's Economy Classic," *Motor Trend*, July 1953, 38: Les Viland, "How Ford Got More Miles Per Gallon," *Motor World*, May 22, 1953, 16; Bill Schroeder, "1953 Mobilgas Economy Run," *Motorsport*, August 1953, 20; Andrew R. Boone, "Champ Gas Savers Show How to Run Your Own Economy Run," *Popular Science*, June 1953, 170; "Is the Economy Run Honest?" *Speed Age*, July 1953, 56; Keith Monroe, "The Long Long Run for the Big Big money," *True's Automobile Yearbook* No. 2, 1953, 17.
79. "Ford Wins 1953 Economy Run," *The Autocar*, May 15, 1953, 675.
80. "Is The Economy Run Honest?" *Speed Age*, July 1953, 56.
81. "Briefs ... by the Editors," *Motor Trend*, July 1953, 5; "Bonneville Record Runs," *Auto Racing Yearbook 1953*, 82; "Dodge Breaks Speed Mark at El Mirage Lake," *Los Angeles Times*, May 10, 1953.
82. "Rugged Run for Gas Economy," *1954 Mobilgas Economy Run* (Floyd Clymer, 1954), 21; *Automobile Topics*, May 1954, 16.
83. Contest Board of the American Automobile Association, "Official Rules 1954 Mobilgas Economy Run," circa 1953, 12.
84. "The Sealed Trial," *Road & Track*, June 1954, 41.
85. Bob Finlay, "Studebaker Wins Economy Run," *Automotive News*, April 12, 1954, 1.
86. Cliff Gromer, "Rules: Do Not Break, Bend or Mutilate," *Drag Racing Online*, May 9, 2007; accessed April 19, 2012. http://www.dragracingonline.com/columns/cliffnotes/ix_2.1.html.
87. "Analyzing the Economy Run," *Chilton's Motor Age*, June 1954, 54.
88. Don Francisco, "Mobilgas Economy Run: How It Is Won," *Hot Rod Magazine*, August 1954, 18.
89. Russ Catlin, "The Run Viewed by a Newsman," *1954 Mobilgas Economy Run* (Floyd Clymer, 1954), 26.
90. "20 Cars Start Second Lap of Economy Run," *Twin Falls Times-News*, April 7, 1954.
91. Ragnar Berglund, "Economy Run Smorgasbord" *1954 Mobilgas Economy Run* (Floyd Clymer, 1954), 35; *Automobile Topics*, May 1954, 25.
92. "Economy Run Results to Be Unveiled Today," *Long Beach Independent*, April 8, 1954.
93. "20 Drivers Welcomed to Twin Falls on Last Leg of Sun Valley Economy Run," *Twin Falls Times-News*, April 8, 1954.
94. "Economy Run Results to Be Unveiled Today," *Long Beach Independent*, April 8, 1954.

95. Russ Catlin, "The Run Viewed by a Newsman," *1954 Mobilgas Economy Run* (Floyd Clymer, 1954), 26.
96. "Analyzing the Economy Run," *Chilton's Motor Age*, June 1954, 54.
97. Floyd Clymer, "Introduction," *1954 Mobilgas Economy Run* (Floyd Clymer, 1954), 1.
98. Bob Finlay, "Studebaker Wins Economy Run," *Automotive News*, April 12, 1954, 1.
99. "Studebakers Monopolize Rugged '54 Economy Run," *Los Angeles Times*, April 11, 1954.
100. Bob Finlay, "Studebaker Wins Economy Run," *Automotive News*, April 12, 1954, 1.
101. General Petroleum Corporation, "Official Results: 1954 Mobilgas Economy Run," print advertisement, circa 1954.
102. The Studebaker Corporation, "Studebaker Wins 1954 Mobilgas Economy Run Sweepstakes!" print advertisement, circa 1954.
103. The Studebaker Corporation, "Advanced Style Studebakers Win '54 Mobilgas Run," print advertisement, circa 1954.
104. The Studebaker Corporation, print advertisement, *The Saturday Evening Post*, June 26, 1954, 85.
105. The Studebaker Corporation, "Cut Your Driving Costs with an Out-Ahead '54 Studebaker," print advertisement, circa 1954.
106. The Studebaker Corporation, "Get a Thrifty '54 Studebaker, Canada's No. 1 Economy Car," print advertisement, circa 1954.
107. The Studebaker Corporation, 1955 model year brochure, D-217-9-54, 9.
108. Dodge Division, Chrysler Corporation, "Dodge Wins Again!" print advertisement, circa 1954.
109. Lincoln Division, Ford Motor Company, "Lincoln Wins Again! Class G Mobilgas Economy Run," print advertisement, circa 1954.
110. "Lincoln Says So Much About You—So Beautifully," print advertisement, circa 1954.
111. Dodge Division, Chrysler Corporation, 1954 model year, "Winner and Still Champ!" brochure, DMA NO. 2308—4-54.
112. Barney Navarro, "The Economy Run," *Car Life*, August 1954, 12; Don Francisco, "Mobilgas Economy Run: How It Is Won," *Hot Rod Magazine*, August 1954, 18; Bill Schroeder, "Clean Sweep for Studebaker," *Motorsport*, August 1954, 22: Walt Woron, "Stingy Studebaker Snares Sweepstakes," *Motor Trend*, June 1954, 27; Walt Woron, "I Drove the Mobilgas Economy Run," *Motor Trend*, July 1954, 32; "The Sealed Trial," *Road & Track*, June 1954, 41; Sheldon Reed, "Mobilgas Economy Run: It's a Rugged Driving Test," *Speed Age*, July 1954, 16.
113. "Dodge Wins Again!" Chrysler Corporation, *Dodge News* 19, No. 4, 11; "Powerful, Yes—and Economical, Too," *Lincoln-Mercury Times*, Ford Motor Company, July–August 1954, inside back cover.
114. Sheldon Reed, "Mobilgas Economy Run: It's a Rugged Driving Test," *Speed Age*, July 1954, 16.
115. Walt Woron, "I Drove the Mobilgas Economy Run," *Motor Trend*, July 1954, 32.

Chapter 4

1. Contest Board of the American Automobile Association, "Official Rules 1955 Mobilgas Economy Run," circa 1954, 1.
2. "Studebaker Again Wins Economy Run," *Automobile Topics*, April 1955, 18.
3. Randall Gould, "The 1955 Mobilgas Economy Run as Seen Through the Eyes of a Coloradan," *1955 Mobilgas Economy Run* (Floyd Clymer, 1955), 20.
4. Contest Board of the American Automobile Association, "Official Rules 1955 Mobilgas Economy Run," circa 1954, 12.
5. Robert M. Finlay, "Studebaker Keeps Economy Title," *Automotive News*, March 21, 1955, 1.
6. Ibid.
7. Bob Russo, "I Drove in Mobilgas Economy Run," *Speed Age*, July 1955, 22.
8. Russ Catlin, "The Press Had a Few Thousand Words for It," *1955 Mobilgas Economy Run* (Floyd Clymer, 1955), 30.
9. Lynn Rogers, "Studebaker Takes First Day Honors in Gas Run," *Los Angeles Times*, March 15, 1955.
10. Randall Gould, "The 1955 Mobilgas Economy Run as Seen Through the Eyes of a Coloradan," *1955 Mobilgas Economy Run* (Floyd Clymer, 1955), 21.
11. Bob Russo, "I Drove in Mobilgas Economy Run," *Speed Age*, July 1955, 22.
12. Lynn Rogers, "Studebaker Takes First Day Honors in Gas Run," *Los Angeles Times*, March 15, 1955.
13. Bob Russo, "I Drove in Mobilgas Economy Run," *Speed Age*, July 1955, 22.
14. Lynn Rogers, "Nash Wins Day's Honors in Mobilgas Test Run," *Los Angeles Times*, March 16, 1955.
15. "Nash Wins Sweepstakes Leg in Economy Run," *Albuquerque Journal*, March 16, 1955.
16. Russ Catlin, "The Press Had a Few Thousand Words for It," *1955 Mobilgas Economy Run* (Floyd Clymer, 1955), 30.
17. Lynn Rogers, "Studebaker Again Wins Economy Run," *Los Angeles Times*, March 17, 1955.
18. Philiberto Curial de Brevannes, "Indian Trail: 1955," *The Flying Red Horse* (Summer 1955): 12.
19. Robert M. Finlay, "Studebaker Keeps Economy Title," *Automotive News*, March 21, 1955, 1.
20. Ibid.
21. "Competition Keen in Economy Run," *Los Angeles Times*, March 27, 1955.
22. Russ Catlin, "The Press Had a Few Thousand Words for It," *1955 Mobilgas Economy Run* (Floyd Clymer, 1955), 30.
23. "Randall Gould, "The 1955 Mobilgas Economy Run as Seen Through the Eyes of a Coloradan," *1955 Mobilgas Economy Run* (Floyd Clymer, 1955), 21.
24. Robert M. Finlay, "Studebaker Keeps Economy Title," *Automotive News*, March 21, 1955, 1.
25. "Mobilgas Economy Run," *Car Life*, June 1955, 78.
26. Russ Catlin, "The Press Had a Few Thousand Words for It," *1955 Mobilgas Economy Run* (Floyd Clymer, 1955), 30.
27. Walt Woron, "Slushpumps Aren't Spendthrifts," *Motor Trend*, May 1955, 29.
28. Robert M. Finlay, "22 Cars Begin L.A.-Pike's Peak Economy Test," *Automotive News*, March 14, 1955, 1.
29. General Petroleum Corporation, "Official Results of the Mobilgas Economy Run," print advertisement, circa 1955.
30. Borg-Warner, "In 1955 Mobilgas Economy Run—B-W Automatic Transmission Wins Over All Other Automatics for Economy!" print advertisement, circa 1955.
31. Buick Motor Division, General Motors Corporation, "Buick Roadmaster with Variable Pitch Dynaflow Wins First in Its Class," print advertisement, circa 1955.
32. Ford Division, Ford Motor Company, "Ford Tops 'Big 3' in Mobilgas Economy Run," print advertisement, circa 1955.
33. Studebaker Division, Studebaker-Packard Corporation, "Again Studebaker Wins Mobilgas Run Sweepstakes!" print advertisement, circa 1955.
34. Studebaker Division, Studebaker-Packard Corporation, "Studebaker's Advanced Design Wins '55 Mobilgas Run!" print advertisement, circa 1955.
35. Studebaker Division, Studebaker-Packard Corporation, print advertisement, *The Saturday Evening Post*, May 21, 1955, 66.
36. Nash Motors Division, American Motors Corporation, "Nash Wins!" print advertisement, circa 1955.
37. Nash Motors Division, American Motors Corporation, "Only Nash Dealers Have 2 Winners to Sell," print advertisement, circa 1955.
38. American Motors Corporation, "Rambler Smashes Miles Per Gallon Record!" print advertisement, circa 1955.
39. Nash Motors Division, American Motors Corporation, 1955 model year features brochure, NSP-55-2087, 26 and Nash Motors Division, American Motors Corporation, 1956 model year features brochure, NSP-56-5134-100M-2-56, 21.
40. "Studebaker Again Wins Economy Run," *Automobile Topics*, April 1955, 21; "Mobilgas Economy Run," *Car Life*, June 1955, 78; "The Economy Run Winners," *Motor Life*, June 1955, 26; Walt Woron, "Slushpumps Aren't Spendthrifts," *Motor Trend*, May 1955, 29; "Studebaker Wins Gas-Saving Race," *Popular Science*, May 1955, 124; "1955 Economy Run," *Road & Track*, June 1955, 11; Bob Russo, "I Drove in Mobilgas Economy Run," *Speed Age*, July 1955, 22.
41. "Sweepstakes Champion," *1955 Mobilgas Economy Run* (Floyd Clymer, 1955), 12.
42. Russ Catlin, "To the Mountains from the Sea," *Automobile Topics*, April 1956, 26.
43. "1955 LeMans Disaster," Wikipedia, the free encyclopedia, accessed May 12, 2012, http://en.wikipedia.org/w/index.php?title=1955_Le_Mans disaster; "AAA Contest Board," Wikipedia, the free encyclopedia, accessed May 12, 2012, http://en.wikipedia.org/w/index.php?title=AAA_Contest_Board.
44. United States Auto Club, "Official Rules 1956 Mobilgas Economy Run," circa 1955, 1.
45. Floyd Clymer, "1956 Economy Run Highlights," *Automobile Topics*, April 1956, 25.
46. United States Auto Club, "Official Rules 1956 Mobilgas Economy Run," circa 1955, 13.
47. Randall Gould, "The Longest Economy Run," *Automobile Topics*, April 1956, 22.
48. United States Auto Club, "Official Rules 1956 Mobilgas Economy Run," circa 1955, 8.
49. "Imperial Tops Field in Economy Run," *Automobile Topics*, April 1956, 16.
50. Wally Parks, "Operation Featherfoot," *Hot Rod Magazine*, June 1956, 40.

51. Randall Gould, "The Longest Economy Run," *Automobile Topics*, April 1956, 22.
52. "Imperial Tops Field in Economy Run," *Automobile Topics*, April 1956, 16.
53. "Economy Run Statistics," *Pasadena Star News*, April 1, 1956.
54. "Imperial Takes Lead as Mishap Mars Economy Run," *Pasadena Star News*, March 20, 1956; Russ Catlin, "To the Mountains from the Sea," *Automobile Topics*, April 1956, 26.
55. "Nash, Olds Favored in '56 Economy Run," *Oxnard Press-Courier*, March 20, 1956.
56. Russ Catlin, "To the Mountains from the Sea," *Automobile Topics*, April 1956, 26.
57. Ibid.
58. Randall Gould, "The Longest Economy Run," *Automobile Topics*, April 1956, 22.
59. Lynn Rogers, "Imperial Still Leading Mobilgas Economy Run," *Los Angeles Times*, March 21, 1956; Russ Catlin, "To the Mountains from the Sea," *Automobile Topics*, April 1956, 26.
60. Russ Catlin, "To the Mountains from the Sea," *Automobile Topics*, April 1956, 26.
61. Lynn Rogers, "Imperial Still Holding Lead in Economy Run," *Los Angeles Times*, March 22, 1956.
62. "Economy Run Heads Down Home Stretch," *Ogden Standard-Examiner*, March 22, 1956.
63. Lynn Rogers, "Imperial Still Holding Lead in Economy Run," *Los Angeles Times*, March 22, 1956.
64. Randall Gould, "The Longest Economy Run," *Automobile Topics*, April 1956, 22.
65. "Dealer Alsbury Leads All the Way in Run," *Automotive News*, April 2, 1956, 2.
66. "1956 Mobilgas Economy Run," *Road & Track*, June 1956, 23.
67. Floyd Clymer, "1956 Economy Run Highlights," *Automobile Topics*, April 1956, 25.
68. Randall Gould, "The Longest Economy Run," *Automobile Topics*, April 1956, 22.
69. Lynn Rogers, "Imperial Wins Economy Run to Colorado Springs," *Los Angeles Times*, March 23, 1956.
70. Randall Gould, "The Longest Economy Run," *Automobile Topics*, April 1956, 22.
71. "Nash, Olds Favored in '56 Economy Run," *Oxnard Press-Courier*, March 20, 1956.
72. Floyd Clymer, "1956 Economy Run Highlights," *Automobile Topics*, April 1956, 25.
73. Socony-Mobil Oil Company, "Official Winners Mobilgas Economy Run," print advertisement, circa 1956.
74. Socony-Mobil Oil Company, "Testing Ground of Performance and Mileage ... the Mobilgas Economy Run," print advertisement, circa 1956.
75. Mobil Oil "El año que duro cuatro dias!" print advertisement, circa 1956.
76. Chrysler Division, Chrysler Corporation, "Chrysler-Built Imperial *Beats All* Cars in Mobilgas Economy Run!" print advertisement, circa 1956.
77. Ford Division, Ford Motor Company, "Ford Wins in Economy Too!" print advertisement, circa 1956.
78. Ford Division, Ford Motor Company, "Ford wins all 3," print advertisement, circa 1956.
79. Chevrolet Motor Division, General Motors Corporation, 1956 model year salesman's brochure.
80. Oldsmobile Division, General Motors Corporation, "Double Victory! Oldsmobile Rockets to Two First Places in Mobilgas Economy Run," print advertisement, circa 1956.
81. Pontiac Motor Division, General Motors Corporation, "Here's the Big Fact You Need to Know About the Mobilgas Economy Run," print advertisement, circa 1956.
82. American Motors Corporation, "*Again In 1956* ... Rambler Delivers the Most Actual Miles-Per-Gallon in the Mobilgas Economy Run," print advertisement, circa 1956.
83. American Motors Corporation, "Rambler Tops the Field in '56 Mobilgas Economy Run!" print advertisement, circa 1956.
84. "Everybody Seems To Win in Economy Run," *Hudson Owner's Report*, American Motors Corporation, circa 1956, 3.
85. "The Price of Power," *Motor Life*, June 1956, 89.
86. Wally Parks, "Operation Featherfoot," *Hot Rod Magazine*, June 1956, 40, "The Price of Power," *Motor Life*, June 1956, 89, Wesley S. Griswold, "Is Your Right Foot Costing You Money?" *Popular Science*, May 1956, 105, "1956 Mobilgas Economy Run," *Road & Track*, June 1956, 23, "The Economy Run Winner — A One-Picture Story," *Ford Times*, Ford Motor Company, July 1956, 62.

Chapter 5

1. United States Auto Club, "Official Rules 1957 Mobilgas Economy Run," circa 1956, 7.
2. Constance Moon, "Who Says Women Can't Drive?" *The Flying Red Horse* (Summer 1957): 17.
3. United States Auto Club, "Official Rules 1957 Mobilgas Economy Run," circa 1956, 13.
4. United States Auto Club, "Final Detailed Results 1957 Mobilgas Economy Run," circa 1957.
5. General Petroleum Corporation, Mobilgas Economy Run Press Release No. 57–10, for release in Sunday Editions, March 17, 1957, or thereafter.
6. General Petroleum Corporation, "Route Of Runs Always Varied for Competitors," Mobilgas Economy Run Press Release, No. 57–04, circa 1957.
7. Nancy Reynolds, "Feminine Highlights," *1957 Mobilgas Economy Run* (Floyd Clymer, 1957), 42.
8. General Petroleum Corporation, "Cars Like You Drive Compete for Honors on Mobilgas Run," Mobilgas Economy Run Press Release, No. 57–02, circa 1957; General Petroleum Corporation, "Mobilgas Run Cars Have No Fuel-Saving Gadgets," Mobilgas Economy Run Press Release, No. 57–012.
9. William Carroll, "Operation Economy," *Motor Trend*, July 1957, 32.
10. General Petroleum Corporation, Mobilgas Economy Run Press, Release No. 57–22, circa 1957.
11. United States Auto Club, "Official Rules 1957 Mobilgas Economy Run," circa 1956, 13.
12. William Carroll, "Operation Economy," *Motor Trend*, July 1957, 32.
13. Ibid.
14. Nancy Reynolds, "Feminine Highlights," *1957 Mobilgas Economy Run* (Floyd Clymer, 1957), 42.
15. "Tricks Used to Keep Run Course Secret," *Los Angeles Times*, April 28, 1957.
16. General Petroleum Corporation, Mobilgas Economy Run Press Release, No. 57–39 for release Sunday, April 14, 1957.
17. Bob Russo, "How We Won the Economy Run," *Speed Age*, August 1957, 16.
18. Bill Barrett, "Driver of the Month," *1957 Mobilgas Economy Run* (Floyd Clymer, 1957), 35.
19. "1957 Mobilgas Economy Run," *Hot Rod Magazine*, June 1957, 22.
20. *1957 Mobilgas Economy Run* (Floyd Clymer, 1957), 15.
21. Art Stephan, "Mobilgas Auto Run Starts Third Leg," *Long Beach Press-Telegram*, April 17, 1957.
22. *1957 Mobilgas Economy Run* (Floyd Clymer, 1957), 19.
23. Jack Weed, "Economy Run Swept by Chrysler," *Automotive News*, April 29, 1957, 1; United States Auto Club, "Final Detailed Results 1957 Mobilgas Economy Run," circa 1957.
24. United States Auto Club, "Final Detailed Results 1957 Mobilgas Economy Run," circa 1957.
25. Dodge News Bureau, Dodge Division, Chrysler Corporation, "Dodge Tops Mileage Run," press release, April 18, 1957.
26. Homer Fey, "Alsbury Leads All Way to Win Economy Run," *Oakland Tribune*, April 19, 1957.
27. "Victory Repeated by Imperial," *Automobile Topics*, June 1957, 19.
28. Bob Russo, "How We Won the Economy Run," *Speed Age*, August 1957, 16.
29. "1957 Mobilgas Economy Run," *Hot Rod Magazine*, June 1957, 22.
30. "Clean Sweep for Chrysler," *Time Magazine*, April 29, 1957.
31. Jack Weed, "Pro-Male Rule 'Robs' 3 Women in Economy Run," *Automotive News*, April 29, 1957, 2.
32. Bill Barrett, "Driver of the Month," *1957 Mobilgas Economy Run* (Floyd Clymer, 1957), 35.
33. Jack Weed, "Pro-Male Rule 'Robs' 3 Women in Economy Run," *Automotive News*, April 29, 1957, 2.
34. United States Auto Club, "Official Rules 1957 Mobilgas Economy Run," circa 1957, 7.
35. Chrysler Division, Chrysler Corporation, Press Release 47 57, 4–19–57.
36. Jack Weed, "Pro-Male Rule 'Robs' 3 Women in Economy Run," *Automotive News*, April 29, 1957, 2.
37. Socony-Mobil Oil Company, "Official Winners Mobilgas Economy Run," print advertisement, circa 1957.
38. Chrysler Corporation, "The Switch Is On," print advertisement, circa 1957.
39. Chrysler Division, Chrysler Corporation, "Imperial Wins," print advertisement, circa 1957.
40. Plymouth Division, Chrysler Corporation, "Southern California Loves The Mobilgas Economy Run Winner," print advertisement, circa 1957.
41. William J. Losher, "Here's How We Won the Economy Run!" *Dodge News* 22, No. 6, 2.
42. The Electric Auto-Lite Company, "Power Tip 'Fires Up' All Winners as Chrysler-Built Cars Sweep '57 Mobilgas Economy Run." print advertisement, circa 1957.
43. Ford Division, Ford Motor Company, "Ford Beats All Mobilgas Entries in Actual Miles Per Gallon," print advertisement, circa 1957.
44. American Motors Corporation, "Rambler Wins!" print advertisement, circa 1957.
45. American Motors Corporation, "Rambler Does It Again!" print advertisement, circa 1957.

46. American Motors Corporation, "Rambler Sets Economy Record," print advertisement, circa 1957.

47. "Clean Sweep for Chrysler," *Time Magazine*, April 29, 1957.

48. Constance Moon, "Who Says Women Can't Drive?" *The Flying Red Horse* (Summer 1957): 17.

49. "1957 Mobilgas Economy Run," *Hot Rod Magazine*, June 1957, 22; "Operation Economy," *Motor Trend*, July 1957, 32; "The Run and the Women," *Road & Track*, July 1957, 24; Bob Russo, "How We Won the Economy Run," *Speed Age*, August 1957, 16.

50. "The Run and the Women," *Road & Track*, July 1957, 24.

51. United States Auto Club, "Official Rules 1958 Mobilgas Economy Run," circa 1957, 14.

52. "1958 Economy Run," *Motor Life*, July 1958, 76.

53. United States Auto Club, "Official Rules 1958 Mobilgas Economy Run," circa 1957, 7.

54. American Motors Corporation, print advertisement, *Automotive News*, April 14, 1958, 30.

55. "Change in Engine Rule Bars Rambler from Run," *Automotive News*, April 14, 1958, 2.

56. W.N. "Bill" Doushkess, "Run to the Land of the Bluebonnets," *1958 Mobilgas Economy Run* (Floyd Clymer, 1958), 6.

57. Robert C. Scollay, "Feather-Foot Champ," *Motor Trend*, July 1958, 28.

58. United States Auto Club, "Official Rules 1958 Mobilgas Economy Run," circa 1957, 13.

59. Robert C. Scollay, "Feather-Foot Champ," *Motor Trend*, July 1958, 28.

60. W.N. "Bill" Doushkess, "Run to the Land of the Bluebonnets," *1958 Mobilgas Economy Run* (Floyd Clymer, 1958), 6.

61. Ibid.

62. Magnolia Petroleum Company, "Coming to Galveston April 17," print advertisement, circa April 1957.

63. "Great Interest Is Shown in Mobilgas Economy Run," *1958 Mobilgas Economy Run* (Floyd Clymer, 1958), 9.

64. "10 Dealers Enter Autos in Mobilgas Economy Run," *Automotive News*, April 14, 1958, 2.

65. W.N. "Bill" Doushkess, "Run to the Land of the Bluebonnets," *1958 Mobilgas Economy Run* (Floyd Clymer, 1958), 6.

66. Lynn Rogers, "2 Time Winner Takes Lead in Economy Run," *Los Angeles Times*, April 14, 1958.

67. Lynn Rogers. "Mel Alsbury Widens His Economy Run Lead," *Los Angeles Times*, April 15, 1958.

68. ""Great Interest Is Shown in Mobilgas Economy Run," *1958 Mobilgas Economy Run* (Floyd Clymer, 1958), 9.

69. Robert C. Scollay, "Feather-Foot Champ," *Motor Trend*, July 1958, 28.

70. Homer Fey, "Mobilgas Economy Run in Third Day," *Oakland Tribune*, April 15, 1958.

71. Robert C. Scollay, "Feather-Foot Champ," *Motor Trend*, July 1958, 28.

72. *1958 Mobilgas Economy Run* (Floyd Clymer, 1958), 41; United States Auto Club, "Final Detailed Results 1958 Mobilgas Economy Run," circa 1958.

73. Lynn Rogers, "Close Fight Shapes Up in Day's Economy Run," *Los Angeles Times*, April 16, 1958.

74. *1958 Mobilgas Economy Run* (Floyd Clymer, 1958), 29.

75. United States Auto Club, "Final Detailed Results 1958 Mobilgas Economy Run," circa 1958.

76. Lynn Rogers, "DeSoto Passes Edsel on Economy Run Lap," *Los Angeles Times*, April 17, 1958.

77. Homer Fey, "Mobilgas Economy Run in Final Day," *Oakland Tribune*, April 17, 1958.

78. "City to 'Dress Up' For Finish of Run," *Galveston Daily News*, April 16, 1958.

79. W.N. "Bill" Doushkess, "Run to the Land of the Bluebonnets," *1958 Mobilgas Economy Run* (Floyd Clymer, 1958), 6.

80. *1958 Mobilgas Economy Run* (Floyd Clymer, 1958), 29.

81. Barry Hart, "Isle Called Ideal Finishing Place for Mobilgas Event," *Galveston Daily News*, April 18, 1958.

82. "Chrysler Takes Most Run Honors," *Galveston Daily News*, April 18, 1958.

83. Lynn Rogers, "Mel Alsbury Jr. Wins Economy Run Honors," *Los Angeles Times*, April 18, 1958.

84. "Victory for the Heavies," *Time Magazine*, April 28, 1958.

85. W.N. "Bill" Doushkess, "Run to the Land of the Bluebonnets," *1958 Mobilgas Economy Run* (Floyd Clymer, 1958), 6.

86. *1958 Mobilgas Economy Run* (Floyd Clymer, 1958), 40.

87. Socony-Mobil Oil Company, "Official Winners Mobilgas Economy Run," print advertisement, circa 1958.

88. Socony-Mobil Oil Company, "Meet Nicky Griffen," print advertisement, circa 1958.

89. Magnolia Petroleum Company, "29 Cars Averaged 18.7 Miles Per Gallon in '58 Mobilgas Economy Run," print advertisement, circa 1958.

90. Chrysler Division, Chrysler Corporation, "Imperial Wins Again," print advertisement, circa 1958.

91. Chrysler Division, Chrysler Corporation, print advertisement, *Automobile Topics*, June 1958, 11.

92. Chrysler Division, Chrysler Corporation, "The Man Who Owns It Owns *More*!" print advertisement, circa 1958.

93. Plymouth Division, Chrysler Corporation, "Plymouth Wins Again in Gas Economy Tests!" print advertisement, circa 1958.

94. Plymouth Division, Chrysler Corporation, "Plymouth Wins Low Price Class," print advertisement, circa 1958.

95. The Electric Auto-Lite Company, print advertisement, *Motor*, September 1958, 158.

96. Oldsmobile Division, General Motors Corporation, "Travel Test an Olds 88," print advertisement, circa 1958.

97. Chrysler Division, Chrysler Corporation, 1959 model year brochure, CS-422, 18.

98. Chrysler Division, Chrysler Corporation, 1959 model year brochure, 11.

99. "Alsbury-Imperial Combination Wins Mobilgas for Third Time!" *Automobile Topics*, May 1958, 22; "1958 Economy Run," *Motor Life*, July 1958, 76; Robert C. Scollay, "Feather-Foot Champ," *Motor Trend*, July 1958, 76: "Yes, the New Cars *Do* Take More Gas," *Popular Science*, July 1958, 68.

100. Barry Hart, "Isle Called Ideal Finishing Place for Mobilgas Event," *Galveston Daily News*, April 18, 1958.

Chapter 6

1. "Ton Mile Score Fair to All," *1958 Mobilgas Economy Run* (Floyd Clymer, 1958), 42.

2. United States Auto Club, "Official Rules 1959 Mobilgas Economy Run," circa 1958, 1.

3. "1959 Mobilgas Economy Run The Longest Route," *1959 Mobilgas Economy Run* (Floyd Clymer, 1959), 5.

4. General Petroleum Corporation, "Stock Safeguards," Mobilgas Economy Run Press Release, March 15, 1959.

5. United States Auto Club, "Official Rules 1959 Mobilgas Economy Run," circa 1958, 8.

6. Ibid., 7.

7. Griff Borgeson, "Run for the Mileage," *Sports Cars Illustrated*, April 1960, 30.

8. Ibid.

9. Ibid.

10. "1959 Mobilgas Economy Run the Longest Route," *1959 Mobilgas Economy Run* (Floyd Clymer, 1959), 5.

11. General Petroleum Corporation, "Memo to Editors," Mobilgas Economy Run Press Release, circa March 1960.

12. Bill Dredge, "47 Economy Run Entries Flagged Away at Dawn," *Los Angeles Times*, April 5, 1959.

13. General Petroleum Corporation, "Actual Route of Run," Mobilgas Economy Run Press Release, April 3, 1959.

14. Randall Gould, "The Longest—Toughest Mobilgas Economy Run," *Automobile Topics*, June 1959, 8.

15. "Lark V-8 Sets Pace with 22.28 M.P.G. in Mobilgas Economy Run," *Studebaker News*, Studebaker-Packard Corporation, May 1959, 4.

16. "Lensmen in Melee at Beginning of Run," *Los Angeles Times*, April 6, 1959.

17. *1959 Mobilgas Economy Run* (Floyd Clymer, 1959), 17.

18. Griff Borgeson, "Run for the Mileage," *Sports Cars Illustrated*, April 1960, 30.

19. Bill Dredge, "Economy Run Drivers End First Day in Utah," *Los Angeles Times*, April 6, 1959.

20. "Statement by A.C. Pillsbury, Director, United States Auto Club and Chief Steward, Mobilgas Economy Run," *1959 Mobilgas Economy Run* (Floyd Clymer, 1959), 21.

21. United States Auto Club, "Final Detailed Results 1959 Mobilgas Economy Run," circa 1959.

22. *1959 Mobilgas Economy Run* (Floyd Clymer, 1959), 23.

23. Bill Dredge, "Rambler Holds Early Lead in Economy Run," *Los Angeles Times*, April 7, 1959.

24. Randall Gould, "The Longest—Toughest Mobilgas Economy Run," *Automobile Topics*, June 1959, 8.

25. Griff Borgeson, "Run for the Mileage," *Sports Cars Illustrated*, April 1960, 30.

26. Bill Dredge, "Gas Economy Run Hits Snowstorm in Rockies," *Los Angeles Times*, April 8, 1959.

27. Griff Borgeson, "Run for the Mileage," *Sports Cars Illustrated*, April 1960, 30.

28. Bill Dredge, "Snow, Wind Slow Economy Drivers," *Los Angeles Times*, April 9, 1959.

29. Ibid.

30. Randall Gould, "The Longest—Toughest Mobilgas Economy Run," *Automobile Topics*, June 1959, 8.

31. Bill Dredge, "Rambler Takes First in Gas

Economy Event," *Los Angeles Times*, April 10, 1959; Randall Gould, "The Longest — Toughest Mobilgas Economy Run," *Automobile Topics*, June 1959, 8; General Petroleum Corporation, "Class C Low Price Class, 8 cylinder," Mobilgas Economy Run Press Release, circa April 1959.

32. Randall Gould, "The Longest — Toughest Mobilgas Economy Run," *Automobile Topics*, June 1959, 8.

33. "Lark V-8 Sets Pace with 22.28 M.P.G. in Mobilgas Economy Run," *Studebaker News*, Studebaker-Packard Corporation, May 1959, 4.

34. Randall Gould, "The Longest — Toughest Mobilgas Economy Run," *Automobile Topics*, June 1959, 8.

35. General Petroleum Corporation, "Class B Low Price Class, 6 Cylinder," Mobilgas Economy Run Press Release, circa April 1959.

36. Randall Gould, "The Longest — Toughest Mobilgas Economy Run," *Automobile Topics*, June 1959, 8.

37. General Petroleum Corporation, print advertisement, *Motor Trend*, June 1959, 10.

38. American Motors Corporation, "Rambler Wins First Two Places," print advertisement, circa 1959.

39. Plymouth Division, Chrysler Corporation, "Plymouth Proves Its Big Difference," print advertisement, circa 1959.

40. Plymouth Division, Chrysler Corporation, 1959 model year mailer brochure.

41. Dodge Division, Chrysler Corporation, print advertisement, *The Saturday Evening Post*, June 20, 1959, 55.

42. Studebaker-Packard Corporation, "The Lark By Studebaker Scores Most Miles Per Gallon of All V-8's in All Classes in Mobilgas Economy Run," print advertisement, circa 1959.

43. Chevrolet Division, General Motors Corporation, print advertisement, *The Saturday Evening Post*, July 25, 1959, 48.

44. The B.F. Goodrich Company, "Nine Economy Winners Ride on BFG Tires," press release, 4–959.

45. Mel Alsbury, Inc., 1959 model year brochure.

46. "1959 Mobilgas Economy Run Story," *Automobile Topics*, June 1959, 6; "Mobilgas Economy Run Winners," *Motor Trend*, June 1959, 80; "Dodge, Small Rambler Win Economy Laurels," *Popular Science*, June 1959, 80; Griff Borgeson, "Run for the Mileage," *Sports Cars Illustrated*, April 1960, 30; "Victory for Rambler," *Time*, April 20, 1959.

47. "Robert L. Minckler to Retire as Mobil Oil Vice-President," *Torrance Herald*, January 28, 1960.

48. United States Auto Club, "Official Rules 1960 Mobilgas Economy Run," circa 1959, 1.

49. "Economy Run Entries Strictly Stock Cars," *Automobile Topics*, June 1960, 25.

50. "Facts & Forecasts," *Motor Life*, January 1960, 6; "Two Days of Safety Tests Are Added to Pure Oil Thrift Trials," *Automobile Topics*, January 1960.

51. United States Auto Club, "Official Rules 1960 Mobilgas Economy Run," circa 1959, 8.

52. "*Mobilgas Economy Run Fact Book*," Mobil Oil Company, circa 1960, 65.

53. "Ten 'Compacts' in Economy Run," *Torrance Press*, March 3, 1960.

54. "Economy Run Entries Strictly Stock Cars," *Automobile Topics*, June 1960, 25.

55. "Ten 'Compacts' in Economy Run," *Torrance Press*, March 3, 1960.

56. "Economy Run Entries Strictly Stock Cars," *Automobile Topics*, June 1960, 25.

57. United States Auto Club, "Official Rules 1960 Mobilgas Economy Run," circa 1959, 13.

58. United States Auto Club, "Final Detailed Results 1960 Mobilgas Economy Run," circa 1960.

59. Bill Dredge, "56 Cars Start Toughest Economy Run at Dawn," *Los Angeles Times*, April 2, 1960.

60. Bill Dredge, "Flagstaff Greets Weary Economy Run Drivers," *Los Angeles Times*, April 3, 1960.

61. Ibid.

62. Ibid.

63. Bill Dredge, "Rambler Leads Pack in Day's Economy Run," *Los Angeles Times*, April 4, 1960.

64. United States Auto Club, "Final Detailed Results 1960 Mobilgas Economy Run," circa 1960.

65. Floyd Freel, "Compacts Meet Tough Economy Run Competition," *Pasadena Independent*, April 4, 1960.

66. United States Auto Club, "Final Detailed Results 1960 Mobilgas Economy Run," circa 1960.

67. United States Auto Club, "Final Detailed Results 1960 Mobilgas Economy Run," circa 1960; Bill Dredge, "Blustery Winds Slash Economy Run Mileages" *Los Angeles Times*, April 5, 1960.

68. Floyd Freel, "Economy Run in Last Lap," *Pasadena Independent*, April 6, 1960.

69. Bill Dredge, "Farm Wagons, Trucks Impede Economy Run," *Los Angeles Times*, April 6, 1960.

70. Ibid.

71. Bill Dredge, "Rambler Wins Economy Run for 2nd Time," *Los Angeles Times*, April 7, 1960.

72. "Rambler Wins Compact Class in Mobilgas Economy Run — 28.35 mpg," *Automobile Topics*, June 1960, 20.

73. United States Auto Club, "Final Detailed Results 1960 Mobilgas Economy Run," circa 1960.

74. "Economy Spectacular," *Motor Life*, June 1960, 72.

75. "Rambler Wins Compact Class in Mobilgas Economy Run — 28.35 mpg," *Automobile Topics*, June 1960, 20.

76. "Economy Spectacular," *Motor Life*, June 1960, 72.

77. United States Auto Club, "Final Detailed Results 1960 Mobilgas Economy Run," circa 1960.

78. Mobil Oil Company, print advertisement, *Motor Trend*, July 1960, 3.

79. Mobil Oil Company, "Compact Cars Averaged 25.50 Miles Per Gallon on the Mobilgas Economy Run," print advertisement, circa 1960.

80. American Motors Corporation, "Rambler American Custom Wins Over All Compacts!" print advertisement, circa 1960.

81. American Motors Corporation, "Rambler American *Beat All* Cars in Miles Per Gallon," print advertisement, circa 1960.

82. Chevrolet Division, General Motors Corporation, "Economy Proved in Mobilgas Run!" print advertisement, circa 1960; Chrysler Corporation, "More Car and More Economy Go Together," print advertisement, circa 1960.

83. Chevrolet Division, General Motors Corporation, "Economy Proved in Mobilgas Run!" print advertisement, circa 1960.

84. Chrysler Corporation, "Talk About Economy!" print advertisement, circa 1960.

85. Chrysler Corporation, "More Car and More Economy Go Together," print advertisement, circa 1960.

86. Chrysler and Imperial Division, Chrysler Corporation, "Chrysler Wins!" print advertisement, circa 1960.

87. Chrysler and Imperial Division, Chrysler Corporation, "Four-Time Winner!" print advertisement, circa 1960.

88. Plymouth Division, Chrysler Corporation, "4 Straight Wins for Plymouth ... 4 Straight Facts for You ...," print advertisement, circa 1960.

89. The Electric Auto-Lite Company, "Autolite Spark Plugs Sparked More Than Twice as Many Mobilgas Economy Run Winners as All Other Spark Plugs Combined," print advertisement, circa 1960.

90. Studebaker-Packard Corporation, "Studebaker Economy Proved Again in Mobilgas Economy Run!" print advertisement, circa 1960.

91. Studebaker-Packard Corporation, "Paragon of Precision, Performance, Power...," print advertisement, circa 1960.

92. "Rambler Wins Compact Class in Mobilgas Economy Run — 28.35 mpg," *Automobile Topics*, June 1960, 20; "Economy Spectacular," *Motor Life*, June 1960, 72; Bob Russo, "Light Foot Forward," *Motor Trend*, June 1960, 62; "Rambler American Sets Economy Mark in Compact Class," *Popular Science*, June 1960, 66; "Mobil Gas Economy Run," *Small Cars Illustrated*, August 1960, 58; "All We Women Did ..." *Time*, April 18, 1960.

93. Bob Russo, "Light Foot Forward," *Motor Trend*, June 1960, 62.

94. Chevrolet Division, General Motors Corporation, 1960 model year brochure.

Chapter 7

1. United States Auto Club, "Official Rules 1961 Mobilgas Economy Run," circa 1959, 7.

2. Bill Dredge, "The Tough Test," *Mobilgas Economy Run 1961 Fact Book*," Mobil Oil Company, circa 1961, 7.

3. Mobil Oil Company, "1961 Mobilgas Economy Run," Mobilgas Economy Run Press Release, February 6, 1961.

4. United States Auto Club, "Official Rules 1961 Mobilgas Economy Run," circa 1959, 7.

5. "Ramblers Pull Out of Run in Hassle," *Automotive News*, March 20, 1961, 74.

6. United States Auto Club, "Official Rules 1961 Mobilgas Economy Run," circa 1959, 9.

7. Bill Dredge, "Falcon Sets Mark as Economy Run Opens," *Los Angeles Times*, March 12, 1961.

8. "American Motors Protests Decision," *Automobile Topics*, May 1961, 17.

9. United States Auto Club, "Official Rules 1961 Mobilgas Economy Run," circa 1959, 7.

10. Bill Dredge, "The Tough Test," *Mobilgas Economy Run 1961 Fact Book*," Mobil Oil Company, circa 1961, 7.

11. United States Auto Club, "Official Rules 1961 Mobilgas Economy Run," circa 1959, 9; Bob Russo, "Featherfoot Trail," *Motor Life*, June 1961, 56.

12. Bob Cahill with John Sloan, "Mopar and the Mobil Economy Run, Part 3," *High Performance MoPar*," January 1996, 43.

13. Ibid.

14. United States Auto Club, "Official Rules 1961 Mobilgas Economy Run," circa 1959, 11.

15. Ibid., 13.

16. Bob Cahill with John Sloan, "Mopar and

the Mobil Economy Run, Part 3," *High Performance MoPar*," January 1996, 43.

17. Terry D. Dunham, *The Buick: A Compete History*, updated edition (Kutztown, PA: Automobile Quarterly Publications, 1987), 282.

18. "Ramblers Pull Out of Run in Hassle," *Automotive News*, March 20, 1961, 74.

19. Mobil Oil Company, "Mobil Run Start," Mobilgas Economy Run Press Release, March 11, 1961.

20. "Ramblers Pull Out of Run in Hassle," *Automotive News*, March 20, 1961, 74.

21. Bill Dredge, "Falcon Sets Mark as Economy Run Opens," *Los Angeles Times*, March 12, 1961.

22. "Pillsbury Statement," *Automobile Topics*, May 1961, 18.

23. United States Auto Club, "Final Detailed Results 1961 Mobilgas Economy Run," circa 1961.

24. Bill Dredge, "Falcon Leads Mobilgas Run in Second Day," *Los Angeles Times*, March 13, 1961.

25. Bill Dredge, "Race Driver Continues to Lead Economy Run," *Los Angeles Times*, March 14, 1961.

26. Ibid.

27. Slim Barnard, "Falcon Still Leading Economy Run," *Los Angeles Examiner*, March 14, 1961.

28. William Carroll, "Falcon Near Run Record," *Automotive News,* March 20, 1961, 1.

29. Ibid.

30. Slim Barnard, "Falcon Still Leads 1961 Economy Run," *Los Angeles Examiner*, March 15, 1961.

31. Bill Dredge, "Race Driver Continues to Lead Economy Run," *Los Angeles Times*, March 14, 1961.

32. Ibid.

33. Bill Dredge, "The Tough Test," *Mobilgas Economy Run 1961 Fact Book*," Mobil Oil Company, circa 1961, 7.

34. Bill Dredge, "Falcon Keeps Its Economy Run Lead," *Los Angeles Times*, March 15, 1961.

35. William Carroll, "Falcon Near Run Record," *Automotive News,* March 20, 1961, 1; Bill Dredge, "The Tough Test," *Mobilgas Economy Run 1961 Fact Book*," Mobil Oil Company, circa 1961, 7.

36. William Carroll, "Falcon Near Run Record," *Automotive News,* March 20, 1961, 1.

37. Bill Dredge, "Falcon Keeps Its Economy Run Lead," *Los Angeles Times*, March 15, 1961.

38. Bill Dredge, "Auto Economy Drive Slowed by Rainstorm," *Los Angeles Times*, March 16, 1961.

39. Bill Dredge, "The Tough Test," *Mobilgas Economy Run 1961 Fact Book*," Mobil Oil Company, circa 1961, 7.

40. William Carroll, "Falcon Near Run Record," *Automotive News,* March 20, 1961, 1.

41. "Economy Run Driver Unhurt in Accident," *Oakland Tribune*, March 16, 1961.

42. Bob Russo, "Featherfoot Trail," *Motor Life*, June 1961, 56.

43. Bill Dredge, "The Tough Test," *Mobilgas Economy Run 1961 Fact Book*," Mobil Oil Company, circa 1961, 7; Art Stephan, "Long Beach Drivers Take Victories in Economy Run," *Long Beach Independent*, March 17, 1961.

44. Jim Whipple, "Detroit Listening Post," *Popular Mechanics*, May 1961, 76.

45. William Carroll, "Falcon Near Run Record," *Automotive News,* March 20, 1961, 1.

46. "Falcon Files with Economy," *Today's Motor Sports*, June 1961, 38.

47. Mobil Oil Company, print advertisement, *Automotive News*, March 20, 1961, 70.

48. Mobil Oil Company, "$100,000 Mobilgas Economy Run Sweepstakes," print advertisement, circa 1961.

49. Ford Division, Ford Motor Company, "Economy Is Another Reason Why Falcon Had Won the Battle of the Compacts," print advertisement, circa 1961.

50. Ford Division, Ford Motor Company, "1st and 2nd Place Mobilgas Economy Run Drivers ... Reveal Their Secrets," print advertisement, circa 1961.

51. Ford Division, Ford Motor Company, "Forget It!" print advertisement, circa 1961.

52. "New Facts for High Mileage," *Ford Times*, Ford Motor Company, July 1961, 48.

53. Buick Motor Division, General Motors Corporation, "Buick Special Sweeps Class C with 25.09 MPG," print advertisement, circa 1961.

54. Buick Motor Division, General Motors Corporation, "See it ... Drive it ..! The Car That Swept Its Class in the Mobilgas Economy Run," print advertisement, circa 1961.

55. Buick Motor Division, General Motors Corporation, *Buick Dealer Flash Sheet*, 1961 model year salesman's brochure.

56. Buick Motor Division, General Motors Corporation, 1962 model year brochure.

57. Dodge Division, Chrysler Corporation, "Leadfoot Loses Mobilgas Economy Run," print advertisement, circa 1961.

58. Chrysler-Plymouth Division, Chrysler Corporation, "Plymouth Does It Again!" print advertisement, circa 1961.

59. Chrysler-Plymouth Division, Chrysler Corporation, "Plymouth ... Known by the Economy It Keeps," print advertisement, circa 1961.

60. Chrysler-Plymouth Division, Chrysler Corporation, print advertisement, *The Saturday Evening Post*, September 30, 1961, 75.

61. Chrysler-Plymouth Division, Chrysler Corporation, "Newport Wins Class 'F' Mobilgas Economy Run," print advertisement, circa 1961.

62. "Ford Falcon Wins Mobilgas Economy Run; Average 32.68 MPG," *Automobile Topics*, May 1961, 16; Bob Russo, "Featherfoot Trail," *Motor Life*, June 1961, 56; Don Werner, "The Mobilgas Economy Run," *Motor* Trend, April 1961, 24; Barney Navarro, "The Cars in the Economy Run," *Motor Trend*, April 1961, 26; "The Drivers in the Economy Run," *Motor Trend*, April 1961, 28; "What Does the Economy Run Prove?" *Motor Trend*, April 1961, 31; Don Werner, "How Much Economy Run Economy?" *Motor Trend*, June 1961, 64; "Economy-Run Winners Tell You How to Save Gas," *Popular Science*, June 1961, 79; "Falcon Flies with Economy," *Today's Motor Sports*, June 1961, 38.

63. Bob Russo, "Featherfoot Trail," *Motor Life*, June 1961, 56.

64. Don Werner, "The Mobilgas Economy Run," *Motor Trend*, April 1961, 24.

65. "What Does the Economy Run Prove?" *Motor Trend*, April 1961, 31.

66. Don Werner, "How Much Economy Run Economy?" *Motor Trend*, June 1961, 64.

Chapter 8

1. United States Auto Club, "Official Rules 1962 Mobil Economy Run," circa 1961, 1.

2. Mobil Oil Corporation, Mobil News Release, March 6, 1967.

3. Mobil Oil Company, Mobil Economy Run Press Release, October 31, 1961.

4. Mobil Oil Company, Mobil Economy Run Press Release, February 26, 1962.

5. United States Auto Club, "Competitive Divisions for the 1962 Mobil Economy Run," November 20, 1961.

6. Bob Cahill with John Sloan, "MoPar and the Mobil Economy Run, Part 3," *High Performance MoPar*, January 1996, 43.

7. William Carroll, "Run Fun—Life and Hard Times Along the Route" by William Carroll *Automotive News*, April 16, 1962, 42.

8. William Carroll, "Reporters Prove Economy Too," *1962 Mobil Economy Run Post-Run Fact Book*," Mobil Oil Company, circa 1962, 7.

9. Mobil Oil Company, "Big Road Show," Mobil Economy Run Press Release, April 1, 1962.

10. Bill Dredge, "Auto Economy Run to Start in Pasadena," *Los Angeles Times*, March 31, 1962.

11. Ibid.

12. Bill Dredge, "Cars Battle for Economy Run Lead," *Los Angeles Times*, April 2, 1962.

13. William Carroll, "Rambler, Pontiacs Star in Run," *Automotive News*, April 16, 1962, 1.

14. Ibid.

15. Bill Dredge, "Michigan Driver Leads Field in Economy Run," *Los Angeles Times*, April 3, 1962.

16. William Carroll, "Rambler, Pontiacs Star in Run," *Automotive News*, April 16, 1962, 1; United States Auto Club, "Final Detailed Results 1962 Mobil Economy Run," circa 1962.

17. Bill Dredge, "Cars Battle for Economy Run Lead," *Los Angeles Times*, April 2, 1962.

18. William Carroll, "Rambler, Pontiacs Star in Run," *Automotive News*, April 16, 1962, 1; "Thompsons Press On in Gas Run," *Long Beach Independent*, April 3, 1962.

19. *1962 Mobil Economy Run Post-Run Fact Book*," Mobil Oil Company, circa 1962, 17.

20. United States Auto Club, "1962 Mobil Economy Run Third Day Itinerary Tucumcari–Tulsa," April 2, 1962.

21. *1962 Mobil Economy Run Post-Run Fact Book*," Mobil Oil Company, circa 1962, 17.

22. William Carroll, "Rambler, Pontiacs Star in Run," *Automotive News*, April 16, 1962, 1.

23. Wayne Thomas, "Expensive Economy Run," *Car and Driver*, July 1962, 84.

24. United States Auto Club, "1962 Mobil Economy Run Fourth Day Itinerary Tulsa–St. Louis," April 3, 1962.

25. Dan Boone, "Head Winds Slow Economy Run," *Los Angeles Herald-Examiner*, April 4, 1962.

26. William Carroll, "Rambler, Pontiacs Star in Run," *Automotive News*, April 16, 1962, 1.

27. "Rambler Is Champ in Economy Run," *Popular Science*, June 1962, 12.

28. Bill Dredge, "Viland's Rambler Still Leads Economy Run," *Los Angeles Times*, April 5, 1962; "Rambler Is Champ in Economy Run," *Popular Science*, June 1962, 12.

29. Bill Dredge, "Viland's Rambler Still Leads Economy Run," *Los Angeles Times*, April 5, 1962; Wayne Thomas, "Expensive Economy Run," *Car and Driver*, July 1962, 84; Bill Thomas, "A Woman Driver Worth Copying," *Los Angeles Times*, March 31, 1963.

30. Bill Dredge, "Viland's Rambler Still Leads Economy Run," *Los Angeles Times*, April 5, 1962.

31. United States Auto Club, "1962 Mobil Economy Run Sixth Day Itinerary Chicago–Detroit," April 5, 1962.
32. Bill Dredge, "Rambler Wins Honors as Economy Run Ends," *Los Angeles Times*, April 6, 1962.
33. "Annual Mobilgas Lightfoot Derby," *Car Life*, July 1962, 57.
34. Ibid.
35. Mobil Oil Company, "Mobil Economy Run ... Demonstrated Mileage Possible with Highest Megatane-Rated Mobil," print advertisement, circa 1962.
36. American Motors Corporation, "Flash! Rambler American '400' Wins Class-A," print advertisement, circa 1962.
37. Champion Spark Plug Company, "For Top Economy and Performance, Twice as Many Car Manufacturers ...," print advertisement, circa 1962.
38. Ford Division, Ford Motor Company, "Ford Falcon '62 Best Shape Economy's Ever Been In!" print advertisement, circa 1962
39. Ford Division, Ford Motor Company, *1962 News Flash*, vol. 62 CM 40-L6, April 1962.
40. Chevrolet Motor Division, General Motors Corporation, "Chevrolets Demonstrate Outstanding Economy in Mobil Run!" print advertisement, circa 1962.
41. Pontiac Motor Division, General Motors Corporation, "Hey, Little Brother, I Guess We Showed 'Em in the 1962 Mobil Economy Run," print advertisement, circa 1962.
42. Pontiac Motor Division, General Motors Corporation, 1962 model year brochure, SP 1394 4–62.
43. Buick Motor Division, General Motors Corporation, "V-8 Buick Special Racks Up 23.68 M.P.G.," print advertisement, circa 1962.
44. Chrysler-Plymouth Division, Chrysler Corporation, "Plymouth Makes It Six Straight," print advertisement, circa 1962.
45. Chrysler-Plymouth Division, Chrysler Corporation, "Chrysler New Yorker Wins Class "H" in Mobil Economy Run," print advertisement, circa 1962.
46. Chrysler-Plymouth Division, Chrysler Corporation, "Chrysler Scores Advertising First at Mobil Economy Run Finish Line," press release No. 4362; William Carroll, "Rambler, Pontiacs Star in Run," *Automotive News*, April 16, 1962, 42.
47. "Rambler American Economy Run Victor," *Automobile Topics*, May 1962, 21; William Carroll, "Rambler, Pontiacs Star in Run," *Automotive News*, April 16, 1962, 1; Wayne Thomas, "Expensive Economy Run," *Car and Driver*, July 1962, 84; "Annual Mobilgas Lightfoot Derby," *Car Life*, July 1962, 57; "Rambler Is Champ in Economy Run," Popular Science, June 1962, 12.
48. William Carroll, "Reporters Prove Economy Too," *1962 Mobil Economy Run Post-Run Fact Book*," Mobil Oil Company, circa 1962, 7.
49. Bob Thomas, "Restraint Key to Gas Economy," *1963 Mobil Economy Run Fact Book*, Mobil Oil Company, circa 1963, 7.
50. Mobil Oil Company, "1963 Run Announcement Story," Mobil Economy Run Press Release, February 18, 1962.
51. United States Auto Club, "Competitive Divisions for the 1963 Mobil Economy Run," November 26, 1962.
52. Tom Swain, "The Feather-Foot Derby — Some Predictions on the Mobil Economy Run," *Car Life*, April 1963, 44.
53. United States Auto Club, "Official Rules 1963 Mobil Economy Run," circa 1962, 25.
54. Chrysler-Plymouth News Bureau, Chrysler Corporation, "Thumbnail Biographies Chrysler-Plymouth Drivers and Co-Drivers 1963 Mobil Economy Run," circa 1963.
55. *United States Auto Club Yearbook 1963*, United States Auto Club, circa 1964, 6.
56. Leon J. Pinkson, "A History of Motoring Progress in San Francisco," *San Francisco History — History of Motoring*, accessed October 31, 2012; http://www.sfgenealogy,com/sfhistory/hgaut.htm.
57. Jim Whipple, "The Economy Runs ... Do Drivers or Gimmicks Win Them?" *Popular Mechanics*, July 1963, 65.
58. "The Race for Economy: An Inside Look into Preparations for the Annual Mobil Economy Run," *Road & Track*, April 1963, 64.
59. "Extra Economy's There for Skillful Driver, 1963 Mobilgas Economy Run Proves Again," *Auto Club News Pictorial*, Automobile Club of Southern California, June 1963, Part II, 5.
60. "What Price Economy?" *Car Life*, July 1963, 67.
61. Tom Swain, "The Feather-Foot Derby — Some Predictions on the Mobil Economy Run," *Car Life*, April 1963, 44.
62. Bob Thomas, "Viland Takes First Day Lead in Economy Run," *Los Angeles Times*, March 31, 1963.
63. Mobil Oil Company, "Entries Close for Mobil Run," Mobil Economy Run Press Release, March 19, 1963.
64. *1963 Mobil Economy Run Fact Book*, Mobil Oil Company, circa 1963, 16.
65. "Rambler Ahead in Gas Run," *Los Angeles Herald-Examiner*, April 1, 1963 ;W.W. Jackson, "Mobil Oil Economy Run," *Auto Sports*, July 1963, 43.
66. "Rambler Leads Event After First Two Days," *Ogden Standard Examiner*, April 1, 1963.
67. W.W. Jackson, "Mobil Oil Economy Run," *Auto Sports*, July 1963, 43.
68. Ibid.
69. Bob Thomas, "Blue Skies Aid Economy Run Drivers," *Los Angeles Times*, April 2, 1963.
70. Floyd Freel, "Cars Reach Halfway in Mileage Run," *Pasadena Independent*, April 2, 1963.
71. Bob Thomas, "Restraint Key to Gas Economy," *1963 Mobil Economy Run Fact Book*, Mobil Oil Company, circa 1963, 7.
72. Bob Thomas, "Blue Skies Aid Economy Run Drivers," *Los Angeles Times*, April 2, 1963.
73. United States Auto Club, "1963 Mobil Economy Run Fourth Day Itinerary Cheyenne–Omaha," April 2, 1963.
74. Bob Thomas, "12 Economy Run Cars 'Lost' in Route Mixup," *Los Angeles Times*, April 3, 1963.
75. Ibid.; "Freeway Leads 12 Mobil Economy Run Cars Astray," *Los Angeles Herald-Examiner*, April 3, 1963; W.W. Jackson, "Mobil Oil Economy Run," *Auto Sports*, July 1963, 43; Mobil Oil Company, "Army of Workers," Mobil Economy Run Press Release, Marchl 8, 1964.
76. "Freeway Leads 12 Mobil Economy Run Cars Astray," *Los Angeles Herald-Examiner*, April 3, 1963.
77. United States Auto Club, Certificate of Stock Status and Performance, circa 1963.
78. W.W. Jackson, "Mobil Oil Economy Run," *Auto Sports*, July 1963, 43.
79. "Mobil Run Drivers Arrive in Chicago," *Los Angeles Herald-Examiner*, April 4, 1963.
80. W.W. Jackson, "Mobil Oil Economy Run," *Auto Sports*, July 1963, 43.
81. Bob Thomas, "Economy Run Autos Near Finish" *Los Angeles Times*, April 4, 1963.
82. "W.W. Jackson, "Mobil Oil Economy Run," *Auto Sports*, July 1963, 43; "What Price Economy?" *Car Life*, July 1963, 67.
83. "What Price Economy?" *Car Life*, July 1963, 67.
84. "Extra Economy's There for Skillful Driver, 1963 Mobilgas Economy Run Proves Again," *Auto Club News Pictorial*, Automobile Club of Southern California, June 1963, Part II, 5.
85. W.W. Jackson, "Mobil Oil Economy Run," *Auto Sports*, July 1963, 43.
86. "What Price Economy?" *Car Life*, July 1963, 67.
87. W.W. Jackson, "Mobil Oil Economy Run," *Auto Sports*, July 1963, 43.
88. Tom Swain, "The Feather-Foot Derby — Some Predictions on the Mobil Economy Run," *Car Life*, April 1963, 44.
89. Bob Thomas, "Restraint Key to Gas Economy," *1963 Mobil Economy Run Fact Book*, Mobil Oil Company, circa 1963, 7.
90. Mobil Oil Company, "Results 1963 Mobil Economy Run," print advertisement, circa 1963.
91. American Motors Corporation, "Flash! From Mobil Economy Run Rambler American "440" Wins Class C 28.61 Miles Per Gallon," print advertisement, circa 1963.
92. American Motors Corporation, "Go for Fun — Save a Bundle," print advertisement, circa 1963.
93. Ford Division, Ford Motor Company, "Total Performance," print advertisement, circa 1963.
94. Ford Division, Ford Motor Company, 1964 model year performance brochure, 11.
95. Chrysler-Plymouth Division, Chrysler Corporation, "Of *Course* Valiant Scored 26.26 M.P.G. in the Mobil Economy Run — Valiant Is a Great *Economy* Car!" print advertisement, circa 1963.
96. Chrysler-Plymouth Division, Chrysler Corporation, "Camel Gets 13.33 M.P.G. in Sahara Economy Run," print advertisement, circa 1963.
97. Buick Motor Division, General Motors Corporation, "23.81 Average MPG Los Angeles to Detroit," print advertisement, circa 1963.
98. Buick Motor Division, General Motors Corporation, 1963 model year brochure.
99. Chrysler-Plymouth Division, Chrysler Corporation, "Terrific Combination!" print advertisement, circa 1963.
100. Oldsmobile Division, General Motors Corporation, "Oldsmobile's Dynamic 88 Wins Class "G" in Mobil Economy Run," print advertisement, circa 1963.
101. W.W. Jackson, "Mobil Oil Economy Run," *Auto Sports*, July 1963, 43.
102. Ibid.; "Rambler Wins Overall Economy Run Honors," *Automobile Topics*, May 1963, 14; "What Price Economy?" *Car Life*, July 1963, 67; "What Do They Prove? *Motorcade*, Spring 1964, 24; "Mobil Economy Run Winners," *Motor Trend*, June 1963, 80; Jim Whipple, "The Economy Runs ... Do Drivers or Gimmicks Win Them?" *Popular*

Mechanics, July 1963, 65; "The Race for Economy: An Inside Look into Preparations for the Annual Mobil Economy Run," *Road & Track*, April 1963, 64.

103. "What Price Economy?" *Car Life*, July 1963, 67.

104. Jim Whipple, "The Economy Runs ... Do Drivers or Gimmicks Win Them?" *Popular Mechanics*, July 1963, 65.

105. "Daytona Trials," *Motor Trend*, April 1963, 70; Hubert Luckett, "'Keep-'Em-Honest' Car Tests," *Popular Science*, April 1963, 67.

Chapter 9

1. Mobil Oil Company, "Official Results Mobil Economy Run ... America's Mileage Classic," brochure, circa 1964.

2. Chevrolet Motor Division, General Motors Corporation, "The Teen Team: Its Background and Special Schooling for Competition in the Mobil Run," News of the Chevy Teen Team Press Release, March 12, 1964.

3. Chevrolet Motor Division, General Motors Corporation, "Why Chevrolet Is Using Teenage Drivers in the Mobil Economy Run," News of the Chevy Teen Team Press Release, February 26, 1964.

4. Mobil Oil Company, "1964 Announcement Story for Auto Editors," Mobil Economy Run Press Release, February 13, 1964.

5. United States Auto Club, "Competitive Divisions for the 1964 Mobil Economy Run," November 6, 1963.

6. United States Auto Club, "Official Rules 1964 Mobil Economy Run," circa 1963, 10.

7. United States Auto Club, "Final Detailed Results 1964 Mobil Economy Run," circa 1964.

8. Chevrolet Motor Division, General Motors Corporation, "The Teen Team: Its Background and Special Schooling for Competition in the Mobil Run," News of the Chevy Teen Team Press Release, March 12, 1964.

9. Lincoln-Mercury Division, Ford Motor Company, Press Release 64–4019 (rev.), March 13, 1964.

10. Dodge Division, Chrysler Corporation, "New York Cab Driver to Compete in Mobil Economy Run," Dodge News Bureau Press Release, circa 1964.

11. Bob Cahill with John Sloan, "Mopar And The Mobil Economy Run," *High Performance Mopar*, October 1995, 31.

12. Buick Motor Division, General Motors Corporation, publicity photographs of cars and drivers at the impound with billboard in the background, circa 1964.

13. Bob Thomas, "Economy Run Drivers Throttle Down Today," *Los Angeles Times*, April 3, 1964.

14. United States Auto Club, "1964 Mobil Economy Run First Day Itinerary Los Angeles-Phoenix," April 3, 1964.

15. Bob Thomas, "Teen-Age Drivers Best Pros in Economy Run," *Los Angeles Times*, April 4, 1964.

16. Ibid.

17. Bob Thomas, "The Kids Are Great, but Experience Hard to Beat," *Los Angeles Times*, April 12, 1964.

18. United States Auto Club, "Final Detailed Results 1964 Mobil Economy Run," circa 1964.

19. Bob Thomas, "The Kids Are Great, but Experience Hard to Beat," *Los Angeles Times*, April 12, 1964.

20. United States Auto Club, "Final Detailed Results 1964 Mobil Economy Run," circa 1964; Bob Thomas, "Mobil Economy Run Bucks Rough Weather," *Los Angeles Times*, April 5, 1964.

21. John Hawkins, "Economy Run '64," *Auto and Motor Sports Illustrated*, July 1964.

22. Bob Thomas, "Mobil Economy Run Bucks Rough Weather," *Los Angeles Times*, April 5, 1964.

23. United States Auto Club, "Final Detailed Results 1964 Mobil Economy Run," circa 1964.

24. Bob Thomas, "Teen-Agers Lagging in Mobil Run," *Los Angeles Times*, April 6, 1964.

25. Bob Thomas, "The Kids Are Great, but Experience Hard to Beat," *Los Angeles Times*, April 12, 1964.

26. United States Auto Club, "Final Detailed Results 1964 Mobil Economy Run," circa 1964.

27. Bob Thomas, "Teen-Agers Lagging in Mobil Run," *Los Angeles Times*, April 6, 1964.

28. United States Auto Club, "Final Detailed Results 1964 Mobil Economy Run," circa 1964.

29. Ibid.; *United States Auto Club Yearbook 1964*, United States Auto Club, circa 1965, 87.

30. United States Auto Club, "Final Detailed Results 1964 Mobil Economy Run," circa 1964.

31. Bob Thomas, "Factory Teams Follow Along on Economy Run," *Los Angeles Times*, April 7, 1964; *United States Auto Club Yearbook 1964*, United States Auto Club, circa 1965, 87.

32. Bob Thomas, "North Hollywood Girl Shines in Mobil Run," *Los Angeles Times*, April 8, 1964.

33. United States Auto Club, "1964 Mobil Economy Run Sixth Day Itinerary Indianapolis-Harrisburg," April 8, 1964.

34. Bob Thomas, "45 Cars Face Final Leg in Auto Economy Run," *Los Angeles Times*, April 9, 1964.

35. "Chevy's Teen Team Proves Its Skill," *Friends Magazine*, Chevrolet Motor Division, General Motors Corporation, September 1964, 9.

36. United States Auto Club, "1964 Mobil Economy Run Sixth Day Itinerary Indianapolis-Harrisburg," April 8, 1964.

37. Bob Thomas, "The Kids Are Great, but Experience Hard to Beat," *Los Angeles Times*, April 12, 1964.

38. Bob Thomas, "45 Cars Face Final Leg in Auto Economy Run," *Los Angeles Times*, April 9, 1964.

39. The Auto Editors of *Consumer Guide*, *50 Years of American Automobiles 1939–1989* (Lincolnwood, IL: Publications International, 1989).

40. Chevrolet Motor Division, General Motors Corporation, Press Release 3657, April 9, 1964.

41. "Teen-Age Driving Skill," *Hot Rod Magazine*, July 1964, 91.

42. Chevrolet Motor Division, General Motors Corporation, Press Release 3657, April 9, 1964.

43. Bob Thomas, "The Kids Are Great, but Experience Hard to Beat," *Los Angeles Times*, April 12, 1964.

44. Mobil Oil Company, "Here Are the results of the 1964 Mobil Economy Run," print advertisement, circa 1964.

45. American Motors Corporation, "Rambler American 440 Wins Class B with 27.83 M.P.G.," print advertisement, circa 1964.

46. Champion Spark Plug Company, "For Top Economy, Rambler Installs Champion Spark Plugs Right at the Factory," print advertisement, circa 1964.

47. Buick Motor Division, General Motors Corporation, "We Only Won Three Classes," print advertisement, circa 1964.

48. Buick Motor Division, General Motors Corporation, print advertisement, *Motor Trend*, July 1964, 9.

49. Buick Motor Division, General Motors Corporation, 1964 model year brochure.

50. Buick Motor Division, General Motors Corporation, 1964 model year salesman's brochure.

51. Chrysler Corporation, "Most Wins By One Manufacturer in 1964 Mobil Economy Run," print advertisement, circa 1964.

52. Chrysler-Plymouth Division, Chrysler Corporation, 1964 model year performance brochure.

53. Chevrolet Motor Division, General Motors Corporation, "We'd Like to Say Some Nice Things About America's Young Adult Drivers," print advertisement, circa 1964.

54. Chevrolet Motor Division, General Motors Corporation, 1964 model year economy performance brochure.

55. Chevrolet Motor Division, General Motors Corporation, "Chevy's Teen Team Proves Its Skill," *Friends Magazine*, September 1964, 9.

56. John Hawkins, "Economy Run '64," *Auto and Motor Sports Illustrated*, July 1964; "Teen-Age Driving Skill," *Hot Rod Magazine*, July 1964, 91; Earl Schilling, "How I Learned to Be a Teen-Age Gas Saver," *Popular Science*, June 1964, 58.

57. Jim Whipple, "Detroit Listening Post," *Popular Mechanics*, July 1964, 34.

58. John Hawkins, "Economy Run '64," *Auto and Motor Sports Illustrated*, July 1964.

59. Bob Knoll, "Coast to Coast in the Pursuit of Economy," *New York Times*, December 24, 2006.

60. Shell Oil Company, television commercial, circa 1964.

61. Mobil Oil Company, "Longest Economy Run to Field 45 Cars," Mobil Economy Run Press Release, February 15, 1965.

62. United States Auto Club, "Competitive Divisions for the 1965 Mobil Economy Run," November 6, 1964.

63. Mobil Oil Company, "Longest Economy Run to Field 45 Cars," Mobil Economy Run Press Release, February 15, 1965.

64. United States Auto Club, "Competitive Divisions for the 1965 Mobil Economy Run," November 6, 1964.

65. United States Auto Club, "Official Rules 1965 Mobil Economy Run," circa 1964, 10.

66. William Carroll, "Close Margins in Economy Run" *Automotive News*, April 19, 1965, 2.

67. United States Auto Club, "Official Rules 1965 Mobil Economy Run," circa 1964, 7.

68. Lincoln-Mercury Division, Ford Motor Company, Press Release 65–5023, circa 1965.

69. John A. Heitmann, "The Race That Wasn't a Race: The Mobilgas Economy Run 1936–1968," *Automotive History Review*, Fall 2008, 24.

70. Don Royer, "Car 39 Covers 1965 Mobil Economy Run," *The Mobil Dealer News*, April 1965, 2.

71. Wayne Thomas, "Behind the Bedlam of the Featherfoot Fleet," *Motor Trend*, March 1966, 60.
72. Bob Thomas, "'Rookie' Driver Leader of Mobil Economy Run," *Los Angeles Times*, April 5, 1965.
73. Mobil Oil Company, Mobil Economy Run Press Release, April 5, 1965.
74. Bob Thomas, "Girl Driver Gets Stuck in Mud on Economy Run," *Los Angeles Times*, April 6, 1965.
75. Ibid.; United States Auto Club, "Final Detailed Results 1965 Mobil Economy Run," circa 1965.
76. United States Auto Club, "Final Detailed Results 1965 Mobil Economy Run," circa 1965.
77. Daniel Boone, "Flagstaff Snow Greets Economy Run," *Los Angeles Herald-Examiner*, April 6, 1965.
78. United States Auto Club, "1965 Mobil Economy Run Time Schedule and Route—1st Day—April 5th."
79. Don Royer, "Car 39 Covers 1965 Mobil Economy Run," *The Mobil Dealer News*, April 1965, 2.
80. United States Auto Club, "1965 Mobil Economy Run Time Schedule and Route—2nd Day—April 6th."
81. Jean Calvin, "Mobil Economy Run," *Sports Car Graphic*, July 1965, 66.
82. United States Auto Club, "Final Detailed Results 1965 Mobil Economy Run," circa 1965.
83. Ibid.
84. United States Auto Club, 1965 Mobil Economy Run Time Schedule and Route—2nd Day—April 6th."
85. United States Auto Club, "Final Detailed Results 1965 Mobil Economy Run," circa 1965.
86. Bob Thomas, "Secret Car 'Unveiled' During Economy Run," *Los Angeles Times*, April 7, 1965.
87. Daniel Boone, "Mobil Run Cars Hold Leads by Narrow Margins," *Los Angeles Herald-Examiner*, April 8, 1965.
88. Don Royer, "Car 39 Covers 1965 Mobil Economy Run," *The Mobil Dealer News*, April 1965, 2.
89. Jean Calvin, "Mobil Economy Run," *Sports Car Graphic*, July 1965, 66.
90. Bob Thomas, "Economy Run Leaders Only 105 Feet Apart," *Los Angeles Times*, April 8, 1965.
91. Daniel Boone, "Mobil Run Cars Hold Leads by Narrow Margins," *Los Angeles Herald-Examiner*, April 8, 1965.
92. United States Auto Club, "Final Detailed Results 1965 Mobil Economy Run," circa 1965.
93. Don Royer, "Car 39 Covers 1965 Mobil Economy Run," *The Mobil Dealer News*, April 1965, 2.
94. Leroi Tex Smith, "The Kids Put 'Em Down!" *Car Craft*, July 1965 30.
95. Bob Thomas, "Economy Test Driver Runs Out of Gasoline," *Los Angeles Times*, April 10, 1965.
96. United States Auto Club, "Final Detailed Results 1965 Mobil Economy Run," circa 1965.
97. Daniel Boone, "A Fast Pace," *Los Angeles Herald-Examiner*, April 18, 1965.
98. Hugh Randolph, "Ford Jumps Into Lead For Economy Run Honors," *Oakland Tribune*, April 9, 1965.
99. Daniel Boone, "Mobil Field Pared; Entry Out of Gas," *Los Angeles Herald-Examiner*, April 10, 1965.
100. Bob Thomas, "Economy Test Driver Runs Out of Gasoline," *Los Angeles Times*, April 10, 1965.
101. Don Royer, "Car 39 Covers 1965 Mobil Economy Run," *The Mobil Dealer News*, April 1965, 2; United States Auto Club, "Final Detailed Results 1965 Mobil Economy Run," circa 1965.
102. United States Auto Club, "1965 Mobil Economy Run Time Schedule and Route—5th Day—April 9th."
103. United States Auto Club, "1965 Mobil Economy Run Sixth Day Itinerary, Niagara Falls—Albany, April 10, 1965."
104. Ibid.
105. *United States Auto Club Yearbook 1965*, United States Auto Club, circa 1966, 95.
106. Daniel Boone, "Mobil Mileage Run Ends Today; Three Very Tight Races," *Los Angeles Herald-Examiner*, April 11, 1965.
107. Don Royer, "Car 39 Covers 1965 Mobil Economy Run," *The Mobil Dealer News*, April 1965, 2.
108. Bob Thomas, "Economy Title Goes to Rambler American," *Los Angeles Times*, April 12, 1965.
109. William Carroll, "Close Margins in Economy Run," *Automotive News*, April 19, 1965, 2.
110. Terry Robards, "Mobil Run: An Expert Can Run Out of Gas, Too," *New York Herald-Tribune*, April 11, 1965.
111. Bob Thomas, "Economy Run Drivers Scale Down Trip Costs," *Los Angeles Times*, April 12, 1965.
112. Mobil Oil Company, "Mobil Economy Run Official Results," print advertisement, circa 1965.
113. Mobil Oil Company, print advertisement, *Automotive News (1965 Almanac Issue)*, 266.
114. American Motors Corporation, "Rambler American 440 Wins Class B with 25.655 M.P.G.," print advertisement, circa 1965.
115. Chrysler-Plymouth Division, Chrysler Corporation, 1965 model year brochure.
116. Bob Thomas, "Economy Run Drivers Scale Down Trip Costs," *Los Angeles Times*, April 12, 1965.
117. "Roughest, Toughest and Longest Mobil Economy Run," *Auto Topics*, July 1965, 48; Leroi Tex Smith, "The Kids Put 'Em Down!" *Car Craft*, July 1965, 30; Dick Wells, "Mobil's Mileage Misers," *Hot Rod Magazine*, June 1965, 87; John Ethridge, "Mobil's Mileage Melee," *Motor Trend*, July 1965, 64; Jean Calvin, "Mobil Economy Run," *Sports Car Graphic*, July 1965, 66.
118. Leroi Tex Smith, "The Kids Put 'Em Down!" *Car Craft*, July 1965, 30.
119. Wayne Thomas, "Behind the Bedlam of the Featherfoot Fleet," *Motor Trend*, March 1966, 60.
120. Matt Stone, "High Gas Prices a Bummer?" *Motor Trend*, January 2006, 172.
121. "Santa Monica Lad to Drive in Dodge Economy Run," *Jet*, April 1, 1965, 49; "Teen-Age Driver Sets Precedent," *Ebony*, July 1965, 49.
122. Don Royer, "Car 39 Covers 1965 Mobil Economy Run," *The Mobil Dealer News*, April 1965, 2.
123. Dick Wells, "Mobil's Mileage Misers," *Hot Rod Magazine*, June 1965, 87.
124. Hugh Randolph, "Rambler Edges Falcon in Mobil Run," *Oakland Tribune*, April 12, 1965.

Chapter 10

1. Mobil Oil Corporation, Mobil Economy Run Press Release, January 26, 1966.
2. United States Auto Club, "Competitive Divisions for the 1966 Mobil Economy Run," October 22, 1965.
3. United States Auto Club, "Official Rules 1966 Mobil Economy Run," circa 1965, 7.
4. United States Auto Club, "Final Detailed Results 1966 Mobil Economy Run," circa 1966.
5. Mobil Oil Corporation, Mobil Economy Run Press Release, February 21, 1966.
6. Mobil Oil Corporation, "Driver Biographies," 1966 Mobil Economy Run Press Booklet.
7. "Street Hemi, Unofficial Mobil Run Entry, Performs Flawlessly," *Chrysler-Plymouth Division Times*, Chrysler Corporation, April 1966, 2.
8. Mobil Oil Corporation, Mobil Economy Run Press Release, March 11, 1966.
9. Bruce Young, "Nine-Time Winner Leads Economy Run," *Long Beach Press Telegram*, March 15, 1966; Daniel Boone, "Crash Mars Mobil Economy Run," *Los Angeles Herald-Examiner*, March 15, 1966; Bob Thomas, "Auto Economy Run Proves to Be Expensive," *Los Angeles Times*, March 21, 1966; Hugh Randolph, "Injury Accident Mars Economy Run," *Oakland Tribune*, March 15, 1966; "Economy Run Cars Crashed," *Yuma Sun*, March 15, 1966.
10. Daniel Boone, "Crash Mars Mobil Economy Run," *Los Angeles Herald-Examiner*, March 15, 1966.
11. Hugh Randolph, "Mountains Cut In on Mobil Car Run," *Oakland Tribune*, March 16, 1966.
12. United States Auto Club, "1966 Mobil Economy Run Time Schedule and Route—2nd Day—March 15th."
13. Dean Reeder, "50 Cars Resume Trip on Mobil Economy Run," *Albuquerque Tribune*, March 16, 1966.
14. Daniel Boone, "Mountains Take Toll of Economy Run's Fuel Average," *Los Angeles Herald-Examiner*, March 16, 1966.
15. Hugh Randolph, "Mountains Cut In on Mobil Car Run," *Oakland Tribune*, March 16, 1966.
16. Hugh Randolph, "Leaders Keep Pace on Mobil Run," *Oakland Tribune*, March 17, 1966; *United States Auto Club Yearbook 1966*, United States Auto Club, circa 1967, 93.
17. Frank Mower and Bob Dunker, "The Featherfoot Drags," *Car Craft*, July 1966, 62.
18. Bob Cahill with John Sloan, "Mopar and the Mobil Economy Run, Part 3," *High Performance Mopar*, January 1966, 43.
19. Daniel Boone, "Rambler Holds the Lead," *Los Angeles Herald-Examiner*, March 17, 1966.
20. United States Auto Club, "1966 Mobil Economy Run Fourth Day Itinerary, Wichita—St. Louis, March 17, 1966."
21. Daniel Boone, "Rambler Keeping the Lead," *Los Angeles Herald-Examiner*, March 18, 1966.
22. Frank Mower and Bob Dunker, "The Featherfoot Drags," *Car Craft*, July 1966, 62.
23. Bruce Young, "Speed King Arfons Shines in New Role," *Long Beach Press Telegram*, March 18, 1966.
24. "Accident Mars Economy Run Here," *Terra Haute Star*, March 19, 1966.

25. Bob Thomas, "Auto Economy Run Proves to Be Expensive," *Los Angeles Times*, March 21, 1966.
26. *United States Auto Club Yearbook 1966*, United States Auto Club, circa 1967, 93.
27. Frank Mower and Bob Dunker, "The Featherfoot Drags," *Car Craft*, July 1966, 62.
28. Daniel Boone, "Weather Perils Drivers," *Los Angeles Herald-Examiner*, March 19, 1966.
29. Frank Mower and Bob Dunker, "The Featherfoot Drags," *Car Craft*, July 1966, 62.
30. United States Auto Club, "1966 Mobil Economy Run Sixth Day Itinerary, Cleveland — Philadelphia, March 19, 1966."
31. "Cars Pause Here During Economy Run," *Bedford Gazette*, March 20, 1966.
32. United States Auto Club, "1966 Mobil Economy Run Seventh Day Itinerary, Philadelphia — Boston March 20, 1966."
33. Ibid.
34. Evelyn Swift, "The Fifth Wheel," *Los Angeles News-Advertiser*, March 24, 1966.
35. Bob Thomas, "Auto Economy Run Proves to Be Expensive," *Los Angeles Times*, March 21, 1966.
36. Hugh Randolph, "52 Cars Pull Away in Start of Longest Mobil Economy Run," *Oakland Tribune*, March 14, 1966.
37. Mobil Oil Corporation, "How Far Can One Gasoline Go to Show How Far You Can Go?" print advertisement, circa 1966.
38. Mobil Oil Corporation, "1966 Mobil Economy Run Proves Gasoline Mileage Potential in All Cars Like Yours," print advertisement, circa 1966.
39. Mobil Oil Corporation, "Mobil Proves It in Public," print advertisement, circa 1966.
40. American Motors Corporation, "Rambler American Wins Mobil Economy Run — Class A ... and Does It with New Acceleration Champ Engine!" print advertisement, circa 1966.
41. American Motors Corporation, "Who Were the Big Winners in the Mobil Economy Run?" television commercial, circa 1966.
42. Chevrolet Motor Division, General Motors Corporation, "Chevrolet's No. 1 in Mobil Economy Run!" print advertisement, circa 1966.
43. Chevrolet Motor Division, General Motors Corporation, 1966 model year dealer promotional packet.
44. Chevrolet Motor Division, General Motors Corporation, 1966 model year salesman's brochure.
45. Buick Motor Division, General Motors Corporation, "Lady Driver Repeats! Buick LeSabre Wins First Place in Mobil Economy Run — Class G!" print advertisement, circa 1966.
46. Buick Motor Division, General Motors Corporation, 1966 model year brochure.
47. Buick Motor Division, General Motors Corporation, 1967 model year brochure, 22.
48. Oldsmobile Division, General Motors Corporation, "New Honors for the Car-of-the-Year!" print advertisement, circa 1966.
49. "Mobilgas Economy Run" *Auto Topics*, June 1966, 20; Frank Mower and Bob Dunker, "The Featherfoot Drags," *Car Craft*, July 1966, 62.
50. Mobil Oil Corporation, "Mobil Economy Run History," 1966 Mobil Economy Run Press Booklet.
51. John A. Heitmann, "The Race That Wasn't a Race: The Mobilgas Economy Run 1936–1968," *Automotive History Review*, Fall 2008, 25.
52. Mobil Oil Corporation, "Driver Biographies," 1966 Mobil Economy Run Press Booklet.
53. John A. Heitmann, "The Race That Wasn't a Race: The Mobilgas Economy Run 1936–1968," *Automotive History Review*, Fall 2008, 25.
54. Ibid.
55. Griff Borgeson, "Art Pillsbury," *Road & Track*, December 1966, 88.
56. Mobil Oil Corporation, Mobil Economy Run Press Release 67282ce, circa 1967.

Chapter 11

1. Mobil Oil Corporation, "1967 Mobil Economy Run 6-Day, L.A. — Detroit Event," Mobil Economy Run Press Release, # 67172 1-A 1-C, March 9, 1967.
2. Mobil Oil Corporation, Mobil Economy Run Route Press Photo Caption Sheet, 67202.
3. United States Auto Club, "Official Rules 1967 Mobil Economy Run," circa 1966, 7.
4. United States Auto Club, "Competitive Divisions for the 1967 Mobil Economy Run," November 25, 1966.
5. Mobil Oil Corporation, "Driver Biographies," 1967 Mobil Economy Run Press Booklet.
6. J.J. Faerber, Buick Sales-Engineering Coordinator to Chief Steward Art Rene, March 14, 1967.
7. Hugh Randolph, "42 Cars Head Out on First Lap of Mobil Economy Run," *Oakland Tribune*, April 4, 1967.
8. Bob Cahill with John Sloan, "MoPar and the Mobil Economy Run, Part 3," *High Performance MoPar*, January 1996, 43.
9. United States Auto Club, 1967 Mobil Economy Run First Day Itinerary Los Angeles-Lake Tahoe, April 4, 1967.
10. United States Auto Club, "1967 Mobil Economy Run Time Schedule and Route with Changes in Course Itinerary — 1st Day — April 4th."
11. Hugh Randolph, "Gas Misers Fight Rain," *Oakland Tribune*, April 5, 1967.
12. United States Auto Club, "Final Detailed Results 1967 Mobil Economy Run," circa 1967.
13. Mobil Oil Corporation, "The 1967 Mobil Economy Run — A Hazardous Record Breaker," Mobil Economy Run Press Release, dated April 9, 1967.
14. Hugh Randolph, "Gas Misers Fight Rain," *Oakland Tribune*, April 5, 1967.
15. United States Auto Club, "Final Detailed Results 1967 Mobil Economy Run," circa 1967.
16. United States Auto Club, "Final Detailed Results 1967 Mobil Economy Run," circa 1967.
17. Hugh Randolph, "3-Car Fight for Mileage Title Developing in Mobil Run," *Oakland Tribune*, April 6, 1967.
18. United States Auto Club, "Final Detailed Results 1967 Mobil Economy Run," circa 1967.
19. Hugh Randolph, "Chrysler Leads 4 Classes in the Mobil Economy Run," *Oakland Tribune*, April 7, 1967.
20. United States Auto Club, "Final Detailed Results 1967 Mobil Economy Run," circa 1967.
21. Daniel Boone, "Economy Run Cars Near Chicago Goal," *Los Angeles Herald-Examiner*, April 8, 1967.
22. Mobil Oil Corporation, "The 1967 Mobil Economy Run — A Hazardous Record Breaker," Mobil Economy Run Press Release, April 9, 1967.
23. Daniel Boone, "More Miles to the Gallon," *Los Angeles Herald-Examiner*, April 10, 1967, *United States Auto Club Yearbook 1967*, United States Auto Club, circa 1968, 97; Mobil Oil Corporation, "The 1967 Mobil Economy Run — A Hazardous Record Breaker," Mobil Economy Run Press Release, April 9, 1967.
24. United States Auto Club, "Final Detailed Results 1967 Mobil Economy Run," circa 1967.
25. United States Auto Club, "Run into Chicago," April 8, 1967.
26. Daniel Boone, "Economy-Run Cars Near Detroit Home," *Los Angeles Herald-Examiner*, April 9, 1967.
27. Hugh Randolph, "Three Plymouths Win Mobil Run," *Oakland Tribune*, April 10, 1967.
28. United States Auto Club, "Final Detailed Results 1967 Mobil Economy Run," circa 1967.
29. Mobil Oil Corporation, "Mobil Run Mileage Average Jumps, Six-Year Downward Trend Reversed," Mobil Economy Run Press Release, April 9, 1967.
30. Daniel Boone, "More Miles to the Gallon," *Los Angeles Herald-Examiner*, April 10, 1967.
31. Mobil Oil Corporation, "The 1967 Mobil Economy Run — A Hazardous Record Breaker," Mobil Economy Run Press Release, April 9, 1967.
32. Hugh Randolph, "Three Plymouths Win Mobil Run," *Oakland Tribune*, April 10, 1967.
33. Sheryl James, "Fuel-Efficient Fun! The Mobil Economy Run," *Michigan History*, September–October 2012.
34. Mobil Oil Corporation, "How Does the Mileage in the 1967 Mobil Economy Run Compare with Yours? Read It and Weep," print advertisement, circa 1967.
35. Mobil Oil Corporation, "Official Results 1967 Mobil Economy Run," brochure, circa 1967.
36. Chrysler-Plymouth Division, Chrysler Corporation, "Plymouth Stuns Automotive World in Mobil Economy Run," print advertisement, circa 1967.
37. Chrysler-Plymouth Division, Chrysler Corporation, "Suggested Ideas for Merchandising — Chrysler-Plymouth's Win in the Mobil Economy Run," circa 1967.
38. Buick Motor Division, General Motors Corporation, "Marta Retzlaff Drove 2837.8 Miles in the Mobil Run to Demonstrate Buick LeSabre's Winning Economy," print advertisement, circa 1967.
39. Buick Motor Division, General Motors Corporation, 1967 model year brochure.
40. American Motors Corporation, "Rambler American Makes Big News This Year...," television commercial, circa 1967.
41. Brock Yates, "Mobil Economy Run: The Thrill of Victory, the Agony of Defeat," *Car and Driver*, July 1967, 44; Jean Calvin, "The Rally Scene," *Sports Car Graphic*, July 1967.
42. Nancy Kennedy, "Ford, Fashion, and the 1967 Mobil Economy Run," *Ford Times*, April 1967, 28.
43. Mobil Oil Corporation, "1968 Mobil Economy Run Adopts 'Visit Vacationland USA' Theme," Mobil Economy Run Press Release, March 5, 1968.
44. United States Auto Club, "Official Rules 1968 Mobil Economy Run," circa 1967.
45. United States Auto Club, "Competitive Divisions for the 1968 Mobil Economy Run," November 25, 1967.
46. Mobil Oil Corporation, "41 Cars Enter

1968 Mobil Economy Run," Mobil Economy Run Press Release No. 68272.

47. Dodge Division, Chrysler Corporation, "From Lady Leadfoot to Female Featherfoot," Dodge Public Relations Press Release No. 31468-TW, circa 1968.

48. Dodge Division, Chrysler Corporation, "Two Housewife-Racers, Auto Writer-Racer and New York Grandmother Cab Driver in Dodges for Mobil Economy Run," Dodge News Bureau Press Release, circa 1968.

49. United States Auto Club, "Break-In Route #1," February 13, 1968.

50. Bob Cahill with John Sloan, "MoPar and the Mobil Economy Run, Part 3," *High Performance MoPar*, January 1996, 43.

51. Dodge Division, Chrysler Corporation, "'East and West Twain' to Be Joined by Former N.Y. Showgirl and Former California Mayor," Dodge Public Relations Press Release, circa 1968.

52. Don MacDonald, "Cars on Trial — Part Two," *Westways Magazine*, Automobile Club of Southern California, June 1968, 40.

53. Robert Beckman, "Anaheim Driver Moves Out in Whopping Lead," *Long Beach Independent*, April 3, 1968.

54. Mobil Oil Corporation, "Second Day Story," Mobil Economy Run Press Release, April 3, 1968.

55. Dan Boone, "Delayed Economy Run on Third Leg," *Los Angeles Herald-Examiner*, April 4, 1968.

56. Dan Boone, "Field Cut in Economy Run," *Los Angeles Herald-Examiner*, April 5, 1968.

57. Mobil Oil Corporation, "Third Day Story," Mobil Economy Press Release, April 4, 1968.

58. Bob Thomas, "Featherfoots to Step on It in Mobil Economy Run," *Los Angeles Times*, March 31, 1968.

59. *United States Auto Club Yearbook 1968*, United States Auto Club, circa 1969, 107.

60. Mobil Oil Corporation, "Third Day Story," Mobil Economy Press Release, April 4, 1968.

61. *United States Auto Club Yearbook 1968*, United States Auto Club, circa 1969, 107.

62. Mobil Oil Corporation, "Fourth Day Story," Mobil Economy Press Release, April 5, 1968.

63. Robert Beckman, "King Mourning Ends Mobil Run," *Long Beach Independent*, April 6, 1968.

64. United States Auto Club, "Final Detailed Results 1968 Mobil Economy Run," circa 1968.

65. Don MacDonald, "Cars on Trial — Part Two," *Westways Magazine*, Automobile Club of Southern California, June 1968, 40.

66. Mike Cook, "When the Economy Run Ran Out of Gas," *Hemmings Sports & Exotic Cars*, December 2000.

67. Great Scott, "Unusual Factors Hit Shortened Mobil Run," *Competition Press and Autoweek*, April 15, 1968.

68. Mike Cook, "When the Economy Run Ran Out of Gas," *Hemmings Sports & Exotic Cars*, December 2000.

69. Bob Cahill with John Sloan, "MoPar and the Mobil Economy Run, Part 3," *High Performance MoPar*, January 1996, 43.

70. Great Scott, "Unusual Factors Hit Shortened Mobil Run," *Competition Press and Autoweek*, April 15, 1968.

71. Mobil Oil Corporation, "Results of the 1968 Mobil Economy Run. Your Car Can Do It. Can *You*?" print advertisement, circa 1968.

72. Mobil Oil Corporation, "What You Learn from the Mobil Economy Run Could Save You a Long Walk," print advertisement, circa 1968.

73. Mobil Oil Corporation, "Official Results 1968 Mobil Economy Run," brochure, circa 1968.

74. Buick Motor Division, General Motors Corporation, "Fourth Victory for Buick LeSabre in the Mobil Economy Run," print advertisement, circa 1968.

75. Buick Motor Division, General Motors Corporation, "Wouldn't You Really Rather Drive a Winner?" 1968 model year brochure.

76. Ford Division, Ford Motor Company, "Fairlane — Success Car '68 — Tops All Intermediate Sixes in Mobil Economy Run," print advertisement, circa 1968.

77. American Motors Corporation, "We Broke the Rules When We Introduced a Compact Car That Could Compete in Price with the Imports," print advertisement, circa 1968.

78. Chrysler-Plymouth Division, Chrysler Corporation, 1969 Barracuda model year brochure, 8–68 81–505–9025, 9.

79. Chrysler International S.A., 1969 model year brochure, DE-1969, 7.

80. Don MacDonald, "Cars on Trial — Part Two," *Westways Magazine*, Automobile Club of Southern California, June 1968, 40.

Chapter 12

1. Mobil Oil Corporation, Mobil News Release, December 29, 1968.

2. Bob Thomas, "Affluent Drivers Spell Doom for Economy Run," *Los Angeles Times*, December 29, 1968.

3. Mike Cook, "When the Economy Run Ran Out of Gas," *Hemmings Sports & Exotic Cars*, December 2000.

4. "1969 Mobil Economy Run Abandons Directional Aids, Adds Models," *Competition Press and Autoweek*, December 16, 1968.

5. Bob Thomas, "Affluent Drivers Spell Doom for Economy Run," *Los Angeles Times*, December 29, 1968.

6. Hugh Randolph, "Economy Run Cancelled," *Oakland Tribune*, December 29, 1968.

7. Bob Thomas, "Affluent Drivers Spell Doom for Economy Run," *Los Angeles Times*, December 29, 1968.

Chapter 13

1. J. "Kelly" Flory, Jr., *American Cars, 1946–1959* (Jefferson, NC: McFarland, 2008), 9.

2. "1956 Mobilgas Economy Run," *Road & Track*, June 1956, 23.

3. James M. Flammang and the Auto Editors of *Consumer Guide*, *Cars of the Fabulous '50s* (Lincolnwood, IL: Publications International, 1995), 335.

4. Bob Pendergast, "No Time for Balloon-Foots," *Hot Rod Magazine*, January 1959, 24.

5. "The First Economy Run," *Road & Track*, January 1959.

6. "37 Imported Cars Seek Honors in Economy Run," *Bakersfield Californian*, October 12, 1958.

7. "Mobil Mileage Rally for Imported Autos Replete with Thrills," *Automobile Topics*, November 1958.

8. Bob Pendergast, "No Time for Balloon-Foots," *Hot Rod Magazine*, January 1959, 24.

9. Ibid.; Griff Borgeson, "First Mileage Rally," *Sports Cars Illustrated*, May 1958, 46.

10. "Mobil Mileage Rally for Imported Autos Replete with Thrills," *Automobile Topics*, November 1958, "Hilly Run Puts Foreign Cars to Gas Test," *Business Week*, October 25, 1958, 32; Bob Pendergast, "No Time for Balloon-Foots," *Hot Rod Magazine*, January 1959, 24; "The First Economy Run," *Road & Track*, January 1959; Griff Borgeson, "First Mileage Rally," *Sports Cars Illustrated*, May 1958, 46.

11. Bob Pendergast, "No Time for Balloon-Foots," *Hot Rod Magazine*, January 1959, 24.

12. "37 Imported Cars Seek Honors in Economy Run," *Bakersfield Californian*, October 12, 1958.

13. Floyd Freel, "Import Report," *Pasadena Star News*, November 15, 1959.

14. "Tiny British Car Wins Mobile Mileage Rally," *Oxnard Press Courier*, November 9, 1959.

15. British Motor Corporation, "56½ Miles Per Gallon!" print advertisement, circa 1959.

16. John Green Corporation, Renault Distributor, "All America Loves a Winner!" print advertisement, circa 1959.

17. "Sprite Averages 56 MPG to Take Mileage Rally," *Automobile Topics*, January 1960;"The Imports Fight for the Economy Crown," *Motor Life*, January 1960, 77; "56.5 Miles Per Gallon ?!!"; *Motor Trend*, January 1960, 73; "Mobil Mileage Rally," *Small Car Parade*, March 1960, 30.

18. "Mobil Mileage Rally," *Small Car Parade*, March 1960, 30.

19. "NSU Prinz Tops Field in Mobil Mileage Run," *Automobile Topics*, January 1961.

20. Mobil Oil Company, "34 Foreign Cars Average 40.24 Miles Per Gallon in Mobil Mileage Rally," print advertisement, circa 1960.

21. Renault, Inc. "Renault Wins," print advertisement, circa 1960.

22. "NSU Prinz Tops Field in Mobil Mileage Run," *Automobile Topics*, January 1961; Bill Dredge, "The Magic of Mileage," *Motor Trend*, February 1961, 64; "Imports Have Economy Run, Too," *Popular Science*, January 1961.

23. "Foreign Autos Visit County on Mileage Run Sunday," *Oxnard Press Courier*, November 10, 1961.

24. "Recapping 1961 Mobil Mileage Rally," *Automobile Topics*, January 1962.

Bibliography

Newspapers

Albuquerque Journal
Bakersfield Californian
Galveston Daily News
Long Beach Independent
Long Beach Press Telegram
Los Angeles Examiner
Los Angeles Herald-Examiner
Los Angeles Mirror
Los Angeles Times
Oakland Tribune
Ogden Standard Examiner
Oxnard Press Courier
Pasadena Star News
Salt Lake Tribune
San Diego Union
San Francisco Call Bulletin
Seattle Times
Torrance Herald
Twin Falls Times-News

Magazines

Auto Topics: July 1965, June 1966.
Automobile Topics: January 18, 1937; January 24, 1938; January 20, 1941; March 1952, May 1952, May 1953, May 1954, April 1955, April 1956, June 1957, May 1958, November 1958, June 1959, January 1960, June 1960, January 1961, May 1961, January 1962, May 1963.
Automotive News: March 19, 1951; April 27, 1953; April 12, 1954; March 21, 1955; April 2, 1956; April 29, 1957; April 14, 1958; April 28, 1958; March 20, 1961; April 16, 1962; April 19, 1965.
Car & Driver: July 1962, July 1967, February 1981.
Car Craft: July 1965, July 1966.
Car Life: August 1954, June 1955, July 1962, April 1963, July 1963.
Chilton's Motor Age: May 1952, June 1954.
Dodge News: Volume 18, Number 6 (1953); Volume 19, Number 4 (1954); Volume 22, Number 6 (1957).
Flying Red Horse: May-June 1951, May-June 1952, Summer 1954, Summer 1955, Summer 1957, Spring–Summer 1958.
Ford News: February 1938, February 1940, March 1941.
Ford Times: June 1950, September 1951, July 1956, July 1961, April 1967.
Gilmore Cub: February–March 1936, February 1937, March 1938, February 1939, February 1940.
Hot Rod Magazine: August 1954, June 1956, June 1957, January 1959, July 1964, June 1965.
Lincoln-Mercury Times: May–June 1950, May–June 1951, July–August 1952, July–August 1953, July–August 1954.
Motor: April 1936, February 1937, February 1941, May 1952.
Motor Age: February 1939, February 1940, February 1941.
Motor Life: June 1955, June 1956, July 1958, January 1960, June 1960.
Motor Trend: February 1950, March 1950, April 1950, June 1951, July 1952, June 1953, July. 1953, June 1954, July 1954, May 1955, July 1957, July 1958, June 1959, January 1960, June 1960, February 1961, April 1961, June 1961, June 1963, April 1964, July 1965, March 1966.
Motor West: February 1936, February 1937, February 1938, February 1939, February 1940, February 1941, March 1950, April 1951, May 1952.
Motorsport: June 1951, July 1952, August 1953, August 1954.
Popular Mechanics: July 1963, July 1964.
Popular Science: May 1941, June 1951, June 1952, June 1953, June 1954, May 1955, July 1958, June 1959, June 1960, January 1961, June 1961, June 1962, June 1964.
Road & Track: April 1950, June 1954, June 1955, June 1956, July 1957, April 1963, December 1966, September 2003.
Small Car Parade: March 1960.
Speed Age: May 1950, July 1951, July 1952, July 1953, July 1954, July 1955, August 1957.
Sports Car Graphic: July 1965.
Sports Cars Illustrated: May 1959.
The Studebaker Wheel: May 1936, May 1939, February 1940.
Time: February 27, 1950; March 19, 1951; April 28, 1952; May 4, 1953; April 19, 1954; April 2, 1956; April 29, 1957; April 28, 1958; April 20, 1959, April 19, 1960.

Companies (Advertising and Promotional Items)

American Motors Corporation
Buick Motor Division — General Motors Corporation
Chevrolet Motor Division — General Motors Corporation
Chrysler Division — Chrysler Corporation
Chrysler-Plymouth Division — Chrysler Corporation
Dodge Division — Chrysler Corporation
Ford Division — Ford Motor Company
Graham-Paige Motors
Hudson Motor Car Company
Kaiser-Frazer Corporation
Lincoln Division — Ford Motor Company
Lincoln-Mercury Division — Ford Motor Company
Mercury Division — Ford Motor Company
Nash-Kelvinator Company
Nash Motors
Oldsmobile Division — General Motors Corporation
Packard Motor Car Company
Pontiac Motor Division — General Motors Corporation
Studebaker Corporation
Willys-Overland Company

Books

Butler, Don. *The History of Hudson.* Sarasota, FL: Crestline Publishing, 1982.
Dunham, Terry D. *The Buick: A Compete History.* Updated edition. Kutztown, PA: Automobile Quarterly Publications, 1987.
Editors of *Consumer Guide. 50 Years of American Automobiles 1939–1989.* Lincolnwood, IL: Publications International, 1989.
Flammang, James M., and the Auto Editors of *Consumer Guide. Cars of the Fabulous '50s.* Lincolnwood, IL: Publications International, 1995.
Flory, J. "Kelly," Jr. *American Cars, 1946–1959.* Jefferson, NC: McFarland, 2008.

Additional Publications

Clymer, Floyd. *Mobilgas Economy Run*: 1950, 1951, 1952, 1953, 1954, 1955, 1957, 1958, 1959.
United States Auto Club Yearbook: 1956, 1957, 1958, 1959, 1960, 1961, 1962, 1963, 1964, 1965, 1966, 1967, 1968.

Index

Numbers in ***bold italics*** indicate pages with photographs.